SNL G-503

Ordnance Catalog
for the
GPW
series of military vehicles

edited by
Brian Greul

The GPW Model MB, commonly referred to as the Jeep is probably the most ubiquitous American military vehicle ever produced. Simple, rugged, and capable of hard work it began as a World War II vehicle and descendant vehicles are still produced today as passenger vehicles.

This book is intended to support enthusiasts and their restoration efforts by providing a professionally printed, 8.5x11 version of this key manual for this vehicle. There are other books in this series and a book that has all of them.

Every effort has been made to faithfully reproduce the document while cleaning up the pages to make them usable to you the reader. However, we are dealing with original works that have been electronically preserved from nearly 80 years ago. There are a number of artifacts in the source documents. Your understanding is appreciated.

An 8.5x11 3 hole punched loose leaf copy may be purchased for your 3 ring binder. Email books@ocotillopress.com for current information.

Should you have suggestions or feedback on ways to improve this book please send email to Books@OcotilloPress.com

Edited 2021 Ocotillo Press
ISBN 978-1-954285-11-8

Printed in the United States of America

Ocotillo Press
Houston, TX 77017
Books@OcotilloPress.com

Disclaimer: The user of this book is responsible for following safe and lawful practices at all times. The publisher assumes no responsibility for the use of the content of this book. The publisher has made an effort to ensure that the text is complete and properly typeset, however omissions, errors, and other issues may exist that the publisher is unaware of.

THE ORDNANCE CATALOG

STANDARD
NOMENCLATURE
LIST

FOR

TRUCK, ¼-TON, 4 x 4,
COMMAND RECONNAISSANCE

WILLYS-OVERLAND
MOTORS INC., MODEL MB

Parts formerly listed in TM-lo-1512 change No. 3, dated 1 Augurt 1943 and in TM-10-1348 change No. 1, dated 10 March 1943, are now **included in this** Standard Nomenclature List No. Q-503.

Supersedes Addendum, SNL G-503, dated 2 Feb. 1943, and Organizational Spare **parts** and Equipment List, SNL G-503, dated 8 June 1943.

SECTION 1

GENERAL INFORMATION

For full information relative to THE ORDNANCE PROVISION SYSTEM, of which the Standa Nomenclature List is a part, see the Introduction to the Ordnance Catalog (IOC), the Ordnance Provisi System Regulations (OPSR), and the Index to Ordnance Publications (OFSB 1-1). (See OFSB 1-8 for c tribution of publications.)

THE STANDARD NOMENCLATURE LIST of Spare Parts and Equipment is divided into ei§ sections. These are, besides this General Information Section:

Maintenance Parts Procurement List (Section 2)
Vehicular Spare Parts and Equipment List (Section 3)
Organizational Spare Parts and Equipment List (Section 4)
Ordnance Maintenance Unit Stockage List (Section 5)
Depot Stockage List (Section 6)
Geographical or Seasonal Maintenance Parts List (Section 7)
Indexes (Section 8)

MAINTENANCE PARTS PROCUREMENT LIST (Section 2)

This section lists all serviceable parts in the vehicle and also designates which of these parts were p cured for maintenance of the vehicle. Column 6, Quantity Required Per Unit Assembly, shows the quanti of each part on the vehicle used in the location for which the listing is given. Column 7, 12 Mos. Field Main nance, shows the quantity of each part that has been estimated to be required for the third and fourth echelo for the maintenance of one hundred (100) vehicles for twelve months. These figures are exclusive of the qua tities estimated for Organization Maintenance Parts, which appear in Section 4. Column 8, Major Overha (5th echelon), shows the quantity of each part that has been estimated to be required for fifth echelon main nance for one hundred (100) vehicles for twelve months. Column 9, Total First Year Procurement, shows t quantity of each part procured for the total or complete maintenance by all echelons of one hundred (10 vehicles for twelve months. Column 10, Estimated Requirements per 100 Rebuilds, shows the estimat quantity of each part required to rebuild one hundred (100) vehicles.

Using troops performing first and second echelon maintenance may requisition those parts marked a % sign in Column 7 of Section 2. Those parts so marked which are not also listed in the Organization Maintenance Parts and Equipment List (Section 4) are to be requisitioned for replacement purposes on when time and the tactical situation permit, and trained personnel and suitable tools are available. Norma only those parts shown in Section 4 are carried by using troops for first and second echelon maintenance.

VEHICULAR SPARE PARTS AND EQUIPMENT LIST (Section 3)

This section lists all maintenance parts, accessories, equipment, common and special tools, suppli ammunition and sighting equipment that are carried in or on the vehicle. Also included are any other maj items that are issued with, and as an integral part of the vehicle. For example, the armament (guns and gu mounts) is listed in this section, as well as the armament spare parts that are carried on the vehicle.

ORGANIZATIONAL SPARE PARTS AND EQUIPMENT LIST (Section 4)

This section lists all maintenance parts, accessories, common and special tools, equipment and suppli fire control equipment, applicable cleaning and preserving materials, required by, and authorized for usi troop organizations in connection with the use and maintenance of the vehicle. Column 6 through 13 in th section show the estimated quantities of parts required by each company or regiment or separate battali to maintain the various quantities of the vehicle assigned to each respectively. The Organizational Accessori shows the tool sets allocated to these organizations, the figures in the quantity column indicating the quantiti per tool set.

ORDNANCE MAINTENANCE UNIT STOCKAGE LIST (Section 5)

This section lists all maintenance parts, accessories, common and special tools, equipment and suppli required to be carried by Ordnance Maintenance Units. The quantities of the parts shown are those requir for the maintenance of the various quantities of the vehicle on a basis of thirty (30) days field maintenanc

DEPOT STOCKAGE LIST (Section 6)

This section is an explanation of the method to be used for determining the quantities of parts to be stocked by Field, Base and Mobile Depots.

GEOGRAPHICAL OR SEASONAL MAINTENANCE PARTS LIST (Section 7)

This section consists of sub-divisions for each geographical location, season or particular terrain conditions. Each sub-division contains a list of maintenance parts, accessories, common and special tools, equipment and supplies, which, due to the peculiarities of the geographical location, season, or terrain condition, are required to be carried in stock by Ordnance Maintenance Units *in addition to* the stockage prescribed in Section 5. The quantities of parts shown in each sub-division are those of the *additional items* or *additional quantities* (of the items in Section 5) required for the maintenance of one hundred (100) vehicles for twelve months within the particular operating condition involved.

FEATURES OF THIS PUBLICATION

The following paragraphs explain various important details with which the user should be familiar for proper understanding of the catalog and practices peculiar to this individual book.

STANDARDIZED GOVERNMENT GROUPING: The parts listed in this catalog are divided into groups and sub-groups in accordance with the Standardized Government Grouping. This division into groups facilitates the location and identification of each part. The Standardized Government Grouping as adapted to this catalog is outlined on page 8.

ILLUSTRATIONS: Assembly views, exploded views and cross sectional drawings are used throughout this catalog to aid the user in locating and identifying parts. All illustrations are placed as closely as possible to the groups to which they apply. On each illustration or separate reference legend, the noun name, piece-mark and appropriate group number of each part illustrated are given. This promotes the quick identification of each part and accurate reference to the text for complete listing. Each illustration is identified by a figure number and title and in Section 8, Indexes, in the back of the book, will be found an index of all figures and the page on which they appear.

COMPONENTS OF KITS, SETS AND ASSEMBLIES: Whenever any doubt might exist as to what items are included as components of kits, sets or assemblies, an explanatory note or complete listing of such components is shown.

ATTACHING PARTS: Attaching parts such as bolts, nuts, rivets, screws and washers, etc., are indented immediately following the parts which they attach. This promotes quick selection of all parts to be requisitioned for use in installing a bracket, generator, etc. Whenever two similar parts, having the same attaching parts, such as right and left fenders are listed, they are placed one immediately following the other and the attaching parts for both are covered by a single listing under the second part. The quantities shown for the attaching parts are those necessary to attach both items.

STANDARD HARDWARE AND BULK MATERIAL: A complete resume of standard hardware and bulk materials will be found in Group 23, Section 2. For these items, which are also listed throughout Section 2 for each place of use, the quantity required per unit assembly (Column 6) is shown only for each listing in the various groups other than group 23 and the echelon breakdown (Columns 7, 8 and 9) is given only in group 23.

INDEXES: Particular attention is directed to the fact that all items are not included in all indexes. This is covered by explanatory notes at the beginning of each index. Section 8 contains these indexes and in this sequence: Illustrations, Alphabetical, Numerical (Ordnance Drawing Numbers, Manufacturers' and Vendors' Part Numbers, and Official Stockage Numbers).

INTERCHANGEABILITY OF COMPONENTS: The engine, transmission, steering gear, clutch and the differential of the rear axle are the same as or similar to the corresponding units installed in Tractor, Snow, M7 (reviewed in SNL G194).

ORDERING PARTS, ITEM STOCK NUMBERS: All piece marks, item stock number *and* complete nomenclature must be carefully included on the requisition when parts are ordered. This is extremely essential. The prefix letter and number of the item stock number signifies only that the item was originally coded under that major item and should not be interpreted as meaning that the part is used only with the one vehicle. This prefix continues unchanged when the item is listed again in catalogs for other vehicles in which it is used.

HOW TO FIND PARTS

There are several ways to find any part quickly, using the various indexes, illustrations and standardized government grouping. The following instructions and examples are given as a guide:

a. If only the noun name and description of the part is known—

1. Turn to alphabetical index and locate the item by the noun name and description.

2. Turn to the group shown opposite the name. Group numbers appear on the upper outside corner of each page.

3. Refer to the proper sub-heading under which the part is listed and locate the part in alphabetical order under the sub-heading.

b. If only the location or function of the part in the vehicle is known—

1. Turn to the standardized government grouping list, page 8, and determine the number of the group in which the part is most likely to be listed.

2. If found in the list, review the illustrations which apply to group or allied groups. When the proper illustration is found, using the code letter, refer to the legend to determine the ordnance or manufacturers' part number, description of the part, and group number.

3. Turn to the proper group and locate the part, either by part number or by description.

c. If only the item stock number of the part is known—

1. Refer to the item stock number index and locate the proper item stock number.

2. Turn to the group shown opposite the item stock number and locate the part.

d. If only the ordnance and or manufacturers' number of the part is known—

1. Refer to the numerical index of ordnance and or manufacturers' numbers and locate the proper part number.

2. Turn to the group shown opposite the ordnance and or manufacturers' part number and locate the part.

For example: To find information on a brake shoe one may refer to the Standardized Government Grouping List. Under the "Brake Group" major group 12, is listed sub-group 1200 "Brake Assembly." Turning to this group in the catalog section of the brake shoes are found in alphabetical sequence. If some doubt exists as to whether or not the correct shoe has been found, a check can be made against the illustration referred to in the plate number column.

The shoe also could be found by referring to the alphabetical index under Shoe, brake. The same sub-group reference is shown and by following the steps outlined in the paragraph above, the shoes are found under sub-group 1201.

Another way of finding the shoe is to determine the sub-group in which the parts may be found and refer to the illustrations which apply to that group. The part may be found on the illustration and part number, noun name, and group number will be found in the legend.

The quantities specified as first year procurement are based on actual procurement as of 15 January 1944, plus all changes thereto as included in Procurement List No. TRK 780R, dated 28 October 1943. There are a number of new items included in Procurement List No. TRK 780R which will not be available until approximately 1 May 1944.

MANUFACTURERS' IDENTIFICATION SYMBOLS

Listed below are the symbols assigned to the various original manufacturers and used as a prefix to their numbers throughout this catalog.

SYMBOL	MANUFACTURERS' NAME	ADDRESS
AB	Aetna Ball Bearing Mfg. Co.	4600 Schubert Ave., Chicago, Illinois
AC	A. C. Spark Plug Division (General Motors Corp.)	Flint, Michigan
AD	Alemite Division (Stewart Warner Corp.)	Chicago, Illinois
AE	Appleton Electric Co.	1745 Wellington Ave., Chicago, Illinois
AL	Electric Auto-Lite Co.	Mulberry and Champlain St., Toledo, Ohio
AN	Ainsworth Mfg. Co.	2200 Franklin St., Detroit, Michigan
AVM	Atwood Vacuum Mfg. Co.	Rockford, Illinois
BA	Bassick Co. (The)	36 Austin St., Bridgeport, Conn.
B-B	Borg and Beck	Rockford, Illinois
BOW	Bower Roller Bearing Co.	3040 Hart Ave., Detroit, Michigan
BUE	Budd, Edw. G. Mfg. Co.	Philadelphia, Pennsylvania
BX	Bendix Products Corp.	401 Bendix Drive, South Bend, Indiana
CA	Columbus Auto Parts	Columbus, Ohio
CAR	Carter Carburetor Co.	2820-56 N. Spring Ave., St. Louis, Missouri
CCG	Chicago Screw Co.	1020 S. Homan Ave., Chicago, Illinois
CL	Clum Mfg. Co.	601 W. National Ave., Milwaukee, Wisconsin
CN	Copeland Gibson Products Corp.	7451 West 8 Mile Rd., Detroit, Michigan
CNC	Cinch Mfg. Co.	Chicago, Illinois
CP	Champion Spark Plug	Toledo, Ohio
DM	Douglas Mfg. Co. (H. A.)	Bronson, Michigan
DTC	Deutschmann, Tobe, Corp.	Washington Ave., Canton, Massachusetts
EAT	Eaton Mfg. Co. (The)	Cleveland, Ohio
ER	Erie Resistor Corp.	64 W. 12th St., Erie, Pennsylvania
FB	Federal Bearings Co.	Poughkeepsie, New York
FM	Ford Motor Co.	Dearborn, Michigan
FO	Flex-O-Tube Co.	752 14th St., Detroit, Michigan
FYR	Fyr-Fighter Co.	2221 Crane St., Dayton, Ohio
GM	General Motors Corp.	General Motors Bldg., Detroit, Michigan
HH	Oakes Products Division (Houdaille Hershey Corp.)	National Bank Bldg., Detroit, Michigan
HI	Hayes Industries	Jackson, Michigan
HLH	Holland Hitch Co.	Holland, Michigan
HO	Hoover Ball and Bearing Co.	Hoover Ave., Ann Arbor, Michigan
HP	Harris Products Co.	Detroit, Michigan
HR	Harrison Radiator Corp.	Washburn and Walnut Sts., Lockport, New York
HY	Hyatt Bearings Division (General Motors Corp.)	P.O. Box 71, Harrison, New Jersey
KHW	Kelsey Hayes Wheel Co.	Detroit, Michigan
KM	Kent-Moore Organization	485 W. Milwaukee, Detroit, Michigan
KS	King Seeley Corp.	305 1st St., Ann Arbor, Michigan

SYMBOL	MANUFACTURERS' NAME	ADDRESS
LK	Link Belt Co.	307 N. Michigan Ave., Chicago, Illinois
LO	Wagner Electric Co.	6420 Plymouth Ave., St. Louis, Missouri
MAE	Monroe Auto Equipment	1400 E. 1st St., Monroe, Michigan
MAS	Master Engineering Service	
MDR	McCord Radiator and Mfg. Co.	2588 E. Grand Blvd., Detroit, Michigan
MIL	Mallory and Co. (P. R.)	3029 E. Washington, Indianapolis, Indiana
MRC	Marlin-Rockwell Corp.	Jamestown, New York
MSP	Midland Steel Products Co.	Cleveland, Ohio
MZ	"Mazda"-General Electric Co.	Schenectady, New York
ND	New Departure Mfg. Co.	1940 Hughes St., Bristol, Connecticut
NP	New Process Gear, Inc.	500 Plum St., Syracuse, New York
PMA	Pressed Metals of America, Inc.	Port Huron, Michigan
PU	Purolator Products, Inc.	365 Frelinghuysen Ave., Newark, New Jersey
RG	Ross Gear and Tool Co.	Lafayette, Indiana
RMC	Rich Mfg. Co.	Battle Creek, Michigan
RZ	(Rzeppa) Gear Grinding Machine Co.	Detroit, Michigan
SA	Stant Mfg. Co.	Connersville, Indiana
SKF	SKF Industries, Inc.	Front St. and Erie Ave., Philadelphia, Pennsylvania
SOL	Solar Mfg. Co.	588 Avenue A., Bayonne, New Jersey
SP	Spicer Mfg. Corp.	4100 Bennett Rd., Toledo, Ohio
SPR	Sprague Specialties Co.	North Adams, Massachusetts
SPW	Sparks Withington Co.	Jackson, Michigan
SRM	Soreng Manegold Co.	1901 Clybourne Ave., Chicago, Illinois
SV	Schrader's Son, A.	470 Vanderbilt Ave., Brooklyn, New York
SW	Stewart-Warner Corp.	1828 Diversey Pkwy., Chicago, Illinois
SZE	Schwarze Electric Co.	1939 Berry St., Adrian, Michigan
SPS	Spun Steel	Canton, Ohio
TB	Thomas and Betts	36 Butler St., Elizabeth, New Jersey
TM	Timken Roller Bearing Co. (The)	1935 Kelley Ave., Canton, Ohio
TP	Thompson Products Inc.	2196 Clarkwood Rd., Cleveland, Ohio
TR	Torrington Co. (The)	Torrington, Connecticut
TRI	Trico Products Co.	817 Washington St., Buffalo, New York
TRU	American Chain and Cable Co. (TRU-Stop)	Wilkes-Barre, Pennsylvania
TSE	Taylor Sales Engineering Co.	125 E. Lexington Ave., Elkhart, Indiana
USL	USL Battery Corp.	Highland Ave., Niagara Falls, New York
VG	Victor Mfg. and Gasket Co.	Roosevelt Rd., Chicago, Illinois
WB	Willard Storage Battery Co.	E. 131 st and St. Clair St., Cleveland, Ohio
WEB	Warner Electric Brake Co.	Beloit, Wisconsin
WG	Warner Gear Co.	Muncie, Indiana
WH	Weatherhead Co. (The)	30 E. 131st St., Cleveland, Ohio
WIL	Wilcox-Rich	9771 French Rd., Detroit, Michigan
WO	Willys-Overland Motors, Inc.	Toledo, Ohio
YA	Yale and Towne Mfg. Co.	Philadelphia, Pennsylvania

LIST OF ABBREVIATIONS

loy-S.	alloy steel	in.	inch
np.	ampere	incand.	incandescent
hd.	button-head	lgh.	length
	brass	mach.	machine
g.	bearing	mf. d.	micro-farad
	bronze	M. I.	malleable iron
pin	cotter pin	mtg.	mounting
	cadmium	N. C.	National coarse
r.	chromium	N. F.	National fine
I.	cast iron	ni.	nickel
-hd.	countersunk head	No.	number
ntd.	continued	O.D.	outer diameter
p.	copper	ov.	oval
	candlepower	ov-hd.	oval head
le-hd.	double head	Pkzd.	parkerized
am.	diameter	pltd.	plated
l.	drilled	rd.	round
l. f/c-pin	drilled for cotter pin	rd-hd.	round head
grs.	engineers	S.	steel
	for	s-fin.	semi-finished
	filament	sq.	square
hd.	fillister head	sq-hd.	square-head
	finished	thd.	thread
ıd.	flat head	thk.	thick
	foot or feet	tung.	tungsten
ls.	headless	w/	with
x.	hexagonal	w/o	without
x-hd.	hexagonal-head	z.	zinc
Ɔ.	inner diameter		

GOVERNMENT GROUP INDEX

01 ENGINE GROUP

0100	Engine Assembly
0101	Crankcase and Cylinder Block
0101A	Cylinder Head
0102	Crankshaft
0102A	Crankshaft Bearing
0103	Pistons
0103A	Rings
0103B	Piston Pins
0104	Connecting Rods
0104A	Connecting Rod Bearings
0105	Valves, Seats, Guides and Spring
0105A	Tappets
0105C	Valve Covers and Gaskets
0106	Camshaft
0106A	Camshaft Bearings
0106C	Chains and Sprockets
0106D	Chain Case Cover
0107	Oil Pump
0107A	Oil Pan
0107A-1	Oil Float
0107E	Oil Filter
0107F	Oil Filter Attaching Parts
0107G	Oil Filler
0107H	Level and/or Float Gauges
0107J	Oil Distribution System
0107N	Crankcase Ventilator
0108	Manifolds as an Assembly
0108A	Intake Manifold
0108B	Exhaust Manifold
0108C	Heat Controls
0109	Flywheel
0109B	Ring Gear
0109C	Flywheel Housing
0110	Mountings

02 CLUTCH GROUP

0201A	Facings and Rivets
0201B	Clutch Driven Plate
0202	Cover, Pressure Plates, and Springs
0203	Release Yoke, Bearing Fork
0204	Pedal
0204A	Pedal Linkage

03 FUEL GROUP

0300	Fuel Tanks
0300A	Tank Gauge Unit
0301	Carburetor
0301A	Carburetor Attaching Parts
0301B	Choke
0301C	Air Cleaner
0302	Fuel Pump
0302A	Attaching Parts
0303	Accelerator and Linkage
0303A	Hand Throttle and Linkage
0304	Fuel Lines
0306B	Fuel Strainer

04 EXHAUST GROUP

0401	Muffler
0402	Exhaust Pipe

05 COOLING GROUP

0501	Radiator
0502	Thermostat
0503	Water Pump
0503A	Fan and Pulleys
0503B	Fan Belt

06 ELECTRICAL GROUP

0601	Generator
0601A	Generator Attaching Parts
0601B	Generator Regulator
0602	Cranking Motor
0603	Distributor
0604	Ignition Coil
0604A	Ignition Wiring
0604B	Spark Plugs
0604C	Ignition Lock
0605	Instruments
0605B	Instrument Wiring
0605C	Panel Light
0606	Switches
0606B	Chassis Wiring
0606D	Circuit Breakers
0606E	Junction Blocks
0606F	Trailer Electric Coupling
0607	Head Lights
0607A	Sealed Beam
0607B	Lamps
0607D	Blackout Lights
0608	Tail Light
0609	Horns
0609B	Horn Button and Wiring
0610	Battery
0610A	Battery Hangers

07 TRANSMISSION GROUP

0700	Transmission Assembly
0701	Case
0703	Main Drive Gear and Bearings
0704	Main Shaft and Bearings
0704B	Gears
0704C	Countershaft and Reverse Idler Sha
0706A	Gear Shift Lever and Parts
0706B	Gearshifter Rails and Parts
0706C	Gearshifter Yokes or Forks

TRANSMISSION TRANSFER ASSEMBLY

0800	Transmission Transfer Assembly
0801	Case
0802	Drive Gear
0803	Driven Gears, Shafts, Bearings
0804	Idler Gear, Shaft, Bearings
0805	Shifter Rods
0805A	Yokes
0805B	Shift Lever and Parts
0809	Mountings
0810	Speedometer Drive Gears

PROPELLER SHAFT AND UNIVERSAL JOINT

0901	Propeller Shaft Assemblies
0902	Universal Joints

FRONT AXLE GROUP

1000	Front Axle Assembly
1001	Housing
1002	Differential Assembly
1003	Differential Gears, Pinion and Bearing
1006	Steering Knuckle, Flange and Arm
1007	Axle Shaft, Universal Joints

REAR AXLE GROUP

1100	Rear Axle Assembly
1101	Housing Assembly
1102	Axle Drive Shafts
1103	Differential and Carrier Assembly
1104	Differential Pinion Bearing

BRAKE GROUP

1200	Brake Assembly
1201	Hand Brake Emergency Brake Parts
1202	Shoes and Facings
1203	Brake Shoe Support
1203A	Guide Springs
1203B	Adjusting Pin and Anchor Plate
1204	Pedal and Spring
1205	Master Cylinder
1207	Wheel Cylinders
1209B	Hoses
1209C	Tubes and Clips

WHEELS, HUBS AND DRUMS GROUP

1301	Wheel Assembly
1301A	Bearings
1301B	Seals
1301C	Retainers
1302	Hubs and Drums
1302A	Studs and Bolts

14 STEERING GROUP

1401	Steering Connecting Rod
1402	Tie Rod
1403A	Gear Assembly
1404	Wheel Assembly
1405	Brackets

15 FRAME & BRACKETS GROUP

1500	Frame and Brackets
1502	Pintle Hook
1505	Spare Wheel

16 SPRINGS AND SHOCK ABSORBER GROUP

1601	Front Spring
1601A	Rear Springs
1601B	Spring Bumpers
1601C	Torque Reaction Spring
1602	Shackle and Spring Attaching Parts
1603	Shock Absorbers and Mountings

17 HOOD, FENDERS, APRONS GROUP

1701	Fenders
1702	Splash Apron
1704	Hood

18 BODY GROUP

1800	Body
1800A	Body Handles
1801	Body Mounting Parts
1802	Body Pillars
1802A	Body Panels
1803	Floor Mats and Seals
1804	Seat
1804A	Cushions
1809	Instrument Panel and Mounting Parts
1809A	Compartment and Parts
1811	Windshield Assembly
1811A	Windshield Glass
1811B	Windshield Frame, Moulding and Seals
1811D	Windshield Wiper Arm and Blades
1817	Tool Box
1817A	Spare Gas Carrier

21 BUMPERS AND GUARD GROUP

2101	Bumpers
2103	Radiator Guards
2103A	Lamp Guards

22 MISCELLANEOUS BODY CHASSIS AND ACCESSORY GROUP

2201	Tarpaulins
2201A	Bows
2201F	Side Curtains
2202	Rear View Mirrors
2203	Identification and Caution Plates
2203A	Reflectors
2204	Speedometer and Parts

23 GENERAL USE STANDARDIZED PARTS GROUP

2301	Tools
2301A	Pioneering Equipment
2301B	Tire Chains
2302	Brake Fluid
2304B	Bolts
2304C	Chains
2304D	Clamps Hose
2304E	Elbows
2304F	Fittings
2304G	Nuts
2304H	Pins
2304Y	Plugs
2304J	Rivets
2304K	Screws
2304L	Washers
2304M	Wire Locking
2305	Miscellaneous Cotter Keys
2305A	Woodruff Keys
2310	Winterization Kits

24 FIRE EXTINGUISHER SYSTEM GROUP

2402	Portable System and Parts

26 RADIO GROUP

2601	Radio Installation Parts
2603	Radio Suppression System Parts

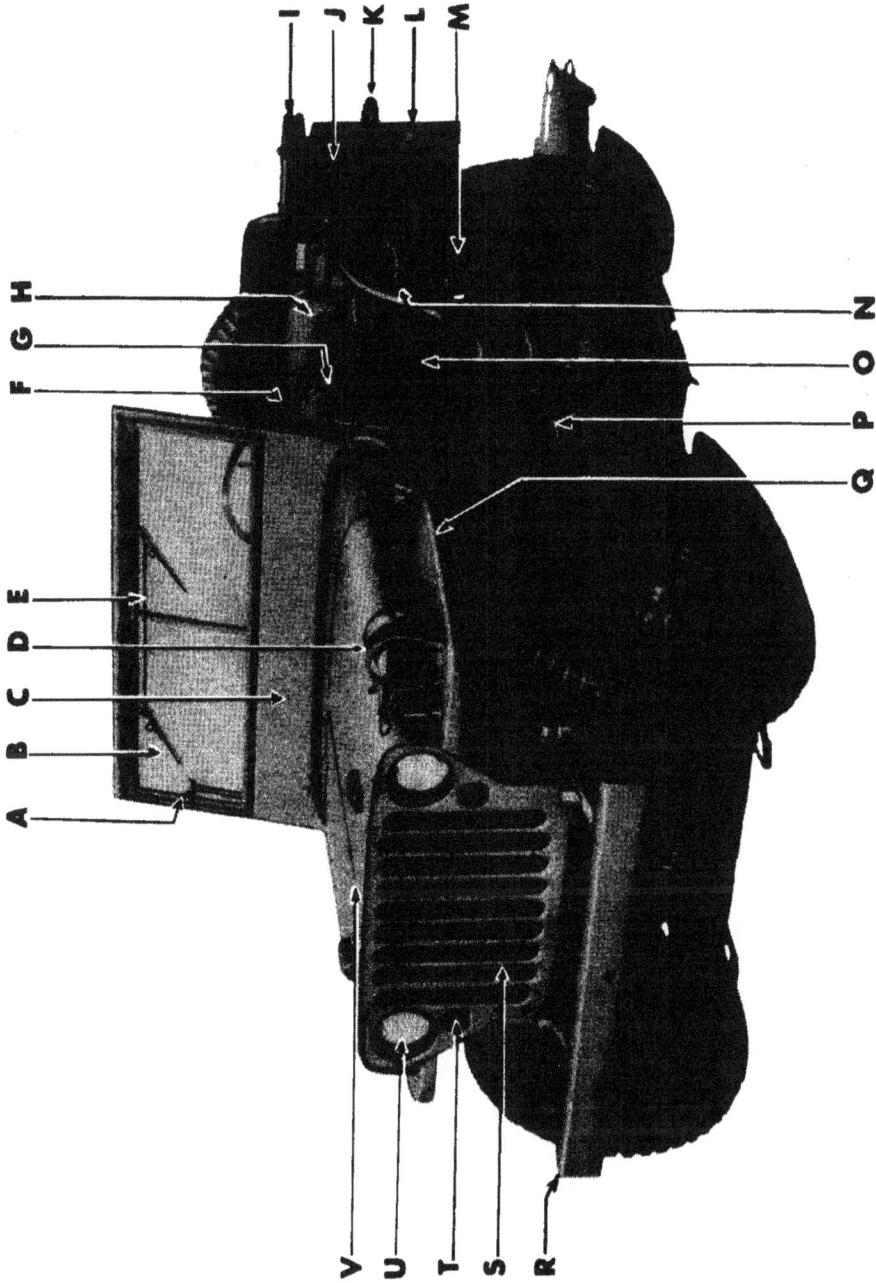

RA PD 305058

Key	Item	Willys Part No.	Ford Part No.	Gov't Group No.
L	REFLECTOR	WO-A-1306	FM-GPW-13380-A	2203A
M	STRAP	WO-A-3139	FM-GPW-1128267	2301A
N	HANDLE	WO-A-2390	FM-GPW-1129670	1800A
O	CUSHION	WO-A-2986	FM-GPW-1162900-B	1804A
P	PANEL	WO-A-3008	FM-GPW-1102039	1802A
Q	FENDER	WO-A-2942	FM-GPW-16006	1701
R	BUMPER	WO-A-1117	FM-GPW-17750	2101
S	GUARD assembly	WO-A-3615	FM-GPW-8307	2103
T	LIGHT	WO-A-1437	FM-GPW-13200	0607D
U	LIGHT	WO-A-1305	FM-GPW-13005	0607
V	HOOD	WO-A-3225	FM-GPW-16610	1704

RA PD 305058-A

FIGURE A—¼ TON 4 x 4 RECONNAISSANCE TRUCK, ¾ FRONT VIEW

Key	Item	Willys Part No.	Ford Part No.	Gov't Group No.
A	WINDSHIELD assembly	WO-A-3210	FM-GPW-1103010	1811
B	GLASS	WO-A-2478	FM-GPW-1103100	1811A
C	PANEL	WO-A-3190	Willys only	1811
D	LAMP assembly	WO-A-6142	FM-GPW-43150	0607D
E	WIPER	WO-A-11433	FM-GPW-17500	1811D
F	WHEEL assembly	WO-A-6858	FM-GPW-3600-A-3	1404
G	MIRROR assembly	WO-A-2934	FM-21CS-17682-B	2202
H	SEAT	WO-A-3107	FM-GPW-1160016-B	1804A
I	BOW	WO-A-2897	FM-GPW-1151266	2201A
J	BRACKET	WO-A-2754	FM-GPW-1153030	2201A
K	HANDLE	WO-A-2389	FM-GPW-1129672	1800A

11

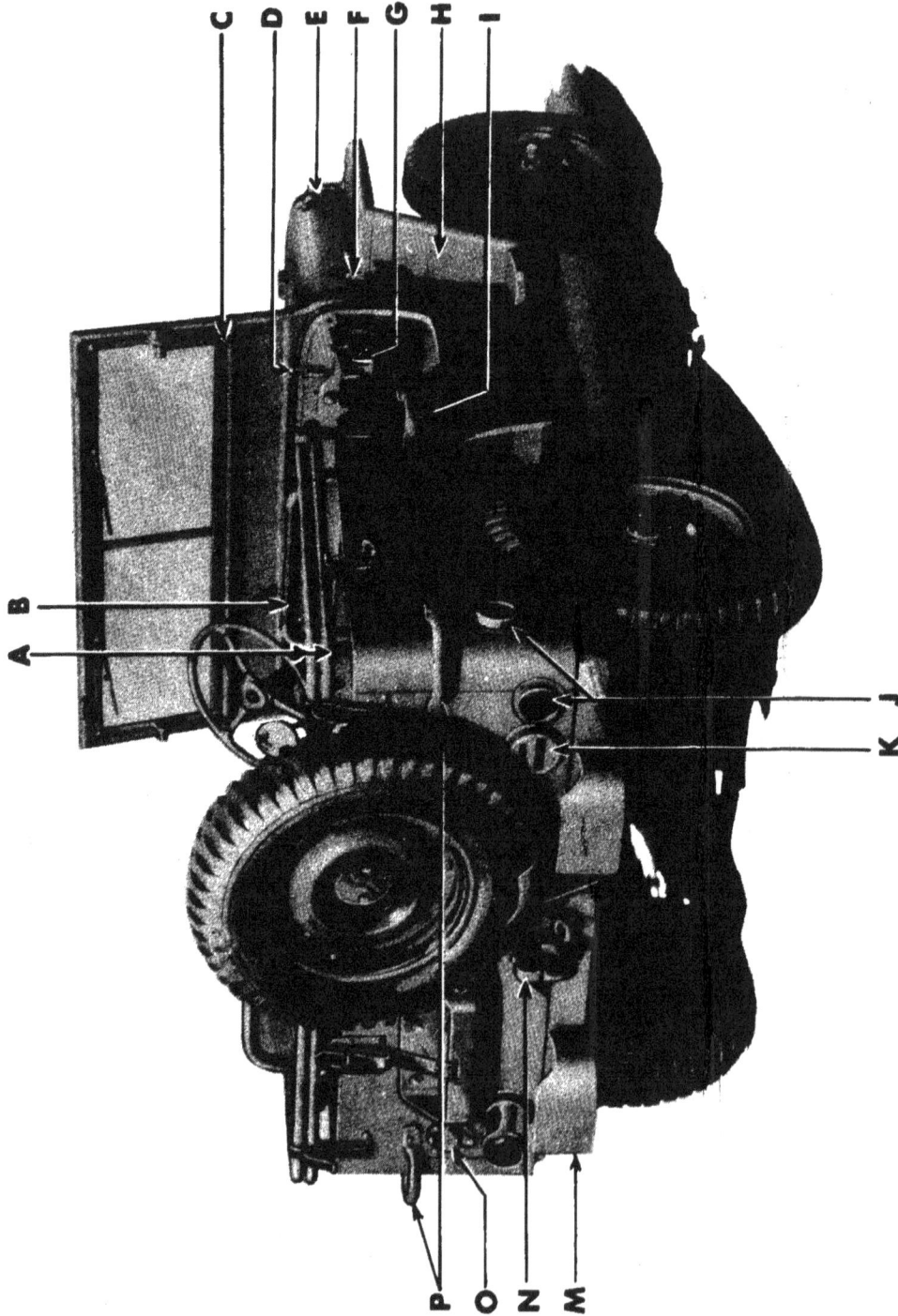

RA PD 305059

FIGURE B—¼ TON 4 x 4 COMMAND RECONNAISSANCE TRUCK, ¾ REAR VIEW

RA PD 305059-A

Key	Item	Willys Part No.	Ford Part No.	Gov't Group No.
A	BOW assembly	WO-A-2897	FM-GPW-1151266	2201A
B	SEAT assembly	WO-A-3107	FM-GPW-1160016-B	1804A
C	ARM	WO-A-2235	FM-GP-1103302	1811
D	CLAMP	WO-A-2227	FM-GP-1103482-A	1811
E	CATCH assembly	WO-A-2896	FM-GPW-16892	1704
F	SCREW	WO-A-2214	Willys only	1811
G	STRAP assembly	WO-A-2883	FM-GPW-1131414	1804
H	FENDER assembly	WO-A-2943	FM-GPW-16005	1701

Key	Item	Willys Part No.	Ford Part No.	Gov't Group No.
I	HANDLE	WO-A-2390	FM-GPW-1129670	1800A
J	REFLECTOR	WO-A-1306	FM-GPW-13380-A	2203A
K	LIGHT assembly	WO-A-1065	FM-GPW-13404-B	0608
M	BUMPERETTE	WO-A-1157	FM-GPW-17775	2101
N	HOOK	WO-A-593	FM-GPW-5182	1502
O	SOCKET assembly	—	FM-11YS-18142-B	0606F
P	HANDLE	WO-A-2389	GPW-1129672	1800A

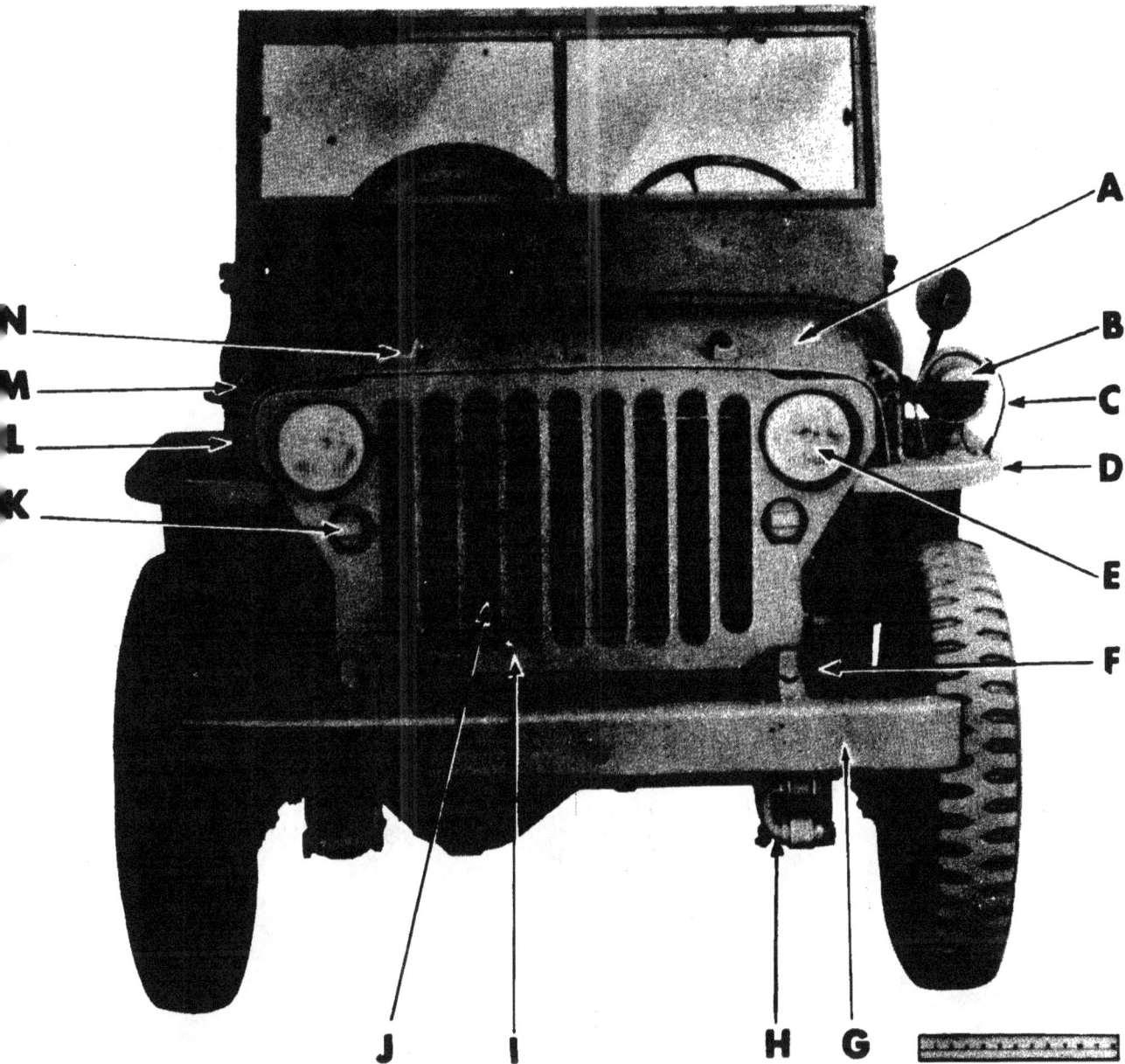

RA PD 305062

FIGURE C—¼ TON 4 x 4 COMMAND RECONNAISSANCE TRUCK,
FRONT VIEW

Key	Item	Willys Part No.	Ford Part No.	Gov't Group No.
A	HOOD assembly	WO-A-3225	FM-GPW-16610	1704
B	LIGHT assembly	WO-A-6142	FM-GPW-13150	0607D
C	GUARD w/SUPPORT	WO-A-4118	FM-GPW-13176	2103A
D	FENDER w/SPLASHER	WO-A-2942	FM-GPW-16006	1701
E	LIGHT assembly	WO-A-1304	FM-GPW-13006	0607
F	SHOCK ABSORBER assembly	WO-A-6902	FM-GPW-18045	1603
G	BUMPER	WO-A-1117	FM-GPW-17750	2101
H	BOLT	WO-A-513	FM-GPW-5778	1602
I	GUARD assembly	WO-A-3615	FM-GPW-8307	2103
J	CORE w/TANK assembly	WO-A-1214	FM-GPW-8005	0501
K	LIGHT assembly	WO-A-1437	FM-GPW-13200	0607D
L	CATCH assembly	WO-A-2896	FM-GPW-16892	1701
M	CATCH assembly	WO-A-3197	FM-GPW-1103027	1704
N	BUMPER assembly	WO-A-4683	(Willys only)	1704

RA PD 305062-A

SNL G-503

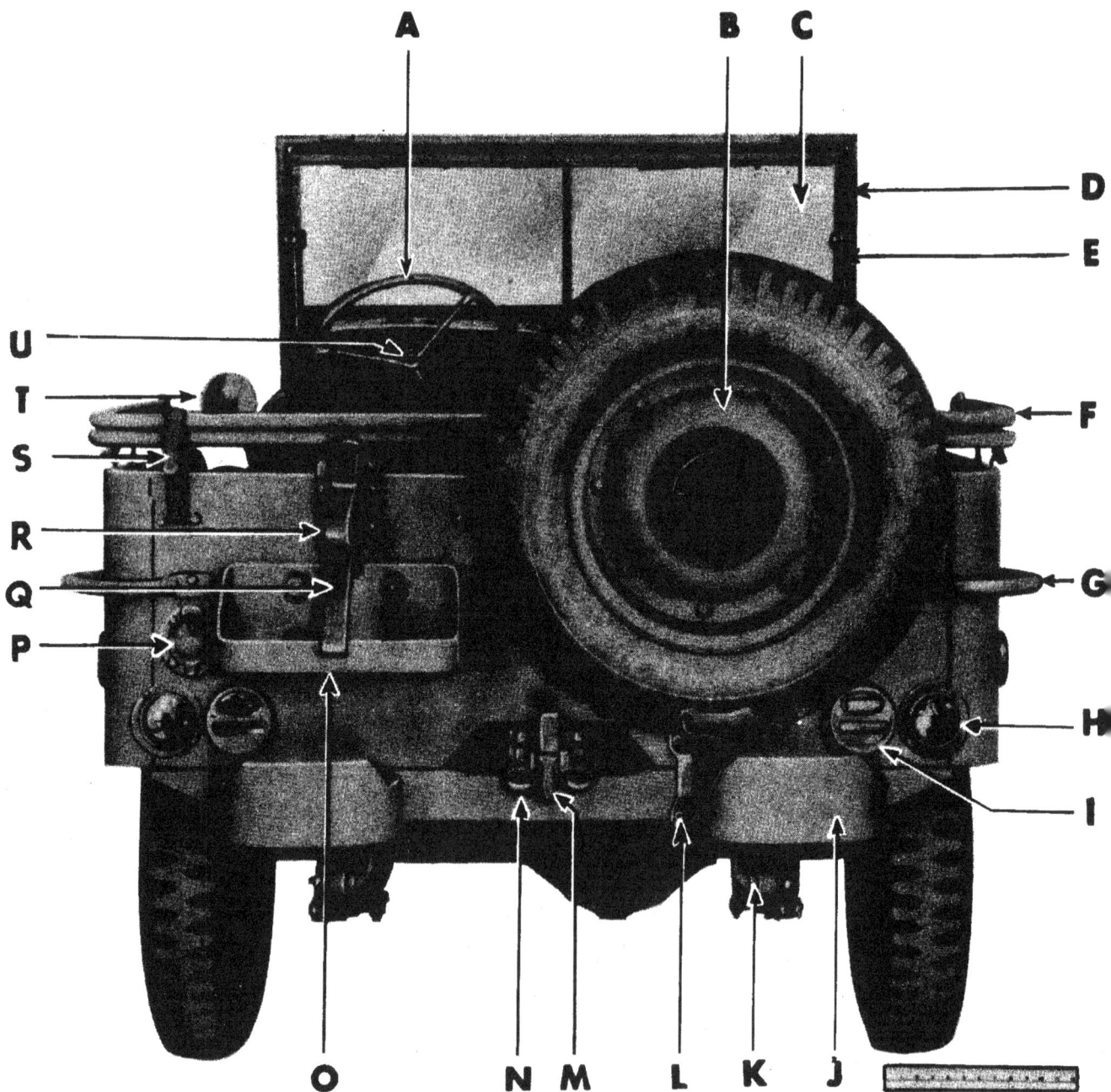

RA PD 305061

FIGURE D—¼ TON 4 x 4 COMMAND RECONNAISSANCE TRUCK, REAR VIEW

Key	Item	Willys Part No.	Ford Part No.	Gov't Group No.
A	WHEEL	WO-A-6858	FM-GPW-3600-A3	1404
B	WHEEL	WO-A-5467	FM-GPW-1025-C	1301
C	GLASS	WO-A-2478	FM-GP-1103100	1811A
D	WINDSHIELD assembly	WO-A-3210	FM-GPW-1103010	1811
E	ARM	WO-A-2235	FM-GP-1103302	1811
F	BOW assembly	WO-A-2897	FM-GPW-1151266	2201A
G	HANDLE	WO-A-2389	FM-GPW-1129672	1800A
H	REFLECTOR	WO-A-1306	FM-GPW-13380-A	2203A
I	LIGHT	WO-A-1065	FM-GPW-13404-B	0608
J	BUMPERETTE	WO-A-1157	FM-GPW-17775	2101
K	SHACKLE	WO-A-6069	FM-GPW-5605	1602
L	BRACKET	WO-A-2359	FM-GPW-1433	1505
M	HOOK	WO-A-593	FM-GPW-5182	1502
N	BOLT	WO-A-6393	FM-GPW-5186	1502
O	BRACKET	WO-A-4123	FM-GPW-1140330	1817A
P	SOCKET	WO-A-6586	FM-11YS-18151-B	0606F
Q	STRAP	WO-A-4127	(Willys only)	1817A
R	STRAP	WO-A-4127		
S	STRAP	WO-A-3110	FM-GPW-1152720	
T	MIRROR	WO-A-2934	FM-21CS-17682-B	
U	BUTTON	WO-A-634	FM-GPW-3627	0609B

RA PD 305061-A

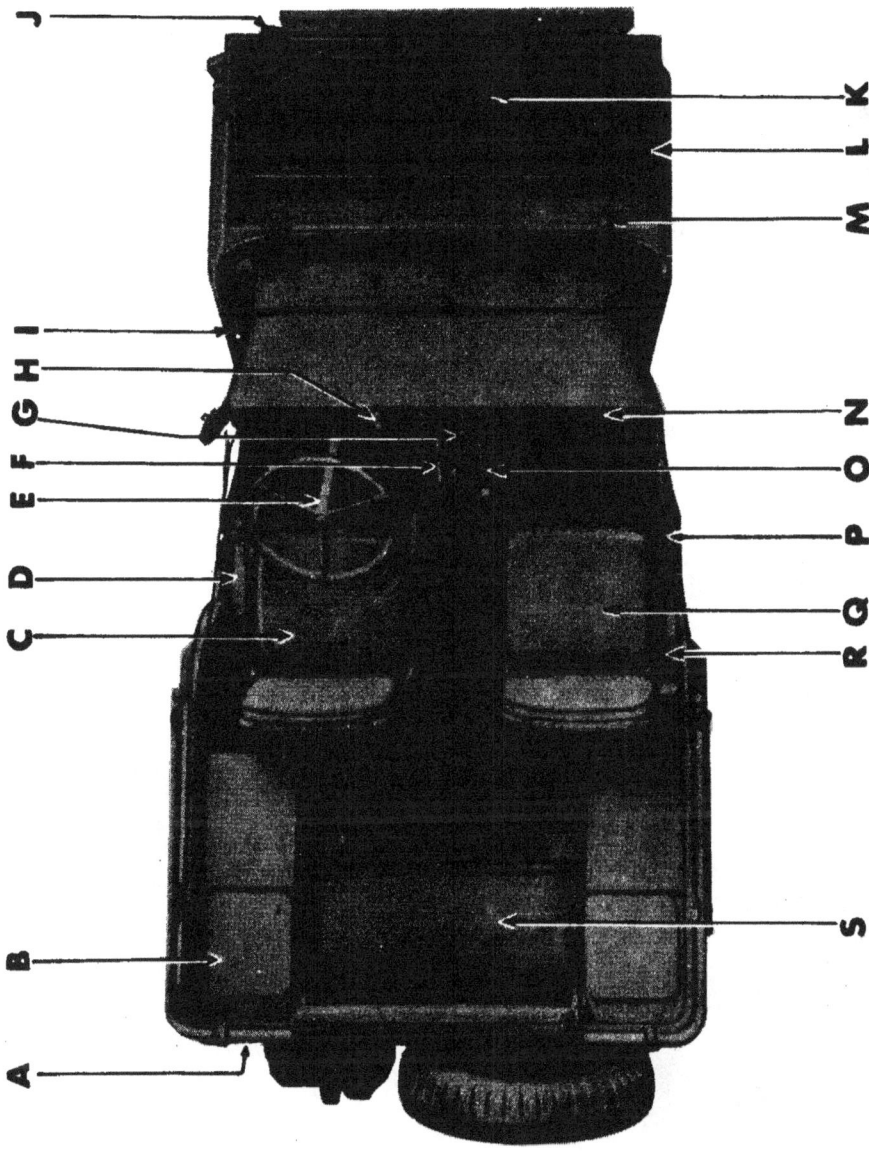

RA PD 305060

FIGURE E—¼ TON 4 x 4 COMMAND RECONNAISSANCE TRUCK, TOP VIEW

Key	Item	Willys Part No.	Ford Part No.	Gov't Group No.
A	BOW assembly	WO-A-2897	FM-GPW-1151266	2201A
B	LID w/HINGE	WO-A-3227	FM-GPW-1146100	1817
C	SEAT assembly	WO-A-3107	FM-GPW-1160016-B	1804A
D	TANK	WO-A-6618	FM-GPW-9002-B	0300
E	GEAR, steering	WO-A-1239	FM-GPW-3504	1403
F	LEVER assembly	WO-A-1380	FM-GPW-7210-A	0706A
G	HANDLE w/TUBE	WO-A-1242	FM-GPW-2780	1201
H	SOCKET assembly	WO-A-1411	FM-GPW-13710	0605C
I	SCREW	WO-A-2214	Willys only	1811
J	RETAINER	WO-A-3203	FM-GPW-1103028	1704

Key	Item	Willys Part No.	Ford Part No.	Gov't Group No.
K	WIPER assembly	WO-A-11433	Willys only	1811D
L	WINDSHIELD assembly	WO-A-3210	FM-GPW-1103010	1811
M	CLAMP assembly	WO-A-2227	FM-GPW-1103482-A	1811
N	CATCH	WO-A-2791	FM-GPW-1104388	1811
O	LEVER	WO-A-1505	FM-GPW-7793	0805B
P	STRAP assembly	WO-A-2883	FM-GPW-1131414	1804
Q	SEAT assembly	WO-A-3106	FM-GPW-1160012-A	1804A
R	PAD assembly	WO-A-3115	FM-GPW-116400	1804A
S	SEAT assembly	WO-3108	FM-GPW-1160026	1804A

RA PD 305060-A

SNL G-503

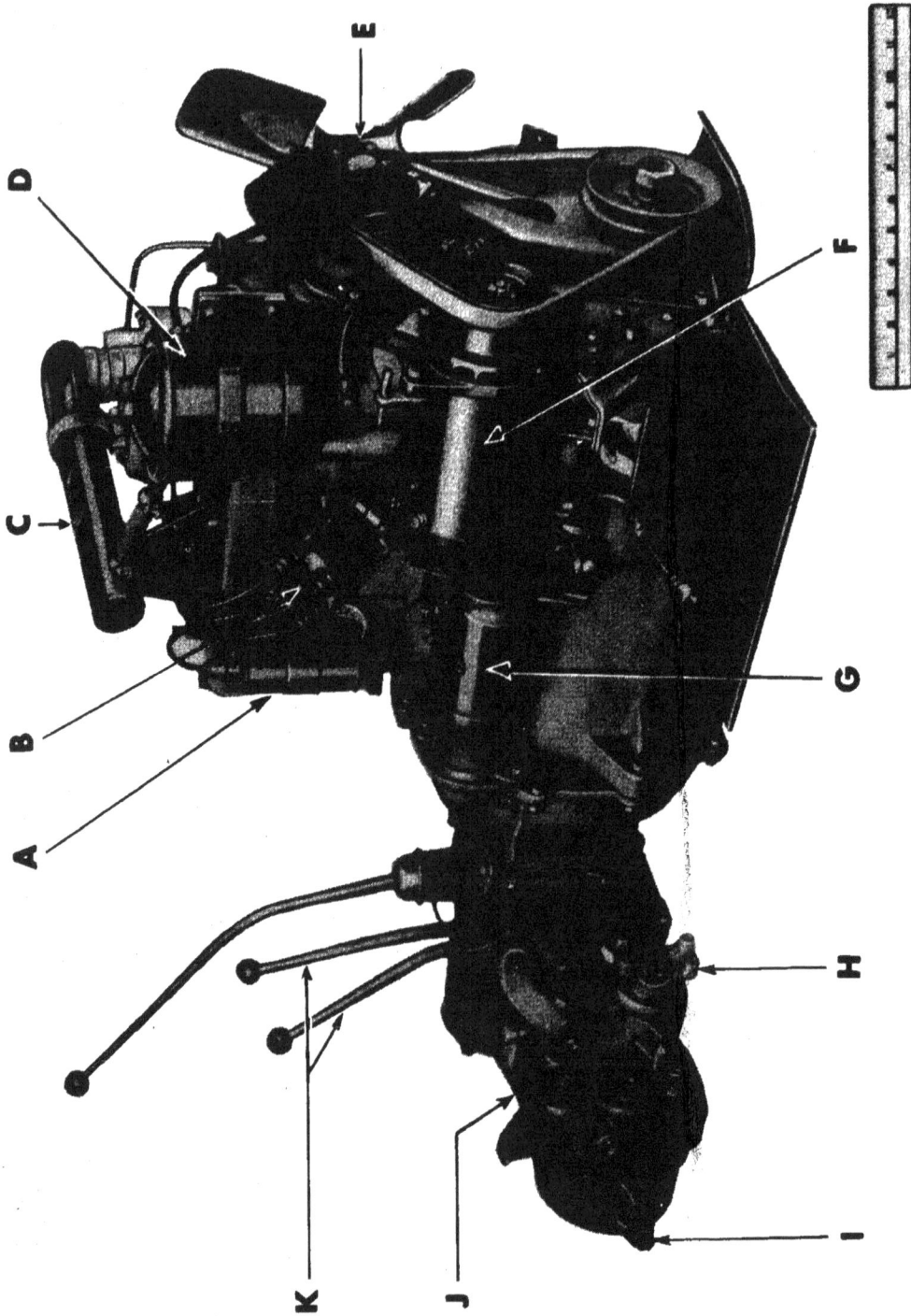

RA PD 305064

FIGURE F—POWER PLANT ASSEMBLY, RIGHT FRONT VIEW

Key	Item	Willys Part No.	Ford Part No.	Gov't Group No.
A	COIL assembly	WO-A-7792	FM-GPW-12000-B	0604
B	DISTRIBUTOR assembly	WO-A-1244	FM-GPW-1200	0603
C	TUBE w/BRACKET assembly	WO-A-6911	FM-GPW-9637-B	0301C
D	FILTER assembly	WO-A-1230	FM-GPW-18660-A	0107E
E	FAN assembly	WO-A-447	FM-GPW-8600	0503A
F	GENERATOR assembly	WO-A-5992	FM-GPW-10000-A	0601

Key	Item	Willys Part No.	Ford Part No.	Gov't Group No.
G	CRANKING motor assembly	WO-A-1245	FM-GPW-11001-A	0602
H	YOKE	WO-A-1106	FM-GP-7709	0803
I	BRAKE assembly	WO-A-1008	FM-GPW-2529	1201
J	TRANSFER assembly	WO-A-1195	FM-GPW-7700	0800
K	LEVER	WO-A-1505	FM-GPW-7793	0805B

RA PD 305064-A

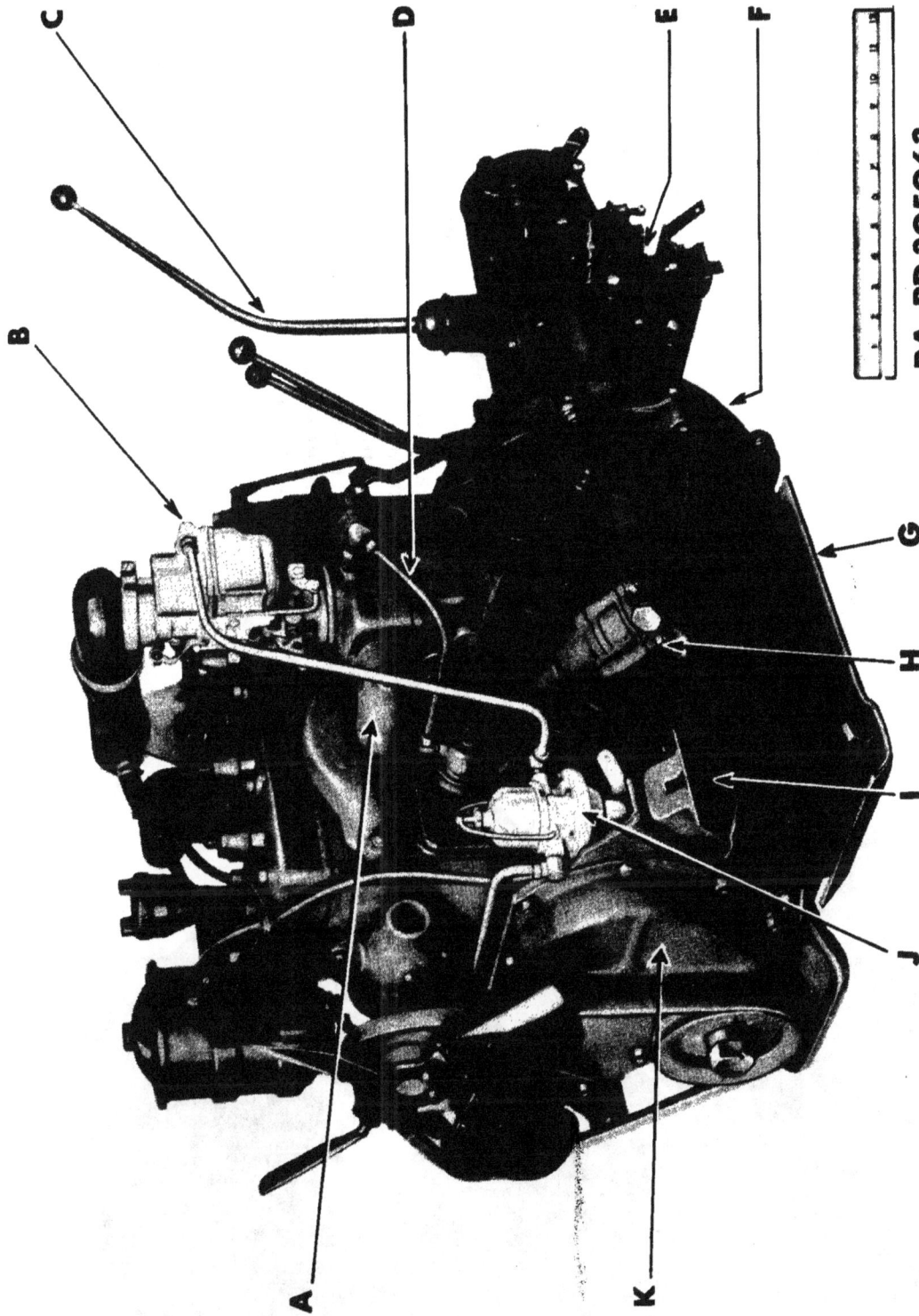

RA PD 305063

FIGURE G—POWER PLANT ASSEMBLY, LEFT FRONT VIEW

Key	Item	Willys Part No.	Ford Part No.	Gov't Group No.
G	PAN assembly	WO-A-7238	FM-GPW-6675	0107A
H	PUMP assembly	WO-637636	FM-GPW-6600	0107
I	INSULATOR assembly	WO-A-7498	FM-GPW-6038-A	0110
J	PUMP assembly	WO-A-8323	FM-GPW-9350-B	0302
K	COVER assembly	WO-A-1190	FM-GPW-6016	0106C

RA PD 305063-A

Key	Item	Willys Part No.	Ford Part No.	Gov't Group No.
A	MANIFOLD assembly	WO-A-1165	FM-GPW-9410-B	0108
B	CARBURETOR assembly	WO-A-1223	FM-GPW-9510	0301
C	LEVER assembly	WO-A-1380	FM-GPW-7210-A	0706A
D	TUBE assembly	WO-A-6922	FM-GPW-6756	0107N
E	TRANSMISSION assembly	WO-A-1145	FM-GPW-7000	0700
F	HOUSING assembly	WO-A-439	FM-GPW-6392	0109C

SNL G-503

FIGURE H—ENGINE, CROSS SECTIONAL, END VIEW

Key	Item	Willys Part No.	Ford Part No.	Gov't Group No.
A	OILER	WO-107128	FM-B-10141	0603
B	DISTRIBUTOR	WO-A-1244	FM-GPW-12100	0603
C	COIL	WO-A-1424	FM-GPW-12000-A	0604
D	GUIDE	WO-375811	FM-GPW-6510-B	0105
E	MANIFOLD	WO-A-1166	FM-GPW-9424-B	0108A
F	COVER assembly	WO-630303	FM-GPW-6520	0105C
G	VALVE	WO-636439	FM-GPW-9460	0108C
H	BAFFLE	WO-630298	FM-GPW-6762	0107N
I	MANIFOLD	WO-A-912	FM-GPW-9428	0108E
J	BODY	WO-A-699	FM-FPW-6758-B	0107N
K	SPRING	WO-637615	FM-GPW-12083	0603
L	GEAR	WO-637425	FM-GPW-6610	0107
M	DISC	WO-636600	FM-GPW-6673	0107
N	PUMP assembly	WO-637636	FM-GPW-6600	0107
O	PINION	WO-343306	FM-GPW-6614	0107
P	RETAINER	WO-630390	FM-GPW-6644	0107
Q	SHIM	WO-630389	FM-GPW-6628	0107
R	SPRING	WO-356155	FM-GPW-6654	0107
S	PLUNGER	WO-630518	FM-GPW-6663	0107
T	SHAFT assembly	WO-636599	FM-GPW-6608	0107
U	OIL PAN assembly	WO-A-7238	FM-GPW-6675	0107A
V	PLUG	WO-639979	FM-GPW-6727	0107A
W	SUPPORT	WO-630397	FM-GPW-6617	0107A-1
X	DOWEL	WO-635377	FM-GPW-6369	0102
Y	BOLT	WO-381519	FM-GPW-6345	0102
Z	FLOAT assembly	WO-630396	FM-GPW-6615	0107A-1
AB	TUBE	WO-A-6915	FM-GPW-6763C	0107G
AC	INDICATOR	WO-A-5168	FM-GPW-6766-B	0107H

RA PD 305045-A

SECTION 2

MAINTENANCE

PARTS

PROCUREMENT

LIST

MAINTENANCE PARTS PROCUREMENT LIST

MAJOR ITEMS AUTHORIZED FOR UNIT REPLACEMENT FOR 100 VEHICLES FOR 12 MONTHS

Truck, ¼-Ton, 4 x 4, Command Reconnaissance - - - - - - - - - - - - 76

| Figure Number | Official Stockage Number | Part Number | | ITEM | Quantity Reqd. per Unit Assy. | Per 100 Major Items | | | Estimated Reqmts. per 100 Rebuilds |
| | | Willys | Ford | | | 12 Mos. Field Maintenance | Major Overhaul (5th Ech) | Total First Year Procurement | |
Col. 1	Col. 2	Col. 3	Col. 4	Col. 5	Col. 6	Col. 7	Col. 8	Col. 9	Col. 10
				GROUP 01—ENGINE					
				0100—ENGINE ASSEMBLY					
		WO-A-5497	FM-GPW-6005	ENGINE assembly (includes CLUTCH, FLYWHEEL HOUSING, and all units ready to run)	1	18	—	18	—
		WO-A-1493		ENGINE assembly (Willys only) (less AIR CLEANER, FLYWHEEL HOUSING, CABLES, CARBURETOR, CLUTCH, COIL, DISTRIBUTOR, FAN, FAN BELT, FUEL PUMP, GENERATOR, OIL FILTER, SPARK PLUGS and CRANKING MOTOR)	1				
F, G		WO-A-5338	FM-GPW-6002	PLANT, power (includes ENGINE, TRANSFER CASE and TRANSMISSION complete, ready to run)	1	—	—	—	—
				0101—CRANKCASE AND CYLINDER BLOCK					
01-2		WO-A-1272	FM-GPW-6010	BLOCK, cylinder, engine, w/BEARING, assembly	1	—	—	—	—
01-12		WO-A-6793	FM-GPW-6009	BLOCK, cylinder, engine, w/BEARING and PISTON, assembly	1	—	3	4	10
01-2		WO-A-1126	FM-9N-8115	COCK, drain, engine cylinder water jacket (¼ in., S., z-pltd., taper pipe thread) (WH-145-A)	1	%12	4	18	13
		WO-A-1536	FM-GPW-18390	GASKET SET, engine overhaul	as req.	—	40	50	105
				(Includes:					
		1 WO-630365	FM-GPW-6288	GASKET, chain cover					
		1 WO-630359	FM-GPW-6020	GASKET, cylinder block					
		1 WO-A-8558	FM-GPW-6051-B	GASKET, cylinder head					
		1 WO-634814	FM-GPW-9450	GASKET, exhaust pipe flange					
		1 WO-638737	FM-GPW-9417	GASKET, fuel pump					
		1 WO-A-6357	FM-GPW-9445	GASKET, insulator and diffuser					
		1 WO-638640	FM-GPW-9448	GASKET, intake and exhaust manifold					
		1 WO-634811	FM-GPW-9435	GASKET, intake to exhaust manifold					
		1 WO-630398	FM-GPW-6627	GASKET, oil float support					
		1 WO-639980	FM-GPW-6710	GASKET, oil pan					
		1 WO-639870	FM-GPW-6659	GASKET, oil pump cover					
		4 WO-630392	FM-GPW-6619	GASKET, oil pump cover					
		1 WO-314338	FM-GPW-6734	GASKET, oil drain plug					
		1 WO-375027	FM-GPW-6625	GASKET, oil pump shaft					

FIGURE 01-1—ENGINE, SIDE VIEW

Key	Item	Willys Part No.	Ford Part No.	Gov't Group No.
A	FAN assembly	WO-A-447	FM-GPW-8600	0503A
B	SHAFT w/BEARING assembly	WO-636297	FM-GPW-8530	0503
C	WASHER	WO-644034	FM-GPW-8557-A	0503
D	SEAL assembly	WO-640031	FM-GAA-8524	0503
E	IMPELLER	WO-639993	FM-GPW-8512	0503
F	PISTON assembly	WO-637041	FM-GPW-6105-A	0103
G	PIN	WO-636961	FM-GPW-6135-A	0103B
H	THERMOSTAT	WO-637646	FM-GPW-8575	0502
I	ELBOW	WO-A-1192	FM-GPW-8250	0101A
J	RETAINER	WO-639651	FM-GPW-8578	0502
K	VALVE	WO-637183	FM-GPW-6505	0105
L	VALVE	WO-637182	FM-GPW-6507	0105
M	HEAD	WO-A-1534	FM-GPW-6050	0101A
N	MANIFOLD assembly	WO-A-912	FM-GPW-9428	0108B
O	SPRING	WO-638636	FM-GPW-6513	0105
P	SCREW	WO-640020	FM-GPW-6549-B	0105A
Q	PLATE assembly	WO-A-5121	FM-GPW-6044	0110
R	CAMSHAFT	WO-637065	FM-GPW-6250	0106
S	GEAR	WO-635394	FM-GPW-6384	0109B
T	PACKING	WO-637237	FM-GPW-6702	0102A
U	PIPE	WO-630294	FM-GPW-6326	0102A
V	BEARING	WO-638733	FM-GPW-6337-A	0102A
W	TAPPET	WO-637047	FM-GPW-6500-A	0105A
X	CRANKSHAFT	WO-A-7568	FM-GPW-18288	0102
Y	BOLT	WO-640070	(Willys only)	0104
Z	SUPPORT	WO-630397	FM-GPW-6167	0107A-1
AA	FLOAT assembly	WO-630396	FM-GPW-6615	0107A-1
AB	BEARING	WO-638731	FM-GPW-6341-A	0102A
AC	ROD	WO-640067	(Willys only)	0104
AD	NUT	WO-52825	FM-356028-S	0104
AE	BEARING	WO-637008	FM-GPW-6338-A	0102A
AF	WASHER	WO-634796	FM-GPW-6308	0102
AG	COVER assembly	WO-A-1190	FM-GPW-6016	0106D
AH	CHAIN	WO-638457	FM-GPW-6260	0106C
AI	SPROCKET	WO-638459	FM-GPW-6306	0106C
AJ	BELT	WO-A-1495	FM-GPW-8260	0503B
AK	PACKING	WO-637098	FM-GPW-6700	0102A
AL	NUT assembly	WO-387633	FM-GPW-6319	0102
AM	PULLEY	WO-638113	FM-GPW-6312	0102
AN	PLUNGER	WO-375907	FM-GPW-6243	0106
AO	BUSHING	WO-639051	FM-GPW-6262-A	0106A
AP	WASHER	WO-375900	FM-GPW-6245	0106
AQ	SPROCKET	WO-638458	FM-GPW-6246	0106C

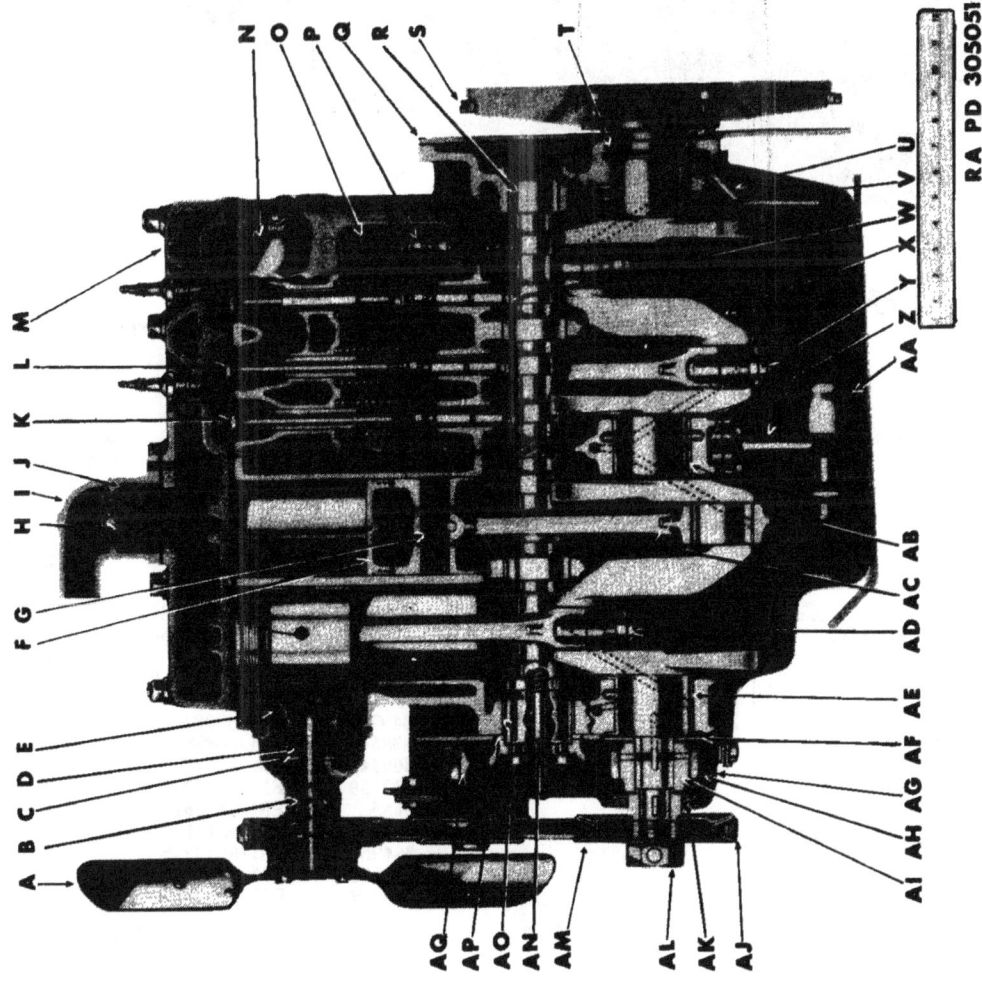

RA PD 305051-A

RA PD 305051

21

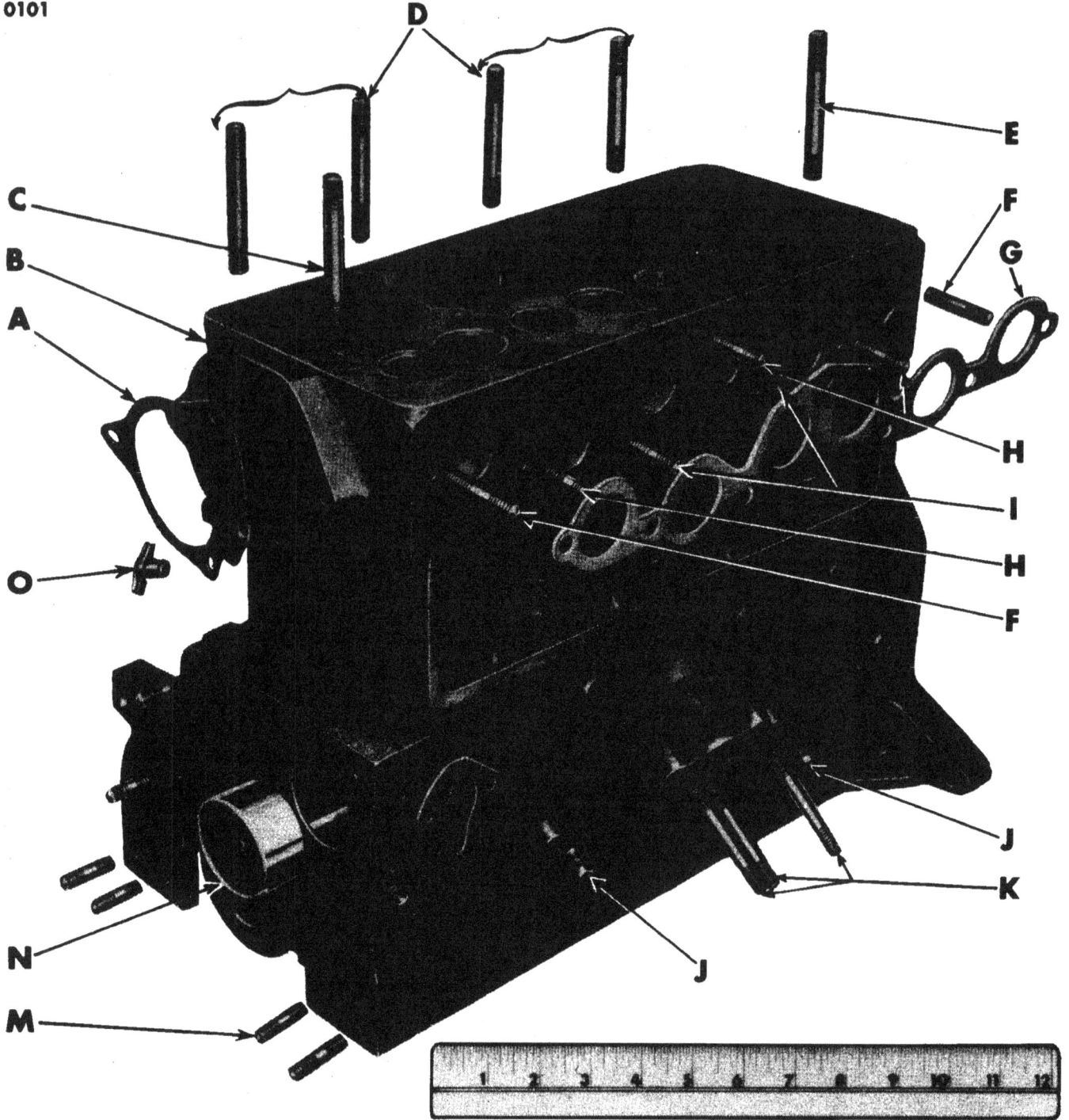

RA PD 305047

FIGURE 01-2—CYLINDER BLOCK ASSEMBLY

Key	Item	Willys Part No.	Ford Part No.	Gov't Group No.
A	GASKET	WO-637053	FM-GPW-8543	0503
B	BLOCK assembly	WO-A-1272	FM-GPW-6010	0101
C	STUD	WO-A-1549		0101A
D	STUD	WO-349368	FM-GPW-6066	0101A
E	STUD	WO-A-1548	FM-GPW-6067	0101A
F	STUD	WO-349712	FM-88082-S	0108
G	GASKET	WO-638640	FM-GPW-9448	0108
H	STUD	WO-300143	FM-88042	0108
I	STUD	WO-632159	FM-88057	0108
J	PLUG	WO-5085	FM-358064-S	0101
K	STUD	WO-375981	FM-88141-S	0107
M	STUD	WO-384958	FM-88022-S	0106D
N	BUSHING	WO-639051	FM-GPW-6262-A	0106A
O	COCK	WO-A-1126	FM-9N-8115	0101

RA PD 305047-A

Figure Number	Official Stockage Number	Part Number: Willys	Part Number: Ford	ITEM	Quantity Reqd. per Unit Assy.	Per 100 Major Items: 12 Mos. Field Maintenance	Per 100 Major Items: Major Overhaul (5th Ech)	Per 100 Major Items: Total First Year Procurement	Estimated Reqmts. per 100 Rebuilds
Col. 1	Col. 2	Col. 3	Col. 4	Col. 5	Col. 6	Col. 7	Col. 8	Col. 9	Col. 10
				0101—CRANKCASE AND CYLINDER BLOCK (Cont'd)					
				1 WO-630394 FM-GPW-6630 GASKET, oil pump to cylinder block					
				1 WO-634813 FM-GPW-6642 GASKET, oil relief spring retainer					
				1 WO-334103 FM-GPW-6353 GASKET, oil slinger					
				4 WO-637863 FM-OIA-12410 GASKET, spark plug					
				1 WO-630305 FM-GPW-6521 GASKET, valve spring cover					
				2 WO-51875 FM-GPW-6555 GASKET, valve cover screw					
				1 WO-630299 FM-GPW-6648 GASKET, ventilator to valve cover					
				1 WO-639650 FM-GPW-8255 GASKET, water outlet elbow					
				1 WO-637053 FM-GPW-8543 GASKET, water pump to cylinder block					
				1 WO-637098 FM-GPW-6700 PACKING, crankshaft					
				2 WO-637237 FM-GPW-6702 PACKING, rear bearing cap					
				2 WO-637790 FM-GPW-6701 PACKING, rear bearing cap					
		WO-A-1537	FM-GPW-18387	GASKET SET, engine, valve job	as req.	80	—	100	—
				(Includes:					
				1 WO-A-8558 FM-GPW-6051-B GASKET, cylinder head					
				1 WO-634814 FM-GPW-9450 GASKET, exhaust pipe flange					
				1 WO-A-6357 FM-GPW-9445 GASKET, insulator and diffuser assembly					
				1 WO-638640 FM-GPW-9448 GASKET, intake and exhaust manifold					
				1 WO-634811 FM-GPW-9435 GASKET, intake to exhaust manifold					
				4 WO-637863 FM-OIA-12410 GASKET, spark plug					
				2 WO-51875 FM-GPW-6555 GASKET, valve cover screw					
				1 WO-630305 FM-GPW-6521 GASKET, valve spring cover					
				1 WO-630299 FM-GPW-6648 GASKET, valve spring cover to ventilator					
				1 WO-639650 FM-GPW-8255 GASKET, water outlet elbow)					
	H6-02-82480	WO-51091	FM-74121-S	PLUG, expansion, S., 1¼ in. (BEDX1AT)	5	% 20	20	50	62
	H6-02-83805	WO-5138	FM-353055-S7	PLUG, pipe, sq-hd., black, ¼ in. (cylinder water jacket, drain) (CPMX1AB)	1	% —	—	—	—
	H6-02-83015	WO-376373	FM-358063-S	PLUG, pipe, alloy-S., ck., ⅜ in. (A164342C) (issue until stock exhausted)	2	% —	—	—	—
01-11	H6-02-83800	WO-5085	FM-358064-S	PLUG, pipe, sq-hd., black, ⅛ in. (CPMX1AA)	4	% —	—	—	—
				0101A—CYLINDER HEAD					
01-3		WO-A-1192	FM-GPW-8250	ELBOW, engine cylinder head water outlet (issue until stock exhausted)	1	% 4	—	—	—
01-3	H1-01-16026	WO-52911	FM-24408-S	BOLT, hex-hd., s-fin., alloy-S., ⅜-16NC-2 x 1⅛ (GM-106331) (elbow to cylinder head) (BANX1CC)	3	% —	—	—	37
01-3	H1-15-18113	WO-5010	FM-34807-S7	WASHER, lock, reg., S., ⅜ in. (13/32 I.D. x 21/32 O.D. x 3/32 thk.) (BECX1K)	3	% —	—	—	—

FIGURE 01-3—CYLINDER HEAD

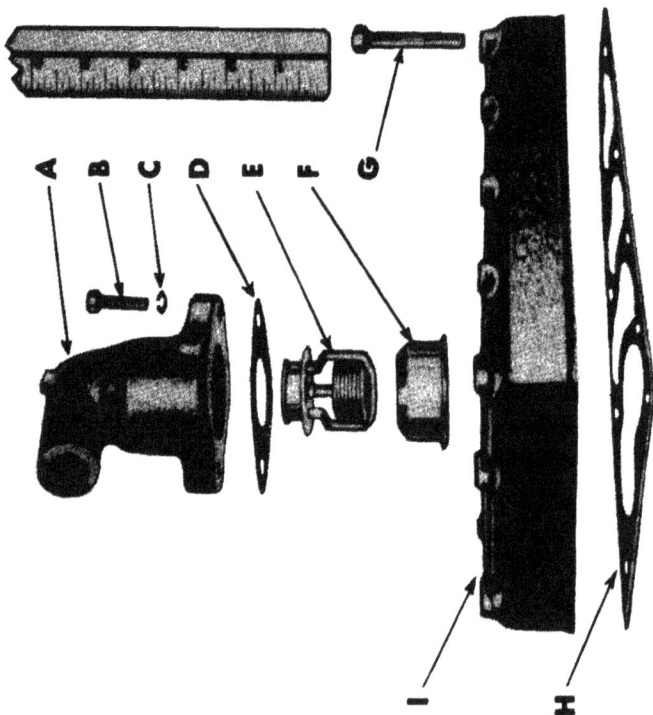

Key	Item	Willys Part No.	Ford Part No.	Gov't Group No.
A	ELBOW	WO-A-1192	FM-GPW-8250	0101A
B	BOLT	WO-52911	FM-24408-S	0101A
C	WASHER	WO-5010	FM-34807-S	0101A
D	GASKET	WO-639650	FM-GPW-8255	0101A
E	THERMOSTAT assembly	WO-637646	FM-GPW-8575	0502
F	RETAINER	WO-639651	FM-GPW-8578	0502
G	BOLT	WO-638635	FM-GPW-6065	0101A
H	GASKET	WO-638678	FM-GPW-6051	0101A
I	HEAD	WO-A-1534	FM-GPW-6050	0101A

RA PD 305046-A

RA PD 305046

GROUP 01—ENGINE (Cont'd)

Figure Number	Official Stockage Number	Part Number		ITEM	Quantity Reqd. per Unit Assy.	Per 100 Major Items				
		Willys	Ford			12 Mos. Field Maintenance	Major Overhaul (5th Ech)	Total FirstYear Procurement	Estimated Reqmts. per 100 Rebuilds	
Col. 1	Col. 2	Col. 3	Col. 4	Col. 5	Col. 6	Col. 7	Col. 8	Col. 9	Col. 10	

0101A—CYLINDER HEAD (Cont'd)

Figure Number	Official Stockage Number	Willys	Ford	ITEM	Quantity Reqd. per Unit Assy.	12 Mos. Field Maintenance	Major Overhaul (5th Ech)	Total FirstYear Procurement	Estimated Reqmts. per 100 Rebuilds
01-3		WO-A-8558	FM-GPW-6051-B	GASKET, engine cylinder head	1	%120	—	150	0
01-3		WO-639650	FM-GPW-8255	GASKET, engine cylinder head water outlet elbow	1	% 8	—	20	—
01-3		WO-A-1534	FM-GPW-6050	HEAD, cylinder engine (ratio 6.48)	1	% 3	3	8	10
01-3	H1-01-16045	WO-638635	FM-GPW-6065	BOLT, hex-hd., s-fin., alloy-S., $\frac{7}{16}$-14NC-2 (rolled thd.) x $2\frac{3}{4}$ (BANX1DL)	9	% 14	14	30	45
	H1-07-16265	WO-638539	FM-351025-S8	NUT, regular, hex., s-fin., alloy-S., $\frac{7}{16}$-20NF-2 (cylinder head stud)		% 16	10	90	30
				before Willys serial 288835)	4				
				after Willys serial 288835)	5				
				all Ford trucks	5				
		WO-A-1550	FM-356025-S	NUT, regular, hex., S., cd-pltd., $\frac{7}{16}$-20NF-2, cylinder head stud 10 and 12, before Willys serial 288835	2	% 8	5	50	15

24

01-2	WO-349368	FM-GPW-6066	STUD, S., 7/16-14NC x 3 3/16 x 7/16-20NF		% 6	6	20	20
12			cylinder head holes, No. 5, 7, 9, 15, before Willys serial 288835	4				
			cylinder head holes, No. 5, 7, 9, 10, 15, after Willys serial 28835, all Ford trucks	5				
01-2	WO-A-1549	FM-GPW-6060	STUD, S., cd-pltd., 7/16-14NC x 3 3/16 x 7/16-20NF					
			cylinder head hole No. 10, before Willys serial 288835 only	1	% 2	2	8	5
01-2	WO-A-1548	FM-GPW-6067	STUD, S., cd-pltd., 7/16-14NC x 3 7/8 x 7/16-20NF (Cylinder head hole No. 12)	1	% 2	2	8	5

0102—CRANKSHAFT

01-12	WO-381519	FM-GPW-6345	BOLT, hex-hd., S., 1/2-13NC x 2 3/8 (crankshaft bearing cap to crankcase)	6	—	10	12	30
01-4	WO-638121	FM-GPW-6303-A	CRANKSHAFT (issue until stock is exhausted, then use WO-A-7568)	1	4	12	—	—
01-4	WO-A-7568	FM-GPW-18288	CRANKSHAFT and DOWEL engine group assembly	1		3	4	10
			(Includes:					
			2 WO-116295 FM-GPW-6390 BOLT					
			1 WO-638121 FM-GPW-6303-A CRANKSHAFT					
			2 WO-52804 FM-33832-S2 NUT					
			2 WO-52330 FM-34909-S2 WASHER)					
01-4	WO-635377	FM-GPW-6369	DOWEL, engine crankshaft bearing	5	12	20	36	62
01-4	WO-334103	FM-GPW-6353	GASKET, engine crankshaft oil slinger	1	10	—	10	—
01-4	WO-5036	FM-74178-S	KEY, Woodruff, No. 9 (pulley to crankshaft)	1	—	8	10	25
01-1	WO-387633	FM-GPW-6319	NUT, engine crankshaft, w/PIN, assembly	1	2	2	6	8
	WO-28025	FM-357574	PIN, clutch, engine crankshaft nut (issue until stock exhausted)	1	4	8	—	—
01-1	WO-638113	FM-GPW-6312	PULLEY, engine crankshaft	1	10	5	18	14
01-12	WO-375920	FM-GPW-6287	RETAINER, engine crankshaft packing	1	8	32	48	100
	WO-630262	FM-GPW-6342-B	SHIM, engine crankshaft (.002 thk.)	as req.	—	32	36	100
01-12	WO-375877	FM-GPW-6310	SLINGER, oil, engine crankshaft	1	—	6	6	20
01-4	WO-634796	FM-GPW-6308	WASHER, thrust, engine crankshaft (S., 3 1/8 O.D. x 1.251 I.D. x .156 thk.)	1	—	3	6	10
01-12	WO-5009	FM-34809-S	WASHER, lock, reg., S., 1/2 (7/8 O.D. x 17/32 I.D. x 1/8 thk.) (bearing cap to crankcase bolt) (BECXIM)	6	% —	—	—	—

0102A—CRANKSHAFT BEARING

01-4	WO-637007	FM-GPW-6333-A	BEARING, engine crankshaft, front upper (std.)	1	—	—	—	—
01-4	WO-637008	FM-GPW-6338-A	BEARING, engine crankshaft, front lower (std.)	1	—	—	—	—
01-4	WO-638730	FM-GPW-6339-A	BEARING, engine crankshaft, center upper (std.)	1	—	—	—	—
01-4	WO-638731	FM-GPW-6341-A	BEARING, engine crankshaft, center lower (std.)	1	—	—	—	—
01-4	WO-638732	FM-GPW-6331-A	BEARING, engine crankshaft, rear upper (std.)	1	—	—	—	—
01-4	WO-638733	FM-GPW-6337-A	BEARING, engine crankshaft, rear lower (std.)	1	—	—	—	—
	WO-637724	FM-GPW-6333-B	BEARING, engine crankshaft, front upper (.010 U.S.)	as req.	—	—	—	—
	WO-637725	FM-GPW-6338-B	BEARING, engine crankshaft, front lower (.010 U.S.)	as req.	—	—	—	—
	WO-639237	FM-GPW-6339-B	BEARING, engine crankshaft, center upper (.010 U.S.)	as req.	—	—	—	—
	WO-639238	FM-GPW-6341-B	BEARING, engine crankshaft, center lower (.010 U.S.)	as req.	—	—	—	—
	WO-639239	FM-GPW-6331-B	BEARING, engine crankshaft, rear upper (.010 U.S.)	as req.	—	—	—	—
	WO-639240	FM-GPW-6337-B	BEARING, engine crankshaft, rear lower (.010 U.S.)	as req.	—	—	—	—
	WO-116522	FM-GPW-6333-C	BEARING, engine crankshaft, front upper (.020 U.S.)	as req.	—	—	—	—
	WO-116524	FM-GPW-6338-C	BEARING, engine crankshaft, front lower (.020 U.S.)	as req.	—	—	—	—
	WO-116526	FM-GPW-6339-C	BEARING, engine crankshaft, center upper (.020 U.S.)	as req.	—	—	—	—
	WO-116528	FM-GPW-6341-C	BEARING, engine crankshaft, center lower (.020 U.S.)	as req.	—	—	—	—
	WO-116530	FM-GPW-6331-C	BEARING, engine crankshaft, rear upper (.020 U.S.)	as req.	—	—	—	—

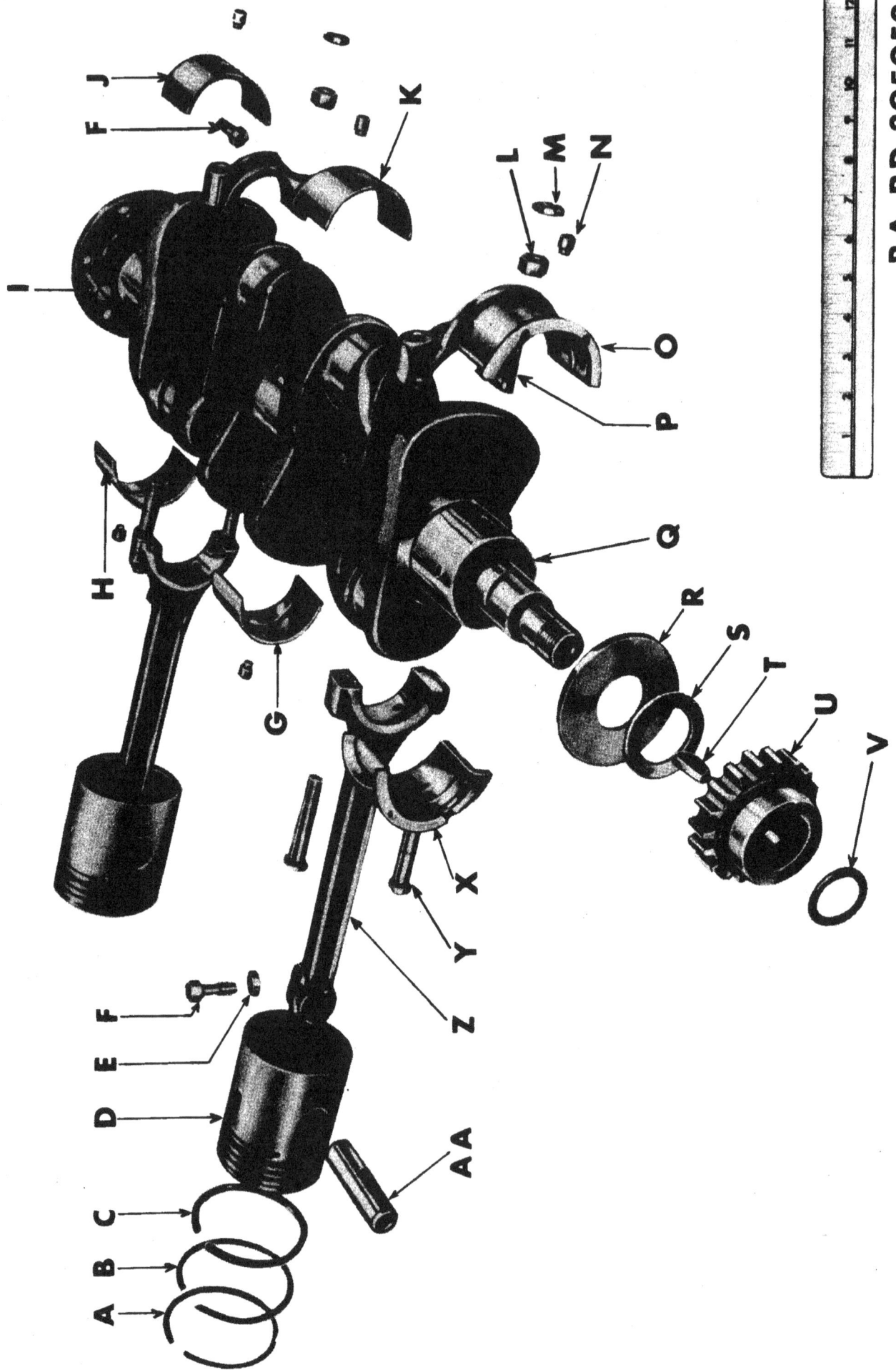

RA PD 305052

Key	Item	Willys Part No.	Ford Part No.	Gov't Group No.
A	RING	WO-639864	FM-GPW-6150-A	0103A
B	RING	WO-116562	FM-GPW-6152-A	0103A
C	RING	WO-116566	FM-GPW-6159-A	0103A
D	PISTON assembly	WO-637041	FM-GPW-6105-A	0103
E	WASHER	WO-5010	FM-34807-82	0103B
F	BOLT	WO-632157	FM-355497	0103B
G	BEARING	WO-638730	FM-GPW-6339-A	0102A
H	BEARING	WO-638732	FM-GPW-6331-A	0102A
I	DOWEL	WO-632156	FM-GPW-6387	0109
J	BOLT	WO-632157	FM-355497	0109
K	BEARING	WO-638733	FM-GPW-6337-A	0102A
L	BEARING	WO-638731	FM-GPW-6341-A	0102A
M	NUT	WO-636962	FM-356021-S	0104
N	NUT	WO-52825	FM-356028-S	0104

Key	Item	Willys Part No.	Ford Part No.	Gov't Group No.
O	DOWEL	WO-635377	FM-GPW-6369	0102
P	BEARING	WO-637008	FM-GPW-6338-A	0102A
Q	BEARING	WO-639862	FM-GPW-6211-A	0104A
R	CRANKSHAFT	WO-A-7568	FM-GPW-18288	0102
S	WASHER	WO-634796	FM-GPW-6308	0102
T	SPACER	WO-630727	FM-GPW-6342-A	0106C
U	KEY	WO-50917	FM-74182-S	0102
V	SPROCKET	WO-638459	FM-GPW-6306	0106C
W	GASKET	WO-334103	FM-GPW-6353	0102
X	BEARING	WO-637007	FM-GPW-6333-A	0102A
Y	BOLT	WO-381519	FM-GPW-6345	0102
Z	ROD assembly	WO-640066	.....	0104
AA	PIN	WO-636961	FM-GPW-6135-A	0103B

RA PD 305052A

GROUP 01—ENGINE (Cont'd)

0102A—CRANKSHAFT BEARING—Cont'd

Figure Number	Official Stockage Number	Part Number — Willys	Part Number — Ford	ITEM	Quantity Reqd. per Unit Assy.	12 Mos. Field Maintenance	Major Overhaul (5th Ech)	Total First Year Procurement	Estimated Reqmts. per 100 Rebuilds
Col. 1	Col. 2	Col. 3	Col. 4	Col. 5	Col. 6	Col. 7	Col. 8	Col. 9	Col. 10
		WO-116532	FM-GPW-6337-C	BEARING, engine crankshaft, rear lower (.020 U.S.)	as req.	—	—	—	—
		WO-A-6798	FM-GPW-18347	BEARING SET, engine crankshaft (std.)	as req.	4	8	15	25
				(Includes:					
				1 WO-637008 FM-GPW-6338-A BEARING					
				1 WO-637007 FM-GPW-6333-A BEARING					
				1 WO-638730 FM-GPW-6339-A BEARING					
				1 WO-638731 FM-GPW-6341-A BEARING					
				1 WO-638732 FM-GPW-6331-A BEARING					
				1 WO-638733 FM-GPW-6337-A BEARING)					
		WO-A-6746	FM-GPW-18348	BEARING SET, engine crankshaft (.010 U.S.)	as req.	2	8	13	25
				(Includes:					
				1 WO-637724 FM-GPW-6333-B BEARING					
				1 WO-637725 FM-GPW-6338-B BEARING					
				1 WO-639237 FM-GPW-6339-B BEARING					
				1 WO-639238 FM-GPW-6341-B BEARING					
				1 WO-639239 FM-GPW-6331-B BEARING					
				1 WO-639240 FM-GPW-6337-B BEARING)					

GROUP 01—ENGINE (Cont'd)

Col. 1 Figure Number	Col. 2 Official Stockage Number	Col. 3 Willys	Col. 4 Ford	Col. 5 ITEM	Col. 6 Quantity Reqd. per Unit Assy.	Col. 7 12 Mos. Field Maintenance	Col. 8 Major Overhaul (5th Ech)	Col. 9 Total First Year Procurement	Col. 10 Estimated Reqmts. per 100 Rebuilds
				0102A—CRANKSHAFT BEARING (Cont'd)					
		WO-A-6747	FM-GPW-18349	BEARING SET, engine crankshaft (.020 U. S.)............	as req.	3	16	22	50
				(Includes:					
				1 WO-116522 FM-GPW-6333-C BEARING					
				1 WO-116524 FM-GPW-6338-C BEARING					
				1 WO-116526 FM-GPW-6339-C BEARING					
				1 WO-116528 FM-GPW-6341-C BEARING					
				1 WO-116530 FM-GPW-6331-C BEARING					
				1 WO-116532 FM-GPW-6337-C BEARING)					
01-12		WO-637098	FM-GPW-6700	PACKING, engine crankshaft, front end............	1	8	—	10	—
01-1		WO-637237	FM-GPW-6702	PACKING, engine crankshaft, rear end (CN-type 310-0).......	2	16	—	20	—
01-12		WO-637790	FM-GPW-6701	PACKING, engine crankshaft rear bearing cap............	2	16	—	20	25
		WO-337112	FM-357420-S	PIN, drain tube to engine crankshaft rear bearing cap (S., $\tfrac{1}{8}$ x $\tfrac{7}{8}$).......	1	—	8	12	—
01-12		WO-630294	FM-GPW-6326	PIPE, drain engine, crankshaft rear bearing cap.	1	—	5	6	15
				0103—PISTONS					
		WO-637037		PISTON, engine, assembly (std.) (conventional type) (Willys only)........	4				
		WO-116017		PISTON, engine, assembly (.020 O.S.) (conventional type) (Willys only).......	as req.				
		WO-116018		PISTON, engine, assembly (.030 O.S.) (conventional type) (Willys only).......	as req.				
		WO-116558		PISTON, engine, assembly (std.) (re-ring type) (Willys only)........	4				
		WO-116560		PISTON, engine, assembly (.020 O.S.) (re-ring type) (Willys only)........	as req.				
		WO-116561		PISTON, engine, assembly (.030 O.S.) (re-ring type) (Willys only)........	as req.				
		WO-116612		PISTON, engine, semi-finished (use until stock is exhausted) (Willys only)	—	16	32	8	
01-4		WO-637041	FM-GPW-6105-A	PISTON, engine, w/PIN, assembly (std. size re-ring)........	4	6			
		WO-116019	FM-GPW-6105-C	PISTON, engine, w/PIN, assembly (.020 O.S. re-ring)........	as req.	3	80	92	250
		WO-116020	FM-GPW-6105-D	PISTON, engine, w/PIN, assembly (.030 O.S. re-ring)........	as req.	3	48	60	150
				0103A—RINGS					
01-4		WO-639864	FM-GPW-6150-A	RING, engine, piston, conventional or re-ring (top groove) (std.).......	4				
		WO-116502	FM-GPW-6150-C	RING, engine, piston, conventional or re-ring (top groove) (.020 O.S.).......	as req.				
		WO-116503	FM-GPW-6150-D	RING, engine, piston, conventional or re-ring type (top groove) (.030 O.S.).......	as req.				
		WO-637042	FM-GPW-6155-A	RING, engine, piston, conventional type (second groove) (std.).......	4				
		WO-116023	FM-GPW-6155-G	RING, engine, piston, conventional type (second groove) (.020 O.S.).......	as req.				
		WO-116024	FM-GPW-6155-H	RING, engine, piston, conventional type (second groove) (.030 O.S.).......	as req.				
		WO-116616	FM-GPW-6156-F	RING, engine, piston, conventional type (bottom groove) (std.).......	4				
		WO-116116	FM-GPW-6156-H	RING, engine, piston, conventional type (bottom groove) (.020 O.S.).......	as req.				
		WO-116117	FM-GPW-6155-J	RING, engine, piston, conventional type (bottom groove) (.030 O.S.).......	as req.				
01-4		WO-116562	FM-GPW-6152-A	RING and SPRING, engine piston, 2nd groove (std., to .009 O.S.).......	as req.				

(Parts list continued from the preceding page — section 0103A)

Ref.	WO No.	FM No.	Description	Qty.	A	B	C	D
	WO-116568	FM-GPW-6159-C	RING and SPRING, engine piston, bottom groove (.020 to .029 O.S.)	as req.	—	—	—	—
	WO-116569	FM-GPW-6159-D	RING and SPRING, engine piston, bottom groove (.030 to .039 O.S.)	as req.	—	—	—	—
	WO-A-6794	FM-GPW-6149-E	RING SET, engine piston, conventional type (std.)	1	—	—	—	—
	WO-A-6796	FM-GPW-6149-G	RING SET, engine piston, conventional type (.020 O.S.)	as req.	70	5	22	3
	WO-A-6797	FM-GPW-6149-H	RING SET, engine piston, conventional type (.030 O.S.)	as req.	45	28	14	2
	WO-116110	FM-GPW-6149-A	RING SET, engine piston (std.) (re-ring type) (issue until stock exhausted)	as req.	—	19	16	8
	WO-116112	FM-GPW-6149-C	RING SET, engine piston (.020 O.S.) (re-ring type) (issue until stock exhausted)	as req.	—	—	8	4
	WO-116113	FM-GPW-6149-D	RING SET, engine piston (.030 O.S.) (re-ring) (issue until stock exhausted)	as req.	—	—	4	4

0103B—PISTON PINS

Ref.	WO No.	FM No.	Description	Qty.	A	B	C	D
01-4	WO-636961	FM-GPW-6135-A	PIN, engine piston (std. size)	4	20	8	6	—
01-4	WO-632157	FM-355497	SCREW, lock, engine piston pin	4	100	62	32	3
01-4	WO-5010	FM-34807-S2	WASHER, lock, reg., S., 3/8 in. (13/32 I.D. x 21/32 O.D. x 3/32 thk.) (piston pin lock screw) (BECX1K)	4	—	—	—	—

0104—CONNECTING RODS

Ref.	WO No.	FM No.	Description	Qty.	A	B	C	D
01-4	WO-640070	FM-GPW-6200	BOLT, engine connecting rod bearing cap (S., 7/16-20NF x 2 3/8) (Willys only)	8	50	30	16	10%
01-4	WO-636962	FM-356021-S	NUT, hex, S., 7/16-20NF-2 (connecting rod bolt)	8	100	100	32	32%
01-4	WO-52825	FM-356028-S	NUT, hex, spring, S., 7/16-20NF-2, stamped (connecting rod bolt) (GM-107381)	8	800	400	256	96%
01-4	WO-640071	FM-GPW-6200	ROD, connecting, engine, No. 1 and 3, assembly (less bearings)	2	5	6	2	2
01-4	WO-640072	FM-GPW-6201	ROD, connecting, engine, No. 2 and 4, assembly (less bearings)	2	5	6	2	2
01-4	WO-640066		ROD, connecting, engine, No. 1 and 3, w/BEARING, assembly (Willys only)	2	—	—	—	—
01-4	WO-640067		ROD, connecting, engine, No. 2 and 4, w/BEARING, assembly (Willys only)	2	—	—	—	—

0104A—CONNECTING ROD BEARINGS

Ref.	WO No.	FM No.	Description	Qty.	A	B	C	D
01-4	WO-639862	FM-GPW-6211-A	BEARING, engine connecting rod (std.)	8	—	91	32	48
	WO-116534	FM-GPW-6211-B	BEARING, engine connecting rod (.010 U.S.)	as req.	—	—	—	—
	WO-116535	FM-GPW-6211-C	BEARING, engine connecting rod (.020 U.S.)	as req.	—	—	—	—
	WO-A-7233	FM-GPW-18330-A	BEARING SET, engine connecting rod (std.) (Includes: 2 WO-639862 FM-GPW-6211-A BEARINGS)	4	—	—	—	—
	WO-A-7234	FM-GPW-18330-B	BEARING SET, engine connecting rod (.010 U.S.) (Includes: 2 WO-116534 FM-GPW-6211-B BEARINGS)	as req.	100	53	32	16
	WO-A-7235	FM-GPW-18330-C	BEARING SET, engine connecting rod (.020 U.S.) (Includes: 2 WO-116535 FM-GPW-6211-C BEARINGS)	as req.	200	99	64	26

0105—VALVES, SEATS, GUIDES AND SPRING

Ref.	WO No.	FM No.	Description	Qty.	A	B	C	D
H	WO-375811	FM-GPW-6510-B	GUIDE, engine exhaust valve stem	4	60	24	19	—
01-5	WO-637045	FM-GPW-6511-B	GUIDE, engine intake valve stem	4	60	24	19	—
01-5	WO-375994	FM-GPW-6546	LOCK, engine valve spring retainer (lower)	16	150	300	48	144
01-5	WO-637044	FM-GPW-6514	RETAINER, engine valve spring (lower)	8	60	50	19	10
01-5	WO-638636	FM-GPW-6513	SPRING, engine valve	8	150	88	48	29
01-1	WO-637183	FM-GPW-6505	VALVE, exhaust, engine (RMC-A-366)	8	100	58	32	19
01-1	WO-637182	FM-GPW-6507	VALVE, intake, engine (RMC-A-365)	8	50	30	16	10

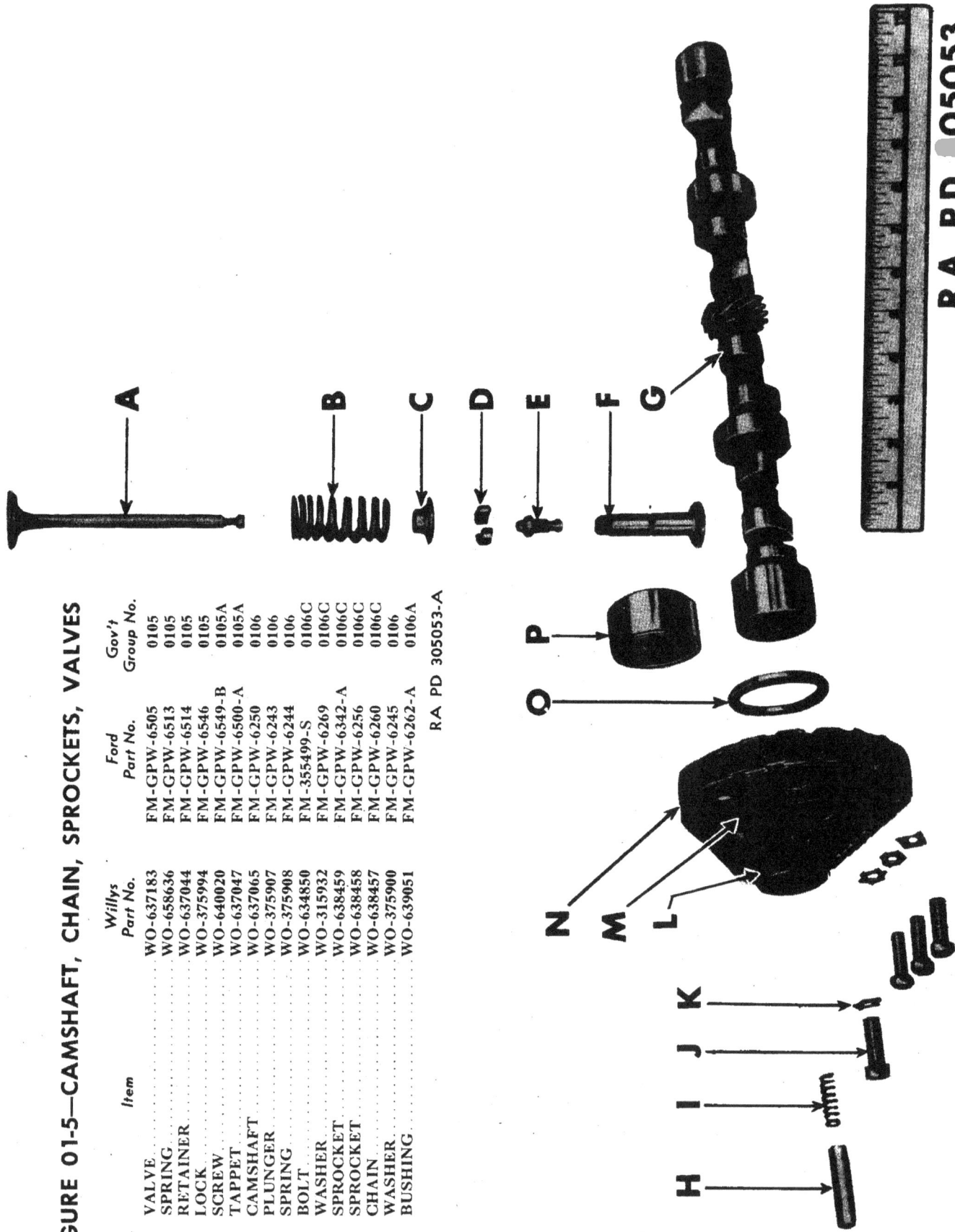

FIGURE 01-5—CAMSHAFT, CHAIN, SPROCKETS, VALVES

Key	Item	Willys Part No.	Ford Part No.	Gov't Group No.
A	VALVE	WO-637183	FM-GPW-6505	0105
B	SPRING	WO-658636	FM-GPW-6513	0105
C	RETAINER	WO-637044	FM-GPW-6514	0105
D	LOCK	WO-375994	FM-GPW-6546	0105
E	SCREW	WO-640020	FM-GPW-6549-B	0105A
F	TAPPET	WO-637047	FM-GPW-6500-A	0105A
G	CAMSHAFT	WO-637065	FM-GPW-6250	0106
H	PLUNGER	WO-375907	FM-GPW-6243	0106
I	SPRING	WO-375908	FM-GPW-6244	0106
J	BOLT	WO-634850	FM-355499-S	0106C
K	WASHER	WO-315932	FM-GPW-6269	0106C
L	SPROCKET	WO-638459	FM-GPW-6342-A	0106C
M	SPROCKET	WO-638458	FM-GPW-6256	0106C
N	CHAIN	WO-638457	FM-GPW-6260	0106C
O	WASHER	WO-375900	FM-GPW-6245	0106
P	BUSHING	WO-639051	FM-GPW-6262-A	0106A

RA PD 305053-A

RA PD 305053

Figure Number Col. 1	Official Stockage Number Col. 2	Willys Col. 3	Ford Col. 4	ITEM Col. 5	Quantity Reqd. per Unit Assy. Col. 6	12 Mos. Field Maintenance Col. 7	Major Overhaul (5th Ech) Col. 8	Total First Year Procurement Col. 9	Estimated Reqmts. per 100 Rebuilds Col. 10
				0105A—TAPPETS					
01-5		WO-640020	FM-GPW-6549-B	SCREW, adjusting, engine valve tappet (self locking)	8	—	19	25	60
01-5		WO-637047	FM-GPW-6500-A	TAPPET, engine valve (WIL-ST-1265) (CCS-W-41-25)	8	—	13	14	40
		WO-115948	FM-GPW-6500-B	TAPPET, engine valve (.004 O.S.) (issue until stock exhausted)	as req.	28	52	—	—
				0105C—VALVE COVERS AND GASKETS					
01-6		WO-630303	FM-GPW-6520	COVER, engine valve spring	1	% 2	2	8	5
01-6		WO-632158	FM-355451-S	BOLT, hex-hd., S., 5/16-18NC-2 x 4 5/8 (front)	1	% 5	3	16	10
01-6		WO-639052	FM-355452-S	BOLT, hex-hd., S., 5/16-18NC-2 x 3 1/16 (rear)	1	% 5	3	16	10
01-6		WO-630305	FM-GPW-6521	GASKET, engine valve spring cover	1	% 32	11	60	—
01-6		WO-51875	FM-GPW-6555	GASKET, engine valve spring cover bolt	2	% 48	—	60	12
				0106—CAMSHAFT					
01-5		WO-637065	FM-GPW-6250	CAMSHAFT	1	—	2	3	5
01-5		WO-375907	FM-GPW-6243	PLUNGER, camshaft thrust	1	5	5	12	10
01-5		WO-375908	FM-GPW-6244	SPRING, camshaft thrust plunger	1	5	5	12	15
01-5		WO-375900	FM-GPW-6245	WASHER, camshaft thrust	1	—	3	5	10
				0106A—CAMSHAFT BEARINGS					
01-2, 5		WO-639051	FM-GPW-6262-A	BUSHING, camshaft, front	1	—	32	35	100
		WO-51460	FM-74127	PLUG, expansion, engine camshaft rear bearing, (1¾ in. O.D.)	1	—	3	5	10
				0106C—CHAINS AND SPROCKETS					
01-4		WO-50917	FM-74182-S	KEY, engine crankshaft sprocket	1	—	8	10	25
01-5		WO-638457	FM-GPW-6260	CHAIN, drive, engine camshaft (½ in. thk. x 1 in. wide) (LK-S-40936)	1	6	6	15	20
01-4		WO-630727	FM-GPW-6342-A	SPACER, engine crankshaft sprocket	1	—	3	3	10
01-4		WO-638459	FM-GPW-6306	SPROCKET, engine crankshaft, S., 18 teeth, 2.870 O.D. x 1.656 thk. (LK-S-35116-1)	1	3	3	9	10
01-5		WO-638458	FM-GPW-6256	SPROCKET, engine camshaft, S., 36 teeth, 5.780 O.D. x 1.013 thk. (LK-S-35117-1)	1	3	3	9	10
01-5	H1-01-16220	WO-634850	FM-355499-S	BOLT, hex-hd., s-fin, alloy-S., 3/8-24NF x 1¼ (to camshaft) (BAOXKD)	4	6	16	40	50
01-5		WO-315932	FM-GPW-6269	WASHER, lock, reg., S., 3/4 in. (0.390 I.D. x .640 O.D. x 3/32 thk.) (BECX1K)	4	16	112	150	350

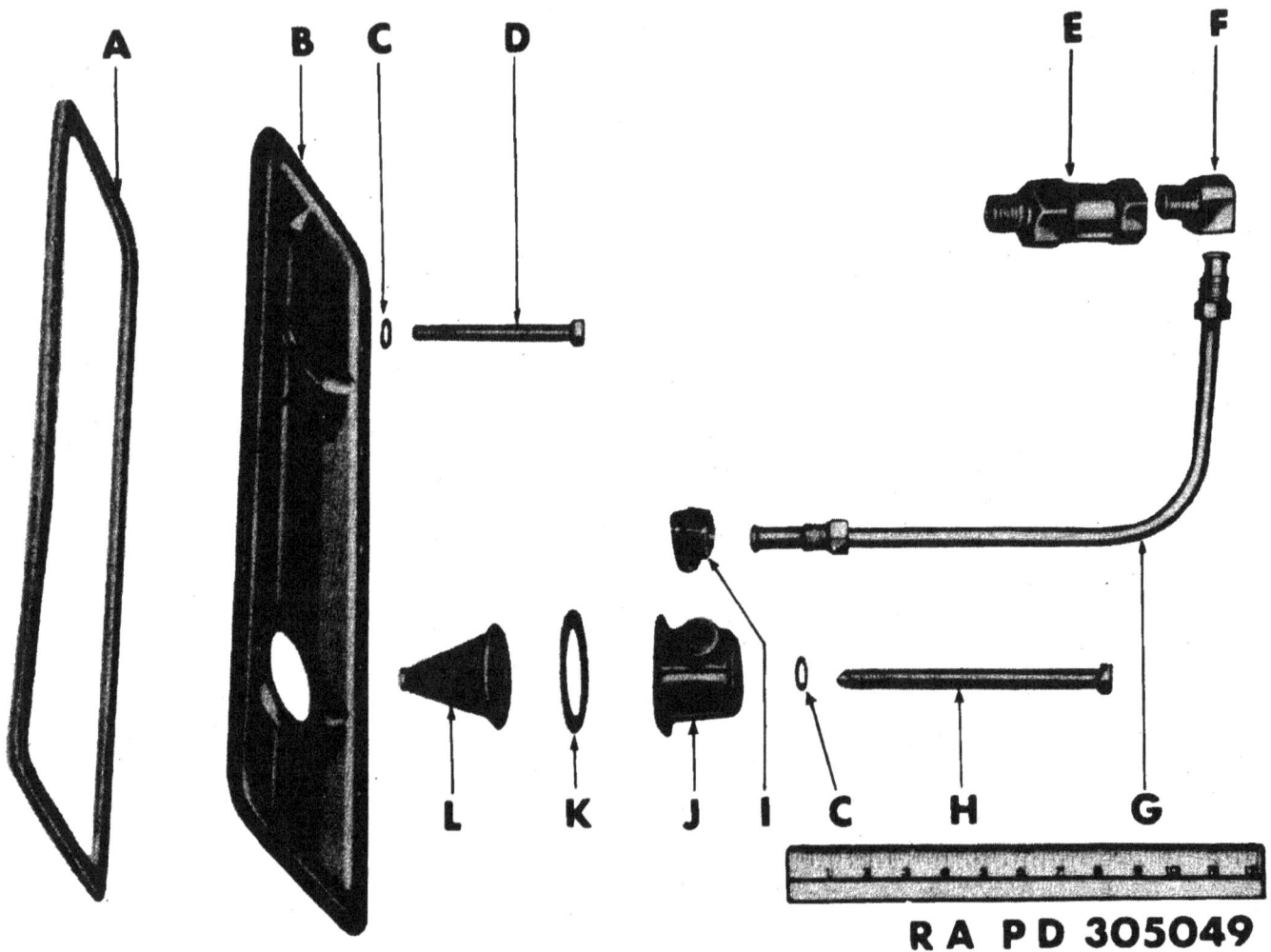

RA PD 305049

FIGURE 01-6—VALVE COVER PLATE WITH CRANKCASE VENTILATOR

Key	Item	Willys Part No.	Ford Part No.	Gov't Group No.
A	GASKET	WO-630305	FM-GPW-6521	0105C
B	COVER	WO-630303	FM-GPW-6519	0105C
C	GASKET	WO-51875	FM-GPW-6555	0105C
D	BOLT	WO-639052	FM-355452	0105C
E	VALVE assembly	WO-A-6895	FM-GPW-6769	0107N
F	ELBOW	WO-A-6885	FM-GPW-6722	0107N
G	TUBE assembly	WO-A-6922	FM-GPW-6756	0107N
H	BOLT	WO-632158	FM-355451-S	0105C
I	ELBOW	WO-384549	FM-GPW-9268	0107N
J	BODY assembly	WO-A-6919	FM-GPW-6758-B	0107N
K	GASKET	WO-630299	FM-GPW-6648	0107N
L	BAFFLE	WO-630298	FM-GPW-6762	0107N

RA PD 305049-A

GROUP 01—ENGINE (Cont'd)

Figure Number	Official Stockage Number	Part Number (Willys)	Part Number (Ford)	ITEM	Quantity Reqd. per Unit Assy.	Per 100 Major Items — 12 Mos. Field Maintenance	Per 100 Major Items — Major Overhaul (5th Ech)	Total First Year Procurement	Estimated Reqmts. per 100 Rebuilds
Col. 1	Col. 2	Col. 3	Col. 4	Col. 5	Col. 6	Col. 7	Col. 8	Col. 9	Col. 10
				0106D—CHAIN CASE COVER					
G	H1-01-16217	WO-A-1190	FM-GPW-6016	COVER, engine timing chain, assembly	1	—	2	3	5
		WO-50163	FM-20349-S2	BOLT, hex-hd, s-fin, alloy-S, 3/8-24NF-3 x 3/4 (GM-100025) (cover to engine plate)	3	—	—	—	30
	H1-01-16219	WO-5919		BOLT, hex-hd, S, 3/8-24NF-3 x 1 (GM-100026) (cover to engine plate) (BAOX1CC) (Willys only)	1	—	—	—	—
			FM-350356-S2	BOLT, hex-hd, s-fin, alloy-S, 3/8-24NF-3 x 3/4 (Ford only)	1	—	—	—	—
		WO-5901	FM-33800-S2	NUT, reg, hex, s-fin, alloy-S, 3/8-24NF-2 (BBB1CA)	10	—	—	—	—
01-2 12	H1-07-19003	WO-384958	FM-88022-S	STUD, S, 3/8-16NC x 1 5/16 x 3/8-24NF (GM-103026) (issue until stock exhausted) (4 used plate and chain cover to cylinder block 2 used plate to cylinder block)	6	92	188	—	—
		WO-5010	FM-34807-S2	WASHER, lock, S, 3/8 in. (2 1/32 O.D. x 1 3/32 I.D. x 3/64 thk.) (BECX1K) Willys	10	—	—	—	—
				Ford	6	—	—	—	—
01-12		WO-630365	FM-GPW-6288	GASKET, engine timing chain cover	1	8	—	10	30
01-12		WO-630359	FM-GPW-6020	GASKET, engine cylinder block, front (issue until stock exhausted)	1	40	80	—	15
		WO-375917	FM-GPW-6286	RING, retaining, engine timing chain cover packing (issue until stock exhausted)	1	4	4	—	—
		WO-630364	FM-GPW-6285	STUD, engine camshaft thrust plunger (issue until stock exhausted)	1	4	4	—	—
				0107—OIL PUMP					
01-7		WO-630384	FM-GPW-6604	BODY, engine oil pump	1	—	—	—	—
01-7		WO-630387	FM-GPW-6664	COVER, engine oil pump, assembly	1	—	4	—	—
01-7	H1-1026108	WO-51819	FM-310079-S	SCREW, machine, oval fil-hd, S, No. 12 (.216)-24NC-2 x 5/8 (cover to body) (BCFX2CG)	1	—	—	—	—
01-7		WO-636600	FM-GPW-6673	DISC, engine oil pump rotor	1	10	10	24	30
01-7		WO-630392	FM-GPW-6619	GASKET, engine oil pump cover	1	3	5	12	15
01-7		WO-639870	FM-GPW-6659	GASKET, engine oil pump cover (vellumoid)	1	—	—	—	—
01-7		WO-380197	FM-355262	GASKET, engine oil pump cover to body (copper) (issue until stock exhausted)	1	32	64	—	—
01-7		WO-375927	FM-GPW-6625	GASKET, engine oil pump shaft	1	—	—	—	—
01-7		WO-630394	FM-GPW-6630	GASKET, engine oil pump to cylinder block	1	6	—	10	—
01-7		WO-634813	FM-GPW-6642	GASKET, engine oil pump relief spring retainer (paper)	1	—	—	—	—
01-7		WO-637425	FM-GPW-6610	GEAR, driven, engine oil pump (12 teeth, 1.1369 O.D. x 3/16 thk.)	1	—	—	—	—
01-7		WO-A-6750	FM-GPW-18380	GASKET SET, service, engine oil pump (Includes: 1 WO-380197 FM-355262-S GASKET 1 WO-630398 FM-GPW-6627 GASKET, oil float support)	as req.	10	37	50	115

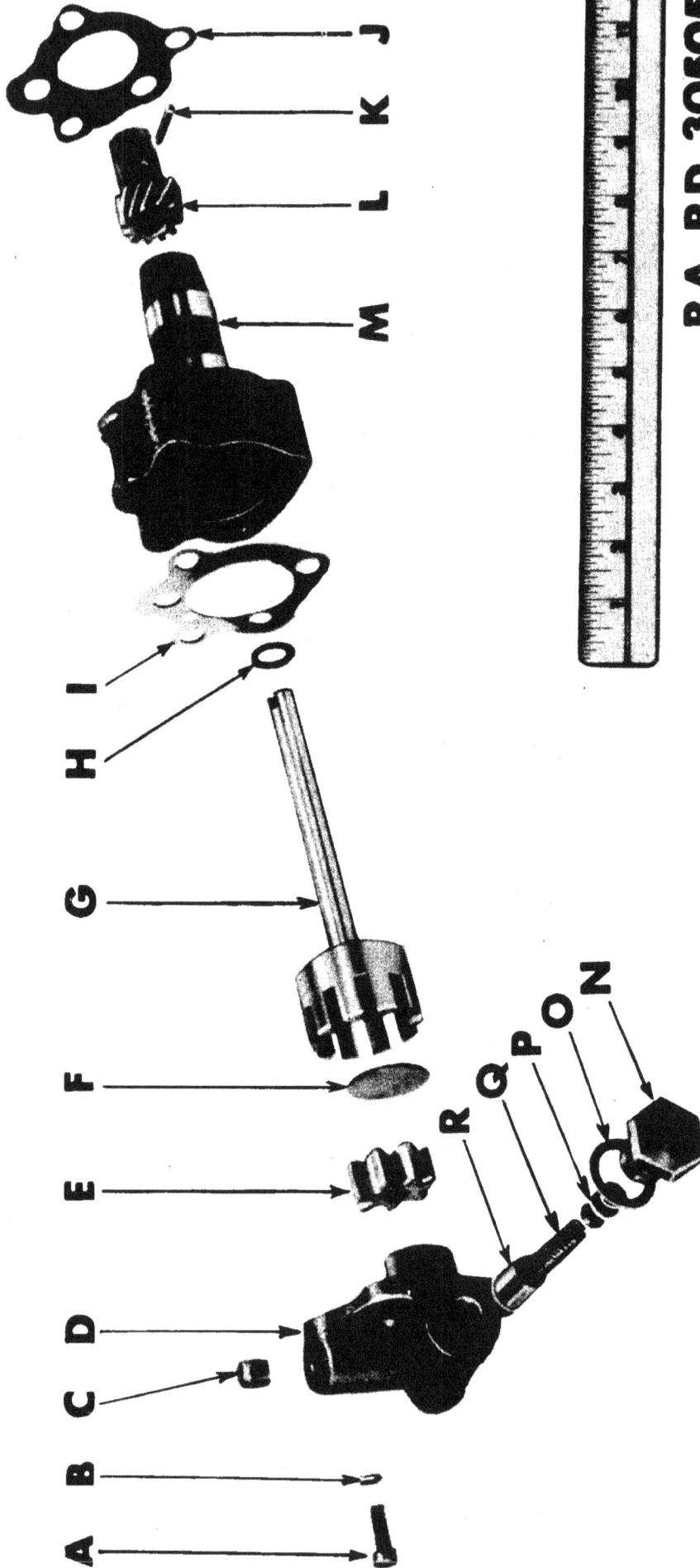

RA PD 305054

FIGURE 01-7—OIL PUMP ASSEMBLY

Key	Item	Willys Part No.	Ford Part No.	Gov't Group No.
A	SCREW	WO-15819	FM-31079-S	0107
B	GASKET	WO-380197	FM-355262	0107
C	PLUG	WO-52525	FM-353052	0107
D	COVER assembly	WO-630387	FM-GPW-6664	0107
E	PINION	WO-343306	FM-GPW-6614	0107
F	DISC	WO-636600	FM-GPW-6673	0107
G	SHAFT assembly	WO-636599	FM-GPW-6608	0107
H	GASKET	WO-375927	FM-GPW-6625	0107
I	GASKET	WO-639870	FM-GPW-6659	0107

Key	Item	Willys Part No.	Ford Part No.	Gov't Group No.
J	GASKET	WO-630394	FM-GPW-6630	0107
K	PIN	WO-330964	FM-GPW-6684	0107
L	GEAR	WO-637425	FM-GPW-6610	0107
M	BODY	WO-630384	FM-GPW-6604	0107
N	RETAINER	WO-630390	FM-GPW-6644	0107
O	GASKET	WO-634813	FM-GPW-6642	0107
P	SHIM	WO-630389	FM-GPW-6628	0107
Q	SPRING	WO-356155	FM-GPW-6654	0107
R	PLUNGER	WO-630518	FM-GPW-6663	0107

RA PD 305054-A

Figure Number	Part Number (Willys)	Part Number (Ford)	Official Stockage Number	ITEM	Quantity Reqd. per Unit Assy.	Per 100 Major Items (12 Mos. Field Maintenance)	Per 100 Major Items (Major Overhaul 5th Ech)	Total First Year Procurement	Estimated Reqmts. per 100 Rebuilds
Col. 1	Col. 3	Col. 4	Col. 2	Col. 5	Col. 6	Col. 7	Col. 8	Col. 9	Col. 10
				0107—OIL PUMP (Cont'd)					
				1 WO-639870 FM-GPW-6659 GASKET, oil pump cover					
				4 WO-630392 FM-GPW-6619 GASKET, oil pump cover					
				1 WO-630394 FM-GPW-6630 GASKET, oil pump to cylinder block					
				1 WO-375927 FM-GPW-6625 GASKET, oil pump shaft					
				1 WO-634813 FM-GPW-6642 GASKET, oil relief spring retainer					
				1 WO-630299 FM-GPW-6648 GASKET, ventilator to valve spring cover					
	WO-A-6749	FM-GPW-18379		4 WO-630389 FM-GPW-6628 SHIM, oil relief spring) KIT, repair, engine oil pump............ (Includes:	as req.	—	5	5	15
				1 WO-637425 FM-GPW-6610 GEAR, driven					
				1 WO-330964 FM-GPW-6684 PIN, driven gear					
				1 WO-343306 FM-GPW-6614 PINION					
				1 WO-636599 FM-GPW-6608 SHAFT)					
01-7	WO-330964	FM-GPW-6684		PIN, engine oil pump driven gear, straight ($\frac{5}{32}$ x $\frac{31}{32}$).......	1	—	8	9	30
01-7	WO-343306	FM-GPW-6614		PINION, engine oil pump (7 teeth).......	1	3	—	5	—
01-7	WO-52525	FM-353052-S		PLUG, pipe, slotted, ⅛ in. (oil pump cover plug) (issue until stock exhausted)	1	16	24	—	10
01-7	WO-637636	FM-GPW-6663		PLUNGER, engine oil pump relief	1	—	3	4	—
H	WO-630518	FM-GPW-6600		PUMP, oil, engine, assembly.....	1	3	2	8	5
01-12	WO-5910	FM-33798-S2		NUT, reg., hex., s-fin., alloy-S., $\frac{5}{16}$-24NF-2 (pump assembly to block) (BBBX1BA)............	3	—	—	—	—
01-2, 12	WO-375981	FM-88141-S		STUD, S., $\frac{5}{16}$-18NC-2 x 2¾ x $\frac{5}{16}$-24NF-2 (issue until stock exhausted)...	3	20	40	—	—
01-8, 12	WO-51833	FM-34806-S2		WASHER, lock, reg., S., $\frac{5}{16}$ in. ($\frac{19}{32}$ O.D. x $\frac{11}{32}$ I.D. x $\frac{1}{16}$ thk.) (BECX1H)..	3	—	3	3	10
01-7	WO-630390	FM-GPW-6644		RETAINER, engine oil pump oil relief spring.	1	—	—	—	—
01-7	WO-636599	FM-GPW-6608		SHAFT, engine oil pump, assembly....	1	—	—	—	—
01-7	WO-630389	FM-GPW-6628		SHIM, engine oil pump oil relief spring.	as req.	10	—	12	—
01-7	WO-356155	FM-GPW-6654		SPRING, engine oil pump oil relief plunger (S., wire, 13 turns, .360 O.D. x 1⅛ free lgth.)..	1	—	6	6	20
				0107A—OIL PAN					
01-8	WO-639980	FM-GPW-6710		GASKET, engine oil pan....	1	198	—	260	—
01-8	WO-314338	FM-GPW-6734		GASKET, engine oil pan drain plug...	1	77	—	100	—
	WO-A-1538	FM-GPW-18512	G503-01-94020	GASKET SET, engine oil pan...... (Includes:	as req.	—	—	—	—
				1 WO-630398 FM-GPW-6627 GASKET, oil float support					
				1 WO-639980 FM-GPW-6710 GASKET, oil pan					
				1 WO-314338 FM-GPW-6734 GASKET, oil pan drain plug					
	WO-A-1167			PAN, oil, engine, assembly (issue until stock exhausted, then use WO-A-7238)	—	4	4	—	—

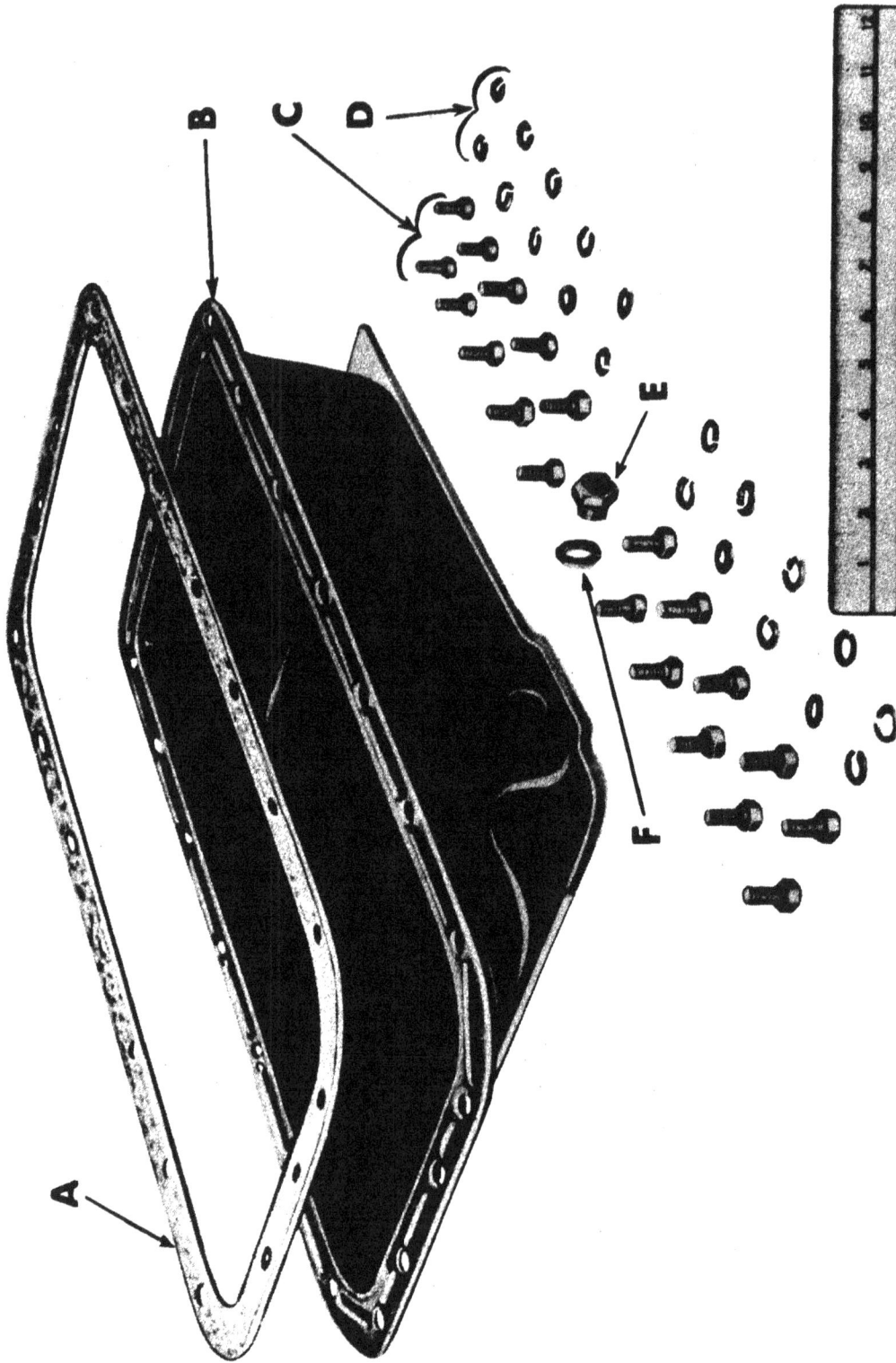

RA PD 305055

FIGURE 01-8—OIL PAN ASSEMBLY

Key	Item	Willys Part No.	Ford Part No.	Gov't Group No.
A	GASKET	WO-639980	FM-GPW-6710	0107A
B	PAN assembly	WO-A-7238	(Willys only)	0107A
C	BOLT	WO-51485	See text	0107A
D	WASHER	WO-51833	See text	0107A
E	PLUG	WO-639979	FM-GPW-6727	0107A
F	GASKET	WO-314338	FM-GPW-6734	0107A

Figure Number Col. 1	Official Stockage Number Col. 2	Willys Col. 3	Ford Col. 4	ITEM Col. 5	Quantity Reqd. per Unit Assy. Col. 6	Per 100 Major Items — 12 Mos. Field Maintenance Col. 7	Major Overhaul (5th Ech) Col. 8	Total First Year Procurement Col. 9	Estimated Reqmts. per 100 Rebuilds Col. 10
				0107A—OIL PAN (Cont'd)					
01-8		WO-A-7238	FM-GPW-6675	PAN, oil, engine, assembly (after Willys serial 297089) (was WO-A-1167)	1	2	2	7	5
01-8		WO-51485		BOLT, hex-hd., 3/8-18NC-2 x 5/8 (6 used oil pan and fan pulley shield to cylinder block) (Willys only) (14 used oil pan to cylinder block and front engine cover) (Willys only)	20	—	—	—	—
01-8			FM-355403-S2	BOLT, hex-hd., S., 5/16-18NC-2 x 11/16, w/WASHER, assembly (Ford only) (BCYX51FG)	20	—	—	—	—
01-8		WO-51833		WASHER, lock, reg., S., 5/16 in. (19/32 O.D. x 11/32 I.D. x 1/16 thk.) (BECX1H) (Willys only)	20	—	—	—	—
01-8		WO-639979	FM-GPW-6727	PLUG, drain, engine oil pan (7/8 in.)	1	% 19	10	36	30
				0107A-1—OIL FLOAT					
H		WO-630396	FM-GPW-6615	FLOAT, strainer, oil, engine, assembly (TSE-215-B)	1	2	—	3	5
		WO-5108	FM-72053	PIN, cotter, S., 1/8 x 1 1/4 (float to support)	1	—	—	—	—
		WO-630398	FM-GPW-6627	GASKET, engine oil strainer float support	1	—	—	—	—
01-1		WO-630397	FM-355396-S	SUPPORT, engine oil strainer float	1	6	6	16	20
		WO-636796	FM-34806-S2	BOLT, support to crankcase (hex-hd., 5/16-18NC-2 x 3/4)	2	8	8	24	25
		WO-51833		WASHER, lock, reg., S., 5/16 in. (19/32 O.D. x 11/32 I.D. x 1/16 thk.) (BECX1H)	2	—	—	—	—
				0107E—OIL FILTER					
01-9		WO-A-6818		CASE, engine oil filter, assembly (Willys only)	1	—	—	—	—
01-9		WO-A-1231	FM-GPW-18687-A	COVER, engine oil filter, assembly (PU-25791)	1	% 2	2	8	5
01-9		WO-A-1232	FM-GPW-18691-A	BOLT, oil filter cover (hex-hd., S., 5/16-20NF-2) (PU-25755)	1	% 5	2	10	5
01-9		WO-A-1236	FM-GPW-18662-A	ELEMENT, engine oil filter, assembly (PU-26637)	1	% 280	40	400	125
F	G503-01-94033	WO-A-1230	FM-GPW-18660-A	FILTER, oil, engine assembly (PU-27078)	1	% 5	2	10	5
01-9		WO-A-1235	FM-GPW-18688-A	GASKET, engine oil filter cover (PU-25802)	1	% 86	32	120	100
01-9		WO-A-1233	FM-GPW-18675-A	GASKET, engine oil filter cover bolt (PU-25756)	1	% 58	10	120	30
01-9		WO-A-1237	FM-358040-S	PLUG, drain, engine oil filter (S., hex., 5/16, 1/4-18 thd.) (PU-25795)	1	% 19		36	10
01-9		WO-A-1234	FM-GPW-18685-A	SPRING, engine oil filter cover bolt (S-wire, tapered, I.D. 13/16 to 17/32) (PU-25757)	1	3	3	3	
				0107F—OIL FILTER ATTACHING PARTS					
01-9		WO-51396	FM-24347-S2	BOLT, hex-hd., s-fin., alloy-S., 5/16-24NF-2 x 3/4 (oil filter to bracket) (BAOX1BA)	3	—	—	—	—
		WO-A-1247	FM-GPW-18663	BRACKET, engine oil filter, assembly (PU-27081)	1	—	4	—	—
01-9		WO-A-1251	FM-GPW-18644-A	CLAMP, engine oil filter, assembly (PU-27081) (Includes: BOLT, NUT)	2	—	—	—	—

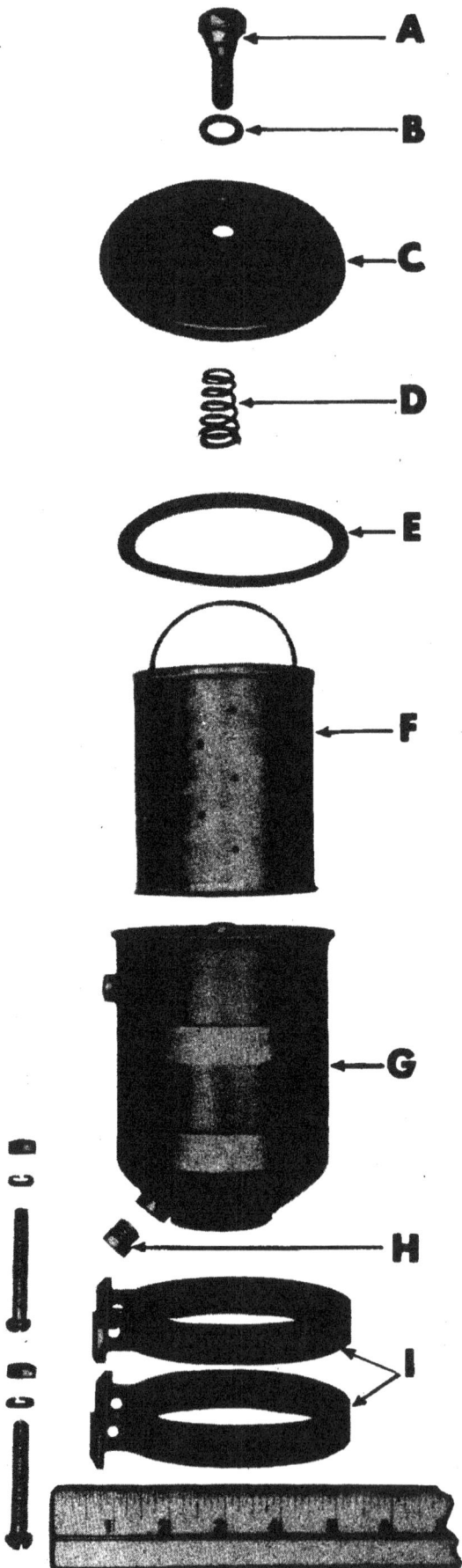

FIGURE 01-9—OIL FILTER ASSEMBLY

Key	Item	Willys Part No.	Ford Part No.	Gov't Group N
A	BOLT........	WO-A-1232	FM-GPW-18691-A	0107
B	GASKET.....	WO-A-1233	FM-GPW-18675-A	0107
C	COVER......	WO-A-1231	FM-GPW-18687-A	0107
D	SPRING.....	WO-A-1234	FM-GPW-18685-A	0107
E	GASKET.....	WO-A-1235	FM-GPW-18688-A	0107
F	ELEMENT assembly...	WO-A-1236	FM-GPW-18662-A	0107
G	CASE assembly...	WO-A-6818	(Willys only)	0107
H	PLUG........	WO-A-1237	FM-358040-S	0107
I	CLAMP, w/ BOLT and NUT.......	WO-A-1251	FM-GPW-18644-A	0107

RA PD 305056

RA PD 305056

GROUP 01—ENGINE (Cont'd)

Figure Number	Official Stockage Number	Part Number		ITEM	Quantity Reqd. per Unit Assy.	Per 100 Major Items			Estimated Reqmts. per 100 Rebuilds
		Willys	Ford			12 Mos. Field Maintenance	Major Overhaul (5th Ech)	Total First Year Procurement	
Col. 1	Col. 2	Col. 3	Col. 4	Col. 5	Col. 6	Col. 7	Col. 8	Col. 9	Col. 10
				0107F—OIL FILTER ATTACHING PARTS (Cont'd)					
		WO-52274	FM-34746-S2	WASHER, plain, S., S.A.E. std., $5/16$ in. (filter to bracket nut) (BEBX1H)......	4	—	—	—	—
		WO-51833	FM-34806-S2	WASHER, lock, reg., S., $5/16$ in. ($1\frac{1}{2}$ O.D. x $11/32$ I.D. x $1/16$ thk.) (BECX1H)......	4	—	—	—	—
				0107G—OIL FILLER					
		WO-A-5105	FM-GPW-6770	BRACKET, support, engine oil filler tube.	1	—	—	—	—
		WO-639655	FM-GPW-6763-A	TUBE, engine oil filler (before Willys engine 114550 only) (For straight type).	1	—	—	—	—

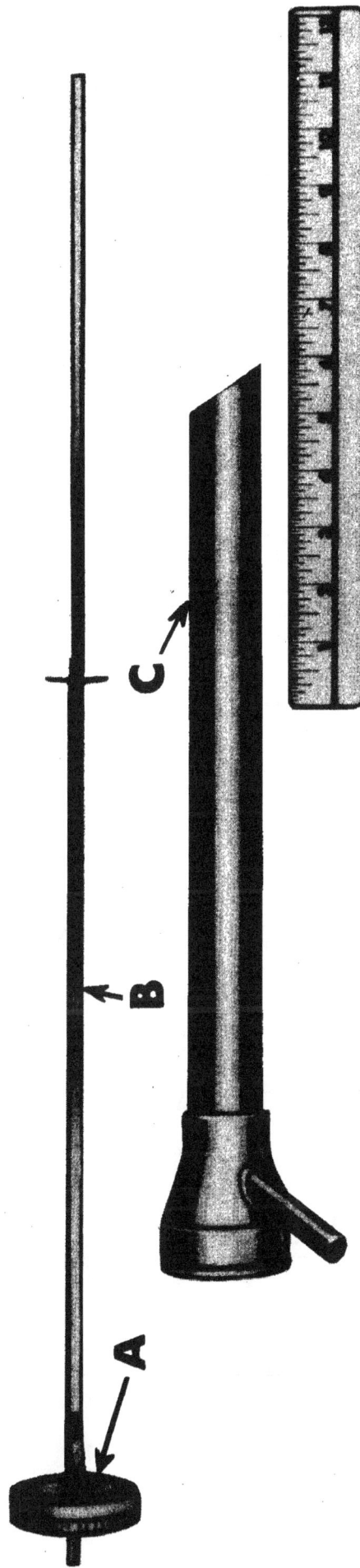

RA PD 305057

FIGURE 01-10—OIL FILLER TUBE AND LEVEL INDICATOR

Key	Item	Willys Part No.	Ford Part No.	Gov't Group No.
A	GASKET	WO-A-7280	FM-GPW-6789	0107H
B	INDICATOR assembly	WO-A-6525	FM-GPW-6766-C	0107H
C	TUBE assembly	WO-A-6915	FM-GPW-6763-C	0107G

RA PD 305057-A

GROUP 01—ENGINE (Cont'd)

Figure Number	Official Stockage Number	Part Number		ITEM	Quantity Reqd. per Unit Assy.	Per 100 Major Items			Estimated Reqmts. per 100 Rebuilds
		Willys	Ford			12 Mos. Field Maintenance	Major Overhaul (5th Ech)	Total First Year Procurement	
Col. 1	Col. 2	Col. 3	Col. 4	Col. 5	Col. 6	Col. 7	Col. 8	Col. 9	Col. 10
				0107G—OIL FILLER (Cont'd)					
01-10		WO-A-5165	FM-GPW-6763-B	TUBE, engine oil filler, assembly (after Willys engine 114550 only) (For funnel type).........	1	—	—	—	—
		WO-A-6915	FM-GPW-6763-C	TUBE, engine oil filler, assembly (after Willys serial 208437 only).........	1	—	2	3	5
				0107H—LEVEL AND/OR FLOAT GAUGES					
01-10		WO-A-7280	FM-GPW-6789	GASKET, engine oil filler cap.........	1	% 8	8	20	25
		WO-639556	FM-GPW-6766-A	INDICATOR and breather CAP, engine oil filler, assembly (before Willys engine 114550 only).........	1	—	—	—	—
H	G503-01-32455	WO-A-5168	FM-GPW-6766-B	INDICATOR and breather CAP, engine oil filler, assembly (after Willys engine 114550 only).........	1	—	—	—	—
01-10		WO-A-6525	FM-GPW-6766-C	INDICATOR and breather CAP, engine oil filler, assembly (after Willys serial 208437 only).........	1	% 6	2	15	5
				0107J—OIL DISTRIBUTION SYSTEM					
		WO-A-1289	FM-GPW-14585	CLIP, tube to front engine plate (7/16 in. S, closed).........	1	—	—	—	—
		WO-A-5449	FM-GPW-2281	CLIP, oil gauge tube to dash (S, 5/16 in. 7/32 bolt hole).........	1	—	—	—	—
		WO-5182	FM-355158-S2	SCREW, rd-hd, S, No. 10 (.190)-24NC-2 x 3/8.........	1	—	—	—	—
		WO-52221	FM-34803-S7	WASHER, lock, S, No. 10 (.190 in.).........	1	—	—	—	—
		WO-387891	FM-9N-18679	CONNECTOR, flared tube, br., 1/4 in. (tube to oil filter).........	1	% 6	3	12	10
		WO-384569	FM-9N-18686	ELBOW, (inverted flared tube, br., 1/4 in.) (1 used flexible connection to engine; 1 used tube to cylinder block).........	2	% 13	6	25	20
		WO-345961	FM-GPW-13434-A	GROMMET, rubber, 13/32 in. (issue until stock exhausted).........	1	% 16	24	—	—
	G503-02-17815	WO-A-1197	FM-GPW-18667	TUBE, engine oil filter inlet, assembly.........	1	% 66	14	108	45
	G503-02-17816	WO-A-1198	FM-GPW-18666	TUBE, flexible (1 used oil filter outlet; 1 used oil gauge tube).........	2	% 85	14	130	45
		WO-A-1450	FM-GPW-9316	TUBE, 1/4 in., oil gauge, assembly.........	1	% 3	—	6	—
		WO-A-1456	FM-GPW-9323	UNION, inverted flared tube, br., 1/4 in. (flexible connection to tube assembly).........	1	% 6	3	12	10
				0107N—CRANKCASE VENTILATOR					
01-6		WO-630298	FM-GPW-6762	BAFFLE, engine crankcase ventilator.........	1	3	3	9	10
01-6		WO-A-6919	FM-GPW-6758-B	BODY, engine crankcase ventilator, assembly (after Willys engine 204040).........	1	2	2	7	5
01-6		WO-384549	FM-GPW-9268	ELBOW, pipe, 1/8 in. (engine crankcase ventilator to tube) (after Willys engine 204040).........	1				
01-6		WO-A-6885	FM-GPW-6722	ELBOW, pipe, 1/4 in. (engine crankcase ventilator tube to valve) (after Willys	1	% 6	3	10	10

Usage	Ord. stock No.	W.O. No.	FM-GPW No.	Description	Units per assy				
01-6		WO-A-6922	FM-GPW-6756	TUBE, engine crankcase ventilator valve, assembly (after Willys engine 204040)	1	% 3	3	9	10
01-6		WO-A-6895	FM-GPW-6769	VALVE, engine crankcase ventilator, assembly (after Willys engine 204040)	1	% 3	3	9	10
		WO-A-1061	FM-GPW-6758	VENTILATOR, engine crankcase, assembly (before Willys engine 204040)	1				

0108—MANIFOLDS AS AN ASSEMBLY

Usage	Ord. stock No.	W.O. No.	FM-GPW No.	Description	Units per assy				
01-2	G503-01-94021	WO-638640	FM-GPW-9448	GASKET, engine intake and exhaust manifold (issue until stock exhausted)	1	% 36	68	46	
01-11		WO-634811	FM-GPW-9435	GASKET, engine intake to exhaust manifold	1	% 38			
		WO-A-7835	FM-GPW-18323	GASKET SET, engine manifold (Includes: 1 WO-638640 FM-GPW-9448 GASKET, intake and exhaust manifold; 1 WO-634811 FM-GPW-9435 GASKET, intake to exhaust manifold; 1 WO-634814 FM-GPW-9450 GASKET, exhaust pipe flange)	as req.				
G	H1-07-19003	WO-A-1165	FM-GPW-9410-B	MANIFOLD, engine intake and exhaust, assembly	1	%			
01-2, 12		WO-53288	FM-33800-S2	NUT, regular, hex, s-fin, alloy-S, Seez Pruf 3/8-24NF-2 (intake and exhaust manifold)	7	% 20	40		
01-11		WO-349712	FM-88082-S	STUD, S, 3/8-16NC x 1 15/16 x 3/8-24NF-2 (issue until stock exhausted)	2				
01-2, 12		WO-632159	FM-88057	STUD, S, 3/8-16NC x 1 1/4 x 3/8-24NF-2	2	% 48	92	90	150
		WO-300143	FM-88042	STUD, S, 3/8-16NC x 1 1/2 x 3/8-24NF-2 (issue until stock exhausted)	3				
		WO-344732	FM-GPW-9443	WASHER, manifold clamp, S, 1 5/16 O.D. x 1 3/16 I.D. x 1/8 to 5/16 thk. at I.D.	2	% 32	48		

0108A—INTAKE MANIFOLDS

Usage	Ord. stock No.	W.O. No.	FM-GPW No.	Description	Units per assy				
01-11	H6-02-83805	WO-A-1166	FM-GPW-9424-B	MANIFOLD, intake, engine assembly	1	% 1	1	4	21
01-11		WO-51486	FM-24386-S2	BOLT, hex-hd, s-fin, alloy-S, 5/16-18NC-2 x 1 (intake to exhaust manifold) (BANX1BC)	4				
01-11		WO-52428	FM-34906-S	WASHER, lock 5/16 in.	4				
01-11		WO-5138	FM-353055-S	PLUG, pipe, sq-hd, black, 1/4 in. (engine intake manifold) (CPMX1AB)	3				

0108B—EXHAUST MANIFOLDS

Usage	Ord. stock No.	W.O. No.	FM-GPW No.	Description	Units per assy				
01-11		WO-A-912	FM-GPW-9428	MANIFOLD, exhaust, engine assembly	1	% 2	2	8	5

0108C—HEAT CONTROLS

Usage	Ord. stock No.	W.O. No.	FM-GPW No.	Description	Units per assy				
01-11		WO-636438	FM-GPW-9462	BEARING, engine heat control valve shaft	1	3	10	15	30
01-11		WO-637211	FM-GPW-9465	KEY, engine heat control lever	1	13	13	30	40
01-11		WO-637210	FM-GPW-9458	LEVER, engine heat control valve counterweight	1	2	5	10	15
01-11	H1-07-16540	WO-6352	FM-355836-S7	NUT, machine screw, hex, S, No. 10 (.190)-24NC-1 (clamp lever to shaft) (BBKX2C)	1				
01-11	H1-10-30102	WO-5272	FM-355160-S2	SCREW, machine, rd-hd, S, No. 10 (.190)-24NC-2 x 3/4 (BCNX2AG)	1				
01-11		WO-637206	FM-GPW-9456	SHAFT, engine heat control valve	1	2	5	10	15
01-11		WO-637208	FM-GPW-9467-A	SPRING, engine heat control valve	1	13	13	30	40

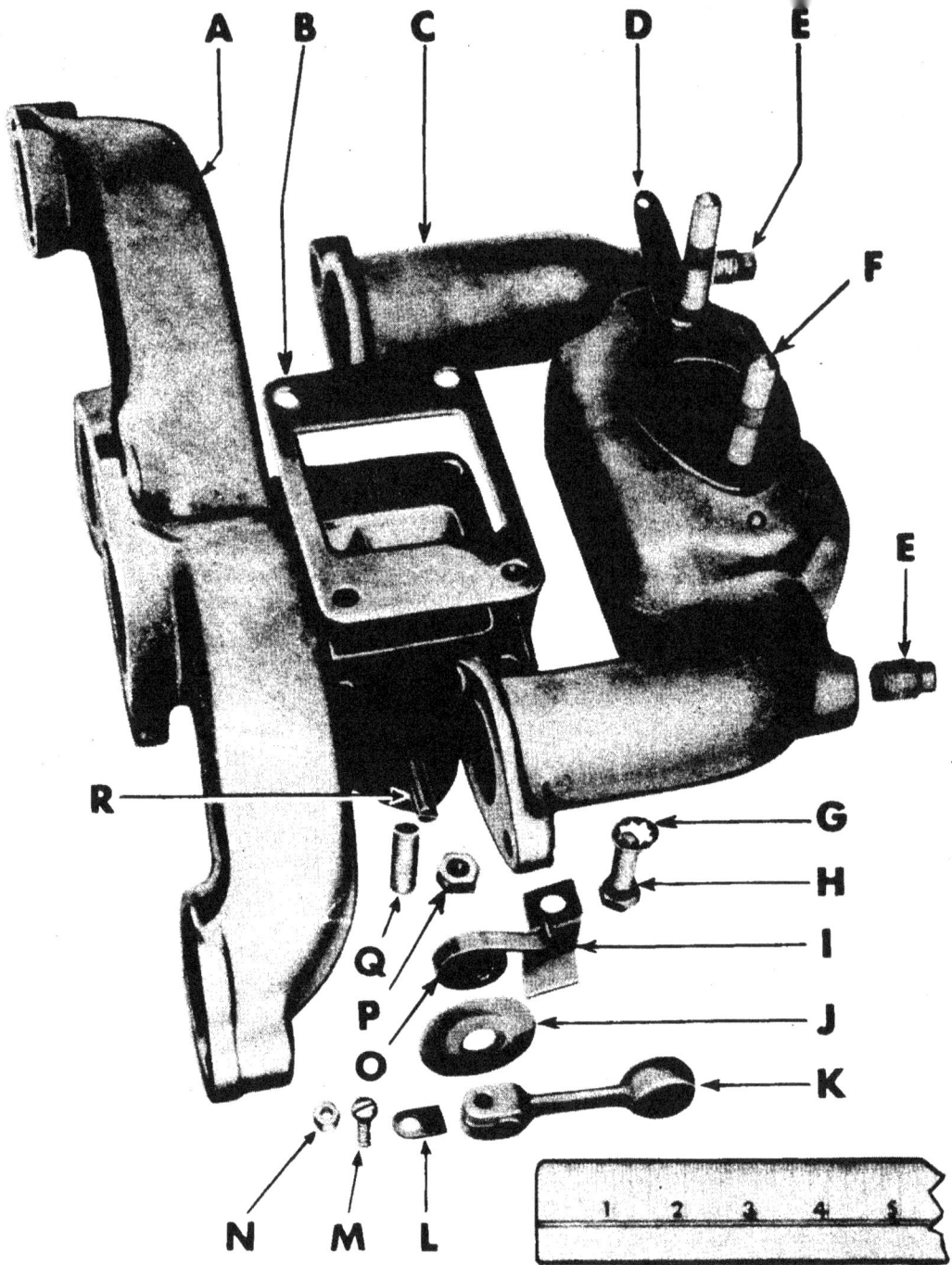

RA PD 305048

FIGURE 01-11—MANIFOLD ASSEMBLY

Key	Item	Willys Part No.	Ford Part No.	Gov't Group No.
A	MANIFOLD	WO-A-912	FM-GPW-9428	0108B
B	GASKET	WO-634811	FM-GPW-9435	0108
C	MANIFOLD	WO-A-1166	FM-GPW-9424-B	0108A
D	CLIP	WO-A-1173	FM-GPW-9751	0303A
E	PLUG	WO-5138	FM-353055-S	0108A
F	STUD	WO-632159	FM-88057	0301A
G	WASHER	WO-52428	FM-34906-S	0108A
H	BOLT	WO-51486	FM-24386-S2	0108A
I	STOP	WO-639743	FM-GPW-9463	0108C
J	WASHER	WO-637209	FM-GPW-9484	0108C
K	LEVER	WO-637210	FM-GPW-9458	0108C
L	KEY	WO-637211	FM-GPW-9465	0108C
M	SCREW	WO-5272	FM-355160-82	0108C
N	NUT	WO-6352	FM-355836-S7	0108C
O	SPRING	WO-637208	FM-GPW-9467-A	0108C
P	NUT	WO-6352	FM-355836-S7	0108C
Q	STUD	WO-332515	FM-88032-S	0402
R	SHAFT	WO-637206	FM-GPW-9456	0108C

RA PD 305048-A

Figure Number	Official Stockage Number	Part Number: Willys	Part Number: Ford	ITEM	Quantity Reqd. per Unit Assy.	Per 100 Major Items: 12 Mos. Field Maintenance	Per 100 Major Items: Major Overhaul (5th Ech)	Per 100 Major Items: Total FirstYear Procurement	Estimated Reqmts. per 100 Rebuilds
Col. 1	Col. 2	Col. 3	Col. 4	Col. 5	Col. 6	Col. 7	Col. 8	Col. 9	Col. 10
				0108C—HEAT CONTROLS (Cont'd)					
01-11		WO-639743	FM-GPW-9463	STOP, engine heat control valve spring	1	13	13	30	40
H		WO-636439	FM-GPW-9460	VALVE, engine heat control	1	2	5	10	15
01-11		WO-637209	FM-GPW-9484	WASHER, engine heat control valve spring (S., special, 1½ O.D. x .310 I.D. x ⅛ thk. at opening)	1	6	6	15	20
				0109—FLYWHEEL					
01-4		WO-632157	FM-355497-S	BOLT, hex-hd., S., ⅜-24NF-2 x 1$\frac{5}{32}$ (flywheel to crankshaft)	4	10	29	50	90
		WO-116295	FM-GPW-6390	BOLT, special hd., S., ½-20NF-2 x 1$\frac{3}{16}$ (flywheel to crankshaft dowel)	2	—	6	12	20
		WO-639578	FM-GPW-7600	BUSHING, flywheel to crankshaft	1	3	3	12	10
01-4		WO-632156	FM-GPW-6387	DOWEL, flywheel to crankshaft	2	—	14	15	45
		WO-A-1443	FM-GPW-6375	FLYWHEEL, engine (issue until stock is exhausted, then use WO-A-7503)	—	4	4	—	—
		WO-A-7503	FM-GPW-18289	FLYWHEEL and DOWEL, engine, group assembly (was WO-A-1443) (Includes: 2 WO-116295 FM-GPW-6390 BOLT, dowel; 1 WO-A-1443 FM-GPW-6375 FLYWHEEL; 2 WO-52804 FM-33832-S2 NUT, dowel bolt; 2 WO-52330 FM-34909-S2 WASHER, lock)	1	3	3	8	10
				NOTE: Parts numbered WO-116295—FM-GPW-6390, WO-52804—FM-33832-S2, WO-52330—FM-34909-S2, are required for straight doweling of flywheel to crankshaft. Eliminates need of special tapered reamer.					
	H1-07-19003	WO-5901	FM-33800-S2	NUT, reg., hex., s-fin., alloy-S., ⅜-24NF-2 (flywheel dowel and bolt) (GM-103026) (BBBX1CA)	6	—	—	—	—
	H1-07-16025	WO-52804	FM-33832-S2	NUT, hex., S., ½-20NF-2 (flywheel dowel bolt) (BBDX1E)	2	—	—	—	—
		WO-52332	FM-34907-S2	WASHER, lock, S., internal teeth, ⅜ in. ($\frac{25}{64}$ I.D. x $\frac{11}{16}$ O.D. x .035 thk.) (BEAX1L)	6	—	—	—	—
		WO-52330	FM-34909-S2	WASHER, lock, S., external teeth, ½ in. (.520 I.D. x ⅞ O.D. x .045 thk.) (BEAX4L)	2	—	—	—	—
				0109B—RING GEAR					
		WO-635394	FM-GPW-6384	GEAR, ring, engine flywheel (S., 97 teeth, 10.980 I.D. x 12.185 O.D. x ⅜ wide)	1	2	3	8	10
				0109C—FLYWHEEL HOUSING					
		WO-630101	FM-355597	BOLT, dowel (transmission to engine plate to cylinder block)	2	10	10	25	30
		WO-630103	FM-GPW-7518	COVER, transmission case inspection	1	—	3	5	10
		WO-51763	FM-24308-S	BOLT, hex-hd., s-fin., alloy-S., ¼-20NC-2 x ½ (cover to case) (BANX4AA)	2	—	—	—	—
		WO-52706	FM-34805-S	WASHER, lock, reg., S., ¼ in. (cover to case bolt) (BECX1G)	2	—	—	—	—

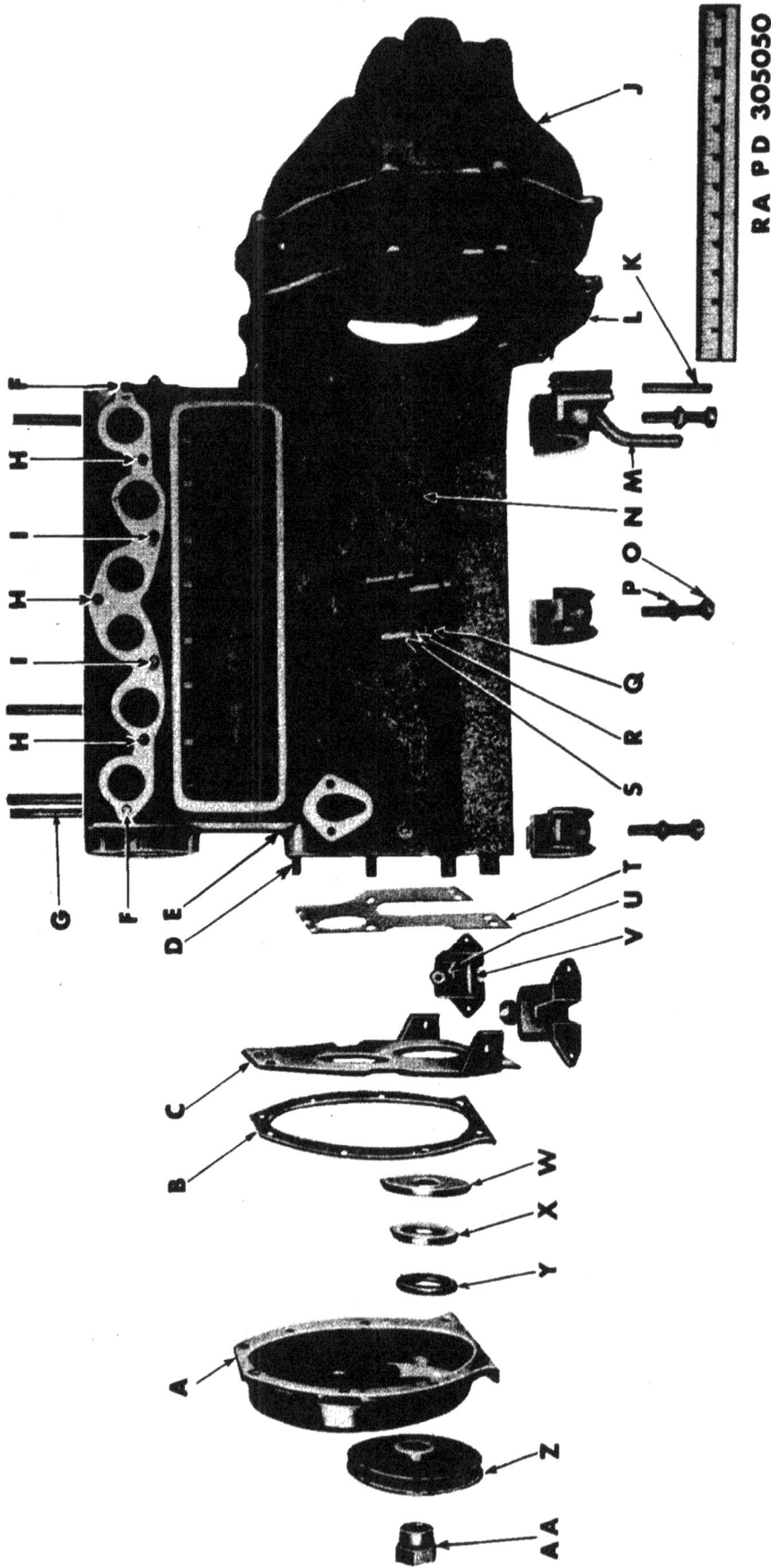

RA PD 305050

FIGURE 01-12—TIMING CHAIN COVER, CYLINDER BLOCK AND FLYWHEEL HOUSING

Key	Item	Willys Part No.	Ford Part No.	Gov't Group No
A	COVER assembly	WO-A-1190	FM-GPW-6016	0106D
B	GASKET	WO-630365	FM-GPW-6288	0106D
C	PLATE assembly	WO-A-1463	FM-GPW-6031-A3	0110
D	STUD	WO-384958	FM-88022-S	0106D
E	BLOCK assembly	WO-A-1272	FM-GPW-6010	0101
F	STUD	WO-349712	FM-88082-S	0108
G	STUD	WO-349368	FM-GPW-6066	0101A
H	STUD	WO-300143	FM-88042	0108
I	STUD	WO-632159	FM-88057	0108
J	HOUSING assembly	WO-A-439	FM-GPW-6392	0109C
K	PACKING	WO-637790	FM-GPW-6701	0102A
L	PLATE	WO-A-5121	FM-GPW-6044	0110
M	PIPE	WO-630294	FM-GPW-6326	0102A
N	PLUG	WO-5085	FM-358064-S	0101

Key	Item	Willys Part No.	Ford Part No.	Gov't Group No.
O	BOLT	WO-381519	FM-GPW-6345	0102
P	WASHER	WO-5009	FM-34809-S	0102
Q	NUT	WO-5910	FM-33798-S2	0107
R	WASHER	WO-51833	FM-34806-S2	0107
S	STUD	WO-375981	FM-88141-S2	0107
T	GASKET	WO-630359	FM-GPW-6020	0106D
U	NUT	WO-5916	FM-33846-S	0110
V	INSULATOR, oil	WO-A-7498	FM-GPW-6038-A	0110
W	SLINGER	WO-375877	FM-GPW-6310	0102
X	RETAINER	WO-375920	FM-GPW-6287	0102
Y	PACKING	WO-637098	FM-GPW-6700	0102A
Z	PULLEY	WO-638113	FM-GPW-6312	0503A
AA	NUT assembly	WO-387633	FM-GPW-6319	2215

GROUP 01—ENGINE (Cont'd)

0109C—FLYWHEEL HOUSING (Cont'd)

Figure Number (Col. 1)	Official Stockage Number (Col. 2)	Willys (Col. 3)	Ford (Col. 4)	ITEM (Col. 5)	Quantity Reqd. per Unit Assy. (Col. 6)	12 Mos. Field Maintenance (Col. 7)	Major Overhaul (5th Ech) (Col. 8)	Total First Year Procurement (Col. 9)	Estimated Reqmts. per 100 Rebuilds (Col. 10)
	H1-10-30132	WO-375217	FM-GPW-7023	COVER, engine flywheel housing timing hole	1		3	4	10
		WO-52036	FM-26147-S7	SCREW, machine, rd-hd., S., 1/4-20NC-2 x 3/8 (BCNX2CC)	1				
		WO-52706	FM-34805-S2	WASHER, lock, reg., S., 1/4 in. (15/32 O.D. x .260 I.D. x 1/16 thk.) (BECX1G)	1		3	4	10
01-12		WO-A-439	FM-GPW-6392	HOUSING, engine flywheel					
		WO-52379	FM-24489-S	BOLT, hex-hd., s-fin., alloy-S., 3/8-24NF-2 x 1 5/8 (housing to bracket at hand brake cable clip) (BAOX4CH)	3				
	H1-01-16039	WO-6606	FM-24409-S7	BOLT, hex-hd., s-fin., alloy-S., 3/8-24NF-2 x 1 1/8 (housing to brkt.)	4				
		WO-6184	FM-24430-S	BOLT, hex-hd., s-fin., alloy-S., 7/16-14NC-2 x 1 1/4 (housing to transmission) (BANX1DD)	8				
	H1-07-19003	WO-5901	FM-33800-S	NUT, reg., hex., s-fin., alloy-S., 3/8-24NF-2 (BBBX1CA)	8				
	H1-15-18113	WO-5010	FM-34807-S	WASHER, lock, reg., S., 3/8 in. (BECX1K)	8		3	10	10
	H6-02-82440	WO-51921	FM-74113-S	PLUG, expansion, S., 3/4 in. diam. (BEDX1AK)	1				

0110—MOUNTINGS

Figure Number (Col. 1)	Official Stockage Number (Col. 2)	Willys (Col. 3)	Ford (Col. 4)	ITEM (Col. 5)	Quantity Reqd. per Unit Assy. (Col. 6)	12 Mos. Field Maintenance (Col. 7)	Major Overhaul (5th Ech) (Col. 8)	Total First Year Procurement (Col. 9)	Estimated Reqmts. per 100 Rebuilds (Col. 10)
		WO-A-146	FM-GPW-6043	BRACKET, mounting, engine, rear (issue until stock exhausted)	1	8	12		
	H1-01-16024	WO-52189	FM-24328-S2	BOLT, hex-hd., s-fin., alloy-S., 3/8-16NC-2 x 5/8 (bracket to transmission)	3				
		WO-6412	FM-24348-S	BOLT, hex-hd., S., 3/8-16NC-2 x 3/4 (bracket to transmission) (GM-100133) (BANX1CA)	1				
	H1-15-18113	WO-5010	FM-34807-S	WASHER, lock, reg., S., 3/8 in. (BECX1K)	4				
01-12	H1-07-19003	WO-A-5125	FM-GPW-6044	CABLE, engine stay, assembly	1	% 2	2	8	5
		WO-5901	FM-33800-S2	NUT, reg., hex., s-fin., alloy-S., 3/8-24NF-2 (GM-103026) (BBBX1CA) — Willys	2				
				Ford	4				
		WO-52909		NUT, hex., S., stamped 3/8-24NF-2 (type "B") (GM-107322) (Willys only)	1				
		WO-52101	FM-34747-S2	WASHER, plain, S., S.A.E. std., 3/8 in. (13/16 O.D. x 13/32 I.D. x 1/16 thk.) (BEBX1K)	1				
		WO-5010	FM-34807-S2	WASHER, lock, reg., S., 3/8 in. (BECX1K)	2				
01-12		WO-A-7498	FM-GPW-6038-A	INSULATOR, engine support, front, assembly (11/16 O.D. x 13/32 I.D. x 1/2 thk.) (BAOX1BC) (Willys only)	2	% 10		20	1
	H1-01-16192	WO-5934		BOLT, hex-hd., s-fin., alloy-S., 5/16-24NF-2 x 1 (insulator to engine front support bracket) (GM-100014) (BAOX1BC) (Willys only)	4				
	H1-01-16193		FM-24427-S7	BOLT, hex-hd., s-fin., alloy-S., 5/16-24NF-2 x 1 1/4 (insulator to engine front support bracket) (Ford only) (BAOX1BD)	4				
	H1-07-19002	WO-5910	FM-33798-S2	NUT, hex., reg., s-fin., alloy-S., 5/16-24NF-2 (GM-103025) (BBBX1BA)	4				
	H1-07-19005	WO-5916	FM-33846-S	NUT, hex., S., 1/2-20NF-2 (insulator to front engine plate) (GM-103028) (BBBX1EA)	2				
		WO-52274	FM-34746-S2	WASHER, plain, S., S.A.E. std., 5/16 in. (BEBX1H)	4				
		WO-51833	FM-34806-S2	WASHER, lock, reg., S., 5/16 in. (11/32 O.D. x 11/32 I.D. x 1/16 thk.) (BECX1H)	4				

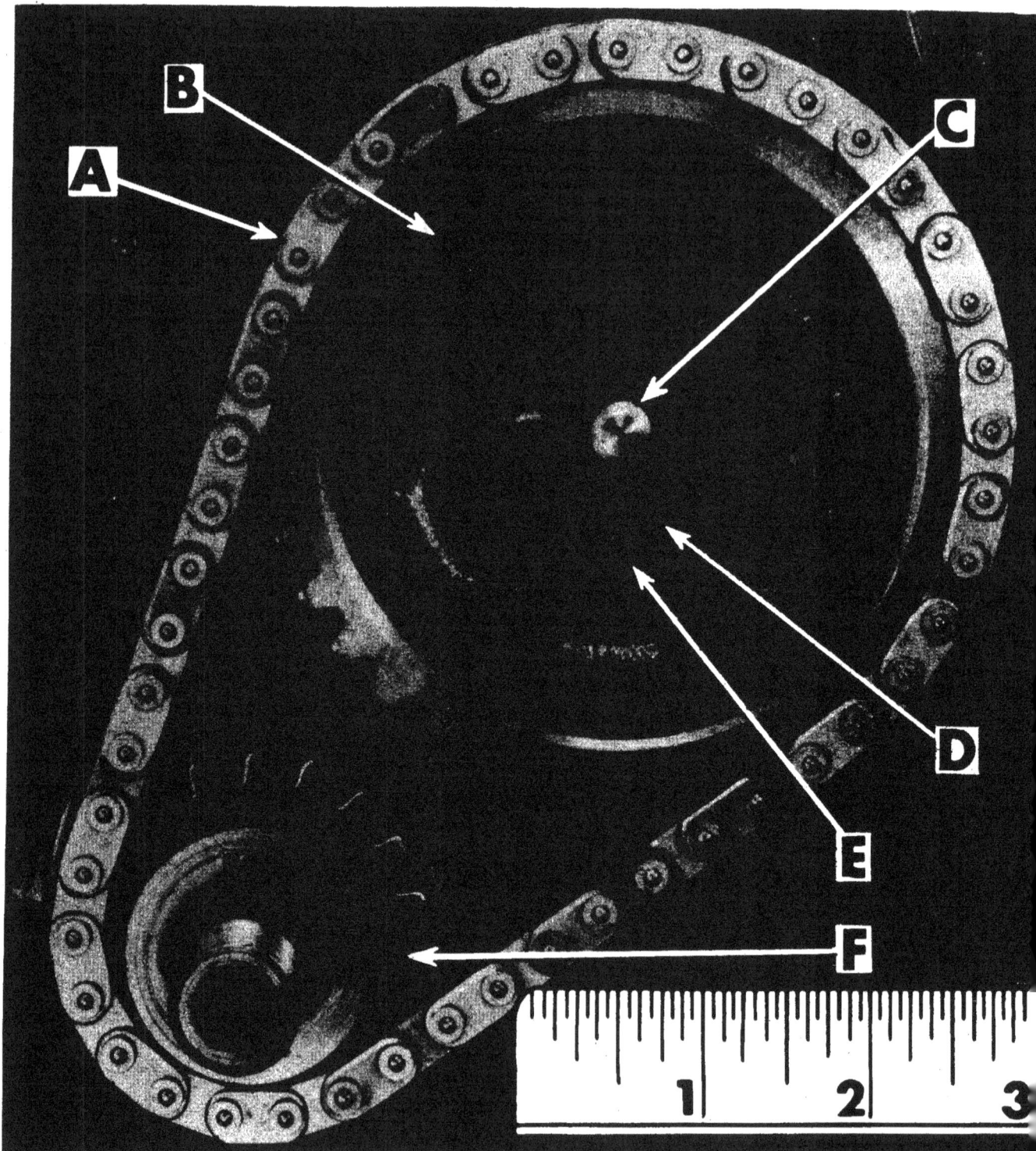

RA PD 305156

FIGURE 01-13—CRANKSHAFT CHAIN AND SPROCKETS

Key	Item	Willys Part No.	Ford Part No.	Gov't Group No.
A	CHAIN	WO-638457	FM-GPW-6260	0106C
B	SPROCKET	WO-638458	FM-GPW-6256	0106C
C	PLUNGER	WO-375907	FM-GPW-6243	0106
D	BOLT	WO-634850	FM-355499-S	0106C
E	WASHER	WO-315932	FM-GPW-6269	0106C
F	SPROCKET	WO-638459	FM-GPW-6306	0106C

RA PD 305156-A

Figure Number	Official Stockage Number	Part Number: Willys	Part Number: Ford	ITEM	Quantity Reqd. per Unit Assy.	Per 100 Major Items: 12 Mos. Field Maintenance	Per 100 Major Items: Major Overhaul (5th Ech)	Per 100 Major Items: Total First Year Procurement	Estimated Reqmts. per 100 Rebuilds
Col. 1	Col. 2	Col. 3	Col. 4	Col. 5	Col. 6	Col. 7	Col. 8	Col. 9	Col. 10
				0110—MOUNTINGS (Cont'd)					
		WO-5009	FM-34809-S2	WASHER, lock, reg., S., 1/2 in. (7/8 O.D. x 17/32 I.D. x 1/8 thk.) (BECX1M)	2	—	—	—	—
		WO-A-6156	FM-GPW-6040-B	INSULATOR, engine support, rear	1	% 2	—	10	—
		WO-5901	FM-33800-S2	NUT, reg., hex., s-fin., alloy-S., 3/8-24NF-2 (insulator to bracket and cross member) (BBBX1CA) (GM-103026)	4	—	—	—	—
		WO-52101	FM-34747-S2	WASHER, plain, S., S.A.E. std., 3/8 in. (13/16 O.D. x 13/32 I.D. x 1/16 thk.) (BEBX1K)	2	—	—	—	—
		WO-5010	FM-34807-S2	WASHER, lock, reg., S., 3/8 in. (13/32 I.D. x 21/32 O.D. x 3/32 thk.) (BECX1K)	4	—	—	—	—
01-12		WO-A-1463	FM-GPW-6031-A3	PLATE, engine, front, assembly (mounts on cylinder block studs)	1	3	3	8	10
01-12		WO-A-5121	FM-GPW-7007	PLATE, engine, rear, assembly	1	—	2	4	5
		WO-A-1051	FM-GPW-6098	SHIM, engine support insulator (issue until stock exhausted)	as req.	4	4	—	—
				NOTE: For engine front support bracket see Group 1500.					
				GROUP 02—CLUTCH					
				0201A—FACINGS AND RIVETS					
02-2		WO-636778	FM-GPW-7577	FACING, engine clutch, rear (B-B-4324)	1	—	—	—	—
02-2		WO-371567	FM-GPW-7549	FACING, engine clutch, front (B-B-2940)	1	—	—	—	—
02-2		WO-374586	FM-351915-S	RIVET, tubular, S., fl-ck-hd., .143 O.D. x 1/4 lgth. (BMGX1)	12	—	—	—	—
		WO-A-6751	FM-GPW-18358	FACING SET, engine clutch disc (Includes: 1 WO-636778 FM-GPW-7577 FACING; 1 WO-371567 FM-GPW-7549 FACING; 15 WO-374586 FM-351926 RIVET)	as req.	6	13	22	40
				0201B—CLUTCH DRIVEN PLATE					
		WO-638992	FM-GPW-7563	PLATE, pressure, engine clutch, assembly (AVM-TP-28-7-1) Includes: LEVERS, PLATES, SPRINGS)	1	13	3	20	10
		WO-630129	FM-355476-S	BOLT, clutch to flywheel (special, hex-hd., S., 5/16-18NC-2 x 1 1/16)	6	24	24	60	75
		WO-5051	FM-34836-S7	WASHER, lock, hv., S., 5/16 in. (.328 I.D. x .578 O.D. x 3/32 thk.) (BECX3H)	6				
02-2		WO-636755	FM-GPW-7550	PLATE, driven, engine clutch, w/HUB, assembly (B-B-11123)	1	19	6	30	20
				0202—COVER, PRESSURE PLATES, AND SPRINGS					
02-2		WO-638151	FM-GPW-7580	COVER, engine clutch (AVM-TP-2811)	1	—	—	—	—
02-2		WO-638157	FM-GPW-7567	CUP, engine clutch pressure spring (AVM-TP-283)	3	—	—	—	—

FIGURE 02-1—CLUTCH ASSEMBLY—SECTIONAL VIEW

Key	Item	Willys Part No.	Ford Part No.	Gov't Group No.
A	FACING	WO-371567	FM-GPW-7549	0201A
B	PLATE	WO-636755	FM-GPW-7550	0201B
C	FACING	WO-636778	FM-GPW-7577	0201A
D	RIVET	WO-374586	FM-351915-S	0201A
E	PLATE	WO-638152	FM-GPW-7566	0202
F	BEARING	WO-635529	FM-GPW-7580-B	0203
G	CARRIER	WO-639654	FM-GPW-7561	0203
H	SPRING	WO-630117	FM-GPW-7562	0203
I	FULCRUM	WO-630068	FM-GPW-7516	0204A
J	LEVER	WO-630112	FM-GPW-7515	0203
K	LEVER	WO-638158	FM-GPW-7591	0202
L	SPRING	WO-638993	FM-GPW-7572	0202
M	PIN	WO-638159	FM-GPW-7564	0203
N	CABLE	WO-A-5102	FM-GPW-7530	0202
O	SCREW	WO-638154	FM-24325-S	0202
P	NUT	WO-638155	FM-33921-S7	0202
Q	WASHER	WO-638305	FM-34745-S2	0202
R	SPRING	WO-638153	FM-GPW-7590	0202
S	CUP	WO-638157	FM-GPW-7567	0202

RA PD 305154-A

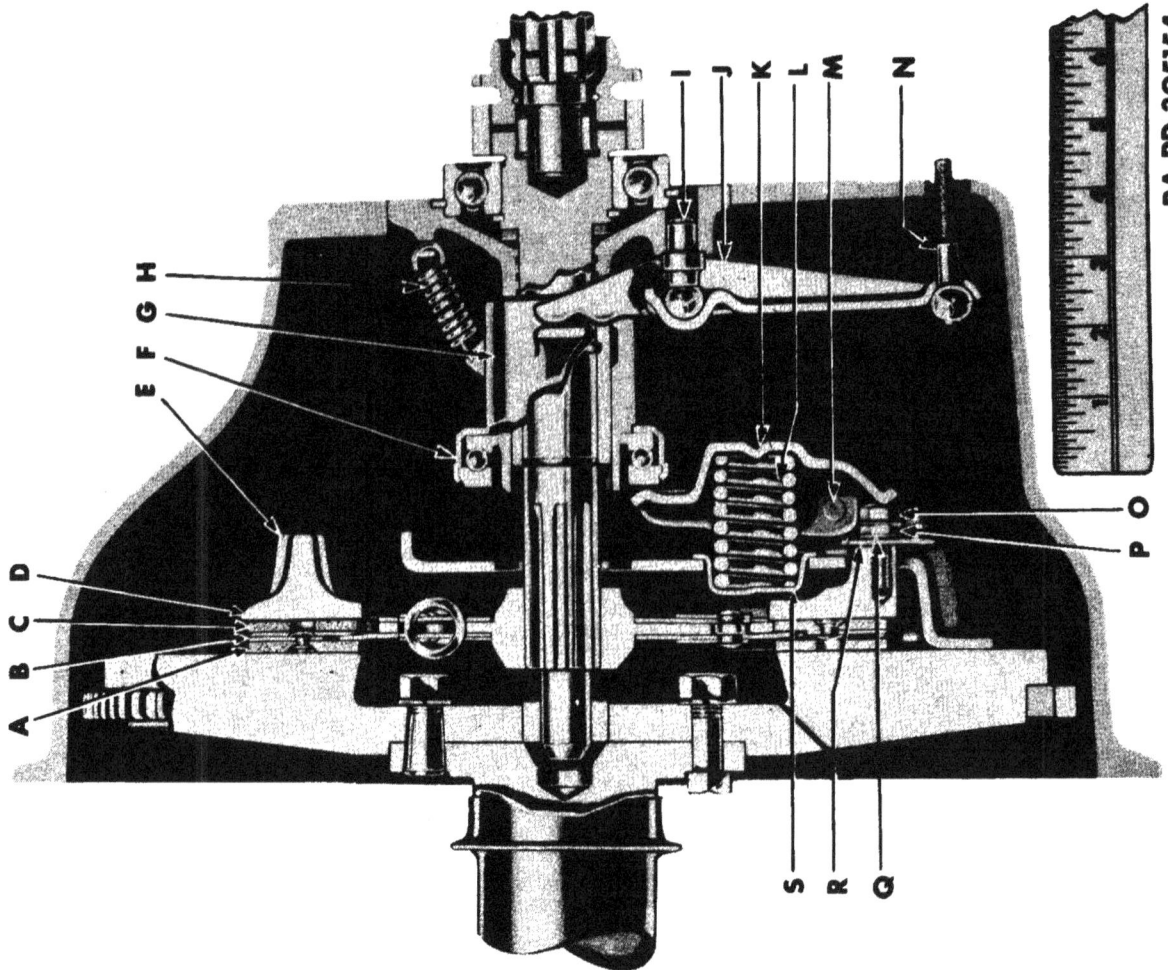

RA PD 305154

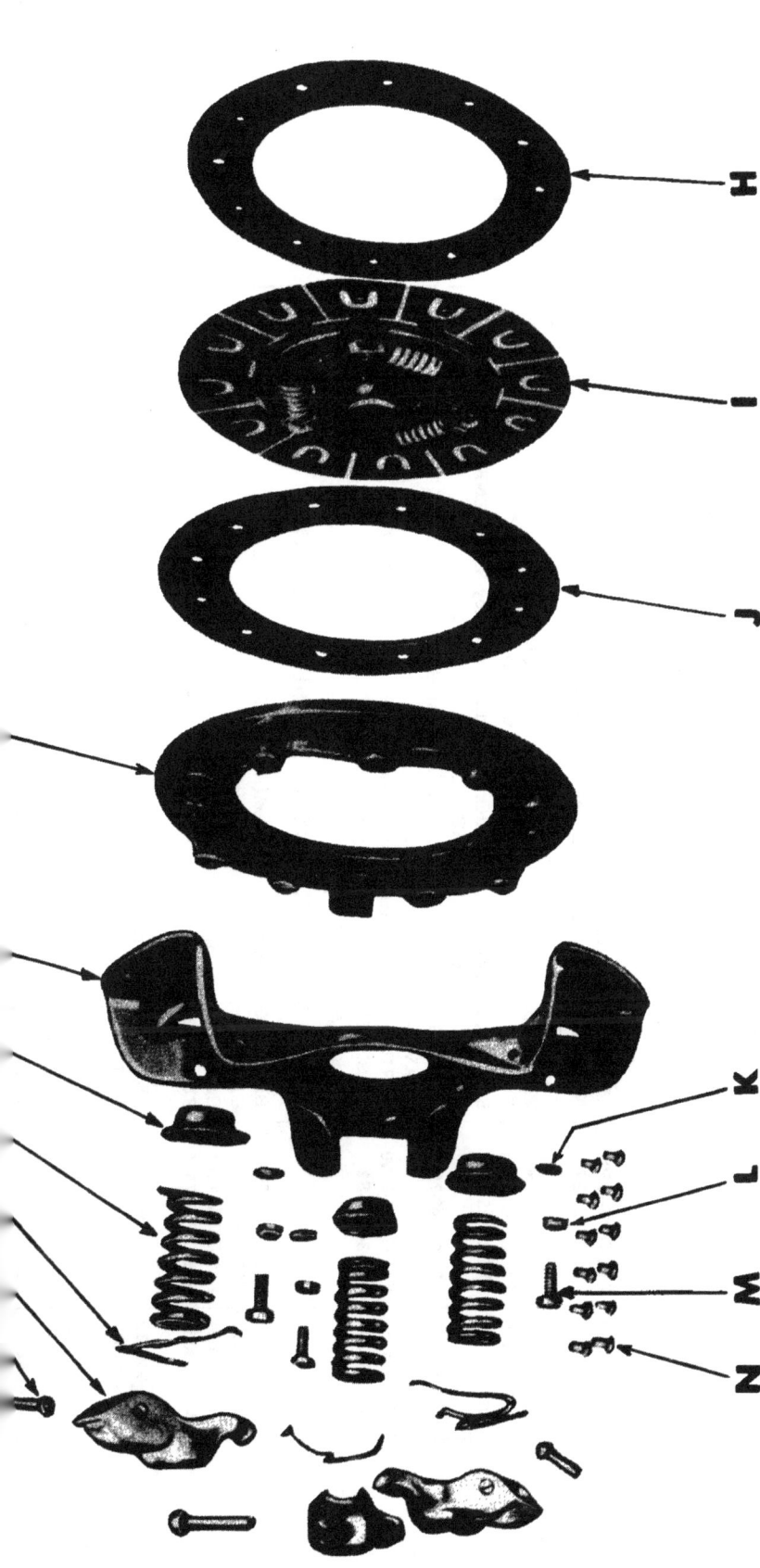

RA PD 305065

FIGURE 02-2—CLUTCH ASSEMBLY

Key	Item	Willys Part No.	Ford Part No.	Gov't Group No.
A	PIN	WO-638159	FM-GPW-7564	0202
B	LEVER	WO-638158	FM-GPW-7591	0202
C	SPRING	WO-638153	FM-GPW-7590	0202
D	SPRING	WO-638993	FM-GPW-7572	0202
E	CUP	WO-68157	FM-GPW-7567	0202
F	COVER	WO-638151	FM-GPW-7570	0202
G	PLATE	WO-638152	FM-GPW-7566	0202

Key	Item	Willys Part No.	Ford Part No.	Gov't Group No.
H	FACING	WO-371567	FM-GPW-7549	0201A
I	PLATE w/HUB and SPRINGS	WO-636775	FM-GPW-7550	0201B
J	FACING	WO-636778	FM-GPW-7577	0201A
K	WASHER	WO-638305	FM-34745-S2	0202
L	NUT	WO-638155	FM-33921-S7	0202
M	SCREW	WO-638154	FM-24325-S	0202
N	RIVET	WO-374586	FM-351926	0201A

RA PD 305065-A

0202

GROUP 02—CLUTCH (Cont'd)

Figure Number Col. 1	Official Stockage Number Col. 2	Part Number Willys Col. 3	Part Number Ford Col. 4	ITEM Col. 5	Quantity Reqd. per Unit Assy. Col. 6	Per 100 Major Items — 12 Mos. Field Maintenance Col. 7	Per 100 Major Items — Major Overhaul (5th Ech) Col. 8	Total First Year Procurement Col. 9	Estimated Reqmts. per 100 Rebuilds Col. 10
				0202—COVER, PRESSURE PLATES, AND SPRINGS (Cont'd)					
		WO-A-6752	FM-GPW-18359	KIT, repair, engine clutch cover (issue until stock is exhausted, then use WO-A-7833).	as req.	6	13	22	20
		WO-A-7833	FM-GPW-18359-B	KIT, repair, engine clutch cover (was WO-A-6752).	as req.	3	6	12	
				(Includes:					
				3 WO-638157 FM-GPW-7567 CUP					
				3 WO-638158 FM-GPW-7591 LEVER					
				3 WO-638155 FM-33921-S7 NUT					
				3 WO-638159 FM-GPW-7565 PIN					
				3 WO-638154 FM-24325-S SCREW					
				3 WO-638993 FM-GPW-7572 SPRING					
				3 WO-638153 FM-GPW-7590 SPRING					
				3 WO-638305 FM-34745-S2 WASHER)					
02-2		WO-638158	FM-GPW-7591	LEVER, release, engine clutch (AVM-TP-2820).	3				
02-1		WO-638155	FM-33921-S7	NUT, lock, engine clutch adjusting screw (hex., s-fin., S., thin $\frac{1}{4}$-28NF-2) (BBBX1B).	3				
02-2		WO-638152	FM-GPW-7566	PLATE, pressure, engine clutch (AVM-TP-2851).	1	3	6	12	20
02-2		WO-638159	FM-GPW-7565	PIN, engine clutch release lever pivot (AVM-TP-287).	3				
02-1		WO-638154	FM-24325-S	SCREW, adjusting, engine clutch (hex-hd, s-fin, alloy-S, $\frac{1}{4}$-28NF-2 x $\frac{5}{8}$) (AVM-TP-2818) (BAOX4AB).	3				
02-2		WO-638993	FM-GPW-7572	SPRING, pressure, engine clutch (spring-steel, 1 in. O.D. x $1\frac{9}{16}$ in. lgth.) (AVM-TP-2831).	3				
02-2		WO-638153	FM-GPW-7590	SPRING, engine clutch pressure plate, return (AVM-TP-2817).	3				
02-1		WO-638305	FM-34745-S2	WASHER, engine clutch adjusting screw, (S., flat, $\frac{5}{8}$ O.D. x $\frac{9}{32}$ I.D. x .050 thk.) (AVM-TP-2827).	3				
				0203—RELEASE YOKE, BEARING FORK					
02-1, 3		WO-635529	FM-GPW-7580-B	BEARING, engine clutch release (AB-A-935-1).	1	10	10	23	30
02-1, 3		WO-A-5102	FM-GPW-7530	CABLE, engine clutch control lever.	1	6		8	
02-3		WO-639654	FM-GPW-7561	CARRIER, engine clutch release bearing.	1	2	2	9	5
02-1		WO-630068	FM-GPW-7516	FULCRUM, engine clutch release lever (TRU-SA-2844-1).	1	3		4	
02-1		WO-630112	FM-GPW-7515	LEVER, clutch control.	1	%5	2	9	
02-3		WO-5910	FM-33798-S	NUT, lock, engine clutch control lever yoke end (S., hex., $\frac{5}{16}$-24NF-2) (GM-103025) (BBBX1BA).	1				
02-3		WO-339043	FM-73880-S	PIN, clevis, engine clutch yoke end to control cable.	1	%8		10	
02-1		WO-630117	FM-GPW-7562	SPRING, engine clutch release bearing carrier.	1	%19		24	
02-3		WO-632177	FM-GPW-7532	YOKE, adjusting, engine (clutch control lever cable).	1	%6		12	

0204—PEDAL

Fig.	Ord. No.	Mfg. Part No.	Description	Qty			
02-3	WO-638792	FM-353043-A-S7	FITTING, grease, straight, 1/4-28NF tapered thd. (AD-1641)	1			
02-3		FM-GPW-2476-A	GROMMET, engine clutch pedal shaft (Ford only)	2			
02-3	WO-5036	FM-74178-S	KEY, Woodruff, No. 9 (pedal to shaft)	1			
02-3	WO-A-1360	FM-GPW-7525	PAD, engine clutch pedal, assembly (issue until stock exhausted)	1		4	4
02-3	WO-A-6359		PAD, engine clutch pedal shank (Willys only) (issue until stock exhausted)	2			8
02-3	WO-A-405	FM-GPW-7250	PEDAL, engine clutch (issue until stock exhausted)	1			4
02-3	WO-50992	FM-24505-S7	BOLT, hex-hd., s-fin., alloy-S., 5/16-24NF-2 x 1 3/4 (pedal clamp) (BAOX1BF)	1			
02-3	WO-6157	FM-24426	BOLT, hex-hd., s-fin., alloy-S., 5/16-18NC-2 x 1 1/4 (shank to pedal) (BANX1BD)	2			
02-3	WO-5910	FM-33798-S2	NUT, reg., hex., s-fin., S., 5/16-24NF-2 (GM-103025) (BBBX1BA)	1			
02-3	WO-51833	FM-34806-S2	WASHER, lock, reg., S., 5/16 in. (19/32 O.D. x 11/32 I.D. x 1/16 thk.) (BECX1H)	1			
02-3	WO-52944	FM-72063-S	PIN, cotter, S., 5/32 x 1 3/8 (shaft to pedals) (issue until stock exhausted)	3		20	40
02-3	WO-A-495	FM-GPW-2473	SHAFT, engine clutch pedal, assembly (issue until stock exhausted)	1		4	4
02-3	WO-650684		SPRING, engine clutch pedal shank draft pad (escutcheon, black enamel, free lgth. 3 1/2 coils) (Willys only) (issue until stock exhausted)	2		19	8
02-3	WO-630593		SPRING, retracting, engine clutch pedal (free length 5 7/8 in., 40 coils)	2		10	30
02-3	WO-A-498	FM-356561-S	WASHER, engine clutch pedal shaft (special, S., 1 1/2 O.D. x 1 1/32 I.D. x 1/8 thk.)	3		4	40
02-3	WO-A-6360		WASHER, engine clutch pedal shank pad (S., 1 3/4 O.D. x 17/32 I.D.) (Willys only) (issue until stock exhausted)	2			8

0204A—PEDAL LINKAGE

Fig.	Ord. No.	Mfg. Part No.	Description	Qty			
02-3	WO-A-180	FM-GPW-7508	BRACKET, frame, engine clutch control (included w/ball STUD assembly)	1			
02-3	WO-52132	FM-24327-S2	BOLT, hex-hd., s-fin., alloy-S., 5/16-24NF-2 x 5/8 (bracket to frame) (GM-106279)	2			
02-3	WO-5910	FM-33798-S2	NUT, S., hex., 5/16-24NF-2 (GM-103025) (BBBX1BA)	2			
02-3	WO-51833	FM-34806-S2	WASHER, lock, S., 5/16 in. (19/32 O.D. x 11/32 I.D. x 1/16 thk.) (BECX1H)	2			
02-3	WO-A-1355	FM-GPW-7503	LEVER, engine clutch control, w/TUBE, assembly	1	4	2	
02-3	WO-5354	FM-72004-S	PIN, cotter, S., 3/8 x 1/2 (engine clutch pedal rod)	2			
02-3	WO-5108	FM-72053	PIN, cotter, S., 1/8 x 1 1/4 (engine clutch control tube) (BFAX1DH)	1			
02-3	WO-A-176	FM-GPW-7539	PAD, engine clutch control tube (felt, 5/8 O.D. x 1/4 thk.) (issue until stock exhausted)	2			
02-3	WO-A-177	FM-GPW-7517	RETAINER, engine clutch control tube spring (S., 19/32 O.D. x 3/32 thk.) (issue until stock exhausted)	2		52	108
02-3	WO-A-499	FM-GPW-7521	ROD, engine clutch release pedal	1		8	16
02-3	WO-887	FM-GPW-7512	SEAL, engine clutch control tube dust (1 in. O.D.)	2		3	4
02-3	WO-A-178	FM-GPW-7545	SPRING, engine clutch control tube (S., 9/16 O.D. x 1 3/4 free lgth.)	1		6	12
02-3	WO-A-181	FM-GPW-7514	STUD, engine clutch control, ball (S., 7/16-14NC)	1		6	12
02-3	WO-A-179	FM-GPW-7507	STUD, engine clutch control, ball, w/BRACKET, assembly	1		3	6
02-3	WO-5336	FM-33801-S2	NUT, reg., hex., s-fin., S., 7/16-14NC-2 (stud to bracket) (BBAX1D)	1		3	50
02-3	WO-5059	FM-34808-S2	WASHER, lock, S., 7/16 in. (7/8 O.D. x 17/32 I.D. x 1/8 thk.) (BECX1L)	1			

GROUP 03—FUEL

0300—FUEL TANKS

Fig.	Ord. No.	Mfg. Part No.	Description	Qty			
02-3	WO-A-1738	FM-GPW-9069	ANTI-SQUEAK, fuel tank (1 x 6 in. long)	2			
02-3	WO-A-1739	FM-GPW-9079	ANTI-SQUEAK, fuel tank (1 x 7 1/4 in. long)	2			

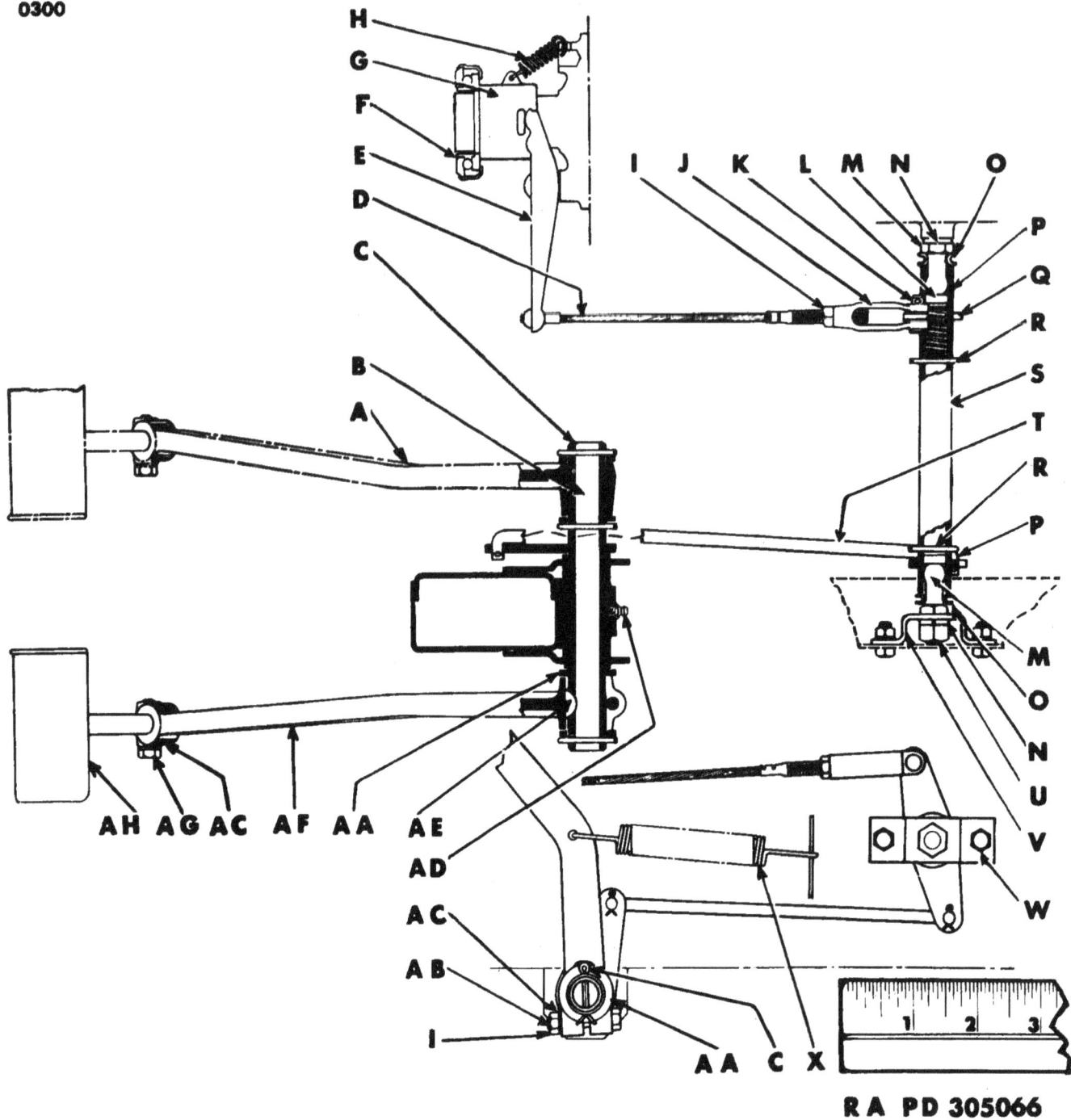

RA PD 305066

FIGURE 02-3—CLUTCH CONTROLS

Key	Item	Willys Part No.	Ford Part No.	Gov't Group No.	Key	Item	Willys Part No.	Ford Part No.	Gov't Group No.
A	PEDAL	WO-A-8253	FM-GPW-2452-B	1204	S	LEVER assembly	WO-A-1355	FM-GPW-7503	0204A
B	SHAFT	WO-A-495	FM-GPW-2473	0204	T	ROD	WO-A-499	FM-GPW-7521	0204A
C	PIN	WO-52944	FM-72063-S	0204	U	NUT	WO-5336	FM-33801-S2	0204A
D	CABLE	WO-A-5102	FM-GPW-7530	0203	V	BRACKET	WO-A-180	FM-GPW-7508	0204A
E	LEVER	WO-630112	FM-GPW-7512	0203				(See WO-A-179)	
F	BEARING	WO-635529	FM-GPW-7580-B	0203	W	SCREW	WO-52132	FM-24327-S2	0204A
G	CARRIER	WO-639654	FM-GPW-7561	0203	X	SPRING	WO-630593	FM-GPW-7523	0204
H	SPRING	WO-630117	FM-GPW-7562	0203	Y	WASHER	WO-A-498	FM-356561	0204
I	NUT	WO-5910	FM-33798-S2	0204	Z	BOLT	WO-50992	FM-24505-S7	0204
J	YOKE	WO-632177	FM-GPW-7532	0203	AA	WASHER	WO-51833	FM-34806-S2	0204
K	PIN	WO-339043	FM-73880-S	0203	AB	GREASE FITTING	WO-638792	FM-353043-A-S7	0204
L	RETAINER	WO-A-177	FM-GPW-7517	0204A					
M	STUD	WO-A-181	FM-GPW-7514	0204A	AC	KEY	WO-5036	FM-74178-S	0204
N	WASHER	WO-5059	FM-34808-S2	0204A	AD	PEDAL	WO-A-405	FM-GPW-7250	0204
O	SEAL	WO-A-887	FM-GPW-7512	0204A	AE	BOLT	WO-6157	FM-24426	0204
P	PAD	WO-A-176	FM-GPW-7539	0204A	AF	PAD	WO-A-1360	FM-GPW-7525	0204
Q	SPRING	WO-A-178	FM-GPW-7545	0204A					
R	PIN	WO-5108	FM-72053	0204A					

RA PD 305066-A

RA PD 305072

FIGURE 03-1—FUEL TANK ASSEMBLY

Key	Item	Willys Part No.	Ford Part No.	Gov't Group No.
A	CAP assembly	WO-A-6333	FM-GPW-9030-B	0300
B	EXTENSION	WO-A-6424	FM-GPW-9034-A	0300
C	STRAP assembly	WO-A-1472	FM-GPW-9095	0300
D	TANK assembly	WO-A-6618	FM-GPW-9002-B	0300
E	PROTECTOR	WO-A-1763	FM-355095-S7	0300A
F	SCREW	WO-5556	(Willys only)	0300A
G	WASHER	WO-52705		0300A
H	GAUGE assembly	WO-A-1292	FM-GPW-9275	0300A

Key	Item	Willys Part No.	Ford Part No.	Gov't Group No.
I	STRAP assembly	WO-A-1472	FM-GPW-9095	0300
J	PLUG	WO-A-5120	FM-353055-S7	0300
K	STRAP assembly	WO-A-1477	FM-GPW-9057	0300
L	NUT	WO-52891	(Willys only)	0300
M	NUT	WO-5910	FM-33798-S2	0300
N	BOLT	WO-A-1483	FM-24505-S7	0300
O	GASKET	WO-A-1293	FM-GPW-9276	0300A
P	CLAMP assembly	WO-A-1481	FM-GPW-9066	0300

RA PD 305072-A

GROUP 03—FUEL (Cont'd)

| Figure Number | Official Stockage Number | Part Number | | ITEM | Quantity Reqd. per Unit Assy. | Per 100 Major Items | | | Estimated Reqmts. per 100 Rebuilds |
| | | Willys | Ford | | | 12 Mos. Field Maintenance | Major Overhaul (5th Ech) | Total First-Year Procurement | |
Col. 1	Col. 2	Col. 3	Col. 4	Col. 5	Col. 6	Col. 7	Col. 8	Col. 9	Col. 10
				0300—FUEL TANKS (Cont'd)					
		WO-A-1740	FM-GPW-9075	ANTI-SQUEAK, fuel tank (1 x 9½ in. long)	2	—	—	—	—
		WO-A-1741	FM-GPW-9071	ANTI-SQUEAK, fuel tank (1 x 18 in. long)	2	—	—	—	—
		WO-A-1480	FM-GPW-9078	ANTI-SQUEAK, fuel tank hold down strap	2	—	—	—	—
		WO-A-1476	FM-GPW-9074	ANTI-SQUEAK, fuel tank hold down strap, inner	2	—	—	—	—
		WO-A-2954	FM-GPW-9063	BRACKET, fuel tank hold down	2	—	—	—	—
		WO-51977	FM-62218-S	RIVET, rd-hd, S., ¼ x 7/16 (BMCX1)	4	—	—	—	—
		WO-A-2953	FM-GPW-9065	BRACKET, fuel tank hold down, front inner	1	—	—	—	—
		WO-52168	FM-20344-S2	BOLT, hex-hd, s-fin, alloy-S, pkzd, ¼-20NC-2 x ⅝ (to front floor pan) (GM-121668) (BANX4AE-15)	2	—	—	—	—
		WO-52217	FM-33795-S2	NUT, reg, hex, S., pkzd, ¼-20NC-2 (GM-118613) (BBAX1A-15)	2	—	—	—	—
		WO-52031	FM-34805-S2	WASHER, reg, lock, S., pkzd, ¼ in. (BECX1G-15)	2	—	—	—	—
		WO-A-2952	FM-GPW-9062	BRACKET, fuel tank hold down, rear inner	1	—	—	—	—
		WO-A-2970	FM-GPW-9051	BRACKET, fuel tank hold down, rear inside, w/SUPPORT, assembly	1	—	—	—	—
		WO-A-3055	FM-GPW-1111322	CAP, fuel tank well drain (issue until stock exhausted)	2	% 4	8	—	—
03-1		WO-A-1254	FM-GPW-9030-A	CAP, fuel tank, w/safety CHAIN and BAR (2⅞ in. O.D.) (before Willys serial 174739) (AC-850042)	1	—	—	—	—
03-1		WO-A-6333	FM-GPW-9030-B	CAP, fuel tank, w/safety CHAIN (4⅝ in. O.D.) (after Willys serial 174739)	1	% 10	—	20	—
03-1		WO-A-1481	FM-GPW-9066	CLAMP, fuel tank hold down, assembly	1	—	—	—	—
03-1		WO-A-1483	FM-24505-S7	BOLT, hex-hd, S., 5/16-24NF-2 x 1¾ (BCBX1BF)	2	—	—	—	—
03-1		WO-5910	FM-33798-S2	NUT, reg, hex, S., 5/16-24NF-2 (GM-103025) (BBBX1BA)	2	%	—	—	—
		WO-52891		NUT, hex, S., stamped, 5/16-24NF-2 (after serial 104433) (Willys only) (GM-147510)	2	—	—	—	—
03-1		WO-A-6424	FM-GPW-9034-A	EXTENSION, fuel tank filler tube, assembly (after Willys serial 174739) (STN-6935-A)	2	—	—	—	—
		WO-A-1275	FM-GPW-9035-B	GASKET, fuel tank cap (before Willys serial 174739) (AC-850024)	1	—	—	—	—
		WO-A-3056	FM-GPW-1111323	NECK, fuel tank well drain hole	1	—	—	—	—
		WO-A-5120	FM-353055-S7	PLUG, pipe ¼ (fuel tank drain plug)	2	—	—	—	—
03-1		WO-A-2930	FM-GPW-1140474	SHIELD, front seat to rear floor fuel tank (driver)	1	% 7	—	9	—
		WO-52170	FM-24308-S2	BOLT, hex-hd, s-fin, alloy-S, pkzd, ¼-20NC-2 x ½ (shield to floor) (GM-119860) (BANX4AA-15)	1	—	—	—	—
		WO-52165	FM-34129-S2	NUT, hex, S., pkzd, ¼-20NC-2 (GM-118643)	3	—	—	—	—
		WO-52702	FM-34745-S	WASHER, plain, S, pkzd, ¼ in.	3	—	—	—	—
		WO-52031	FM-34805-S2	WASHER, reg, lock, S., pkzd, ¼ in. (BECX1G-15)	3	—	—	—	—
03-1		WO-A-1472	FM-GPW-9095	STRAP, hold down, fuel tank, inner, assembly (issue until stock exhausted)	2	% 4	8	—	—
03-1		WO-A-1477	FM-GPW-9057	STRAP, hold down, fuel tank outer, assembly	1	—	—	—	—
		WO-A-1221	FM-GPW-9002-A	TANK, fuel, assembly (before Willys serial 174739) (use WO-A-6897 FM-GPW-9001)	1	—	—	—	—
03-1		WO-A-6618	FM-GPW-9002-B	TANK, fuel, assembly (after Willys serial 174739)	1	—	—	—	—

0300A—TANK GAUGE UNIT

0301—CARBURETOR

Ref	WO No.	FM-GPW No.	Description	No. used				
03-1	WO-A-6897	FM-GPW-9001	TANK, fuel, w/EXTENSION and CAP, assembly (as required to replace tanks on trucks built prior to Willys serial 174739) (issue until stock exhausted)	1			4	%4
03-1	WO-A-3497	FM-GPW-111324	WELL, fuel tank, assembly	1				
			0300A—TANK GAUGE UNIT					
03-1	WO-A-1293	FM-GPW-9276	GASKET, fuel tank gauge (cork) (2⅝ O.D. x 1⅝ I.D. x ⅛ thk.)	1			30	%19
03-1	WO-A-1292	FM-GPW-9275	GAUGE, fuel, tank unit, assembly (AL-9979-A)	1			14	%10
03-1	WO-5556	FM-355095-S7	SCREW, machine, rd-hd., S., No. 8 (.164 in.)-32NC-2 x ⅜ (gauge to tank) (BCNX1FE)	5				
03-1	WO-52705		WASHER, lock, S., No. 8 (.164 in.) (.2627 O.D. x .169 I.D. x 3/64 thk.) (BECX1D) (Willys only)	5				
03-1	WO-A-1763	FM-GPW-9211	PROTECTOR, fuel gauge, tank unit	1			4	%3
	WO-5253		SCREW, machine, rd-hd., S., No. 8 (.164)-32NC-2 x ¼ (fuel gauge tank unit terminal) (BCNX1FC) (Willys only)	1				
	WO-52705		WASHER, lock, S., No. 8 (.164 in.) (.2627 O.D. x .169 I.D. x 3/64 thk.) (fuel gauge tank unit terminal screw) (BECX1D) (Willys only)	1				
			0301—CARBURETOR					
03-2	WO-116181	FM-GPW-9528	ARM, carburetor pump, w/COLLAR, assembly (CAR-53A-168S)	1	10	8	3	%3
03-2	WO-116197	FM-GPW-9583	ARM, carburetor throttle shaft, w/SCREW (CAR-114-21S)	1	5	8	2	%2
03-2	WO-116543	FM-GPW-6333-D	BRACKET, carburetor choke lever, assembly (before 4000 Willys trucks) (CAR-62-108S)	1				
G	WO-301232	FM-31037-S7	SCREW, choke lever bracket (CAR-101-6)	1	8	8	2	%2
03-2	WO-A-1223	FM-GPW-9510	CARBURETOR assembly (CAR-539S)	1	10	13	3	%6
03-2	WO-116204	FM-GPW-9594	CHECK, carburetor discharge disc, assembly (CAR-122-47S)	1				
03-2	WO-116205	FM-GPW-9576	CHECK, carburetor intake ball, assembly (CAR-122-64S)	1				
	WO-116548		CLAMP, carburetor tube, assembly (before Willys serial 103468) (CAR-62-131S) (Willys only)	1				
03-2	WO-116587	FM-GPW-9595	CLAMP, carburetor tube, assembly (after Willys serial 103468) (CAR-62-134) (issue until stock exhausted)	1	5	8	4	
03-2	WO-116208	FM-GPW-9515	COVER, carburetor bowl, w/PIN, assembly (CAR-146-95S)	1			2	2
03-2	WO-116213	FM-31061-S8	SCREW, carburetor bowl cover attaching (CAR-101-82)	4				
03-2	WO-6330	FM-34802-S2	WASHER, lock, S., 294 O.D. x .169 I.D. x 3/64 thk (carburetor bowl cover screw) (CAR-86-10)	4				
03-2	WO-116206	FM-GPW-9905	DISC, carburetor metering rod (CAR-129-15)	1	30	24	10	10
03-2	WO-116584	FM-GPW-9518	FLANGE, carburetor body, assembly (CAR-1-407)	1				
03-2	WO-116215	FM-31662-S	SCREW, carburetor body flange attaching (CAR-101-122)	1				
03-2	WO-5045	FM-34805-S2	WASHER, lock, S., 15/32 O.D. x 9/32 I.D. x 1/16 thk, (carburetor body flange screw) (CAR-86-11)	1				
03-2	WO-116172	FM-GPW-9550	FLOAT, carburetor, w/LEVER, assembly (CAR-21-74S)	1	7	6	2	2
03-2	WO-116202	FM-GPW-9516	GASKET, carburetor body flange (CAR-121-56)	2				
03-2	WO-116203	FM-GPW-9519	GASKET, carburetor bowl cover (CAR-121-73)	1				
03-2	WO-116169	FM-GPW-9608	GASKET, carburetor metering rod jet and plug (CAR-20-26)	3				
03-2	WO-116168	FM-GPW-9569	GASKET, carburetor needle seat and plug (CAR-20-22)	3				
03-2	WO-116171	FM-GPW-9926	GASKET, carburetor nozzle (CAR-20-72)	1				
03-2	WO-116170	FM-GPW-9574	GASKET, carburetor strainer plug (CAR-20-61)	1				
03-2	WO-A-6837	FM-GPW-18352	GASKET SET, carburetor	as req.		45		29

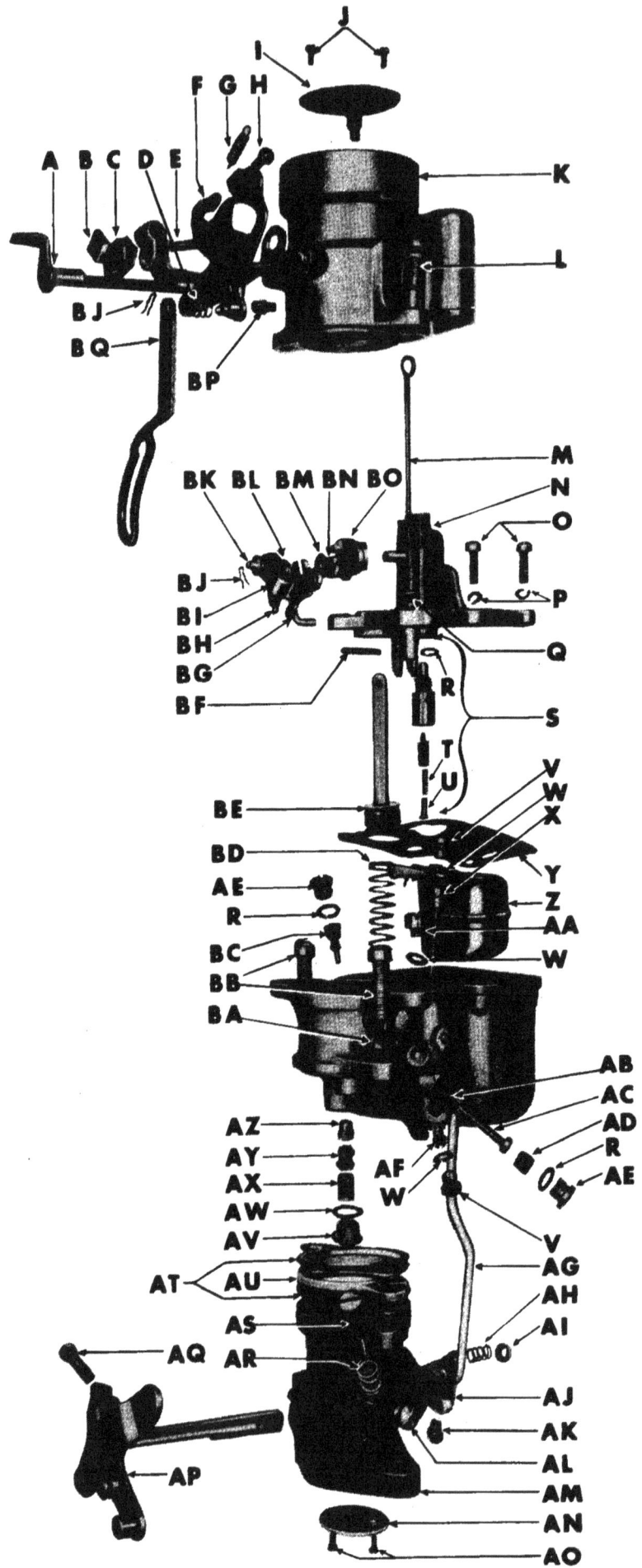

RA PD 305068

FIGURE 03-2—CARBURETOR

Key	Item	Willys Part No.	Ford Part No.	Gov't Group No.
A	SHAFT assembly	WO-116545	FM-GPW-9546	0301
B	NUT	WO-116219	FM-355858-S	0301
C	CLAMP	WO-116587	FM-GPW-9595	0301
D	SPRING	WO-116189	FM-GPW-9587	0301
E	SCREW	WO-116588	FM-355152-S	0301
F	LEVER assembly	WO-116586	FM-GPW-9526	0301
G	SPRING	WO-116184	FM-GPW-9624	0301
H	SCREW	WO-116211	FM-31032-S7	0301
I	VALVE assembly	WO-116157	FM-GPW-9549	0301
J	SCREW	WO-116216	FM-GPW-9586	0301
K	HORN assembly	WO-116544	FM-GPW-9520	0301
L	SCREW	WO-116385	FM-355200-S7	0301
M	ROD	WO-116540	FM-GPW-9906	0301
N	COVER	WO-116208	FM-GPW-9515	0301
O	SCREW	WO-116213	FM-31061-S8	0301
P	WASHER	WO-6630	FM-34802-S2	0301
Q	DISC	WO-116206	FM-GPW-9905	0301
R	GASKET	WO-116168	FM-GPW-9569	0301
S	GASKET	WO-116174	FM-GPW-9567	0301
T	SPRING	WO-116191	FM-GPW-9935	0301
U	PIN	WO-116177	FM-GPW-9566	0301
V	PASSAGE	WO-116165	FM-GPW-9543	0301
W	GASKET	WO-116169	FM-GPW-9608	0301
X	JET	WO-116539	FM-GPW-9533	0301
Y	GASKET	WO-116203	FM-GPW-9519	0301
Z	FLOAT assembly	WO-116172	FM-GPW-9550	0301
AA	JET w/GASKET	WO-116541	FM-GPW-9914	0301
AB	GASKET	WO-116171	FM-GPW-9926	0301
AC	NOZZLE	WO-116166	FM-GPW-9922	0301
AD	PLUG	WO-116161	FM-GPW-9562	0301
AE	PLUG w/GASKET	WO-116164	FM-GPW-9928	0301
AF	JET	WO-116179	FM-GPW-9544	0301
AG	ROD	WO-116198	FM-GPW-9531	0301
AH	SPRING	WO-116185	FM-GPW-9615	0301
AI	RETAINER	WO-116194	FM-GPW-9614	0301
AJ	ARM assembly	WO-116197	FM-GPW-9583	0301
AK	SCREW	WO-52290	FM-31588-S7	0301
AL	PLUG	WO-116162	FM-GPW-9579	0301
AM	FLANGE assembly	WO-116584	FM-GPW-9518	0301
AN	VALVE	WO-116154	FM-GPW-9585	0301
AO	SCREW	WO-116217	FM-GPW-9588	0301
AP	SHAFT assembly	WO-116585	FM-GPW-9581	0301
AQ	SCREW	WO-116651	FM-GPW-9610	0301
AR	SPRING	WO-116183	FM-GPW-9578	0301
AS	SCREW	WO-116176	FM-GPW-9541	0301
AT	GASKET	WO-116202	FM-GPW-9516	0301
AU	INSULATOR	WO-116210	FM-GPW-9554	0301
AV	PLUG, w/GASKET	WO-116163	FM-GPW-9696	0301
AW	GASKET	WO-116170	FM-GPW-9574	0301
AX	STRAINER	WO-116175	FM-GPW-9575	0301
AY	CHECK assembly	WO-116205	FM-GPW-9576	0301
AZ	CHECK assembly	WO-116204	FM-GPW-9594	0301
BA	WASHER	WO-5045	FM-34805-S2	0301
BB	SCREW	WO-116215	FM-31662-S	0301
BC	JET	WO-116180	FM-GPW-9940	0301
BD	SPRING	WO-116188	FM-GPW-9636	0301
BE	PLUNGER assembly	WO-116195	FM-GPW-9631	0301
BF	PIN	WO-116173	FM-GPW-9558	0301
BG	LINK	WO-116199	FM-GPW-9527	0301
BH	SPRING	WO-116187	FM-GPW-9570	0301
BI	LEVER assembly	WO-116537	FM-GPW-9529	0301
BJ	SPRING	WO-116178	FM-GPW-9599	0301
BK	PIN	WO-116209	FM-GPW-9930	0301
BL	SPRING	WO-116538	FM-GPW-9907	0301
BM	WASHER	WO-116207	FM-34711-S	0301
BN	NUT	WO-52615	FM-34051-S7	0301
BO	ARM assembly	WO-116181	FM-GPW-9528	0301
BP	SCREW	WO-116384	FM-355067-S7	0301
BQ	LINK	WO-116542	FM-GPW-9598	0301

RA PD 305068-A

SNL G-503

GROUP 03—FUEL (Cont'd)

Col. 1 Figure Number	Col. 2 Official Stockage Number	Part Number Col. 3 Willys	Part Number Col. 4 Ford	Col. 5 ITEM	Col. 6 Quantity Reqd. per Unit Assy.	Col. 7 12 Mos. Field Maintenance	Col. 8 Major Overhaul (5th Ech)	Col. 9 Total First Year Procurement	Col. 10 Estimated Reqmts. per 100 Rebuilds
				0301—CARBURETOR (Cont'd)					
				(Includes:					
				2 WO-116202 FM-GPW-9516 GASKET, body flange					
				1 WO-116203 FM-GPW-9519 GASKET, bowl cover					
				1 WO-A-6357 FM-GPW-9445 GASKET, diffuser					
				1 WO-116171 FM-GPW-9926 GASKET, nozzle					
				3 WO-116169 FM-GPW-9608 GASKET, jet					
				3 WO-116168 FM-GPW-9569 GASKET, seat and plug					
				1 WO-116170 FM-GPW-9574 GASKET, strainer)					
03-2		WO-116544	FM-GPW-9520	HORN, air, carburetor, assembly (CAR-6-312S)	1	2	2	8	5
03-2		WO-116385	FM-355200-S7	SCREW, carburetor airhorn attaching, w/WASHER (CAR-101-150S)	2	10	10	24	30
03-2		WO-116210	FM-GPW-9554	INSULATOR, carburetor to choke valve (CAR-183-19)	1	% 2	2	8	5
03-2		WO-116179	FM-GPW-9544	JET, carburetor idle well (CAR-43-67)	1	—	—	—	—
03-2		WO-116539	FM-GPW-9533	JET, carburetor, low speed assembly (CAR-11-180S)	1	—	—	—	—
03-2		WO-116180	FM-GPW-9940	JET, carburetor pump (CAR-48-84)	1	—	—	—	—
03-2		WO-116541	FM-GPW-9914	JET, carburetor metering rod, w/GASKET, assembly (CAR-120-151S)	1	—	—	—	—
		WO-A-6840	FM-GPW-18357-B	KIT, repair, carburetor	as req.				
				(Includes:					
				1 WO-116197 FM-GPW-9583 ARM					
				1 WO-116204 FM-GPW-9594 CHECK, discharge disc					
				1 WO-116205 FM-GPW-9576 CHECK, intake ball					
				1 WO-116206 FM-GPW-9905 DISC, metering rod					
				2 WO-116202 FM-GPW-9516 GASKET, body flange					
				1 WO-116203 FM-GPW-9519 GASKET, bowl cover					
				1 WO-A-6357 FM-GPW-9445 GASKET, diffuser					
				1 WO-116171 FM-GPW-9926 GASKET, nozzle					
				1 WO-116179 FM-GPW-9544 JET, idle well					
				1 WO-116539 FM-GPW-9533 JET, low speed assembly					
				1 WO-116541 FM-GPW-9914 JET, metering rod					
				1 WO-116180 FM-GPW-9940 JET, pump					
				1 WO-116199 FM-GPW-9527 LINK, pump connector					
				1 WO-116166 FM-GPW-9922 NOZZLE					
				1 WO-52615 FM-34051-S7 NUT, hex					
				1 WO-116209 FM-GPW-9930 PIN					
				1 WO-116178 FM-GPW-9599 PIN					
				4 WO-116161 FM-GPW-9562 PLUG					
				1 WO-116162 FM-GPW-9579 PLUG					
				5 WO-116160 FM-GPW-9523 PLUG					

Fig. / Group	Federal Stock No.	Ford No.	Qty	Description	Unit	1	2	3	4
			1	*...FM-GPW-9528 PLUG and GASKET*					
	WO-116164	FM-GPW-9928	2	PLUG and GASKET					
	WO-116165	FM-GPW-9543	2	PLUG and GASKET					
	WO-116195	FM-GPW-9631	1	PLUNGER and ROD					
	WO-116194	FM-GPW-9614	1	RETAINER					
	WO-116198	FM-GPW-9531	1	ROD					
	WO-116540	FM-GPW-9906	1	ROD, metering					
	WO-116216	FM-GPW-9586	2	SCREW					
	WO-116217	FM-GPW-9588	2	SCREW					
	WO-116213	FM-31061-S8	2	SCREW					
	WO-116215	FM-31662-S	1	SCREW					
	WO-116385	FM-355200-S7	1	SCREW					
	WO-116184	FM-GPW-9624	1	SPRING					
	WO-116185	FM-GPW-9615	1	SPRING					
	WO-116187	FM-GPW-9570	1	SPRING					
	WO-116188	FM-GPW-9636	1	SPRING					
	WO-116189	FM-GPW-9587	1	SPRING					
	WO-116538	FM-GPW-9907	1	SPRING					
	WO-116174	FM-GPW-9567	1	SPRING and SEAT					
	WO-116175	FM-GPW-9575	1	STRAINER					
	WO-6330	FM-34802-S2	2	WASHER					
	WO-5045	FM-34805-S7	1	WASHER					
	WO-116207	FM-34711-S	1	WASHER)					
	WO-A-5501			KIT, repair, carburetor throttle shaft and lever (Willys only) (CAR-3-4664).. as req. (Includes:	as req.				
	WO-116587	FM-GPW-9595	1	CLAMP, tube					
	WO-116219	FM-355858-S7	1	NUT, clamp					
	WO-116540	FM-GPW-9906	1	ROD, metering					
	WO-116588	FM-355132-S	1	SCREW, clamp					
	WO-116217	FM-GPW-9588	2	SCREW, attaching, valve					
	WO-116585	FM-GPW-9581	1	SHAFT and LEVER, throttle)					
03-2	WO-116537	FM-GPW-9529	1	LEVER, operating, carburetor pump, assembly (CAR-53A-251S)	% 3	3	8		10
03-2	WO-116586	FM-GPW-9526	1	LEVER, carburetor choke, w/BRACKET assembly (after 4000 trucks) (CAR-62-135-S)	% 2	2	8		8
03-2	WO-116545	FM-GPW-9546	1	LEVER, carburetor choke control, w/SHAFT, assembly (CAR-14-246S) (issue until stock exhausted)	% 4	4			
03-2	WO-116542	FM-GPW-9598	1	LINK, carburetor choke (CAR-117-106)	—	—	—		—
03-2	WO-116199	FM-GPW-9527	1	LINK, connector, carburetor pump (CAR-117-58)	—	—	—		—
03-2	WO-116166	FM-GPW-9922	1	NOZZLE, carburetor (CAR-12-255) (issue until stock exhausted)	4	4	—		—
03-2	WO-50922	FM-33800-S2	1	NUT, carburetor flange stud (CAR-105A-13)	—	—	—		—
03-2	WO-52615	FM-34051-S7	1	NUT, hex, S., (carburetor metering rod pin) (CAR-105A-19)	—	—	—		—
03-2	WO-116219	FM-355858-S	1	NUT, carburetor tube clamp (CAR-105A-8) (issue until stock exhausted)	4	8	—		—
03-2	WO-116165	FM-GPW-9543	1	PASSAGE, carburetor low speed jet and idle well, w/GASKET, assembly (CAR-11B-129S)	—	—	—		—
03-2	WO-116173	FM-GPW-9558	1	PIN, carburetor float lever (CAR-24-23)	10		12		15
03-2	WO-116177	FM-GPW-9566	1	PIN, carburetor intake needle (CAR-150-98)	5	5	12		—
03-2	WO-116209	FM-GPW-9930	1	PIN, carburetor metering rod (CAR-150-97)	—	—	—		—
03-2	WO-116204	FM-GPW-119594	1	PLUG, carburetor discharge disc check, assembly (CAR-122-47S)	—	—	—		—
03-2	WO-116162	FM-GPW-9579	1	PLUG, carburetor idle port rivet (CAR-11B-108)	—	—	—		—
03-2	WO-116161	FM-GPW-9562	1	PLUG, carburetor nozzle retainer (CAR-11B-105)	—	—	—		—

GROUP 03—FUEL (Cont'd)

0301—CARBURETOR (Cont'd)

Figure Number	Official Stockage Number	Part Number Willys	Part Number Ford	ITEM	Quantity Reqd. per Unit Assy.	Per 100 Major Items — 12 Mos. Field Maintenance	Per 100 Major Items — Major Overhaul (5th Ech)	Per 100 Major Items — Total First Year Procurement	Estimated Reqmts. per 100 Rebuilds
Col. 1	Col. 2	Col. 3	Col. 4	Col. 5	Col. 6	Col. 7	Col. 8	Col. 9	Col. 10
03-2		WO-116160	FM-GPW-9523	PLUG, carburetor rivet (CAR-11B-79)	1	—	—	—	—
03-2		WO-116159	FM-GPW-9522	PLUG, carburetor rivet (CAR-11B-35)	2	—	—	—	—
03-2		WO-116164	FM-GPW-9928	PLUG, carburetor pump jet and nozzle passage, w/GASKET (CAR-11B-127-S)	2	10	—	12	—
03-2		WO-116163	FM-GPW-9696	PLUG, carburetor strainer passage, w/GASKET (CAR-11B-125S)	1	5	—	8	—
03-2		WO-116195	FM-GPW-9631	PLUNGER, carburetor pump, w/ROD, assembly (CAR-64-62S)	1	—	—	—	—
03-2		WO-116194	FM-GPW-9614	RETAINER, carburetor spring, (CAR-63-35)	1	19	—	24	—
03-2		WO-116540	FM-GPW-9906	ROD, metering carburetor, (CARK75-547)	1	10	—	12	—
03-2		WO-116198	FM-GPW-9531	ROD, connector carburetor throttle (CAR-115-59)	1	—	—	—	—
03-2		WO-116176	FM-GPW-9541	SCREW, adjustment idler carburetor (CAR-30A-39)	1	6	3	12	10
03-2		WO-116651	FM-GPW-9610	SCREW, adjustment carburetor throttle lever (CAR-101-121) (issue until stock exhausted)	1	%	—	—	—
03-2		WO-52290	FM-31588-S7	SCREW, carburetor throttle shaft arm clamp (CAR-101-28)	1		8	—	—
03-2		WO-116218	FM-355132-S	SCREW, carburetor tube clamp (before Willys serial 103468) (CAR-105-11) (Willys only)	1	—	—	—	—
03-2		WO-116588		SCREW, carburetor tube clamp (after Willys serial 103468) (CAR-105-13) (issue until stock exhausted)	1	—	—	—	—
03-2		WO-116211	FM-31032-S7	SCREW, carburetor wire clamp (CAR-101-10)	1	40	80	20	10
03-2		WO-116384	FM-355067-S7	SCREW, fill-hd. No. 8 (.164)-32NC, w/WASHER (carburetor choke valve) (issue until stock exhausted)	1	6	3	—	—
03-2		WO-116545	FM-GPW-9546	SHAFT, carburetor choke control, w/LEVER, assembly	1	—	—	—	—
03-2		WO-116585	FM-GPW-9581	SHAFT, carburetor throttle, w/LEVER, assembly (after Willys serial 103468) (CAR-3-465-S)	1	4	8	12	12
03-2		WO-116189	FM-GPW-9587	SPRING, carburetor connector link (CAR-61-190)	1	%	4	12	—
03-2		WO-116185	FM-GPW-9615	SPRING, carburetor connector rod (CAR-61-128)	1	10	—	12	—
03-2		WO-116184	FM-GPW-9624	SPRING, carburetor choke pull back (CAR-61-119)	1	10	—	9	10
03-2		WO-116183	FM-GPW-9578	SPRING, carburetor idle adjustment screw (CAR-61-57)	1	3	3	—	—
03-2		WO-116191	FM-GPW-9935	SPRING, carburetor intake needle (CAR-61-207)	1	—	—	—	—
03-2		WO-116538	FM-GPW-9907	SPRING, carburetor metering rod (CAR-61-272)	1	19	—	24	—
03-2		WO-116178	FM-GPW-9599	SPRING, carburetor pin (CAR-150-A-10)	1	—	—	—	—
03-2		WO-116186	FM-GPW-9650	SPRING, carburetor plunger (CAR-61-143)	1	—	—	—	—
03-2		WO-116188	FM-GPW-9636	SPRING, carburetor pump (CAR-61-171)	1	—	—	—	—
03-2		WO-116187	FM-GPW-9570	SPRING, carburetor pump arm (CAR-61-169)	1	6	—	8	—
03-2		WO-116174	FM-GPW-9567	SPRING, carburetor, needle, w/SEAT, assembly (CAR-25-94S)	1	8	—	12	—
03-2		WO-116175	FM-GPW-9575	STRAINER, carburetor pump check (CAR-30-20)	1	5	—	8	—
03-2		WO-116157	FM-GPW-9549	VALVE, choker, carburetor, assembly (CAR-7-116S)	1	—	—	—	—
03-2		WO-116216	FM-GPW-9586	SCREW, attaching, carburetor choker valve (CAR-39-10)	2	—	—	—	—
03-2		WO-116154	FM-GPW-9585	VALVE, throttle, carburetor (CAR-2-89)	1	—	—	—	—

Fig.	Ord. Part No.	Mfr. Part No.	Name and description	Units	Col 1	Col 2	Col 3	Col 4
			0301A—CARBURETOR ATTACHING PARTS					
	WO-5010	FM-34807-S2	hausted). WASHER, S., lock, carburetor flange stud ($^{21}/_{32}$ O.D. x$1^3/_8$ I.D. x $^5/_{32}$ thk.) (CAR-86-15).	1	—	—	12	8
01-11	WO-A-6357	FM-GPW-9445	GASKET, insulator, carburetor and intake manifold diffuser, assembly (GM-114942)	1	—	—	—	% 10
	WO-50922	FM-33800-S2	NUT, hex., S., $^3/_8$-24NF-2 (carburetor to manifold stud) (BBBX1CA).	1	1	13	—	—
	WO-632159	FM-88057-S7	STUD, S., $^3/_8$-16NC x $1^{11}/_{16}$ x $^3/_8$-24NF (carburetor to manifold) (issue until stock exhausted)	2	—	—	40	% 20
			0301B—CHOKE					
	WO-345961	FM-GPW-13434-A2	BUSHING, rubber, carburetor choke control through dash (issue until stock exhausted).	2	—	—	24	% 16
03-5	WO-A-7517	FM-GPW-9775-B	CONTROL, carburetor choke or throttle, assembly	1	—	—	8	% 6
	WO-5901	FM-33925-S7	NUT, reg., hex., S., $^3/_8$-24NF-2 (choke control to dash) (GM-103026) (BBBX1CA).	2	—	—	—	%
	WO-52332	FM-34907-S2	WASHER, lock, S., internal teeth, $^3/_8$ in. ($^{11}/_{16}$ O.D. x $^{25}/_{64}$ I.D. x .035 thk.) (BEAX3C).	2	—	—	—	—
18-1	WO-A-1307	FM-GPW-97303	KNOB, carburetor choke throttle w/PLUNGER and WIRE, assembly (issue until stock is exhausted, then use WO-A-7518).	1	—	12	—	% 10
	WO-A-7518	FM-GPW-9778	KNOB, carburetor choke or throttle w/PLUNGER and WIRE, assembly (was WO-A-1307).	2	—	—	—	—
			0301C—AIR CLEANER					
03-3	WO-A-1315		BASE, air cleaner, assembly (before Willys serial 124309) (AC-1542510) (Willys only).	1	—	—	—	—
	WO-A-5629	FM-GPW-9609	BODY, air cleaner, assembly (after Willys serial 124309) (HH-613455) (issue until stock exhausted).	1	—	—	4	—
	WO-A-1313		BOLT, air cleaner cover (S., wing) (before Willys serial 124309) (AC-1542508) (Willys only).	1	—	—	—	—
	WO-A-1279	FM-GPW-9657	BRACKET, support, air cleaner, left, assembly.	1	—	—	—	—
	WO-A-1278	FM-GPW-9656	BRACKET, support, air cleaner, right, assembly.	1	—	—	—	—
	WO-52132	FM-20027-S2	BOLT, hex-hd., s-fin., alloy-S., $^5/_{16}$-24NF-3 x $^5/_8$ (GM-106279).	6	—	—	—	—
	WO-51833	FM-34806-S2	WASHER, lock, S., $^5/_{16}$ in. ($^{11}/_{16}$ O.D. x $^{11}/_{32}$ I.D. x $^1/_{16}$) (BECX1H).	6	—	—	—	—
	WO-53025	FM-356309-S7	WASHER, lock, S., cd-pltd., internal, external tooth, $^5/_{16}$ in. (GM-178532).	3	—	—	—	—
03-3	WO-A-1451	FM-GPW-9686-A	BUSHING, air cleaner tube to carburetor (3 ply rad. hose, $2^1/_4$ x $1^1/_8$).	1	—	12	—	% 10
	WO-A-281	FM-GPW-9628	CLAMP, carburetor air horn, w/BOLT and NUT (clamp, S., $2^{13}/_{32}$ I.D. x $^1/_2$) (bolt and nut cd-pltd.) (before serial 104310) (issue until stock exhausted).	1	—	—	4	—
03-3	WO-53108	FM-GPW-6772	CLAMP, hose, S., $^3/_4$ in. (after Willys serial 208437).	2	—	—	20	% 6
03-3	WO-635097	FM-GPW-9653	CLAMP, hose S., $2^9/_{32}$ in., I.D. (USL-461389).	4	—	—	—	% 6
	WO-A-1515	FM-GPW-2250	CLAMP, hose S., $2^1/_2$ in., I.D. (after Willys serial 104310) (1 used for WO-A-1451 BUSHING: 1 used carburetor to air cleaner tube).	2	—	—	—	% 10
	WO-6273	FM-34114-S2	NUT, hex., S., No. 10 (.190)-32NF-2.	2	—	—	—	—
	WO-6383	FM-27145-S2	SCREW, machine, rd-hd., S., No. 10 (.190)-32NF-2 x 1 (GM-100768) (BCOX1.1AL).	2	—	—	—	—
03-3	WO-A-5621	FM-GPW-18205-B	CLEANER, air, assembly (after Willys serial 124309) (HH-613300).	1	—	10	5	% 3
	WO-115905	FM-355253-S	SCREW, wing, $^1/_4$-20NC-2 x $^3/_4$ (cleaner to brkt.).	4	—	—	—	% 13

FIGURE 03-3—OIL BATH AIR CLEANER

Key	Item	Willys Part No.	Ford Part No.	Gov't Group No.
A	HOSE	WO-A-1311	FM-GPW-9652	0301C
B	TUBE w/BRACKET	WO-A-6911	FM-GPW-9637-B	0301C
C	BUSHING	WO-A-1451	FM-GPW-9686-A	0301C
D	CLAMP	WO-A-1515	FM-GPW-2250	0301C
E	HORN	WO-A-463	FM-GPW-9632	0301C
F	BODY assembly	WO-A-5629	FM-GPW-9609	0301C
G	ELEMENT assembly	WO-A-5630	FM-GPW-9617	0301C
H	CUP	WO-A-5631	FM-GPW-9658	0301C
I	GASKET	WO-A-5633	FM-GPW-9623	0301C
J	GASKET	WO-A-5632	FM-GPW-9621	0301C
K	CLAMP	WO-635097	FM-GPW-9653	0301C

RA PD 305070-A

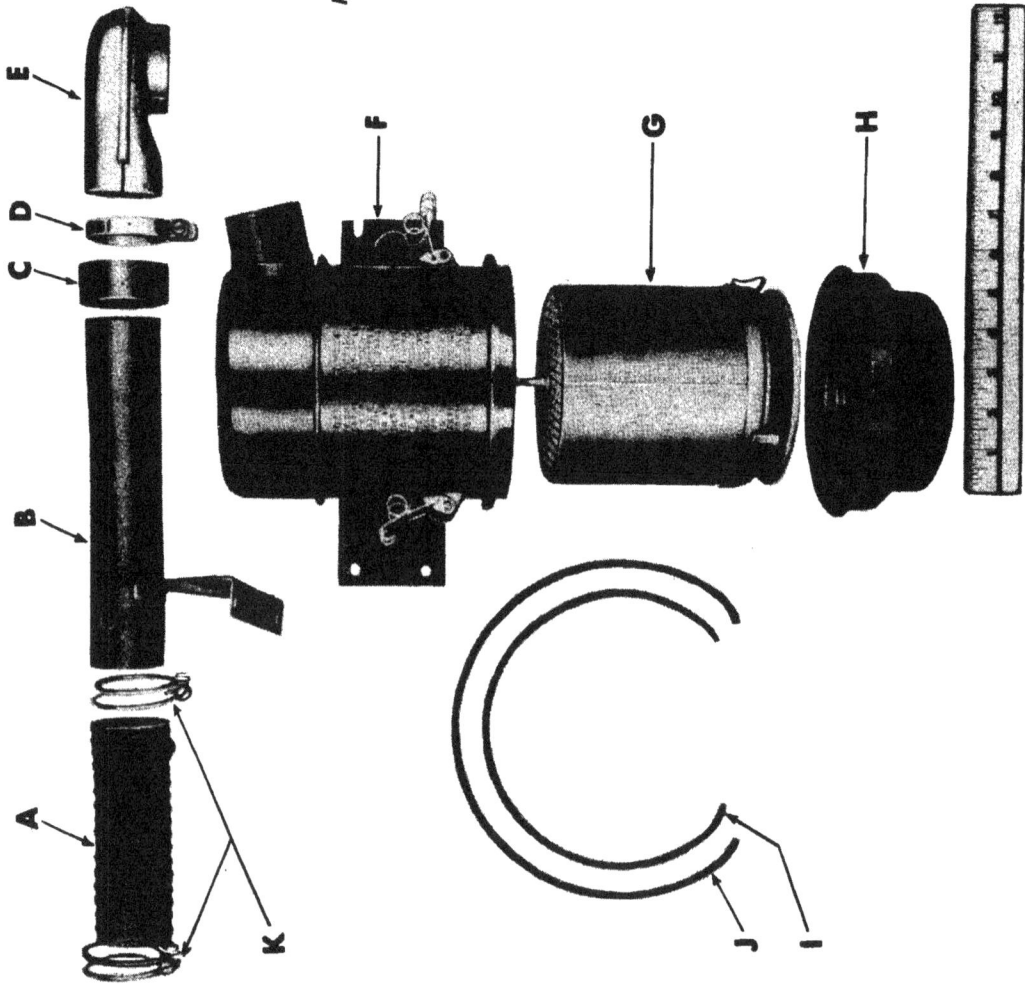

RA PD 305070

0301C—AIR CLEANER (Cont'd)

Figure Number Col. 1	Official Stockage Number Col. 2	Part Number Willys Col. 3	Part Number Ford Col. 4	ITEM Col. 5	Quantity Reqd. per Unit Assy. Col. 6	Per 100 Major Items — 12 Mos. Field Maintenance Col. 7	Per 100 Major Items — Major Overhaul (5th Ech) Col. 8	Per 100 Major Items — Total First-Year Procurement Col. 9	Estimated Reqmts. per 100 Rebuilds Col. 10
		WO-52702	FM-34745-S	WASHER, S., 1/4 in. (5/8 O.D. x 9/32 I.D. x 1/16 thk.)	4	—	—	—	—
		WO-53024		WASHER, lock, S., 1/4 in., cd-pltd., internal, external, 1/4 in. (GM-174916)	2	—	—	—	—
03-3		WO-A-5631	FM-GPW-9658	CUP, oil, air cleaner (after Willys serial 124309) (HH-613306)	1	% 3	—	4	—
		WO-A-1314		ELEMENT, filter, air cleaner, assembly (after Willys serial 124309) (AC-1542509) (Willys only)	1	—	—	—	—
03-3		WO-A-5630	FM-GPW-9617	ELEMENT, air cleaner, w/wing BOLT (after Willys serial 124309) (HH-613387)	1	% 6	—	8	—
03-3		WO-A-5632	FM-GPW-9621	GASKET, air cleaner, body upper (after Willys serial 124309) (HH-613313)	1	% 8	—	15	—
03-3		WO-A-5633	FM-GPW-9623	GASKET, air cleaner, oil cup, lower (after Willys serial 124309) (HH-613314)	1	% 8	—	15	—
03-3		WO-A-463	FM-GPW-9632	HORN, air cleaner	1	% 5	—	8	—
		WO-A-6918	FM-GPW-6771	HOSE, air cleaner tube to oil filler tube	1	% 10	—	12	—
03-3		WO-A-1311	FM-GPW-9652	HOSE, flexible, air cleaner tube to cleaner	1	% 19	—	24	—
		WO-1224		OUTLET, air cleaner (first 24309 Willys trucks) (Willys only) (BCAX1BA) (Willys only)	1	—	—	—	—
		WO-51523		SCREW, hex-hd, cap, S., 5/16-18NC-2 x 3/4 (outlet to brkt.) (GM-100121)	2	—	—	—	—
		WO-A-642	FM-GPW-9647	SHIELD, air cleaner, assembly	1	—	—	—	—
		WO-51732	FM-20309-S7	BOLT, hex-hd, s-fin, alloy-S., 1/4-28NF-2 x 1/2 (shield to frame) (BAOX4AA)	3	—	—	—	—
		WO-5914	FM-33796-S2	NUT, hex, S., 1/4-28NF-2 (GM-103024) (BBBX1AA)	3	—	—	—	—
		WO-5121	FM-34706-S2	WASHER, S., 5/16 in. (3/4 O.D. x 5/16 I.D. x 1/16 thk.)	3	—	—	—	—
		WO-52706	FM-34805-S2	WASHER, lock, S., 1/4 in. (-BECX1G)	3	—	—	—	—
		• WO-A-7191	FM-GPW-9612	SPRING, air cleaner toggle (after Willys serial 124309) (HH-613380)	2	% 3	—	6	—
		WO-A-1290	FM-GPW-9637-A	TUBE, air cleaner, w/BRACKET, assembly (before Willys serial 208437)	1	—	—	—	—
03-2		WO-A-6911	FM-GPW-9637-B	TUBE, air cleaner, w/BRACKET, assembly (after Willys serial 208437)	1	% 3	—	4	—

0302—FUEL PUMP

Figure Number Col. 1	Official Stockage Number Col. 2	Part Number Willys Col. 3	Part Number Ford Col. 4	ITEM Col. 5	Quantity Reqd. per Unit Assy. Col. 6	Per 100 Major Items — 12 Mos. Field Maintenance Col. 7	Per 100 Major Items — Major Overhaul (5th Ech) Col. 8	Per 100 Major Items — Total First-Year Procurement Col. 9	Estimated Reqmts. per 100 Rebuilds Col. 10
03-4		WO-115641	FM-GPW-9399	ARM, rocker, fuel pump, assembly (AC-1521960)	1	% 3	3	10	10
03-4		WO-115657	FM-GPW-9387	BAIL, fuel pump strainer, assembly (AC-1523231)	1	% 3	—	5	—
03-4		WO-A-1045	FM-GPW-9386	BODY, fuel pump priming, w/LEVER, assembly (AC-1537812)	1	% 3	—	4	—
03-4		WO-A-1494	FM-GPW-9355	BOWL, fuel pump (metal) (AC-1537065)	1	% 3	—	—	—
03-4		WO-115653	FM-11A-9361	CLAMP, fuel pump valve (AC-1521956)	1	% 5	5	12	15
03-4		WO-51546	FM-26466-S7	SCREW, fuel pump valve clamp (AC-132629)	2	—	—	—	—
03-4		WO-115650	FM-GPW-9354	COVER, fuel pump, top (AC-132696)	1	—	—	—	—
03-4		WO-113439	FM-31628-S7	SCREW, fill-hd, No. 10 (.190)—32NF-2 x 1/2 (AC-855493)	6	—	—	—	—
03-4		WO-113440	FM-34803-S7	WASHER, lock, S., No. 10 (.190 in.) (AC-855064)	6	—	—	—	—
03-4		WO-116695	FM-GPW-9398	DIAPHRAGM, fuel pump, w/PULL ROD, assembly (AC-1538205)	1	—	—	—	—
03-4		WO-115656	FM-GPW-9364	GASKET, fuel pump bowl (AC-1523096)	1	% 48	—	70	—

FIGURE 03-4—FUEL PUMP ASSEMBLY

Key	Item	Willys Part No.	Ford Part No.	Gov't Group No.
A	BAIL assembly	WO-115657	FM-GPW-9387	0302
B	SEAT	WO-113460	FM-GPW-9388	0302
C	BOWL	WO-A-1494	FM-GPW-9355	0302
D	SCREEN assembly	WO-115654	FM-GPW-9365	0302
E	GASKET	WO-115656	FM-GPW-9364	0302
F	COVER	WO-115650	FM-GPW-9354	0302
G	SCREW	WO-113439	FM-31628-S7	0302
H	WASHER	WO-113440	FM-34803-S7	0302
I	DIAPHRAGM w/PULL ROD	WO-116695	Willys only	0302
J	CLAMP	WO-115653	FM-11A-9361	0302
K	VALVE assembly	WO-115651	FM-11A-9352	0302
L	SCREW	WO-51546	FM-26466-S7	0302
M	GASKET	WO-115652	FM-GPW-9362	0302
N	SPRING	WO-116694	Willys only	0302
O	SPRING	WO-115643	FM-GPW-9380	0302
P	ARM assembly	WO-115641	FM-GPW-9399	0302
Q	LINK	WO-115880	FM-INC-9381	0302
R	PIN	WO-A-1046	FM-GPW-9378	0302
S	BODY w/LEVER	WO-A-1045	FM-GPW-9386	0302
T	SEAL	WO-115870	FM-GPW-19469	0302
U	WASHER	WO-115869	FM-GPW-9468	0302

RA PD 305069-A

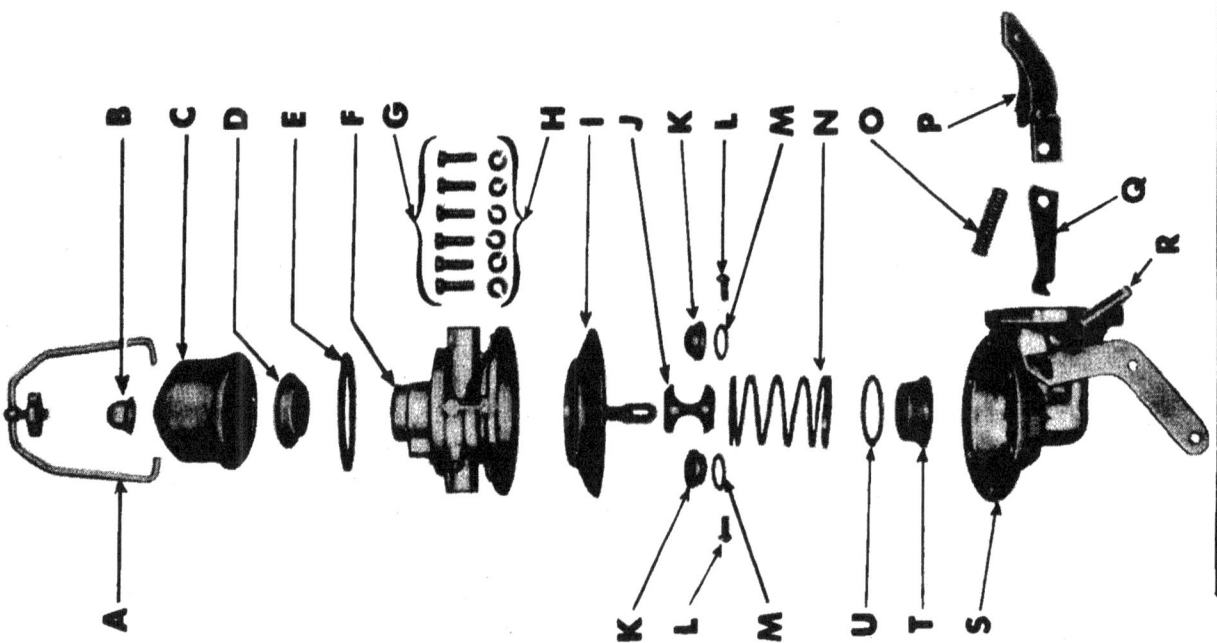

0302—FUEL PUMP (Cont'd)

Figure Number Col. 1	Part Number			ITEM Col. 5	Quantity Reqd. per Unit Assy. Col. 6	Per 100 Major Items				Estimated Reqmts. per 100 Rebuilds Col. 10
	Official Stockage Number Col. 2	Willys Col. 3	Ford Col. 4			12 Mos. Field Maintenance Col. 7	Major Overhaul (5th Ech) Col. 8	Total First Year Procurement Col. 9		
03-4		WO-115652	FM-GPW-9363	GASKET, fuel pump valve (AC-1521953)	2	—	—	—	—
		WO-A-7834	FM-GPW-18373-D	KIT, repair, fuel pump	as req.	26	32	65	100
				(Includes:					
				1 WO-116695 FM-GPW-9398 DIAPHRAGM, w/ROD					
				1 WO-115656 FM-GPW-9364 GASKET, bowl					
				1 WO-638737 FM-GPW-9417 GASKET, fuel pump					
				2 WO-115652 FM-GPW-9363 GASKET, valve					
				1 WO-115880 FM-INC-9381 LINK, rocker arm					
				1 WO-A-1046 FM-GPW-9378 PIN, rocker arm					
				1 WO-115870 FM-GPW-9469 SEAL					
				1 WO-115654 FM-GPW-9365 SCREW, filtering, assembly					
				3 WO-51546 FM-26466-S7 SCREW, clamp					
				3 WO-113439 FM-31628-S7 SCREW, cover					
				1 WO-116694 FM-GPW-9396 SPRING, diaphragm					
				1 WO-115643 FM-GPW-9380 SPRING, rocker arm					
				2 WO-115651 FM-11A-9352 VALVE assembly					
				3 WO-113440 FM-34803-S7 WASHER, lock					
				1 WO-A-1047 FM-GPW-9377 WASHER, pin					
		WO-A-6754		KIT, repair, fuel pump (issue until stock is exhausted, then use WO-A-7956)	as req.			—	—
		WO-A-7956		KIT, repair, fuel pump (was WO-A-6754)	as req.	4	8	—	—
				(Includes:					
				1 WO-115657 FM-GPW-9387 BAIL assembly					
				1 WO-115656 FM-GPW-9364 GASKET					
				1 WO-113461 FM-GPW-9373 NUT					
				1 WO-115654 FM-GPW-9365 SCREW, filtering					
				1 WO-113460 FM-GPW-9388 SEAT, bowl)					
03-4		WO-115880	FM-INC-9381	LINK, fuel pump rocker arm (AC-1521708)	1				
03-4		WO-113461	FM-GPW-9373	NUT, thumb, fuel pump strainer (AC-855763)	1				
03-4		WO-A-1046	FM-GPW-9378	PIN, fuel pump rocker arm (AC-1521578)	1				
Fig. 6		WO-A-8323	FM-GPW-9350	PUMP, fuel, assembly (AC-1538312)	1	10		13	
03-4		WO-115870	FM-GPW-9469	SEAL, fuel pump pull rod (rubber) (AC-1521880)	1	26	32	60	100
03-4		WO-113460	FM-GPW-9388	SEAT, fuel pump strainer bowl (AC-854005)	1	3	3	4	
03-4		WO-115654	FM-GPW-9365	SCREEN, filtering, fuel pump, assembly (AC-1523099)	1	5		10	
03-4		WO-116694	FM-GPW-9396	SPRING, fuel pump diaphragm (AC-1523068)	1				
03-4		WO-115643	FM-GPW-9380	SPRING, fuel pump rocker arm (AC-1522046)	1				
03-4		WO-115651	FM-11A-9352	VALVE, fuel pump, assembly (AC-1523106)	1				
03-4		WO-115869	FM-GPW-9468	WASHER, fuel pump diaphragm spring seat (AC-1521985)	1	26	32	60	100
03-4		WO-A-1047	FM-GPW-9377	WASHER, fuel pump rocker arm pin (AC-1521288)	1				

GROUP 03—FUEL (Cont'd)

Col. 1 Figure Number	Col. 2 Official Stockage Number	Col. 3 Willys	Col. 4 Ford	Col. 5 ITEM	Col. 6 Quantity Reqd. per Unit Assy.	Col. 7 12 Mos. Field Maintenance	Col. 8 Major Overhaul (5th Ech)	Col. 9 Total First Year Procurement	Col. 10 Estimated Reqmts. per 100 Rebuilds
				0302A—FUEL PUMP ATTACHING PARTS					
		WO-638737	FM-GPW-9417	GASKET, fuel pump to block (AC-38263)	1	—	—	—	—
		WO-6428	FM-20366-S2	SCREW, hex-hd, cap, S., 5/16-18NC-2 x 7/8 (fuel pump to cylinder block) (GM-106325) (BCAX1BB)	2	% 38	—	60	—
		WO-51833	FM-34806-S2	WASHER, lock, S., 5/8 in. (fuel pump to cylinder block bolt) (BECX1H)	2	—	—	—	—
				0303—ACCELERATOR AND LINKAGE					
03-5		WO-633013	FM-GPW-9752	BLOCK, adjusting, accelerator throttle rod	1	10	—	12	—
03-5		WO-639607	FM-GPW-9728	BRACKET, accelerator cross shaft	1	—	—	—	—
		WO-5914	FM-33796-S2	NUT, hex, S., 1/4-20NC-2 (GM-103024) (BBBX1AA) (issue until stock exhausted)	2	—	—	—	—
		WO-337304	FM-88350-S	STUD, S., 1/4-20NC-2 x 7/8 (cross shaft bracket)	2	% 8	16	—	—
		WO-52768	FM-356305-S	WASHER, S., plain, 21/64 (at each end of cross shaft)	2	—	—	—	—
		WO-52706	FM-34805-S2	WASHER, lock, S., 1/4 in. (BECX1G)	2	—	—	—	—
03-5		WO-A-1173	FM-GPW-9751	CLIP, throttle rod retracting spring (on manifold) (issue until stock exhausted) (GM-110633)	1	% 4	4	—	—
		WO-6352	FM-355836-S8	NUT, hex, S., No. 10 (.190)-24NC-2 (to manifold) (BBKX2BC)	1	—	—	—	—
		WO-5067	FM-72003-S	PIN, cotter, 1/16 x 1/2 (BFAX1BC)	1	—	—	—	—
		WO-51662	FM-356201-S	WASHER, S., 13/64	1	—	—	—	—
03-5		WO-639610	FM-GPW-9745	CLIP, throttle rod retracting spring (on throttle rod)	1	% 10	—	25	—
		WO-5067	FM-72003-S	PIN, cotter, 1/16 x 1/2 (BFAX1BC)	1	—	—	—	—
		WO-51662	FM-356201-S	WASHER, S., 13/64	1	—	—	—	—
		WO-651298	FM-GPW-9731-B	CONNECTION, accelerator pedal (before 116790 trucks)	1	—	—	—	—
		WO-650482	FM-GPW-9711	HINGE, accelerator pedal (female)	1	—	—	—	—
		WO-A-1174	FM-GPW-9727	LINK, connecting, accelerator pedal (before Willys serial 225209)	1	—	—	—	—
03-5		WO-A-6710	FM-GPW-9719	LINK, connecting, accelerator pedal (after Willys serial 225209)	1	% 5	—	6	—
		WO-5067	FM-72003-S	PIN, cotter, S., 1/16 x 1/2 (BFAX1BC)	1	—	—	—	—
		WO-52702	FM-34745-S2	WASHER, plain, S., S.A.E. std., 1/4 in. (BEBX1G)	1	% 3	—	4	—
		WO-A-1910	FM-GPW-9726	LEVER, accelerator cross shaft, w/BRACKET, assembly	1	% 8	12	—	—
		WO-650484	FM-74019-S	PIN, accelerator pedal hinge (issue until stock exhausted)	1	—	—	—	—
		WO-5067	FM-72003-S	PIN, cotter, 1/16 x 1/2 (BFAX1BC)	2	—	—	—	—
03-5		WO-A-1225	FM-GPW-9716	REST, foot, accelerator (issue until stock exhausted)	1	% 4	4	—	—
		WO-52954	FM-33909-S	NUT, jam, hex, s-fin., alloy-S., 5/16-24NF-2 (foot rest) (GM-114493) (BBDX1BA)	1	—	—	—	—
03-5		WO-A-1084	FM-GPW-9732-A	PEDAL, accelerator, assembly (before Willys 116,790 trucks)	1	—	—	5	—
		WO-A-6851	FM-GPW-9735-B	PEDAL, accelerator, w/HINGE and LINK, assembly	1	% 5	—	6	—

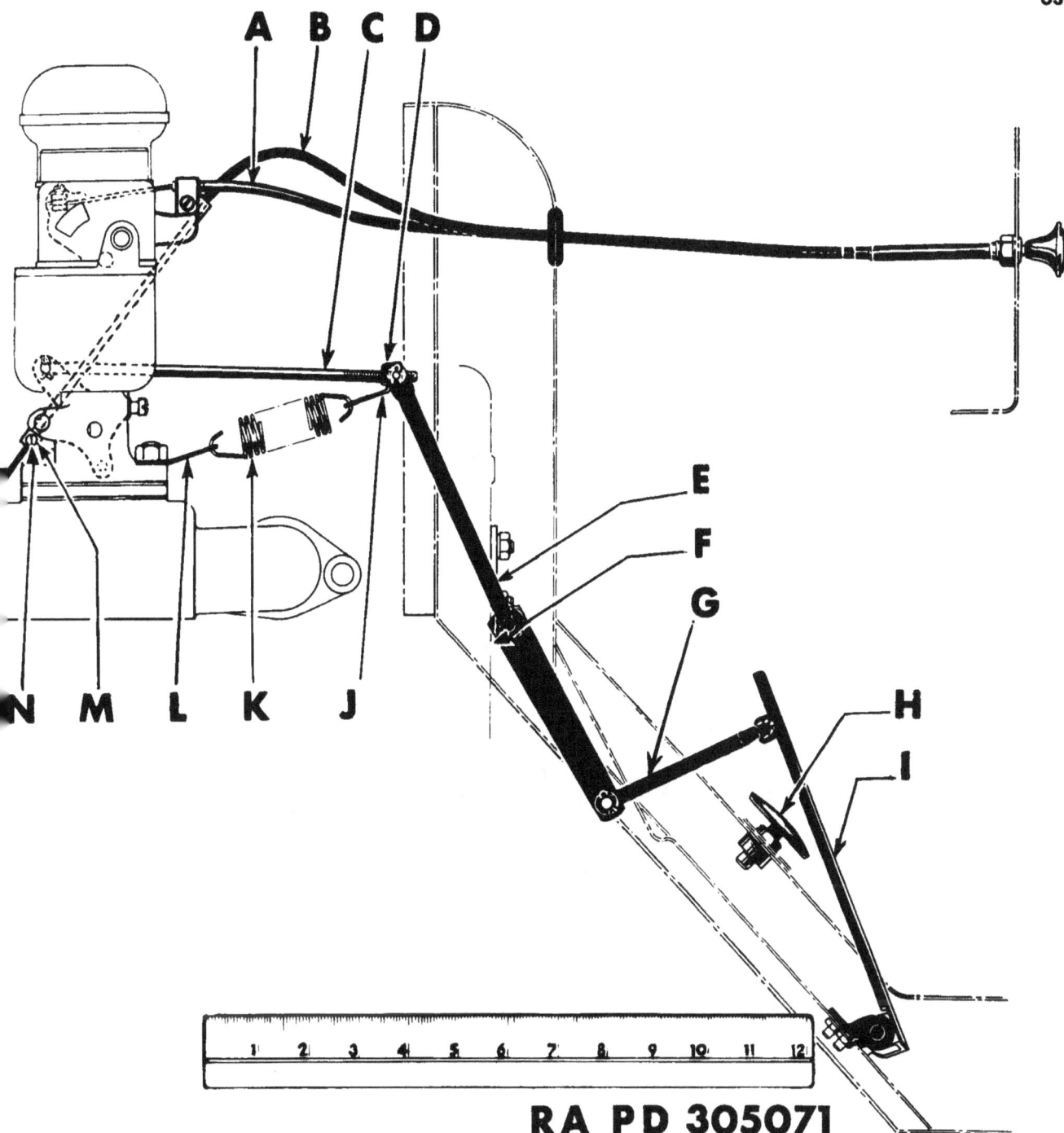

RA PD 305071

FIGURE 03-5—ACCELERATOR THROTTLE AND CHOKE CONTROLS

Key	Item	Willys Part No.	Ford Part No.	Gov't Group No.
A	CHOKE	WO-A-7517	FM-GPW-9775-B	0301B
B	THROTTLE assembly	(See WO-A-7517)	FM-GPW-9775-B	0301B
C	ROD	WO-A-1175	FM-GPW-9742	0303
D	BLOCK	WO-633013	FM-GPW-9752	0303
E	SHAFT w/LEVER	WO-A-1243	FM-GPW-9739	0303
F	BRACKET	WO-639607	FM-GPW-9728	0303
G	LINK	WO-A-6710	FM-GPW-9719	0303
H	REST assembly	WO-A-1225	FM-GPW-9716	0303
I	PEDAL w/LINK	WO-A-6851	FM-GPW-9735-B	0303
J	CLIP	WO-639610	FM-GPW-9745	0303
K	SPRING	WO-633011	FM-GPW-9799	0303
L	CLIP	WO-A-1173	FM-GPW-9751	0303
M	STOP	WO-372438	FM-GPW-11474	0303A
N	SCREW	WO-51040	FM-26457-S	0303A

RA PD 305071-A

SNL G-503

GROUP 03—FUEL (Cont'd)

Col. 1 Figure Number	Col. 2 Official Stockage Number	Part Number — Col. 3 Willys	Part Number — Col. 4 Ford	Col. 5 ITEM	Col. 6 Quantity Reqd. per Unit Assy.	Per 100 Major Items — Col. 7 12 Mos. Field Maintenance	Per 100 Major Items — Col. 8 Major Overhaul (5th Ech)	Per 100 Major Items — Col. 9 Total First Year Procurement	Col. 10 Estimated Reqmts. per 100 Rebuilds
				0303—ACCELERATOR AND LINKAGE (Cont'd)					
03-5		WO-633011	FM-GPW-9799	SPRING, retracting, accelerator	1	% 29	—	33	—
		WO-632174	FM-GPW-9737	SPRING, accelerator cross shaft (issue until stock exhausted)	1	% 4	6	—	—
		WO-650483	FM-GPW-9795	SPRING, accelerator pedal hinge	1	% 29	—	35	—
		WO-6352	FM-355836-S8	NUT, hex, S., No. 10 (.190)-24NC-2 (accelerator pedal assembly to toe board) (GM-128854) (BBKX2BC)	2	—	—	—	—
		WO-51537	FM-355162-S2	SCREW, fl-hd, S., cd-pltd., No. 10 (.190)-24NC-2 x ½ (GM-123854) (BCKX2AE-15)	2	% 10	—	100	—
		WO-52221	FM-34803-S7	WASHER, lock, S., No. 10 (.190 in.)	2	—	—	—	—
				0303A—HAND THROTTLE AND LINKAGE					
				NOTE: For parts not shown here, see Group 0301B					
03-5		WO-A-1302	FM-GPW-9775-B	CONTROL, throttle, assembly (before Willys serial 103468)	1	—	—	—	—
18-1		WO-A-7517		KNOB, throttle, assembly (was WO-A-5106)	1	—	—	—	—
		WO-A-1308	FM-GPW-9778	KNOB, carburetor choke or throttle, w/PLUNGER and WIRE, assembly (Use until stock is exhausted, then use WO-A-7518)	—	% 10	—	12	—
		WO-372438	FM-GPW-11474	STOP, throttle wire (issue until stock is exhausted, then issue WO-A-8834)	1	% 4	8	—	—
03-5		WO-A-8834	FM-GPW-11495	STOP, throttle wire, w/SCREW, assembly (was WO-372438)	1	% 19	—	24	—
03-5		WO-51040	FM-26457-S	SCREW, rd-hd., S., No. 8 (.164)-32NC-2 x ¼ (throttle wire stop screw) (issue until stock is exhausted, then issue WO-A-8834)	1	% 68	132	—	—
				0304—FUEL LINES					
03-6		WO-A-1265	FM-GPW-9154	BUSHING, reducing, ¼ pipe thd. to ⅛ tapered pipe thd.	2	% 10	—	12	—
		WO-A-1325	FM-GPW-9288	CONNECTION, flexible (in fuel line) (FO-HA-8031)	1	% 29	—	40	—
		WO-384549	FM-GPW-9268	CONNECTOR, inverted flare tube, 5/16 in.	2	% 10	—	10	—
		WO-387249	FM-GPW-9267-A	CONNECTOR, inverted flared tube, br., 5/16 in.	3	% 20	—	22	—
		WO-384549	FM-GPW-9268	ELBOW, inverted flared tube (br.)	3	% 16	—	20	—
		WO-A-1368	FM-GPW-9289	TUBE, flexible connection to fuel pump, assembly	1	% 3	—	6	—
		WO-A-1367	FM-GPW-9282	TUBE, fuel filter to flexible connection, assembly	1	% 3	—	4	—
23-3		WO-A-1289	FM-GPW-14585	CLIP, closed, 7/16, bolt hole 13/32 (tube to front engine plate, under generator brace bolt)	1	—	—	—	—
		WO-A-1366	FM-GPW-9237	CLIP, 7/16 in., open, S., 7/32 dia. hole	1	% 3	—	4	—
		WO-A-5450	FM-GPW-9295	TUBE, fuel tank to filter, assembly (73⅝ long x 5/16 O.D.)	5	% 8	12	—	—
				(1 used tube to floor pan front cross sill					
				2 used tube to floor pan rear cross sill					
				1 used tube to fender splasher					

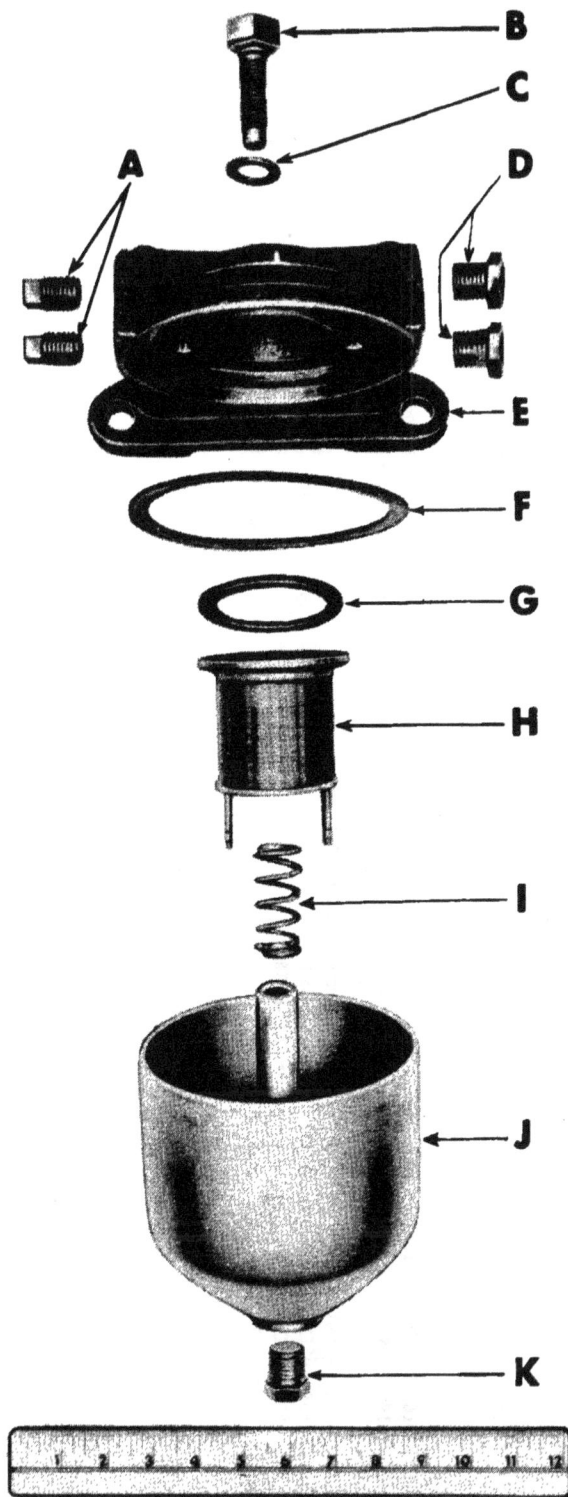

RA PD 305067

FIGURE 03-6—FUEL FILTER ASSEMBLY

Key	Item	Willys Part No.	Ford Part No.	Gov't Group No.
A	PLUG	WO-5138	FM-353055	0306B
B	SCREW	WO-A-1256	FM-GPW-9183	0306B
C	GASKET	WO-A-1257	FM-GPW-9184	0306B
D	BUSHING	WO-A-1265	FM-GPW-9154	0304
E	COVER	WO-A-1258	FM-GPW-9149	0306B
F	GASKET	WO-A-1259	FM-GPW-9160	0306B
G	GASKET	WO-A-1260	FM-GPW-9186	0306B
H	STRAINER-UNIT	WO-A-1261	FM-GPW-9140	0306B
I	SPRING	WO-A-1262	FM-GPW-9182	0306B
J	BOWL w/STUD assembly	WO-A-1263	FM-GPW-9162	0306B
K	PLUG	WO-A-1264	FM-GPW-9185	0306B

RA PD 305067-A

SNL G-503

GROUP 03—FUEL (Cont'd)

Col. 1 Figure Number	Col. 2 Official Stockage Number	Col. 3 Willys	Col. 4 Ford	Col. 5 ITEM	Col. 6 Quantity Reqd. per Unit Assy.	Col. 7 12 Mos. Field Maintenance	Col. 8 Major Overhaul (5th Ech)	Col. 9 Total First Year Procurement	Col. 10 Estimated Reqmts. per 100 Rebuilds
				0304—FUEL LINES (Cont'd)					
		WO-6352	FM-355836-S8	NUT, hex, S., No. 10 (.190)-24NC-2 (clip attaching) (GM-110633) (BBKX2BC)	2	—	—	—	—
		WO-52889	FM-32924	SCREW, binding-hd., S., type A, No. 10 (.190) x ⅝ (clip to cross sill) (GM-140363) (Willys only)	3	% 19	19	100	—
		WO-5113	FM-355130-S7	SCREW, rd-hd., mach., S., No. 10 (.190)-24NC-2 x ½ (clip to dash, clip to fender splasher) (GM-110500) (BCNX2AE)	2	—	—	—	—
		WO-52221	FM-34803-S	WASHER, lock, S., No. 10 (.190 in.) (BECX1E)	2	—	—	—	—
		WO-A-1369	FM-GPW-9369	TUBE, fuel pump to carburetor, assembly	1	% 6	—	8	—
				0306B—FUEL STRAINER					
03-6		WO-A-1263	FM-GPW-9162	BOWL, fuel strainer, w/STUD, center, assembly (AC-1504117)	1	—	—	—	—
03-6		WO-A-1258	FM-GPW-9149	COVER, mounting, fuel strainer (AC-1504212) (issue until stock is exhausted)	1	—	4	6	—
03-6		WO-A-1256	FM-GPW-9183	SCREW, attaching, fuel strainer cover (fill-hd., S.)	1	5	—	—	—
03-6		WO-A-1257	FM-GPW-9184	GASKET, fuel strainer cover cap screw (AC-853562)	1	—	—	—	—
03-6		WO-A-1259	FM-GPW-9160	GASKET, fuel strainer bowl (AC-853558)	1	—	—	—	—
03-6		WO-A-1260	FM-GPW-9186	GASKET, fuel strainer unit (AC-853572)	1	—	—	—	—
		WO-A-6883	FM-GPW-18337	GASKET SET, fuel strainer (Includes: 1 WO-A-1257 FM-GPW-9184 GASKET, cover cap screw; 1 WO-A-1259 FM-GPW-9160 GASKET, strainer bowl; 1 WO-A-1260 FM-GPW-9186 GASKET, strainer unit)	as req	% 77	—	120	—
03-6		WO-A-1264	FM-GPW-9185	PLUG, drain, fuel strainer (AC-127951)	1	% 5	—	—	—
03-6		WO-5138	FM-353055-S	PLUG, pipe, I., ¼ in. std.	2	—	—	6	—
03-6		WO-A-1262	FM-GPW-9182	SPRING, fuel strainer unit (AC-1504118)	1	% 5	—	6	—
		WO-A-7850	FM-GPW-9155-A2	STRAINER, fuel, assembly (issue until stock is exhausted, then use WO-A-7850)	—	% 2	—	4	—
		WO-A-1255	FM-GPW-9155	STRAINER, fuel, assembly (AC-1595235) (was WO-A-1255)	1	—	—	—	—
		WO-5919	FM-24389-S2	BOLT, hex-hd., S., ⅜-24NF-3 x ½ (fuel strainer to dash) (GM-100026)	2	—	—	—	—
		WO-5010	FM-34807-S2	WASHER, lock, S., ⅜ in. (²¹/₃₂ O.D. x ¹³/₃₂ I.D. x ³/₃₂ thk.) (BECX1K)	2	% 5	—	—	—
03-6		WO-A-1261	FM-GPW-9140	STRAINER, fuel, assembly (AC-1595823)	1	—	—	9	—
				GROUP 04—EXHAUST					
				0401—MUFFLER					
		WO-A-655	FM-GPW-5264-A	CLAMP, support, muffler (before Willys serial 143507) (rd-type)	2	—	—	—	—
04-1		WO-A-5753	FM-GPW-5264-B	CLAMP, support, muffler (after Willys serial 143507) (oval type) (issue until stock exhausted)	1	% 4	14	—	—

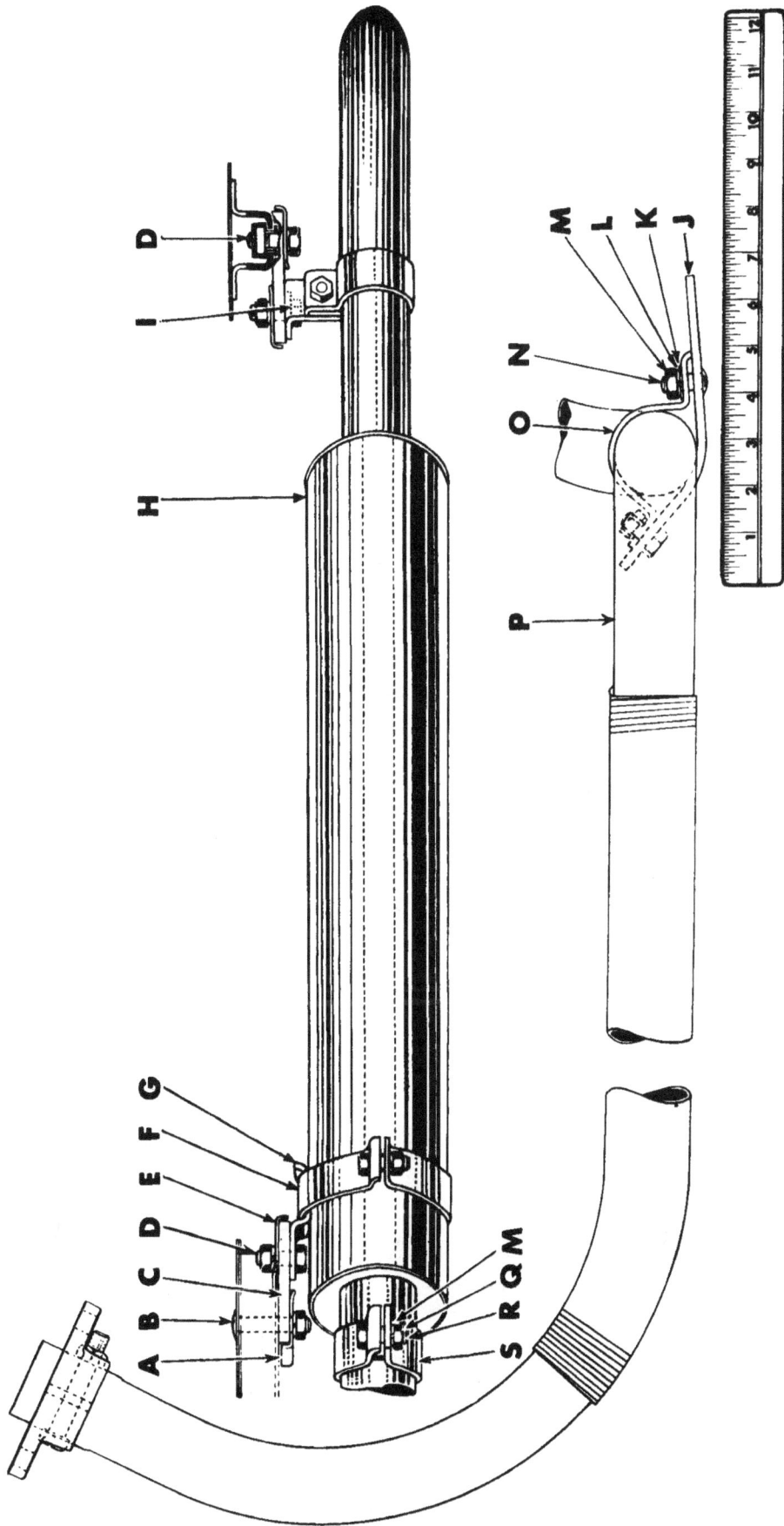

RA PD 305105

FIGURE 04-1—EXHAUST SYSTEM

Key	Item	Willys Part No.	Ford Part No.	Gov't Group No.
K	WASHER	WO-52274	FM-34746-S2	0401
L	WASHER	WO-51833	FM-34806-S2	0401
M	NUT	WO-6167	FM-33797-S2	0402
N	BOLT	WO-52983	FM-23393-S2	0401
O	CLAMP	WO-A-1300	FM-GPW-5251	0402
P	PIPE assembly	WO-A-1296	FM-GPW-5246	0402
Q	NUT	WO-53285	See FM-33798-S2	0401
R	BOLT	WO-5922	FM-24407-S	0402
S	CLAMP	WO-636004	FM-GPW-5270	0402

RA PD 305105-A

Key	Item	Willys Part No.	Ford Part No.	Gov't Group No.
A	PLATE	WO-638058	FM-GPW-5274	0401
B	BOLT	WO-52372	FM-23498-S2	0401
C	INSULATOR	WO-A-658	FM-GPW-5283	0401
D	BOLT	WO-50929	FM-24367-S2	0401
E	PLATE	WO-638058	FM-GPW-5274	0401
F	CLAMP	WO-A-5753	FM-GPW-5264-B	0401
G	STRAP	WO-A-657	FM-GPW-5262	0401
H	MUFFLER assembly	WO-A-6118	FM-GPW-5230-B	0401
I	CLAMP	WO-A-6119	FM-GPW-5298	0401
J	PLATE	WO-A-1253	FM-GPW-9251-B	0402

SNL G-503

GROUP 04—EXHAUST (Cont'd)

Figure Number	Official Stockage Number	Part Number Willys	Part Number Ford	ITEM	Quantity Reqd. per Unit Assy.	12 Mos. Field Maintenance	Major Overhaul (5th Ech)	Total First Year Procurement	Estimated Reqmts. per 100 Rebuilds
Col. 1	Col. 2	Col. 3	Col. 4	Col. 5	Col. 6	Col. 7	Col. 8	Col. 9	Col. 10
				0401—MUFFLER (Cont'd)					
04-1		WO-5934	FM-20387-S2	BOLT, hex-hd, s-fin, alloy-S, 5/16-24NF-2 x 1 (GM-100014) (BAOX1BC)	2	—	—	—	—
04-1		WO-53285	FM-33798-S2	NUT, hex, reg, S, 5/16-24NF-2 (Seez-Pruf) (Willys only)	2	—	—	—	—
		WO-51833	FM-34806-S2	NUT, hex, reg, S, 5/16-24NF-2 (BBBX1BA) (Ford only)	2	—	—	—	—
04-1		WO-A-6119	FM-GPW-5298	WASHER, lock, S, 5/16 in. (19/32 O.D. x 11/32 I.D. x 1/16 thk.) (BECX1H)	2	—	—	—	—
04-1				CLAMP, muffler tail pipe (after Willys serial 143507, oval type muffler) (issue until stock exhausted)	1	% 4	12	—	—
		WO-5922	FM-24407-S	BOLT, hex-hd, s-fin, alloy-S, 5/16-24NF-2 x 1 1/8 (GM-106281) (BAOX4BC)	1	—	—	—	—
04-1		WO-5437	FM-34706-S2	WASHER, plain, S, 5/16 in. (GM-106262)	2	% 6	—	9	—
04-1		WO-A-658	FM-GPW-5283	INSULATOR, muffler support	2	—	—	—	—
		WO-52372	FM-23498-S2	BOLT, carriage, sq-nk, S, 5/16-18NC-2 x 1 3/4 (insulator to support strap and body still) (BADX1CF)	1	% 80	—	100	—
04-1		WO-50929	FM-24367-S2	BOLT, hex-hd, S, 5/16-24NF-2 x 7/8 (GM-106280) (BAOX1BB)	3	—	—	—	—
04-1		WO-53285	FM-33798-S2	NUT, hex, reg, S, 5/16-24NF-2 (Seez-Pruf) (Willys only)	2	—	—	—	—
		WO-51833	FM-34806-S2	NUT, hex, reg, S, 5/16-24NF-2 (Ford only) (BBBX1BA)	2	—	—	—	—
04-1		WO-A-1146	FM-GPW-5230-A	WASHER, lock, S, 5/16 in. (23/32 O.D. x 11/32 I.D. x 1/16 thk.) (BECX1H)	4	—	—	—	—
04-1		WO-A-6118	FM-GPW-5230-B	MUFFLER assembly (rd-type) (before Willys serial 143507) (use WO-A-6118 and attaching parts for replacement)	1	% 80	12	—	—
04-1		WO-638058	FM-GPW-5274	MUFFLER assembly (oval type) (after Willys serial 143507)	1	—	16	94	—
04-1				PLATE, muffler support insulator (issue until stock exhausted)	4	% 4	—	—	—
04-1		WO-A-657	FM-GPW-5262	STRAP, muffler support round type muffler) (issue until stock exhausted) (oval type muffler)	2	% 12	—	—	—
				0402—EXHAUST PIPE					
04-1		WO-A-1300	FM-GPW-5251	CLAMP, exhaust pipe extension (issue until stock exhausted)	1	% 4	12	—	—
04-1		WO-52983	FM-23393-S2	BOLT, carriage, sq-nk, S, 5/16-18NC-2 x 7/8 (pipe extension to skid plate) (GM-119045)	1	% 40	—	50	—
04-1		WO-51523	FM-20346-S2	BOLT, hex-hd, s-fin, alloy-S, 5/16-18NC-2 x 3/4 (GM-100121) (BCAX1BA)	1	—	—	—	—
04-1		WO-6167	FM-33797-S2	NUT, hex, S, 5/16-18NC (GM-102634) (BBAX1B)	2	—	—	—	—
04-1		WO-52274	FM-34746-S2	WASHER, S, 5/16 in. (11/16 I.D. x 1 1/16 O.D. x 1/16 thk.) (BEBX1H)	2	—	—	—	—
04-1		WO-51833	FM-GPW-5270	WASHER, lock, S, 5/16 in.	1	—	—	—	—
04-1		WO-636004		CLAMP, exhaust pipe to muffler (issue until stock exhausted)	1	% 4	12	—	—
		WO-5922	FM-24407-S	BOLT, hex-hd, s-fin, alloy-S, 5/16-24NF-2 x 1 1/8 (GM-106281) (BAOX4BC)	1	—	—	—	—
		WO-53285	FM-33798-S2	NUT, hex, S, 5/16-24NF-2 (Seez-Pruf) (Willys only)	1	—	—	—	—
		WO-51833	FM-34806-S2	NUT, hex, S, 5/16-24NF-2 (BBBX1BA) (Ford only)	1	—	—	—	—
		WO-51833		WASHER, lock, S, 5/16 in. (19/32 O.D. x 11/32 I.D. x 1/16 thk.) (BECX1H)	1	—	—	—	—
		WO-630526		FLANGE, exhaust pipe (Willys only)	1	% 4	12	—	—

Ref.	Stock No.	Federal Stock No.	Description	Qty				
04-1	WO-6486		BOLT, hex-hd, s-fin, alloy-S, 3/8-16NC-2 x 2 1/4 (clamp bolt) (GM-106333) (BCAX1BB) (Willys only)	1	—	—	—	—
04-1	WO-50878	FM-24429-S7	BOLT, hex-hd, s-fin, alloy-S, 3/8-24NF-2 x 1 1/4 (flange to manifold) (GM-100027) (BCBX1CD)	1	—	—	—	—
	WO-53289		NUT, hex, S, 3/8-16NC-2 (Seez-Pruf) (Willys only)	1	—	—	—	—
	WO-53287	FM-33799	NUT, hex, reg, S, 3/8-16NC-2 (Ford only)	1	—	—	—	—
			NUT, hex, S, 3/8-24NF-2 (Seez-Pruf) (Willys only)	2	—	—	—	—
		FM-33800-S2	NUT, hex, S, 3/8-24NF-2 (Ford only)	1	—	—	—	—
	WO-5010	FM-34807-S7	WASHER, lock, reg, S, 3/8 in. (2 5/32 I.D. x 1 3/8 O.D. x 3/32 thk.) (BECX1K)	1	—	100	% 80	—
	WO-634814	FM-GPW-9450	GASKET, exhaust pipe flange	1	—	—	—	—
04-1	WO-A-1253	FM-GPW-5291-B	PLATE, skid, muffler guard, under frame	7	—	—	—	—
	WO-52945		BOLT, carriage, sq-nk, S, 3/8-16NC-2 x 7/8 (plate to cross member) (Willys only)	4	—	—	—	—
	WO-5544	FM-33799-S7	NUT, hex, S, 3/8-16NC-2 (GM-102635) (BBAX1C)	4	—	—	—	—
	WO-52101	FM-34747-S2	WASHER, plain, S, S.A.E. std, 3/8 in. (13/16 O.D. x 13/32 I.D. x 1/16 thk.) (BEBX1K)	4	—	80	% 40	—
	WO-A-1296	FM-GPW-5646	PIPE, exhaust, w/bond STRAP, assembly (issue until stock exhausted)	1	—	—	—	—
	WO-A-10198	FM-GPW-5246-B	PIPE, exhaust, w/o bond strap, assembly (for new radio suppression)	1	—	—	—	—
	WO-332515	FM-88032-S	STUD, S, 3/8-16NC-2 x 1 3/8 x 3/8-24NF-2 (exhaust pipe to manifold)	2	—	—	—	—

GROUP 05—COOLING

0501—RADIATOR

Ref.	Stock No.	Federal Stock No.	Description	Qty				
05-1	WO-A-1215	FM-GPW-8100-A	CAP, filler, radiator, assembly (AC-846709) (STN-6455A)	1	20	—	% 16	—
05-1	WO-A-1126	FM-9N-8115	COCK, drain, radiator (1/4 in.) (WH-145-A) (EAT-IE-1212)	1	6	—	% 6	—
05-1	WO-A-1214	FM-GPW-8005	CORE, radiator, w/TANK and SHROUD, assembly	1	12	—	% 10	—
05-1	WO-A-1546	FM-34084-S	NUT, radiator to support bracket (hex, S, cd-pltd, 7/16-20NF-2) (BBBX1DA-15) (issue until stock exhausted)	4	—	52	% 28	—
05-1	WO-A-1547	FM-34708-S	WASHER, plain, cd-pltd, S, 7/16 in. (1/2 I.D. x 1 1/4 O.D. x 5/64 thk.) (issue until stock exhausted)	2	—	24	% 16	—
	WO-53027	FM-356314-S7	WASHER, lock, S, internal, external tooth, 7/16 in.	2	—	—	—	—
05-1	WO-A-1216	FM-K-7129-B	GASKET, radiator filler cap (AC-846732)	1	60	—	% 38	—
	WO-A-1217	FM-GPW-8133	ROD, brace, radiator (issue until stock exhausted)	1	—	4	% 4	—
	WO-5910	FM-33798-S2	NUT, hex, S, 7/16-24NF-2 (GM-103025) (BBBX1BA)	3	—	—	—	—
	WO-51833	FM-34806-S2	WASHER, lock, S, 7/16 in. (1 1/32 O.D. x 1 1/32 I.D. x 1/16 thk.) (BECX1H)		—	—	—	—
05-1	WO-A-4413	FM-GPW-8125-A	SHIM, radiator to support bracket (issue until stock exhausted)	as req.	—	40	% 20	—

0502—THERMOSTAT

Ref.	Stock No.	Federal Stock No.	Description	Qty				
01-3	WO-639651	FM-GPW-8578	RETAINER, thermostat	1	6	—	% 3	—
01-3	WO-637646	FM-GPW-8575	THERMOSTAT assembly (HR-3108628)	1	25	8	% 10	25

0503—WATER PUMP

Ref.	Stock No.	Federal Stock No.	Description	Qty				
05-2	WO-636297	FM-GPW-8530	BEARING, water pump, w/SHAFT, assembly (double row, annular) (ND-885141) (MRC-D2-13567) (HO-88541)	1	—	—	—	—
	WO-640032	FM-GPW-8548	BELLOWS, water pump seal	1	—	—	—	—
05-2	WO-637052	FM-GPW-8505	BODY, water pump seal	1	—	—	—	—

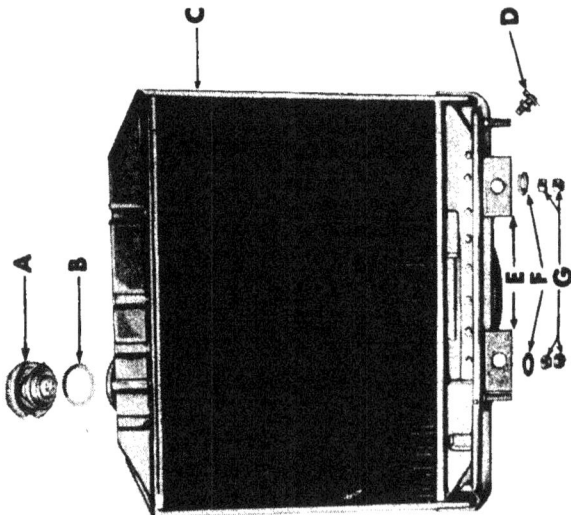

FIGURE 05-1—RADIATOR

Key	Item	Willys Part No.	Ford Part No.	Gov't Group No.
A	CAP	WO-A-1215	FM-GPW-8100-A	0501
B	GASKET	WO-A-1216	FM-GPW-8578	0501
C	CORE assembly	WO-A-1214	FM-GPW-8005	0501
D	COCK	WO-A-1126	FM-9N-8115	0501
E	SHIM	WO-A-4413	FM-GPW-8125-A	0501
F	WASHER	WO-A-1547	FM-34708-S	0501
G	NUT	WO-A-1546	FM-34084-S	0501

RA PD 305174-A

RA PD 305174

GROUP 05—COOLING (Cont'd)

Figure Number	Official Stockage Number	Part Number		ITEM	Quantity Reqd. per Unit Assy.	Per 100 Major Items			Estimated Reqmts. per 100 Rebuilds
		Willys	Ford			12 Mos. Field Maintenance	Major Overhaul (5th Ech)	Total First Year Procurement	
Col. 1	Col. 2	Col. 3	Col. 4	Col. 5	Col. 6	Col. 7	Col. 8	Col. 9	Col. 10
				0503—WATER PUMP (Cont'd)					
05-2		WO-637053	FM-GPW-8543	GASKET, water pump to cylinder block	1	% 38	—	45	—
05-2		WO-639993	FM-GPW-8512	IMPELLER, water pump	1	6	13	21	40
		WO-A-6839	FM-GPW-18515-B	KIT, repair, water pump	as req.	10	32	48	100
				(Includes:					
				1 WO-636297 FM-GPW-8530 BEARING, w/SHAFT					
				1 WO-637053 FM-GPW-8543 GASKET, pump to block					
				1 WO-640031 FM-GPW-8524-A2 SEAL assembly					
				1 WO-640034 FM-GPW-8557-A WASHER					
				1 WO-636298 FM-GPW-8576 SPRING, retaining)					
05-2		WO-639992	FM-GPW-8501	PUMP, water, assembly	1	% 13	16	—	—

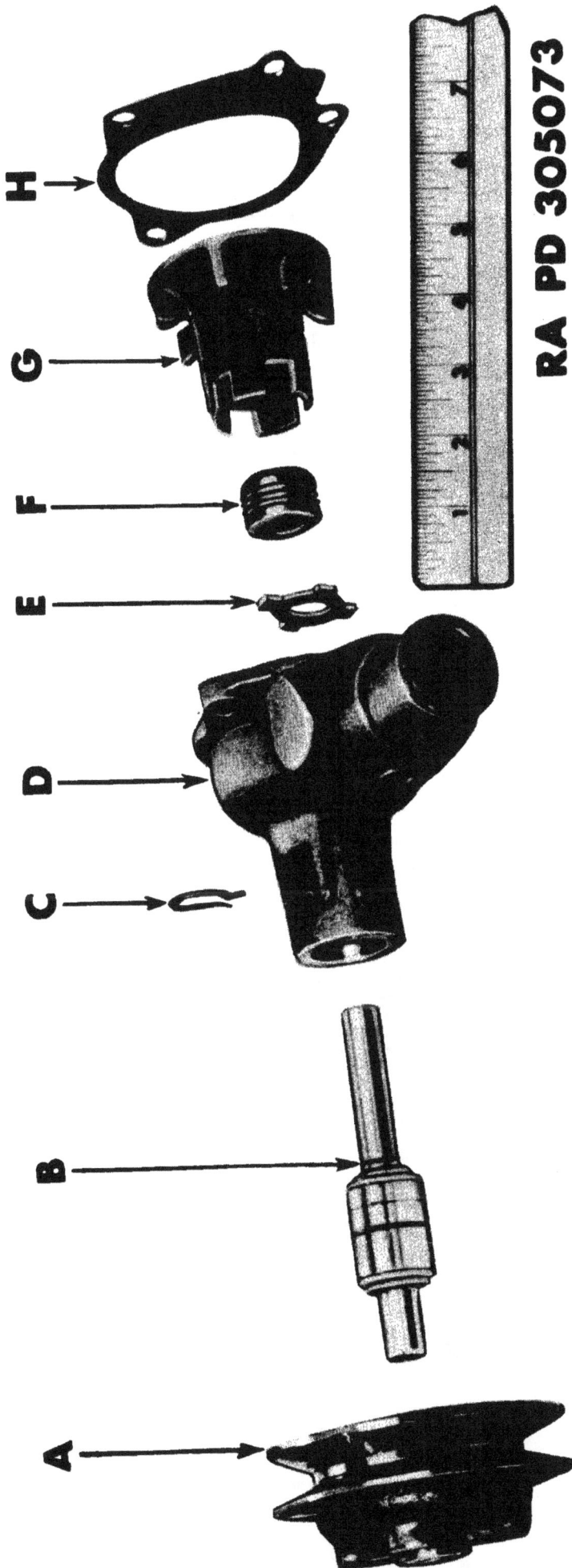

RA PD 305073

FIGURE 05-2—WATER PUMP ASSEMBLY

Key	Item	Willys Part No.	Ford Part No.	Gov't Group No.
A	PULLEY	WO-636299	FM-GPW-8509-A	0503A
B	SHAFT, w/BEARING	WO-636297	FM-GPW-8530	0503
C	SPRING	WO-636298	FM-GPW-8576	0503
D	BODY	WO-637052	FM-GPW-8505	0503
E	WASHER	WO-640034	FM-GPW-8557-A	0503
F	SEAL	WO-640031	FM-GAA-8524	0503
G	IMPELLER	WO-639993	FM-GPW-8512	0503
H	GASKET	WO-637053	FM-GPW-8543	0503

RA PD 305073-A

GROUP 05—COOLING (Cont'd)

Figure Number	Official Stockage Number	Willys	Ford	ITEM	Quantity Reqd. per Unit Assy.	12 Mos. Field Maintenance	Major Overhaul (5th Ech)	Total First Year Procurement	Estimated Reqmts. per 100 Rebuilds
Col. 1	Col. 2	Col. 3	Col. 4	Col. 5	Col. 6	Col. 7	Col. 8	Col. 9	Col. 10
				0503—WATER PUMP (Cont'd)					
		WO-6428	FM-20366-S	BOLT, hex-hd., s-fin., alloy-S., 5/16-18NC-2 x 7/8 (pump to block) (GM-106325) (BANX1BB)	3	—			—
		WO-51858	FM-355442-S	BOLT, hex-hd., s-fin., alloy-S., 5/16-18NC x 2½ (pump to block) (GM-100127) (BANX1BK)	1	—			
		WO-51833	FM-34806-S2	WASHER, lock, reg., S., 5/16 in. (1 19/32 O.D. x 11/32 I.D. x 1/16) (BECX1H)	4	—			
05-2		WO-640031	FM-GPW-8524-A2	SEAL, water pump, assembly (issue until stock exhausted)	1	50	100		
05-2		WO-636298	FM-GPW-8576	SPRING, retaining, water pump bearing	1	—			
		WO-640033	FM-GPW-8572-B	SPRING, water pump seal	1	—			
05-2		WO-640034	FM-GPW-8557-A	WASHER, water pump seal	1	—			
				0503A—FAN AND PULLEYS					
Fig. 6		WO-A-447	FM-GPW-8600	FAN assembly (HI-88632)	1	3	2	8	5
		WO-51514	FM-20324-S2	BOLT, hex-hd., s-fin., alloy-S., 1/4-20NC-2 x 5/8 (fan to pulley) (GM-106319) (BANX4AE)	4	—			
		WO-52706	FM-34805-S2	WASHER, lock, S., 1/4 in. (BECX1G)	4	—			
05-2		WO-636299	FM-GPW-8509-A	PULLEY, fan and water pump	1	3	3	10	10
		WO-A-1124	FM-GPW-8240	SHIELD, fan pulley	1	2	2	6	5
		WO-A-1125	FM-35676-S	WASHER, fan pulley shield (S., 5/8 O.D. x 11/32 I.D.)	6	10	10	40	30
		WO-A-6701		WASHER, fan pulley shield retaining bolt (S., 9/16 O.D. x 1/4 I.D. x .005 thk.)	6	10	10	25	30
				0503B—FAN BELT					
01-1		WO-A-1495	FM-GPW-8620-A1	BELT, drive, fan and generator (issue until stock exhausted then use WO-A-9490)	1	% 48	32	120	100
		WO-A-9490	FM-GPW-8620-A2	BELT, drive, fan and generator (Goodrich Neoprene) (was WO-A-1495)	1	—	—	—	—
				0505—ENGINE WATER FITTINGS AND HOSE					
		WO-52226	FM-60-8287	CLAMP, hose, 1 13/16 in. dia. (used on all radiator hose)	8	—	—	—	—
		WO-A-6373	FM-GPW-8285	HOSE, radiator, upper (issue until stock exhausted)	2	% 136	280	260	—
		WO-A-592	FM-GPW-8284	HOSE, radiator water outlet, lower	1	% 200	—	—	—
		WO-630512	FM-IGT-8260	HOSE, radiator water outlet, upper	1	% 64	132	—	—
		WO-A-6374	FM-GPW-8290	TUBE, connecting, water inlet	1	% 6	—	10	—
		WO-636109	FM-GPW-8269	TUBE, connecting, water outlet	1	% 3	—	5	—

GROUP 06—ELECTRICAL

0601—GENERATOR

Ref	WO No.	FM No.	Description	Qty				
06-1	WO-A-1629	FM-GPW-10105	ARM, generator brush (AL-CCE-54)	2	—	3	6	10
06-1	WO-A-1637	FM-GPW-10005	ARMATURE, generator, assembly (AL-GEG-2120-F)	1	3	3	9	10
06-1	WO-A-1649	FM-GPW-10142	BAND, generator, head (AL-GCE-24)	1	%3	3	8	10
06-1	WO-51248	FM-GPW-10094	BEARING, ball, generator (before Willys engine 158007) (AL-X-295) (ND-1203)	2	—	—	—	—
06-1	WO-A-6299	FM-B-10094	BEARING, ball, generator commutator end (after Willys engine 158007) (AL-X-1655) (ND-77503)	2	6	6	15	20
06-1	WO-A-1623	FM-355260-S7	BOLT, retaining, generator commutator end bearing (hex-hd., S., 1/4-20NF x 5/8) (AL-DA-60) (issue until stock exhausted)	1	8	16	—	—
06-1	WO-A-1630	FM-GPW-10069	BRUSH, main, generator (AL-GCE-1012) (issue until stock exhausted)	2	132	268	34	30
06-1	WO-A-1651	FM-GPW-18274	BRUSH SET, generator (AL-GCE-2012S) (Includes: 2 WO-A-1630 FM-GPW-10069 BRUSH)	as req.	19	10	—	—
06-1	WO-A-1599	FM-GPW-10206-A	BUSHING, insulating, generator field (.203 x .305 x 5/16) (AL-GCY-25)	1	—	—	—	—
06-1	WO-A-1598	FM-GPW-10104	BUSHING, insulating, generator field (.250 x .312 x 5/16) (AL-GCT-25) — Willys 1, Ford 3	1 / 3	—	—	—	—
06-1	WO-A-1604	FM-GPW-10175	COIL, field, generator left, right, assembly (AL-GEB-1005A) (issue until exhausted)	1	4	4	—	—
06-1	WO-A-1606	FM-GPW-10192	COIL, field, generator left, assembly (AL-GEB-1007A)	1	—	—	—	—
06-1	WO-A-1607	FM-GPW-10191	COIL, field, generator right, assembly (AL-GEB-1008B)	1	—	—	—	—
06-1	WO-A-1626	FM-GPW-10118	COVER, generator commutator end cap (AL-GBJ-32A) (issue until stock exhausted)	1	4	4	—	—
06-1	WO-A-1625	FM-GPW-10119	GASKET, generator commutator end cap cover (AL-GBJ-25) (issue until stock exhausted)	1	—	—	—	—
F	WO-A-5992	FM-GPW-10000-A	GENERATOR assembly (AL-GED-5002D) (For attaching parts see Group 0601A)	1	40	80	13	—
06-1	WO-A-1622	FM-GPW-10124	GUARD, oil, generator (flat, S., .986 x 1.570 x .020) (AL-DA-39)	1	%10	—	—	—
06-1	WO-A-6298	FM-GPW-10050	HEAD, generator, commutator end, assembly (after Willys engine 158007) (AL-GCE-2118A)	1	—	—	—	—
06-1	WO-A-6301	FM-GPW-10139	HEAD, generator, drive end (after Willys engine 158007) (AL-GCE-125A)	1	2	2	6	5
06-1	WO-A-6300	FM-GPW-10138	HEAD, generator, drive end, assembly (after engine 158007) (AL-GCE-1125A)	1	2	2	6	5
06-1	WO-A-1592	FM-01A-10193	INSULATOR, generator field connection (AL-GAL-44)	1	—	—	—	—
06-1	WO-A-1595	FM-GPW-10208-A	INSULATOR, generator field terminal post, inner (AL-GBW-67)	1	—	—	—	—
06-1	WO-A-1594	FM-GPW-10202-C	INSULATOR, generator armature terminal post, inner (AL-GBW-66)	1	—	—	—	—
06-1	WO-A-1591	FM-GPW-10202-A	INSULATOR, generator armature terminal post top (17/64 x 3/4 x .062) (AL-GAA-32)	1	5	5	20	15
06-1	WO-A-1641	FM-74144-S	KEY, Woodruff No. 5, S. (generator armature assembly) (AL-X-260)	1	2	2	5	5
06-1	WO-A-7840	FM-GPW-18342	KIT, repair, generator field coil (Includes: 1 WO-A-1599 FM-GPW-10206-A BUSHING; 1 WO-A-1598 FM-GPW-10104 BUSHING; 1 WO-A-1604 FM-GPW-10175 COIL assembly; 1 WO-A-1592 FM-01A-10193 INSULATOR; 1 WO-A-1595 FM-GPW-10208-B INSULATOR)	as req.	—	—	—	—

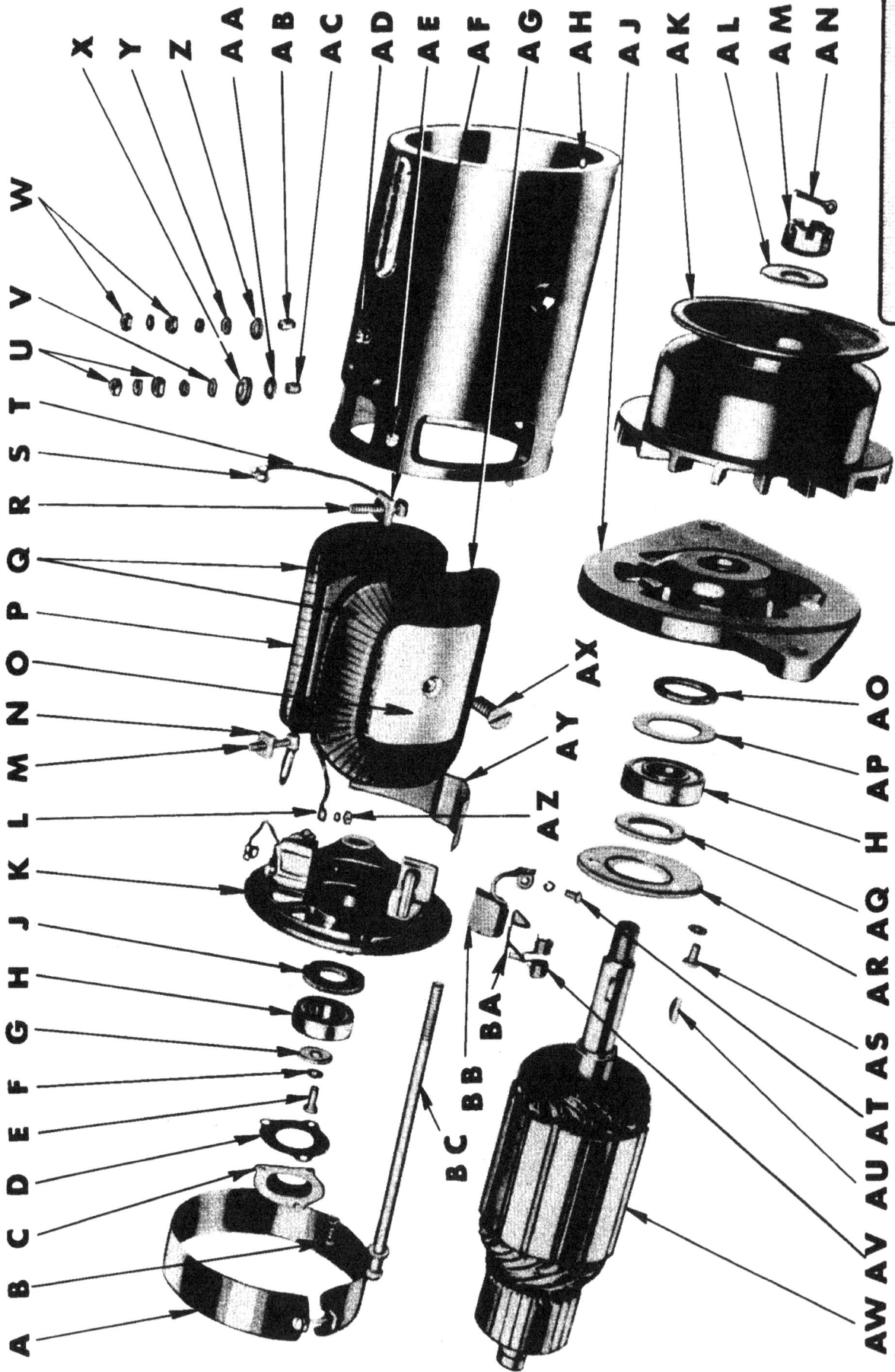

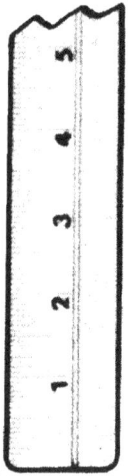

RA PD 305075

A B C D E F G H J K L M N O P Q R S T U V W X Y Z AA AB AC AD AE AF AG AH AJ AK AL AM AN

BC BB BA

AZ AY AX

AW AV AU AT AS AR AQ H AP AO

FIGURE 06-1—GENERATOR ASSEMBLY

Key	Item	Willys Part No.	Ford Part No.	Gov't Group No.	Key	Item	Willys Part No.	Ford Part No.	Gov't Group No.
A	BAND	WO-A-1649	FM-GPW-10142	0601	AC	BUSHING	WO-A-1598	FM-GPW-10104	0601
B	SCREW	WO-A-1636	FM-31588-S	0601	AD	SCREW	WO-A-6297	FM-37789-S7	0601
C	COVER	WO-A-1626	FM-GPW-10118	0601	AE	SCREW	WO-A-1618	FM-36009-S	0601
D	GASKET	WO-A-1625	FM-GPW-10119	0601	AF	INSULATION	WO-A-1594	FM-GPW-10106	0601
E	BOLT	WO-A-1623	FM-31588-S	0601	AG	COIL	WO-A-1607	FM-GPW-10191	0601
F	WASHER	WO-5045	FM-34805-S2	0601	AH	PIN	WO-A-1609	FM-GPW-1008-B	0601
G	WASHER	WO-A-1621	FM-GPW-10099	0601	AJ	HEAD assembly	WO-A-6300	(Willys only)	0601
H	BEARING	WO-A-6299	FM-B-10094	0601	AK	PULLEY	WO-A-1639	FM-GPW-10130	0601
J	WASHER	WO-A-1624	FM-GPW-10116	0601	AL	WASHER	WO-A-1638	FM-GPW-10134	0601
K	HEAD assembly	WO-A-6298	FM-GPW-10050	0601	AM	NUT	WO-A-1640	FM-34032-S7	0601
L	TERMINAL	WO-A-1608	FM-GPW-10218	0601	AN	PIN	WO-A-1642	FM-72043-S	0601
M	POST	WO-A-1605	FM-GPW-10210	0601	AO	WASHER	WO-A-1646	FM-GPW-10122	0601
N	INSULATION	WO-A-1595	FM-GPW-10208-A	0601	AP	GUARD	WO-A-1622	FM-GPW-10124	0601
O	POLE PIECE	WO-A-1600	FM-GPW-10041	0601	AQ	WASHER	WO-A-1644	FM-GPW-10212	0601
P	COIL	WO-A-1606	FM-GPW-10192	0601	AR	RETAINER	WO-106313	FM-GPW-10098	0601
Q	COIL assembly	WO-A-1604	FM-GPW-10175	0601	AS	SCREW	WO-A-1647	FM-36800-S	0601
R	POST	WO-A-1602	FM-GPW-10211-A	0601	AT	SCREW	WO-A-1632	FM-26457-S7	0601
S	TERMINAL	WO-A-1603	FM-GPW-10216	0601	AU	KEY	WO-A-1641	FM-74144-S	0601
T	LEAD assembly	WO-A-1601	FM-GPW-10100	0601	AV	SPRING	WO-A-1628	FM-GPW-10057	0601
U	NUT	WO-A-1611	FM-355883-S	0601	AW	ARMATURE	WO-A-1637	FM-GPW-10005	0601
V	WASHER	WO-A-1616	FM-34705-S2	0601	AX	SCREW	WO-A-1596	FM-355486-S7	0601
W	NUT	WO-A-1617	FM-350853-S7	0601	AY	INSULATION	WO-A-1592	FM-01A-10193	0601
X	INSULATOR	WO-A-1591	FM-GPW-10202-A	0601	AZ	NUT	WO-A-1610	FM-34051-S7	0601
Y	WASHER	WO-A-1615	FM-34703-S7	0601	BA	ARM	WO-A-1629	FM-GPW-10105	0601
Z	WASHER	WO-A-1593	FM-GPW-10206-A	0601	BB	BRUSH	WO-A-1630	(Willys only)	0601
AA	WASHER	WO-A-1597	(Willys only)	0601	BC	SCREW	WO-A-1590	FM-GPW-10120	0601
AB	BUSHING	WO-A-1599	(Willys only)	0601					

RA PD 305075-A

SNL G-503

GROUP 06—ELECTRICAL (Cont'd)

Figure Number	Official Stockage Number	Part Number Willys	Part Number Ford	ITEM	Quantity Reqd. per Unit Assy.	12 Mos. Field Maintenance	Major Overhaul (5th Ech)	Total First Year Procurement	Estimated Reqmts. per 100 Rebuilds
Col. 1	Col. 2	Col. 3	Col. 4	Col. 5	Col. 6	Col. 7	Col. 8	Col. 9	Col. 10
				0601—GENERATOR (Cont'd)					
			FM-GPW-10202-C	WO-A-1594 INSULATOR (bottom)	1				
			FM-GPW-10202-B	WO-A-1591 INSULATOR (top)	1				
			FM-GPW-10100	WO-A-1601 LEAD assembly	1				
			FM-34051-S7	WO-A-1610 NUT (grd. screw)	1				
			FM-350853-S7	WO-A-1617 NUT (terminal)	2				
			FM-355883-S	WO-A-1611 NUT	2				
			FM-36009-S7	WO-A-1618 SCREW (grd.)	1				
			FM-355486-S7	WO-A-1596 SCREW (pole)	2				
			FM-34703-S7	WO-A-1615 WASHER	1				
			FM-GPW-10208-A	WO-A-1597 WASHER (insulating)	1				
			FM-GPW-10206-A	WO-A-1593 WASHER (insulating)	1				
			FM-34803-S7	WO-A-1614 WASHER, lock	1				
			FM-34705-S2	WO-A-1616 WASHER, lock	1				
			FM-356263-S7	WO-A-1612 WASHER, lock	1				
			FM-34801-S7	WO-A-1613 WASHER, lock	1				
		WO-A-7895	FM-GPW-18363-B	KIT, repair, generator (issue until stock is exhausted, then use WO-A-9055)	as req.	16	24	—	—
		WO-A-9055	FM-GPW-18363-C	KIT, repair, generator	as req.	6	6	15	20
				(Includes:					
			FM-GPW-10069	WO-A-1630 BRUSH	2				
			FM-GPW-10119	WO-A-1625 GASKET, cover	1				
			FM-34141-S7	WO-A-1589 NUT	1				
			FM-31588-S7	WO-A-1636 SCREW, cover	3				
			FM-27161-S7	WO-A-1650 SCREW	1				
			FM-26457-S7	WO-A-1632 SCREW, lead	2				
			FM-36800-S	WO-A-1647 SCREW, retainer	3				
			FM-355260-S7	WO-A-1623 SCREW, retaining	1				
			FM-GPW-10057	WO-A-1628 SPRING, brush	2				
			FM-GPW-10116	WO-A-1624 WASHER, felt	1				
			FM-78-10212-A	WO-A-1644 WASHER, felt	1				
			FM-GPW-10212	WO-A-1646 WASHER, felt	1				
			FM-34803-S7	WO-A-1635 WASHER, lock	3				
			FM-34805-S7-8	WO-5045 WASHER, lock	1				
			FM-34806-S7	WO-A-5288 WASHER, lock	1				
			FM-34802-S2	WO-51532 WASHER, lock	2				
			FM-34803-S2	WO-52960 WASHER, lock	3				
			FM-356208-S7	WO-A-1621 WASHER, retaining)	1				
06-1		WO-A-1601	FM-GPW-10100	LEAD, generator armature terminal, assembly (AL-GEB-44) (issue until stock exhausted)	1	12	16	—	—

Group	Ord part No.	Mfr part No.	Description	Units				
06-1	WO-A-1610	FM-34051-S7	NUT, generator field ground screw (hex, S, cd-pltd, No. 6 (.138)-32NC-V (AL-8X-140) (BBKX2A-15) (issue until stock exhausted)	1	8	16	—	—
06-1	WO-A-1617	FM-350853-S7	NUT, generator field terminal post (hex, S, cd-pltd, No. 10 (.190)-32NF-1) (BBMX1C-15) (issue until stock exhausted)	2	16	32	—	—
06-1	WO-A-1611	FM-355883-S	NUT, generator armature terminal (hex, S, cd-pltd, No. 14 (.242)-24NC) (AL-8X-177)	2	—	—	—	—
06-1	WO-A-1589	FM-34141-S7	NUT, generator head band screw (sq-hd, S, cd-pltd, No. 10 (.190)-32NF) (AL-8X-794)	1	—	—	—	—
06-1	WO-302347	FM-B-10141	OILER, generator (press in type) (before Willys engine 158007) (AL-X-489)	2	—	—	—	—
06-1	WO-A-1600	FM-GPW-10041	PIECE, pole, generator (AL-GEB-29)	2	—	—	—	—
06-1	WO-A-1609	FM-GPW-10218	PIN, dowel, generator commutator end plate, assembly (S, 1/8 x 7/16) (AL-MN-21)	2	—	—	—	—
06-1	WO-A-6301	FM-GPW-10139	PLATE, generator, commutator end, assembly (after Willys engine 158007) (AL-GCE-125A)	1	—	—	—	—
06-1	WO-A-1645	FM-GPW-10138	PLATE, generator, drive end, assembly (before Willys engine 158007) (AL-GCE-1125)	1	—	—	—	—
06-1	WO-A-1605	FM-GPW-10211-B	POST, generator field terminal (No. 10 (.190)-32NF) (AL-GEB-58)	1	—	—	—	—
06-1	WO-A-1602	FM-GPW-10211-A	POST, generator armature terminal (No. 14 (.242)-24NC) (AL-GEB-27)	1	—	—	—	—
06-1	WO-A-1639	FM-GPW-10130	PULLEY, drive, generator, w/SPACER, assembly (Willys only)	1	2	2	6	5
06-1	WO-A-9492		PULLEY, drive, generator (AL-SP-484-A)	1	—	—	—	—
06-1	WO-A-1640	FM-34032-S7	NUT, generator pulley to shaft (hex, heavy, alloy-S, slotted, 1/2-20NF-2) (AL-X-156) (BBHX1AA) (issue until stock exhausted)	1	4	8	—	—
06-1	WO-A-1642	FM-72034-S	PIN, generator drive pulley nut (cotter, S, 3/32 x 1) (AL-X-404) (BFAX1CG) (issue until stock exhausted)	1	—	—	—	—
06-1	WO-106313	FM-GPW-10098	RETAINER, generator drive end bearing (AL-CG-6) (issue until stock exhausted)	1	4	4	—	—
06-1	WO-A-1618	FM-36009-S	SCREW, ground, generator (machine fl-hd, S, cd-pltd, No. 6 (.138)-32NC-2 x 3/8) (AL-8X-1420) (issue until stock exhausted)	1	8	16	—	20
06-1	WO-A-1636	FM-31588-S	SCREW, generator commutator end cover (fill-hd, S, cd-pltd, No. 10 (.190)-32NF-2 x 5/16) (AL-8X-870) (issue until stock exhausted)	3	40	80	—	—
06-1	WO-A-1632	FM-26457-S7	SCREW, generator brush lead (rd-hd, S, No. 8 (.164)-32NC-2 x 1/4) (AL-8X-305) (issue until stock exhausted)	2	16	32	—	—
06-1	WO-A-1647	FM-36800-S	SCREW, generator drive end bearing retainer (rd-hd, S, cd-pltd, No. 10 (.190)-32NF-2 x 3/8) (AL-X-311) (issue until stock exhausted)	3	20	40	—	—
06-1	WO-A-6297	FM-37789-S7	SCREW, generator ground (rd-hd, S, cd-pltd, No. 10 (.190)-32NF-2 x 7/16) (after Willys engine 158007) (AL-X-1368)	1	6	6	20	20
06-1	WO-A-1650	FM-27161-S7	SCREW, generator head band (rd-hd, S, cd-pltd, No. 10 (.190)-32NF-2 x 1 1/4) (AL-8X-715) (issue until stock exhausted)	2	8	16	—	—
06-1	WO-A-1590	FM-GPW-10120	SCREW, generator frame (AL-DK-23)	2	—	3	3	10
06-1	WO-A-1596	FM-355486-S7	SCREW, generator pole piece (AL-GBY-38A) (Willys only)	2	12	20	—	—
06-1	WO-A-8842		SPACER, generator shaft (AL-GEG-31) (Willys only)	1	—	—	—	—
06-1	WO-A-1628	FM-GPW-10057	SPRING, generator brush (AL-GCE-53) (issue until stock exhausted)	2	32	68	—	—
06-1	WO-A-1608		TERMINAL, generator field (br, tinned, .175 hole) (AL-X-959) (Willys only) (issue until stock exhausted)	1	8	16	—	—
06-1	WO-A-1603		TERMINAL, generator armature lead (br, tinned, .175 hole) (AL-X-847) (Willys only) (issue until stock exhausted)	1	8	16	—	—
06-1	WO-A-1615	FM-34703-S7	WASHER, S, plain, No. 10 (.190) (generator field terminal) (AL-8X-349) (issue until stock exhausted)	1	20	40	—	—
06-1	WO-A-1616	FM-34705-S2	WASHER, S, cd-pltd, 1/4 in. (generator armature terminal) (AL-8X-361)	1	—	—	—	—
06-1	WO-A-1638	FM-GPW-10134	WASHER, S, cd-pltd, 17/32 x 1 1/4 x .095 (generator armature shaft) (AL-GEW-31) (issue until stock exhausted)	1	5	5	20	15

GROUP 06—ELECTRICAL (Cont'd)

Col. 1	Col. 2	Col. 3	Col. 4	Col. 5	Col. 6	Col. 7	Col. 8	Col. 9	Col. 10
Figure Number	Official Stockage Number	Part Number — Willys	Part Number — Ford	ITEM	Quantity Reqd. per Unit Assy.	Per 100 Major Items — 12 Mos. Field Maintenance	Per 100 Major Items — Major Overhaul (5th Ech)	Total First Year Procurement	Estimated Reqmts. per 100 Rebuilds
				0601—GENERATOR (Cont'd)					
06-1		WO-A-1624	FM-GPW-10116	WASHER, generator commutator end, inner (felt, ¾ x 1⁹⁄₁₆ x ⅛) (AL-DH-7)	1	—	—	—	—
06-1		WO-A-1644	FM-78-10212-A	WASHER, generator drive end head, inner (felt, 1³⁄₁₆ x 1³⁄₁₆ x ⅛) (AL-GAU-31)	1	—	—	—	—
06-1		WO-A-1646	FM-GPW-10212	WASHER, generator drive end head, outer (felt, ⅞ x 1¼ x ⅛) (AL-GT-78)	1	—	—	—	—
06-1		WO-A-1597	FM-GPW-10208-A	WASHER, insulating, generator armature terminal, outer (¹⁷⁄₆₄ x ⁹⁄₁₆ x ¹⁄₁₆) (AL-GC-26)	1	—	—	—	—
06-1		WO-A-1593	FM-GPW-10206-A	WASHER, insulating, generator field terminal, outer (¹³⁄₆₄ x ⁹⁄₁₆ x ¹⁄₁₆) (AL-GBW-34)	1	—	—	—	—
		WO-A-1614	FM-34803-S7	WASHER, generator terminal and ground screw (lock, S., No. 10) (.190) (AL-12X-196)	1	—	—	—	—
		WO-A-1613	FM-34801-S7	WASHER, lock, generator field ground (S., tinned, No. 6 (.138)) (AL-12X-194) (issue until stock exhausted)	7	—	—	—	—
		WO-51532	FM-34802-S2	WASHER, lock, generator brush lead (S., No. 8 (.164)) (AL-X-195)	1	20	40	—	—
		WO-A-1635	FM-34803-S7	WASHER, lock, generator commutator end cover screw (S., tinned, No. 10 (.190)) (AL-12X-544)	2	—	—	—	—
		WO-5168	FM-34803-S7	WASHER, lock, generator bearing retaining bolt (S., No. 10 (.190)) (AL-X-196)	3	—	—	—	—
		WO-A-1612	FM-355263-S7	WASHER, lock, generator armature terminal (S., tinned, No. 14 (.242)) (AL-12X-193)	3	—	19	50	60
06-1		WO-5045	FM-34805-S8	WASHER, lock, S., ¼ in. (generator bearing retaining screw) (AL-X-199)	1	—	—	—	—
		WO-A-1619	FM-34806-S2	WASHER, lock, ⁵⁄₁₆ in. (before Willys engine 158007) (AL-12X-203)	1	—	—	—	—
		WO-A-5288	FM-34806-S7	WASHER, lock, generator frame screw (S., tinned, ⁵⁄₁₆) (after Willys engine 158007) (AL-12X-1014)	2	—	—	—	—
06-1		WO-A-1621	FM-355208-S7	WASHER, retaining, generator commutator end bearing (S., plain, .254 x ¹³⁄₁₆ x .065) (AL-DA-22) (issue until stock exhausted)	1	4	8	—	—
				0601A—GENERATOR ATTACHING PARTS					
		WO-A-1399	FM-GPW-10143	BRACE, generator, short (issue until stock exhausted)	1	—	—	—	—
		WO-6157	FM-24426-S2	BOLT, hex-hd, s-fin., alloy-S., ⁵⁄₁₆-18NC-2 x 1¼ (brace to generator) (BANX1BD)	1	—	4	4	—
		WO-6167	FM-33797	NUT, reg, hex., S., ⁵⁄₁₆-18NC-2 (GM-102634) (BANX1BD)	1	—	—	—	—
		WO-53025	FM-356309-S7	WASHER, lock, S., cd-pltd, internal, external teeth, ⁵⁄₁₆ in. (GM-178378)	1	—	—	—	—
		WO-A-1491	FM-GPW-10153-A	BRACE, generator, w/HANDLE, assembly	1	% 3	—	4	—
		WO-A-1470	FM-GPW-10177	BUSHING, generator brace locking	1	—	—	—	—
		WO-A-1397	FM-355455-S	BOLT, generator support insulator (hex-hd., S., ⁵⁄₁₆-24NF-3×⁹⁄₃₂) (BBBX1BA)	2	% 10	5	16	15
		WO-5910	FM-33798-S2	NUT, hex, S., ⁵⁄₁₆-24NF-2 (GM-103025) (BBBX1BA)	2	—	—	—	—
		WO-53025	FM-356309-S7	WASHER, lock, S., cd-pltd, internal, external tooth, ⁵⁄₁₆ in. (Bond No. 7) (GM-178378)	2	—	—	—	—

Group	Part No.	Part No.	Description	Qty			
	WO-A-1468	FM-GPW-10176	BOLT, pivot, generator brace (hex-hd., S., 3/8-24NF-2 x 7/8) (B19450CB) (issue until stock exhausted)	1	—	—	% 4
	WO-5901	FM-33800-S2	NUT, hex, S., 3/8-24NF-2 (GM-103026) (BBBX1CA)	1	—	8	—
	WO-5010	FM-34807-S2	WASHER, lock, S., 3/8 in. (29/32 O.D. x 13/32 I.D. x 3/32 thk.) (BECX1K)	1	—	4	—
	WO-A-1400	FM-GPW-10162	GUIDE, adjusting, generator brace (issue until stock exhausted)	1	—	—	—
	WO-6606	FM-24409-S7	BOLT, hex-hd., S., 3/8-24NF-2 x 1 1/8 (guide to brace) (GM-106286) (BAOX4AA)	1	—	—	—
	WO-5901	FM-33800-S2	NUT, hex, S., 3/8-24NF-2 (GM-103026) (BBBX1CA)	1	—	—	—
	WO-5455	FM-34707-S2	WASHER, plain, S., 3/8 in.	1	—	—	—
	WO-5010	FM-34807-S2	WASHER, lock, S., 3/8 in.	1	—	—	—
	WO-A-1395	FM-GPW-10178-A	INSULATOR, generator support (rubber, 1 1/16 O.D. x 1/2 I.D.)	2	8	16	% 6
	WO-A-1469	FM-GPW-10155	SPRING, generator brace (issue until stock exhausted)	1	—	—	8
	WO-A-1392	FM-GPW-10166	SUPPORT, generator, assembly (issue until stock exhausted)	1	24	4	% 19
	WO-633949	FM-633949	BOLT, hex-hd., S., 3/8-16NC-2 x 7/8 (support to block)	2	—	—	—
	WO-5010	FM-34807-S2	WASHER, lock, S., 3/8 in. (29/32 O.D. x 13/32 I.D. x 3/32 thk.) (BECX1K)	2	8	—	—
	WO-A-1401	FM-356371-S7	WASHER, generator support (21/64 hole)	2	—	—	—
	WO-A-1396	FM-356436-S	WASHER, generator support (33/64)	2	—	—	% 6

0601B—GENERATOR REGULATOR

Group	Part No.	Description	Qty
06-2	WO-A-7800	ARMATURE, cutout relay coil, assembly (AL-VRY-1043) (Willys only)	1
06-2	WO-A-7808	ARMATURE, current regulator coil, assembly (AL-VRY-1061B) (Willys only)	1
06-2	WO-A-7809	ARMATURE, voltage regulator coil, assembly (AL-VRY-1080B) (Willys only)	1
06-2	WO-A-7807	BRACKET, current regulator coil, adjusting, assembly (AL-VRA-1060) (Willys only)	2
06-2	WO-A-7802	SCREW, mounting bridge (fil-hd., S., cd-pltd., No. 8 (.164)-32NC x 7/16) (AL-8X-878) (Willys only)	4
06-2	WO-A-7803	SCREW, mounting bridge and bracket (fil-hd., S., cd-pltd., No. 8 (.164)-32NC x 5/16 (AL-8X-888) (Willys only)	8
06-2	WO-A-1666	WASHER, plain, S., cd-pltd., No. 8 (.164 in.) (AL-8X-350) (Willys only)	8
06-2	WO-51532	WASHER, lock, S., No. 8 (.164 in.) (AL-X-195) (Willys only)	12
06-2	WO-52781	BRACKET, mounting, resistor (AL-VRA-67) (Willys only)	1
06-2		SCREW, rd-hd., S., cd-pltd., No. 8 (.164)-32NC x 3/8 (AL-8X-55) (Willys only)	6
06-2	WO-A-1666	WASHER, plain, S., cd-pltd., No. 8 (.164 in.) (AL-8X-350) (Willys only)	6
06-2	WO-51532	WASHER, lock, S., No. 8 (.164 in.) (AL-X-195) (Willys only)	6
06-2	WO-A-7801	BRACKET, stationary, cutout relay, w/CONTACT (AL-VRH-1073) (Willys only)	1
06-2	WO-A-7802	SCREW, mounting, bridge (fil-hd., S., cd-pltd., No. 8 (.164)-32NC x 7/16 (AL-8X-878) (Willys only)	2
06-2	WO-A-7803	SCREW, mounting, bridge (fil-hd., S., cd-pltd., No. 8 (.164)-32NC x 5/16 (AL-8X-888) (Willys only)	4
06-2	WO-A-1666	WASHER, plain, S., cd-pltd., No. 8 (.164 in.) (AL-8X-350) (Willys only)	4
06-2	WO-51532	WASHER, lock, S., No. 8 (.164 in.) (AL-X-195) (Willys only)	6
06-2	WO-A-9044	COIL, cutout relay, assembly (AL-VRY-3035) (Willys only)	1
06-2	WO-A-9051	COIL, current regulator, assembly (AL-VRY-3070D) (Willys only)	1
06-2	WO-A-9049	COIL, voltage regulator, assembly (AL-VRY-3071-B) (Willys only)	1
06-2	WO-A-9053	NUT, mounting, coil, hex, S., cd-pltd., 1/4-20NC (AL-8X-1055) (Willys only)	3
	WO-A-8914	WASHER, lock, S., 1/4 in. (AL-X-535) (Willys only)	3

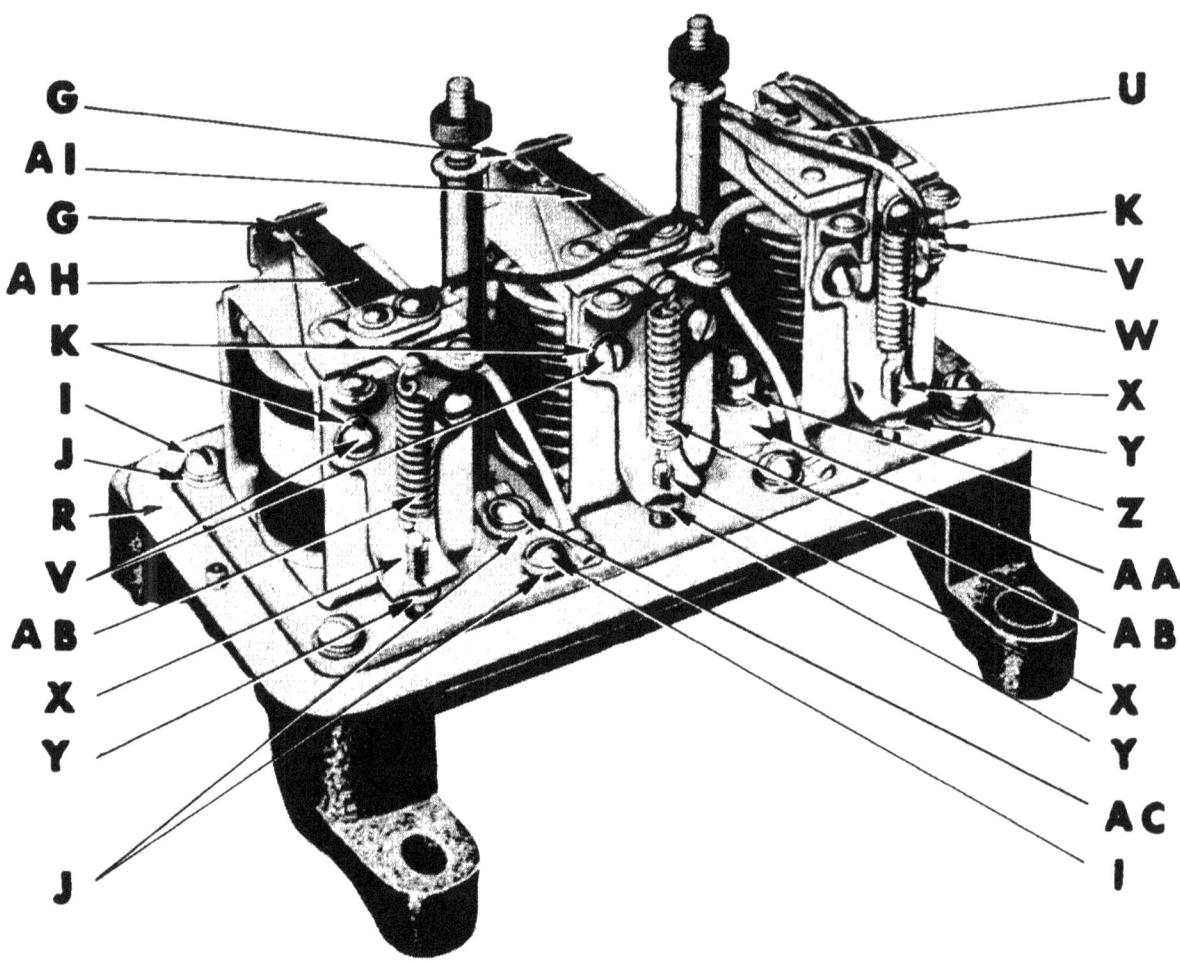

G

AI

G

AH

K

I

J

R

V

AB

X

Y

J

U

K

V

W

X

XY

Z

AA

AB

X

Y

AC

I

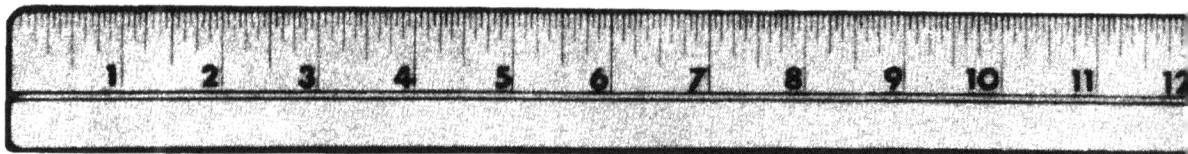

1 2 3 4 5 6 7 8 9 10 11 12

K

AE

AG

AF

K

AE

A

K

A

A

A

A

K

RA PD 30508

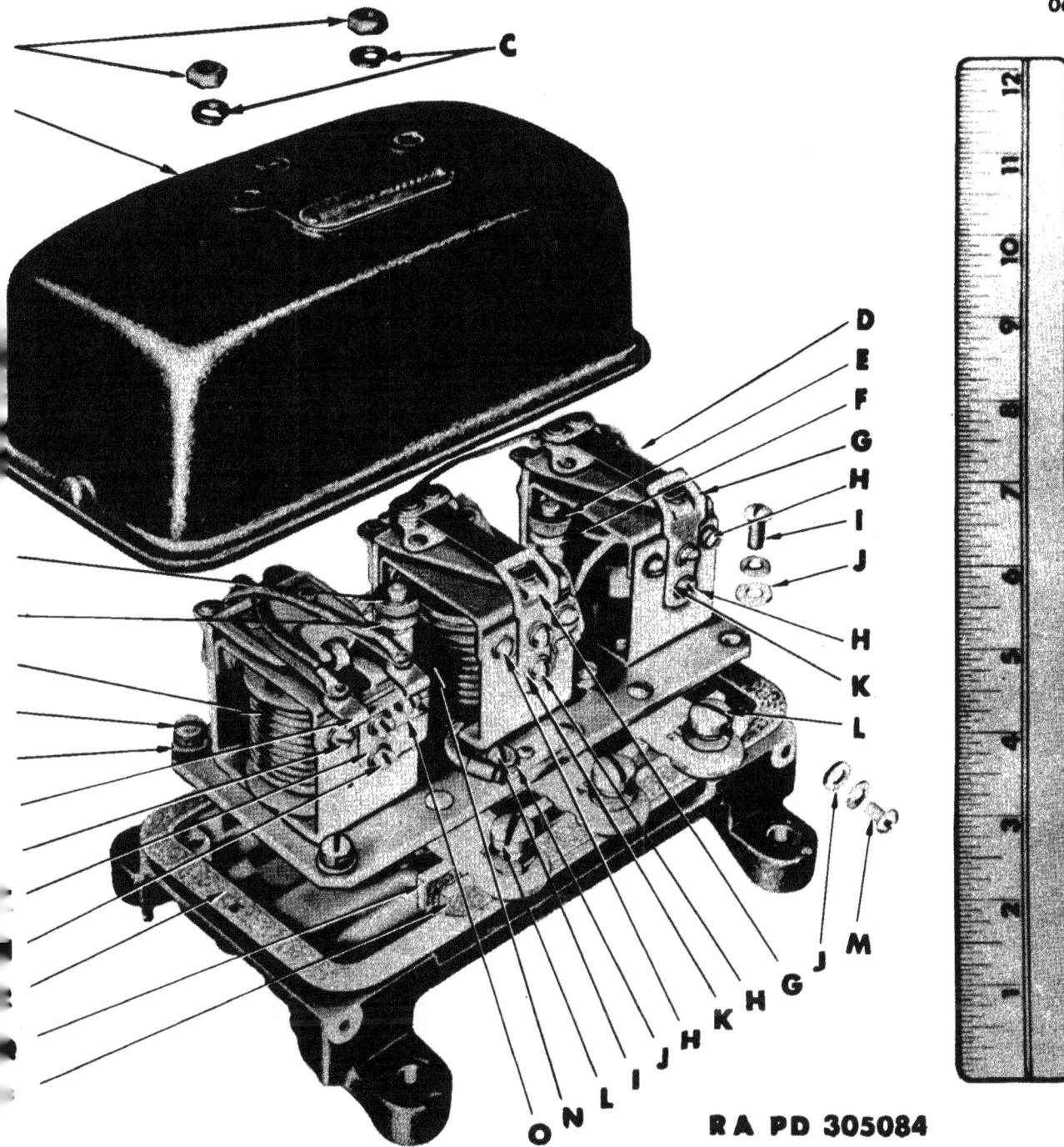

FIGURE 06-2—VOLTAGE REGULATOR

RA PD 305084

Key	Item	Willys Part No.	Ford Part No.	Gov't Group No.	Key	Item	Willys Part No.	Ford Part No.	Gov't Group No.
A	COVER	WO-A-9042	(Willys only)	0601B	S	BRACKET	WO-A-7801	(Willys only)	0601B
B	NUT	WO-A-9040	(Willys only)	0601B	T	COIL assembly	WO-A-9044	(Willys only)	0601B
C	WASHER	WO-A-9041	(Willys only)	0601B	U	ARMATURE	WO-A-7800	(Willys only)	0601B
D	COIL assembly	WO-A-9049	(Willys only)	0601B	V	SCREW	WO-A-7802	(Willys only)	0601B
E	WASHER	WO-A-9043	(Willys only)	0601B	W	SPRING	WO-A-7798	(Willys only)	0601B
F	WASHER	WO-A-1616	(Willys only)	0601B	X	SCREW	WO-A-7797	(Willys only)	0601B
G	BRACKET	WO-A-7807	(Willys only)	0601B	Y	NUT	WO-A-7796	(Willys only)	0601B
H	SCREW	WO-A-7803	(Willys only)	0601B	Z	NUT	WO-A-5260	(Willys only)	0601B
I	SCREW	WO-A-8993	(Willys only)	0601B	AA	CONNECTOR	WO-A-9052	(Willys only)	0601B
J	WASHER	WO-A-5262	(Willys only)	0601B	AB	SPRING	WO-A-7806	(Willys only)	0601B
K	WASHER	WO-A-1666	(Willys only)	0601B	AC	SCREW	WO-A-1620	(Willys only)	0601B
L	SCREW	WO-A-9050	(Willys only)	0601B	AD	RESISTANCE	WO-A-9054	(Willys only)	0601B
M	SCREW	WO-A-1620	(Willys only)	0601B	AE	SCREW	WO-52781	(Willys only)	0601B
N	COIL assembly	WO-A-9051	(Willys only)	0601B	AF	NUT	WO-A-9053	(Willys only)	0601B
O	INSULATION	WO-A-7799	(Willys only)	0601B	AG	RESISTANCE	WO-A-5256	(Willys only)	0601B
P	INSULATION	WO-A-9048	(Willys only)	0601B	AH	ARMATURE	WO-A-7809	(Willys only)	0601B
Q	GASKET	WO-A-9047	(Willys only)	0601B	AI	ARMATURE	WO-A-7808	(Willys only)	0601B
R	GASKET	WO-A-9046	(Willys only)	0601B					

RA PD 305084-A

SNL G-503

GROUP 06—ELECTRICAL (Cont'd)

| Figure Number | Official Stockage Number | Part Number | | ITEM | Quantity Reqd. per Unit Assy. | Per 100 Major Items | | | Estimated Reqmts. per 100 Rebuilds |
| | | Willys | Ford | | | 12 Mos. Field Maintenance | Major Overhaul (5th Ech) | Total First Year Procurement | |
Col. 1	Col. 2	Col. 3	Col. 4	Col. 5	Col. 6	Col. 7	Col. 8	Col. 9	Col. 10
				0601B—GENERATOR REGULATOR (Cont'd)					
06-2		WO-A-9052		CONNECTOR, series coil (AL-VRA-46) (Willys only)	1	—	—	—	—
06-2		WO-A-9042		COVER, voltage regulator, assembly (AL-VRA-1002) (Willys only)	1	—	—	—	—
06-2		WO-A-9040		NUT, mounting, cover, hex, S, cd-pltd, ¼-20NC (AL-8X-163) (Willys only)	2	—	—	—	—
06-2		WO-A-1616		WASHER, plain, S, cd-pltd, ¼ in. (AL-8X-361) (Willys only)	4	—	—	—	—
06-2		WO-A-9041		WASHER, lock, S, tinned, ¼ in. (AL-12X-199) (Willys only)	2	—	—	—	—
06-2		WO-A-9043		WASHER, rubber (VRA-109) (Willys only)	2	—	—	—	—
06-2		WO-A-9046		GASKET, cover (AL-VRA-50) (Willys only)	1	10	10	24	30
06-2		WO-A-9047		GASKET, terminal (AL-VRA-51) (Willys only)	1	5	5	12	15
06-2		WO-A-7799		INSULATION, cutout relay stationary contact bracket (AL-VRA-76) (Willys only)	1	—	—	—	—
06-2		WO-A-9048		INSULATION, terminal (AL-VRA-52) (Willys only)	1	—	—	—	—
		WO-A-7805		KIT, repair, current, regulator	as req.	3	3	9	10
				(Composed of: 1 WO-A-7808 ARMATURE 1 WO-A-7796 NUT, adjusting 1 WO-A-7797 SCREW, adjusting 2 WO-A-7802 SCREW 1 WO-A-7806 SPRING 3 WO-A-1666 WASHER 3 WO-51532 WASHER, lock)					
		WO-A-7794		KIT, repair, generator cutout relay	as req.	3	6	12	20
				(Composed of: 1 WO-A-7800 ARMATURE 1 WO-A-7801 BRACKET and CONTACT 1 WO-A-7799 INSULATION 1 WO-A-7796 NUT, adjusting 1 WO-A-7797 SCREW, adjusting 2 WO-A-7802 SCREW 2 WO-A-7803 SCREW 1 WO-A-7798 SPRING, armature 3 WO-A-1666 WASHER 2 WO-A-7795 WASHER 2 WO-A-7804 WASHER 3 WO-51532 WASHER)					
		WO-A-7810		KIT, repair, voltage regulator	as req.	3	6	12	20
				(Composed of: 1 WO-A-7809 ARMATURE					

Item	Stock No.	Other No.	Description	Qty.				
			1 WO-A-7796 NUT, adjusting	1	—	—	—	—
			1 WO-A-7797 SCREW, adjusting	1	—	—	—	—
			2 WO-A-7802 SCREW	2	10	4	3	—
			1 WO-A-7806 SPRING	1	20	8	6	—
			3 WO-51666 WASHER	3	—	—	—	—
			3 WO-51532 WASHER)	3	—	—	—	—
	WO-A-8997		LEAD, voltage regulator armature to base (AL-VRH-26) (Willys only)	1	—	—	—	—
	WO-A-5260		NUT, series connection (hex, S., cd-pltd, No. 10 (.216)-32NC) (AL-8X-173) (Willys only)	1	—	—	—	—
06-2	WO-A-7796		NUT, adjusting screw (knurled, br., No. 10 (.216)-48NS) (AL-VRA-15) (Willys only)	1	—	—	—	—
06-2	WO-A-5256		RESISTANCE, carbon (marked 7) (AL-TC-51U) (Willys only)	1	—	—	—	% 10
06-2	WO-A-9054		RESISTANCE, carbon (marked 80) (AL-TC-51N) (Willys only)	2	—	13	—	—
	WO-A-1409	FM-GPW-10505	REGULATOR, generator, 6V., 40 amp., assembly (AL-VRY-4203A). (Includes: REGULATOR, current limiting; REGULATOR, voltage; RELAY, cutout)	1	30	20	10	10
	WO-6609	FM-24427-S7	BOLT, hex-hd., s-fin, alloy-S., 5/16-24NF-2 x 1 1/4 (regulator to bracket)	2	—	—	—	—
	WO-5437	FM-34706-S2	WASHER, plain, S., 5/16 in. (7/8 O.D. x 3/8 I.D. x 1/16) (GM-106262)	2	—	—	—	—
	WO-51833	FM-34806-S2	WASHER, lock, S., 5/16 in. (19/32 O.D. x 11/32 I.D. x 1/16 thk.) (BECX1H)	2	—	—	—	—
	WO-A-8498		SEAL, lead and wire (AL-GAG-138) (Willys only)	1	—	—	—	—
06-2	WO-A-7797		SCREW, adjusting (br., No. 10 (.216)-48NS-2 x 19/32) (AL-VRA-16) (Willys only)	3	—	—	—	—
06-2	WO-A-1620		SCREW, rd-hd, S., cd-pltd., No. 10 (.216)-32NF x 5/16 (AL-8X-321) (Willys only) (2 used lead connection, 1 used ground)	3	—	—	—	—
06-2	WO-52781		SCREW, mounting, resistor (rd-hd., S., cd-pltd., No. 8 (.164)-32NC x 3/8) (AL-8X-55) (Willys only)	6	—	—	—	—
06-2	WO-A-8993		SCREW, rd-hd., S., cd-pltd., No. 10 (.216)-32NF x 1/2 (AL-8X-309) (Willys only) (6 used base mounting, 1 used lead connection)	7	—	—	—	—
06-2	WO-A-9050		SCREW, terminal, fil-hd., S., cd-pltd., 5/16-24NF x 1/2 (AL-8X-137) (Willys only)	3	10	20	30	—
06-2	WO-A-7798		SPRING, cut out relay (S., wire, 16 turns) (AL-VRA-17) (Willys only)	1	—	—	—	—
06-2	WO-A-7806		SPRING, voltage regulator and current regulator coil (S., wire, 14 1/2 turns) (AL-VRA-84) (Willys only)	2	—	—	—	—
06-2	WO-A-5262		WASHER, S., cd-pltd., No. 10 (.190 in.) (AL-8X-133A) (Willys only) (6 used base mounting, 1 used ground, 4 used lead connection)	11	—	—	—	—
	WO-5168		WASHER, lock, S., No. 10 (.216 in.) (AL-X-196) (Willys only) (6 used base mounting, 1 used ground screw, 4 used lead connection)	11	—	—	—	—
	WO-A-5288		WASHER, lock, terminal screw (S., tinned 5/16 in.) (AL-12X-1014) (Willys only)	3	—	—	—	—
	WO-A-7795		WASHER, insulating cutout relay stationary contact bracket (.187 x .250 x 1/32) (AL-GAA-35) (Willys only)	2	—	—	—	—
	WO-A-7804		WASHER, insulating cut out relay stationary contact bracket (13/64 x 3/8 x 1/32) (AL-X-1465) (Willys only)	2	—	—	—	—

GROUP 06—ELECTRICAL (Cont'd)

0602—CRANKING MOTOR

Figure Number	Part Number			ITEM	Quantity Reqd. per Unit Assy.	Per 100 Major Items			Estimated Reqmts. per 100 Rebuilds
	Official Stockage Number	Willys	Ford			12 Mos. Field Maintenance	Major Overhaul (5th Ech)	Total First Year Procurement	
Col. 1	Col. 2	Col. 3	Col. 4	Col. 5	Col. 6	Col. 7	Col. 8	Col. 9	Col. 10
06-3		WO-A-1568	FM-GPW-11005	ARMATURE, cranking motor, assembly (AL-MZ-2089)	1	3	3	9	10
06-3		WO-109452	FM-GPW-11077	BAND, head, cranking motor, assembly (AL-MZ-1024G)	1	3	3	8	10
		WO-A-1589	FM-34141-S2	NUT, sq., S., No. 10 (.190)-32NF-2 (AL-8X-794)	1	%			
		WO-A-1588	FM-36954-S7	SCREW, machine, rd-hd., S., No. 10 (.190)-32NF-2 x 1½ (AL-X-714)	1	% 6	6	6	20
		WO-A-1583	FM-GPW-11135	BEARING, cranking motor, .626 x .7535 x .735, absorbent bz. (AL-MG-77A) (issue until stock exhausted)	1	36	74		
06-3		WO-A-1582	FM-GPW-11130	BEARING, cranking motor, intermediate, assembly (AL-MAB-2040A)	1	2	2	6	5
06-3		WO-109431	FM-GPW-11055	BRUSH, insulated cranking motor (AL-MZ-12)	2	6	6	20	20
06-3		WO-109446	FM-GPW-11056	BRUSH, grounded, cranking motor, assembly (AL-MZ-1034)	1				
		WO-A-1552	FM-GPW-18535	BRUSH SET, cranking motor (AL-MZ-2012-S) (issue until stock exhausted) (Includes: 2 WO-109431 BRUSH / 2 WO-109446 BRUSH)	1	72	148		
		WO-639734	FM-GPW-11134	BUSHING, cranking motor (in flywheel housing)	1		3	6	10
		WO-109436	FM-GPW-11107	BUSHING, insulating, cranking motor terminal post (AL-MU-31)	1				
06-3		WO-109427	FM-GPW-11082	COIL, cranking motor, lower left (AL-MZ-1009) (issue until stock exhausted)	1	8	12		5
06-3		WO-109428	FM-GPW-11084	COIL, cranking motor, lower right (AL-MZ-1008) (issue until stock exhausted)	1	8	12		40
06-3		WO-A-1560	FM-GPW-11083	COIL, cranking motor, upper left (AL-MZ-1007)	1				
06-3		WO-A-1563	FM-GPW-11085	COIL, cranking motor, upper right (AL-MZ-1010)	1				
06-3		WO-A-1558	FM-GPW-11090	CONNECTOR, cranking motor field coil (AL-MZ-32) (issue until stock exhausted)	2	4	8		
06-3		WO-A-1573	FM-GPW-11350	DRIVE, Bendix, assembly (AL-EBA-46)	1	3	3	9	10
06-3		WO-A-1576	FM-B-11381	HEAD, driving, cranking motor Bendix (AL-EB-8503) (issue until stock exhausted)	1		4		
		WO-109442	FM-GPW-11061	HOLDER, cranking motor brush (AL-MZ-16)	2				
06-3		WO-A-1585	FM-GPW-11131	HOUSING, cranking motor Bendix drive pinion, assembly (AL-PS-1079A)	1	2	2	8	5
06-3		WO-A-1584	FM-355164-S	BOLT, hex-hd., S., No. 10 (.190)-32NF x ³¹⁄₆₄ (AL-MZ-52)	4	13	13	100	40
		WO-5168	FM-34803-S	WASHER, lock, S., No. 10 (.190 in.) (AL-X-196)	4				
06-3		WO-A-1557	FM-GPW-11089	INSULATION, cranking motor field connection (AL-MZ-30A) (issue until stock exhausted)	1	4	8		15
		WO-5017	FM-74175-S7	KEY, Woodruff, No. 6 (AL-X-261)	1	5	5	20	
06-3		WO-A-6756	FM-GPW-18376	KIT, repair, cranking motor (issue until stock is exhausted, then use WO-A-7836)	as req.	12	20		
		WO-A-7836	FM-GPW-18376-B	KIT, repair, cranking motor (was WO-A-6756) (issue until stock exhausted) (Includes: 1 WO-A-1583 FM-GPW-11135 BEARING / 1 WO-A-1588 FM-36954-S7 SCREW, head band / 1 WO-A-1589 FM-34141-S2 NUT, head band)	as req.	12			

A B C D E F G H J K L M N O P Q R S T U V W X

RA PD 305076

AI AH AG AF AE D AD AC AB V AA Z Y

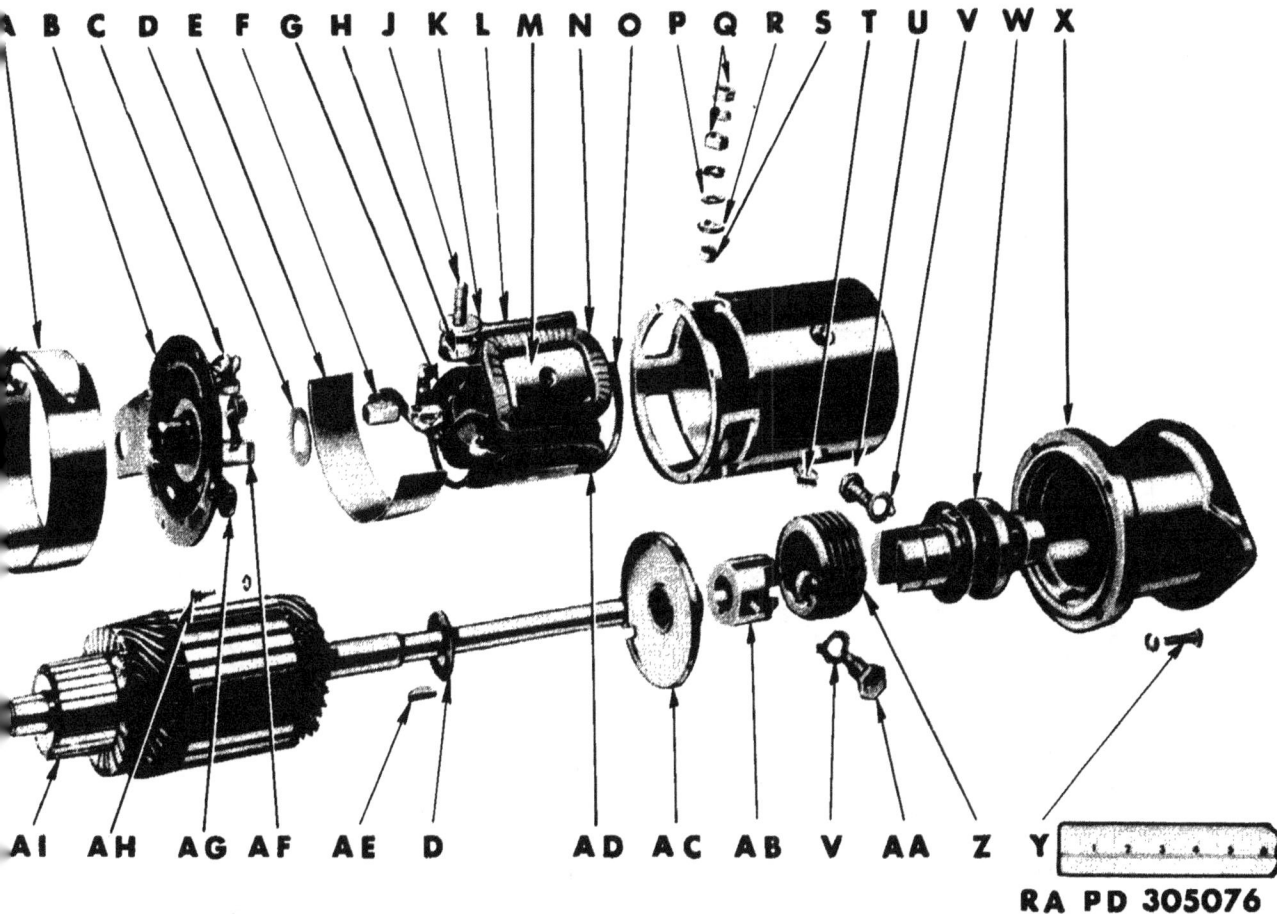

FIGURE 06-3—CRANKING MOTOR ASSEMBLY

Item	Willys Part No.	Ford Part No.	Gov't Group No.	Key	Item	Willys Part No.	Ford Part No.	Gov't Group No.
BAND	WO-109452	FM-GPW-11077	0602	S	BUSHING	WO-109436	FM-GPW-11107	0602
HEAD assembly	WO-A-1566	FM-GPW-11049	0602	T	SCREW	WO-A-1559	FM-355485-S7	0602
BRUSH	WO-109446	FM-GPW-11056	0602	U	SCREW	WO-A-1579	FM-GPW-11382	0602
WASHER	WO-109455	FM-GPW-11036-B	0602	V	WASHER	WO-A-1574	FM-B-11379	0602
INSULA-TION	WO-A-1557	FM-GPW-11089	0602	W	SHAFT	WO-A-1581	FM-GPW-11354	0602
BRUSH	WO-109431	FM-GPW-11055	0602	X	HOUSING assembly	WO-A-1585	FM-GPW-11131	0602
COIL	WO-109427	FM-GPW-11082	0602	Y	BOLT	WO-A-1584	FM-355164-S	0602
TERMINAL	WO-A-1554	FM-GPW-11102	0602	Z	SPRING	WO-A-1577	FM-GPW-11375	0602
TERMINAL	WO-109433	FM-GPW-11103	0602	AA	SCREW	WO-A-1578	FM-GPW-11377	0602
WASHER	WO-109437	FM-GPW-11095	0602	AB	HEAD	WO-A-1576	FM-B-11381	0602
COIL	WO-A-1560	FM-GPW-11083	0602	AC	BEARING	WO-A-1582	FM-GPW-11130	0602
PIECE	WO-A-1556	FM-GPW-11120	0602	AD	COIL	WO-109428	FM-GPW-11084	0602
COIL	WO-A-1563	FM-GPW-11085	0602	AE	KEY	WO-5017	FM-74175-S7	0602
CON-NECTOR	WO-A-1558	FM-GPW-11090	0602	AF	HOLDER	WO-109442	FM-GPW-11060	0602
WASHER	WO-A-1555	FM-34706-S2	0602	AG	SPRING	WO-109445	FM-B-11059	0602
NUT	WO-A-1565	FM-355944-S5	0602	AH	SCREW	WO-A-1572	FM-37364-S7	0602
WASHER	WO-A-1553	FM-GPW-11094	0602	AI	ARMATURE assembly	WO-A-1568	FM-GPW-11005	0602

GROUP 06—ELECTRICAL (Cont'd)

0602—CRANKING MOTOR (Cont'd)

Figure Number Col. 1	Official Stockage Number Col. 2	Part Number Willys Col. 3	Part Number Ford Col. 4	ITEM Col. 5	Quantity Reqd. per Unit Assy. Col. 6	12 Mos. Field Maintenance Col. 7	Major Overhaul (5th Ech) Col. 8	Total First Year Procurement Col. 9	Estimated Reqmts. per 100 Rebuilds Col. 10
				2 WO-A-1572 FM-31596-S SCREW, head commutator					
				2 WO-A-1584 FM-355164-S SCREW, head					
				1 WO-A-1552 FM-GPW-11535 SET, brush					
				4 WO-109445 FM-B-11059 SPRING, brush					
				2 WO-A-1571 FM-34803-S7 WASHER, lock					
				2 WO-52960 FM-34803-S7 WASHER, lock, w/KIT only					
				2 WO-109455 FM-GPW-11036-B WASHER, thrust)					
	WO-A-7842		FM-GPW-18329	KIT, repair, cranking motor Bendix drive	as req.	—	3	3	10
				(Includes:					
				1 WO-A-1576 FM-B-11381 HEAD					
				1 WO-5017 FM-74175-S7 KEY, Woodruff					
				1 WO-A-1578 FM-GPW-11377 SCREW, head spring					
				1 WO-A-1579 FM-GPW-11382 SCREW, shaft spring					
				1 WO-A-1575 FM-B-11357-A SLEEVE					
				1 WO-A-1577 FM-GPW-11375 SPRING, drive					
				1 WO-A-1574 FM-B-11379 WASHER, lock)					
	WO-A-7841		FM-GPW-18319	KIT, repair cranking motor field coil	as req.	2	2	8	5
				(Includes:					
				1 WO-109436 FM-GPW-11107 BUSHING, insulating					
				1 WO-109427 FM-GPW-11082 COIL					
				1 WO-109428 FM-GPW-11084 COIL					
				1 WO-A-1560 FM-GPW-11083 COIL					
				1 WO-A-1563 FM-GPW-11085 COIL					
				2 WO-A-1558 FM-GPW-11090 CONNECTOR					
				1 WO-A-1557 FM-GPW-11089 INSULATOR					
				2 WO-A-1565 FM-355944-S5 NUT					
				1 WO-109433 FM-GPW-11103 POST					
				4 WO-A-1559 FM-355485-S7 SCREW					
				1 WO-A-1554 FM-GPW-11102 TERMINAL					
				2 WO-5051 FM-34806-S2 WASHER					
				1 WO-A-1555 FM-34706-S2 WASHER					
				1 WO-A-1553 FM-GPW-11094 WASHER, insulating					
				1 WO-109437 FM-GPW-11095 WASHER, lock)					
F		WO-A-1245	FM-GPW-11001-A	MOTOR, cranking, assembly (AL-MZ-4113)	1	10	—	13	'
		WO-51406	FM-24428-S	BOLT, hex-hd, s-fin, alloy-S, 3/8-16NC-2 x 1¼ (to bell housing) GM-100135) (BANX1CD)	2	—	—	—	—
		WO-5010	FM-34807-S2	WASHER, lock, S, 3/8 (29/32 O.D. x 13/32 I.D. x 3/32 thk.) (BECX1K)	2	—	—	—	—
06-3		WO-A-1556	FM-GPW-11120	PIECE, pole, cranking motor (AL-MZ-29)	4	—	—	—	—

		WO-A-1586	FM-72798-S7-8	PIN, locating, dowel, cranking motor intermediate bearing (AL-MAB-88)	1	—	—	—	—
06-3		WO-A-1566	FM-GPW-11049	PLATE, cranking motor, commutator end, assembly (AL-MZ-2156)	1	3	3	9	10
06-3		WO-A-1572	FM-37364-S7	SCREW, fill-hd., No. 10 (.190)-32NF x 3/8 (plate screw) (AL-X-902)	4	13	13	40	40
06-3		WO-109433	FM-GPW-11103	POST, terminal, cranking motor (5/16-24NF) (AL-MU-28)	1	—	—	—	—
		WO-A-1565	FM-355944-S5	NUT, hex., 5/16-24NF (AL-5X-1376) (issue until stock exhausted)	2	8	16	—	—
		WO-5051	FM-34806-S2	WASHER, lock, 5/16 in. (AL-X-1014) (BECX3H)	2	—	—	—	—
		WO-A-1580	FM-B-11371	RING, takeup, cranking motor, Bendix drive (AL-EB-8734) (issue until stock exhausted)	1	4	4	—	—
		WO-A-1567	FM-GPW-11069	RIVET, cranking motor, commutator end plate (tubular, ov-hd., 1/8 x 1/4) (AL-X-532)	—	—	—	—	—
				(Willys)	4	—	—	—	—
				(Ford)	8	—	—	—	—
06-3		WO-A-1578	FM-GPW-11377	SCREW, cranking motor Bendix head spring (AL-EB-8506)	1	19	19	45	60
06-3		WO-A-1559	FM-355485-S7	SCREW, cranking motor pole piece (AL-MZ-38A)	4	—	—	—	—
06-3		WO-A-1579	FM-GPW-11382	SCREW, cranking motor Bendix shaft spring (AL-EB-8507)	1	19	19	45	60
06-3		WO-A-1581	FM-GPW-11354	SHAFT, cranking motor Bendix drive w/PINION, assembly (AL-EBA-4611) (issue until stock exhausted)	1	12	16	—	—
		WO-A-1575	FM-B-11357-A	SLEEVE, compression, cranking motor Bendix drive (AL-EB-7819S) (issue until stock exhausted)	1	4	4	—	—
		WO-A-1569	FM-GPW-11053	SPACER, thrust, cranking motor bearing (AL-MZ-51)	1	—	—	—	—
06-3		WO-109445	FM-B-11059	SPRING, cranking motor brush (AL-MZ-19) (issue until stock exhausted)	4	16	24	—	—
06-3		WO-A-1577	FM-GPW-11375	SPRING, Bendix drive, cranking motor (AL-EB-8505)	1	3	—	6	—
		WO-638646	FM-GPW-11140	SUPPORT, cranking motor (issue until stock exhausted)	1	—	4	—	—
		WO-A-1746		BOLT, hex-hd., S., 3/8-16NC x 7/8 (issue until stock exhausted, then use WO-51405)	—	4	4	—	—
		WO-51405		BOLT, hex-hd., s-fin., alloy-S., 3/8-16NC-2 x 1 (to crankcase) (was WO-A-1746) (BCAX1CB) (Willys only)	1	—	—	—	—
		WO-52132	FM-24327-S7	BOLT, hex-hd., S., 5/16-24NF-2 x 5/8 (to cranking motor) (GM-106279)	1	—	—	—	—
		WO-5437	FM-34706-S2	WASHER, S., 5/16 in. (GM-106262)	1	—	—	—	—
		WO-53026	FM-356200-S7	WASHER, lock, S., cd-pltd., internal, external, tooth, 3/8 in. (GM-178551)	2	—	—	—	—
		WO-53025	FM-356309-S7	WASHER, lock, S., cd-pltd., internal tooth, 5/16 in. (GM-178378)	1	—	—	—	—
		WO-51833	FM-34806-S2	WASHER, lock, reg., S., 5/16 in. (BECX1H)	1	—	—	—	—
06-3		WO-A-1554	FM-GPW-11102	TERMINAL, cranking motor field coil (AL-MU-14)	1	—	—	—	—
		WO-A-1555	FM-34706-S2	WASHER, cranking motor terminal post, plain, S., 5/16 in. (AL-MU-37)	1	—	—	—	—
06-3		WO-A-1553	FM-GPW-11094	WASHER, insulating, cranking motor terminal post, outer (AL-MAB-31)	1	—	—	—	—
06-3		WO-109437	FM-GPW-11095	WASHER, insulating, cranking motor terminal post, inner (AL-MU-39) (issue until stock exhausted)	1	8	16	—	—
		WO-A-1571	FM-34803-S7	WASHER, cranking motor, commutator end (lock, No. 10 (.190 in.) (AL-X-544)	1	—	—	—	—
06-3		WO-A-1574	FM-B-11379	WASHER, lock, cranking motor housing screw (AL-EB-108)	1	38	38	84	120
06-3		WO-109455	FM-GPW-11036-B	WASHER, thrust, cranking motor armature, commutator end (fibre, .645 x 1 1/4 x 1/32) (AL-MU-54)	—	—	—	—	—
				(Willys)	2	10	10	20	30
				(Ford)	1	—	—	—	—

0603—DISTRIBUTOR

06-4		WO-A-1674	FM-GPW-12155	ARM, advance control, distributor, assembly (AL-IGS-1080) (issue until stock exhausted)	1	—	4	—	—
06-4		WO-A-1570	FM-GPW-12162	ARM, breaker, distributor, assembly (AL-IGP-3028)	1	—	—	—	—

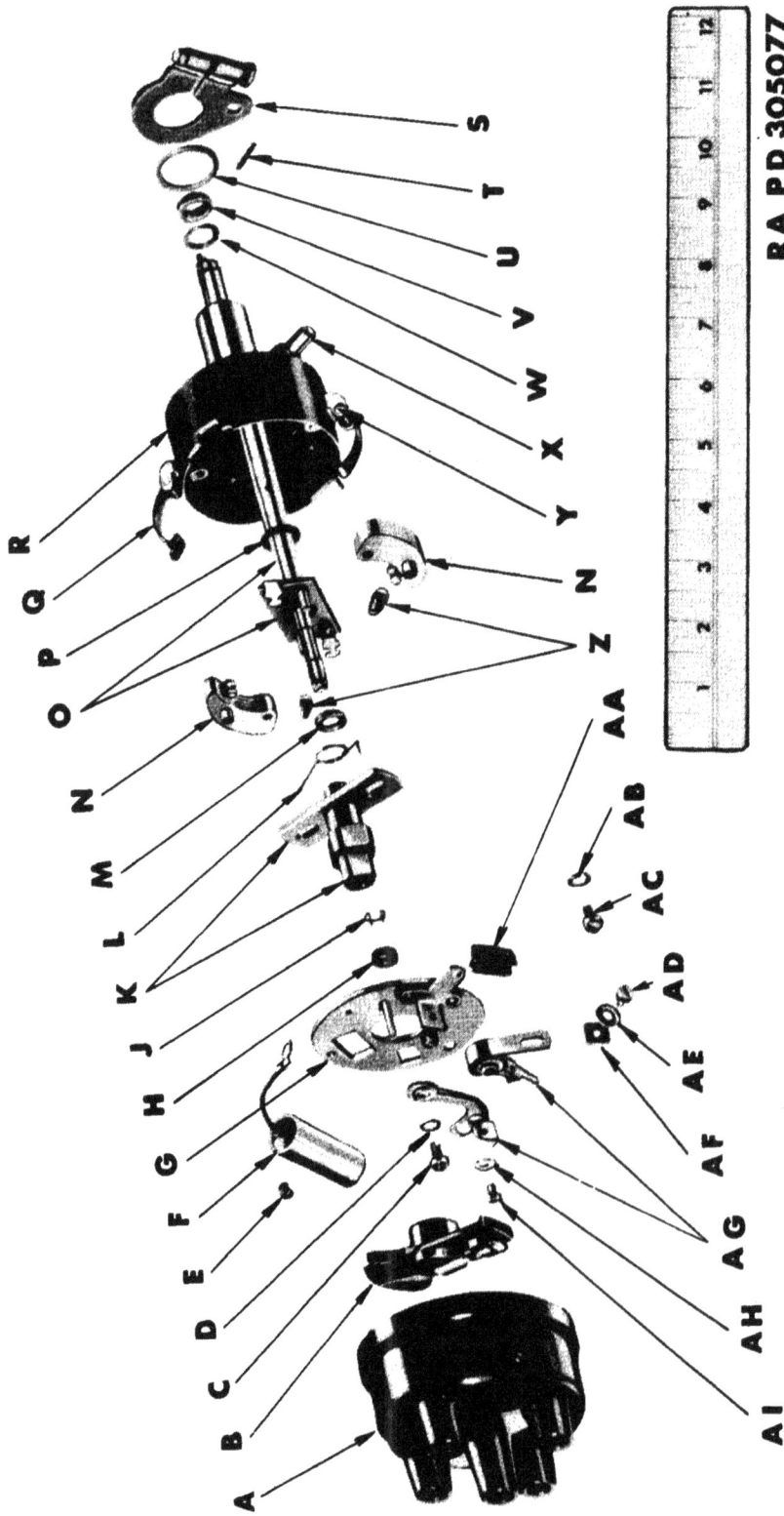

RA PD 305077

FIGURE 06-4—DISTRIBUTOR ASSEMBLY

Key	Item	Willys Part No.	Ford Part No.	Gov't Group No.
A	CAP assembly	WO-A-1655	FM-GPW-12106	0603
B	ROTOR	WO-A-1658	FM-GPW-12200	0603
C	SCREW	WO-A-1636	FM-31588-S7	0603
D	WASHER	WO-109453	FM-34803-S7	0603
E	SCREW	WO-A-1670	FM-31026-S7	0603
F	CONDENSER	WO-A-1631	FM-GPW-12300	0603
G	PLATE	WO-A-1664	FM-GPW-12010	0603
H	WICK	WO-A-1671	FM-GPW-12133	0603
I	WASHER	WO-A-1669	FM-34801-S7	0603
J	SNAP RING	WO-A-1653	FM-GPW-12177	0603
K	PLATE	WO-A-1661	FM-GPW-12176	0603
L	SPRING	WO-A-1684	FM-GPW-12191	0603
M	SPACER	WO-A-1672	FM-GPW-12120	0603
N	WEIGHT	WO-A-1676	FM-GPW-12188	0603
O	SHAFT	WO-A-1678	FM-GPW-12178	0603
P	WASHER	WO-A-1673	FM-GPW-12182	0603
Q	SPRING	WO-A-1682	FM-GPW-12144	0603
R	BASE assembly	WO-A-1679	FM-GPW-12139	0603
S	ARM	WO-A-1674	FM-GPW-12155	0603
T	RIVET	WO-A-1685	FM-72867-S7	0603
U	WASHER	WO-A-1654	FM-GPW-12267	0603
V	COLLAR	WO-A-1659	FM-GPW-12195	0603
W	WASHER	WO-106740	FM-GPW-12193	0603
X	OILER	WO-107128	FM-B-10141	0603
Y	PIN	WO-A-1683	FM-GPW-12145	0603
Z	SPRING	WO-A-1677	FM-GOW-42084	0603
AA	COVER	WO-A-1660	FM-GPW-12174	0603
AB	WASHER	WO-5168	FM-34803-S7	0603
AC	SCREW	WO-A-1686	FM-31583-S7	0603
AD	SCREW	WO-A-1633	FM-36787-S7	0603
AE	WASHER	WO-A-1667	FM-34701-S7	0603
AF	CLIP	WO-A-1663	FM-GPW-12217	0603
AG	ARM	WO-A-1570	FM-GPW-12162	0603
AH	WASHER	WO-A-1666	FM-34702-S7	0603
AI	SCREW	WO-A-1668	FM-31027-S8	0603

RA PD 305077-A

GROUP 06—ELECTRICAL (Cont'd)

0603—DISTRIBUTOR (Cont'd)

Figure Number (Col. 1)	Official Stockage Number (Col. 2)	Willys (Col. 3)	Ford (Col. 4)	ITEM (Col. 5)	Quantity Reqd. per Unit Assy. (Col. 6)	12 Mos. Field Maintenance (Col. 7)	Major Overhaul (5th Ech) (Col. 8)	Total First Year Procurement (Col. 9)	Estimated Reqmts. per 100 Rebuilds (Col. 10)
06-4		WO-A-1679	FM-GPW-12139	BASE, distributor, assembly (AL-IGS-2135)	1	—	—	—	—
06-4		WO-A-1681	FM-GPW-12082	BEARING, distributor, absorbent bronze (AL-IG-579-A) (issue until stock exhausted)	2	16	32	30	15
06-4		WO-A-1655	FM-GPW-12106	CAP, distributor, assembly (AL-IG-1324)	1	19	5	50	15
06-4		WO-A-1663	FM-GPW-12217	CLIP, distributor breaker arm spring (AL-IG-676)	1	19	5		10
06-4		WO-A-1659	FM-GPW-12195	COLLAR, distributor drive shaft (AL-IGB-199)	1		3		
06-4		WO-A-1631	FM-GPW-12300	CONDENSER, distributor, assembly (AL-IGW-2139)	1	% 32	6	45	20
60-4		WO-A-1670	FM-31026-S7	SCREW, fill-hd, No. 6 (.138)-32NC x 5/32 (AL-8X-1546) (issue until stock exhausted)	1	% 32			
		WO-A-1669	FM-34801-S7	WASHER, lock, S., No. 6 (.138) (AL-X-1012)	1	% 96	68		75
06-4		WO-A-1687	FM-GPW-18354	CONTACT SET distributor (AL-1GP-3028FS) (point set). (Includes:	1		24	154	
				1 WO-A-1570 FM-GPW-12162 ARM assembly					
				1 WO-A-1564 FM-GPW-12172 CONTACT, breaker)					
06-4		WO-A-1660	FM-GPW-12174	COVER, distributor, terminal slot (AL-1GC-117)	1	5	5	12	15
F		WO-A-1244	FM-GPW-12100	DISTRIBUTOR assembly (AL-1GC-4705)	1	5	5	14	15
		WO-51514	FM-20308-S2	BOLT, hex-hd, S., 1/4-20NC-2 x 5/8 (to cylinder block)	1	—	—		
		WO-52702	FM-34745-S2	WASHER, S., 1/4 in. (BECX1G)	1	—			
		WO-52706	FM-34805-S2	WASHER, lock, S., 1/4 in. (BECX1G)	1	6	6	16	20
		WO-A-7843	FM-GPW-17343	KIT, repair, distributor (Includes:	as req.				
				1 WO-A-1681 FM-GPW-12082 BEARING					
				1 WO-A-1663 FM-GPW-12217 CLIP, spring					
				1 WO-A-1660 FM-GPW-12174 COVER					
				1 WO-107128 FM-B-10141 OILER					
				2 WP-A-1683 FM-GPW-12145 PIN, hinge					
				1 PLUG (AL-1GS-32)					
				1 WO-A-1653 FM-GPW-12177 RING, lock					
				1 WO-A-1685 FM-72867-S7 RIVET					
				1 WO-A-1670 FM-31026-S7 SCREW, condenser					
				1 WO-A-1668 FM-31027-S8 SCREW					
				2 WO-A-1636 FM-31588-S7 SCREW					
				1 WO-A-1633 FM-36787-S7 SCREW, condenser					
				2 WO-A-1682 FM-GPW-12144 SPRING, cap					
				2 WO-A-1669 FM-34801-S7 WASHER, lock					
				3 WO-5168 FM-34803-S7 WASHER, lock					
				1 WO-A-1672 FM-GPW-12120 WASHER					
				1 WO-A-1667 FM-34701-S7 WASHER, No. 6					

GROUP 06—ELECTRICAL (Cont'd)

0603—DISTRIBUTOR (Cont'd)

```
1   WO-A-1666   FM-34702-S2    WASHER, No. 8
1   WO-A-1673   FM-GPW-12182   WASHER, thrust
1   WO-106740   FM-GPW-12193   WASHER, thrust
1   WO-A-1671   FM-GPW-12133   WICK
```

Figure Number	Official Stockage Number	Willys	Ford	ITEM	Quantity Reqd. per Unit Assy.	12 Mos. Field Maintenance	Major Overhaul (5th Ech)	Total First Year Procurement	Estimated Reqmts. per 100 Rebuilds
Col. 1	Col. 2	Col. 3	Col. 4	Col. 5	Col. 6	Col. 7	Col. 8	Col. 9	Col. 10
06-4		WO-107128	FM-B-10141	OILER, distributor shaft (press in sleeve type) (AL-X-490)	1	5	—	5	—
06-4		WO-A-1662	FM-GPW-12151	PLATE, breaker, distributor, assembly (AL-IGC-2148C)	1	2	24	6	5
06-4		WO-A-1636	FM-31588-S7	SCREW, fill-hd., S., No. 10 (.190)-32NF x 5/16 (AL-SX-870)	2	—	—	—	—
06-4		WO-109453	FM-34803-S7	WASHER, lock, S., No. 10 (.190 in.) (AL-X-1270)	2	2	2	6	5
06-4		WO-A-1664	FM-GPW-12010	PLATE, breaker part, distributor, assembly (AL-IGC-1148)	1	—	12	—	—
06-4		WO-A-1661	FM-GPW-12176	PLATE, distributor cam and stop, 4 cyl., left hand (AL-IGC-1132LB)	1	—	32	—	—
06-4		WO-A-1683	FM-GPW-12145	PIN, distributor cap hinge (AL-X-1448) (issue until stock exhausted)	2	8	32	—	—
06-4		WO-A-1656	FM-GPW-12011	PLUNGER, distributor contact spring (AL-IG-514) (issue until stock exhausted)	1	%16	10	—	—
06-4		WO-A-1564	FM-GPW-12218	POINT, breaker distributor (AL-IGC-1149)	1	16	5	—	—
06-4		WO-A-1685	FM-72867-S7	RIVET, distributor, drive shaft collar (AL-SW-213) (issue until stock exhausted)	1	%29	8	45	30
06-4		WO-A-1658	FM-GPW-12200	ROTOR, distributor (AL-IG-1657R)	1	—	—	12	15
06-4		WO-A-1686	FM-31583-S7	SCREW, condenser terminal (fill-hd., No. 10 (.190)-32NF x 1/4) (AL-8X-872)	1	5	4	—	—
06-4		WO-A-1668	FM-31027-S8	SCREW, locking, distributor, stationary contact (fill-hd., S., No. 8 (.164)-32NC x 3/16) (AL-8X-884) (issue until stock exhausted)	1	%4	—	—	—
06-4		WO-A-1633	FM-36787-S7	SCREW, distributor, breaker arm spring clip (hex-hd., S., slotted, No. 6 (.138)-32NC x 5/16) (AL-IGS-175)	1	—	—	—	—
06-4		WO-A-1678	FM-GPW-12178	SHAFT, drive, distributor assembly (AL-IGS-1134L)	1	—	—	—	—
06-4		WO-A-1675	FM-GPW-12175	SHAFT, drive, distributor, w/GOVERNOR, assembly (AL-IGS-2134L) (issue until stock exhausted)	1	—	—	—	—
06-4		WO-A-1653	FM-GPW-12177	SNAP RING, retaining, distributor cam (AL-IG-680) (issue until stock exhausted)	1	—	4	—	—
06-4		WO-A-1672	FM-GPW-12120	SPACER, distributor cam (AL-IGS-99) (issue until stock exhausted)	1	40	80	—	—
06-4		WO-A-1684	FM-GPW-12191	SPRING, distributor anti-rattle (AL-IGT-69) (issue until stock exhausted)	1	4	8	—	—
06-4		WO-A-1587	FM-GPW-12169	SPRING, distributor breaker arm (AL-IGP-30)	1	4	4	—	—
06-4		WO-A-1657	FM-GPW-12012	SPRING, distributor contact (AL-IG-515) (issue until stock exhausted)	1	16	32	—	—
H		WO-A-1682	FM-GPW-12144	SPRING, distributor cap (AL-IG-694) (issue until stock exhausted)	2	8	12	15	20
06-4		WO-637615	FM-GPW-12083	SPRING, friction distributor shaft	1	5	5	—	—
06-4		WO-A-1677	FM-GPW-12084	SPRING SET, distributor governor weight (AL-IGB-202S) (issue until stock exhausted)	1	—	40	—	—
06-4		WO-5168	FM-34803-S7	WASHER, lock, No. 10 (.190 in.) (condenser terminal screw) (AL-X-196)	1	20	—	—	—
06-4		WO-A-1666	FM-34702-S2	WASHER, distributor stationary contact (lock, No. 8 (.164)) (AL-8X-350) (issue until stock exhausted)	4	—	40	—	—
06-4		WO-A-1667	FM-34701-S7	WASHER, distributor breaker arm spring screw (No. 6 (.138)) (AL-8X-353)	1	20	—	—	—
06-4		WO-A-1654	FM-GPW-12267	WASHER, thrust, distributor advance arm (AL-IG-816C)	1	2	2	6	5

Fig.	Ord. No.	Federal Stock No.	Description	Units				
06-4	WO-A-1673	FM-GPW-12182	WASHER, thrust, distributor shaft, upper (AL-IGS-104) (issue until stock exhausted)	1	—	4	—	—
06-4	WO-106740	FM-GPW-12193	WASHER, thrust, distributor drive shaft, lower (AL-IG-90) (issue until stock exhausted)	1	4	8	—	—
06-4	WO-A-1676	FM-GPW-12188	WEIGHT, distributor governor assembly (AL-IG-2456)	2	3	3	9	10
06-4	WO-A-1671	FM-GPW-12133	WICK, oiling, distributor (felt) (AL-IGH-28) (issue until stock exhausted)	1	4	8	—	—
			0604—IGNITION COIL					
	WO-A-1526	FM-GPW-12030	BRACKET, ignition coil (AL-IG-17980D)	1	—	—	—	—
	WO-5914	FM-33796-S2	NUT, hex., S., 1/4-28NF-2 (bracket to stud) (GM-103024) (BBBX1AA)	2	—	—	—	—
	WO-635886	FM-357689-S	STUD, S., 1/4-20NC x 1 x 1/4-28NF-2	2	% 3	3	10	10
	WO-631105	FM-GPW-12064	WASHER, S., black enamel, 9/32 in. (issue until stock exhausted)	2	% 8	16	—	—
	WO-53024	FM-351274-S7	WASHER, lock, S., cd-pltd., internal, external teeth, 1/4 in. (GM-174916)	4	—	—	—	—
	WO-A-1424	FM-GPW-12000-A	COIL, ignition, w/BRACKET and WASHER, assembly (before Willys serial 288835)	1	—	3	—	—
	WO-A-7792	FM-GPW-12000-B	COIL, ignition, w/BRACKET and WASHER, assembly (after Willys serial 288835) (coil does not have bottom ground connection)	1	% 6	—	13	10
			0604A—IGNITION WIRING					
	WO-301435	FM-GPW-12091-A	BUSHING, spark plug cable support (rubber), 11/16 in. hole) (issue until stock exhausted)	1	% 4	4	—	—
06-5	WO-A-5083	FM-GPW-14321	CABLE, coil primary	1	%	—	—	—
06-5	WO-A-1420	FM-GPW-12298-B	CABLE, coil secondary	1	%	—	—	—
06-5	WO-A-1412	FM-GPW-12287	CABLE, spark plug, No. 1	1	—	—	—	—
06-5	WO-A-1414	FM-GPW-12284	CABLE, spark plug, No. 2	1	—	—	—	—
	WO-A-1416	FM-GPW-12283	CABLE, spark plug, No. 3	1	—	—	—	—
	WO-A-1418	FM-GPW-12286	CABLE, spark plug, No. 4	1	—	—	—	—
	WO-A-6321		INSULATOR, ignition cable (Willys only)	4	—	—	—	—
	WO-A-6757	FM-GPW-18363	KIT, repair, ignition wiring (issue until stock is exhausted. Then use WO-A-7844 for maintenance)	as req.	% 4	8	—	—
	WO-A-7844	FM-GPW-18363-B	KIT, repair, ignition wiring (was WO-A-6757)	as req.	% 5	6	24	20
			(Includes:					
			1 WO-A-1412 FM-GPW-12287 CABLE, No. 1					
			1 WO-A-1414 FM-GPW-12284 CABLE, No. 2					
			1 WO-A-1416 FM-GPW-12283 CABLE, No. 3					
			1 WO-A-1418 FM-GPW-12286 CABLE, No. 4					
			1 WO-A-5083 FM-GPW-14321 CABLE, coil primary					
			1 WO-A-1420 FM-GPW-12298-B CABLE, coil secondary)					
	WO-327257	FM-9N-12113-A	SEAL, weather, spark plug cables, distributor end (rubber)	5	% 10	10	40	30
	WO-A-1652	FM-GPW-12006	TERMINAL, high tension spark plug cables (AL-16-94)	5	% 80	80	200	250
	WO-314369	FM-B-14466	TERMINAL, spark plug cables and coil secondary cable	5	% 19	19	50	60
	WO-307556	FM-B-14463	TIP, coil primary cable	2	—	—	—	—
			0604B—SPARK PLUGS					
	WO-A-1096	FM-11A-12425	CAP, spark plug insulator	4	—	—	—	—
	WO-637863	FM-O1A-12410	GASKET, spark plug (VC-2066C-C1) (MDR-AM-504K)	4	% 40	—	50	—
	WO-A-538	FM-GPW-12405	PLUG, spark, w/GASKET, assembly (14MM) (CP type-AN-7)	4	% 416	64	610	200

GROUP 06—ELECTRICAL (Cont'd)

Figure Number	Official Stockage Number	Part Number (Willys)	Part Number (Ford)	ITEM	Quantity Reqd. per Unit Assy.	Per 100 Major Items — 12 Mos. Field Maintenance	Per 100 Major Items — Major Overhaul (5th Ech)	Per 100 Major Items — Total First Year Procurement	Estimated Reqmts. per 100 Rebuilds
Col. 1	Col. 2	Col. 3	Col. 4	Col. 5	Col. 6	Col. 7	Col. 8	Col. 9	Col. 10
				0604C—IGNITION LOCK					
		WO-A-6814	FM-GPW-3685-B	HANDLE, ignition switch (after 104402 Willys trucks)	1	—	—	—	—
		WO-A-2518	FM-GPW-3685	KEY, ignition switch (before Willys serial 202023)	1	—	—	—	—
		WO-A-6813		NUT, mounting (DM-53170A) (after Willys serial 202023) (Willys only)	1	—	—	—	—
		WO-52131	FM-26457-S7	SCREW, switch terminal (machine rd-hd., S., No. 8 (.164)-32NC-2) (GM-122159)	2	—	—	—	—
18-1		WO-A-6811	FM-GPW-3686-B	SWITCH, keyless ignition, assembly (after Willys serial 202023)	1	% 3	—	5	—
		WO-A-1350	FM-356229-S	WASHER, lock, S., kzd., internal teeth .3350DK-16910D x .018 thk. (before Willys serial 202023)	2	% 19	10	30	—
				0605—INSTRUMENTS					
18-1		WO-A-8186	FM-GPW-10850	AMMETER, 50 amp. (AL-10311A)	1	% 6	—	8	—
		WO-5848		NUT, hex, S., No. 10 (.190)-32NF (to bracket) (AL-8170) (Willys only)	4	—	—	—	—
		WO-A-8129		WASHER, lock, S., No. 10 (.190) (AL-2229) (Willys only)	2	—	—	—	—
		WO-A-8130		BRACKET, mounting, ammeter (AL-9288-A) (Willys only)	1	—	—	—	—
		WO-A-8132		BRACKET, mounting, fuel gauge (AL-10063) (Willys only)	1	—	—	—	—
		WO-A-8131		BRACKET, mounting, oil pressure gauge and heat indicator (AL-21608) (Willys only)	2	—	—	—	—
18-1		WO-A-8184	FM-GPW-9280	GAUGE, fuel (AL-10313-A)	1	% 8	4	1	—
18-1		WO-A-8190	FM-GPW-9273	GAUGE, oil pressure, assembly (AL-10310-A)	1	% 8	4	10	—
		WO-6536		NUT, hex, S., No. 10 (.190)-32NF (to bracket) (AL-2235) (Willys only)	4	—	—	—	—
		WO-A-8129		WASHER, lock, S., No. 10 (.190) (AL-2229) (Willys only)	2	—	—	—	—
		WO-662420	FM-GPW-9319-A	GROMMET, heat indicator through dash (3/16 in., rubber) (issue until stock exhausted) (Willys only)	2	—	—	—	—
18-1		WO-A-8188	FM-GPW-10883	INDICATOR, heat, assembly (Moto Meter) (AL-10312-A)	1	% 4	8	10	—
		WO-6536		NUT, hex, S., No. 10 (.190)-32NF (AL-2235) (gauges to bracket) (Willys only)	1	—	—	—	—
		WO-A-8129		WASHER, lock, S., No. 10 (.190) (AL-2229) (Willys only)	6	—	—	—	—
				0605B—INSTRUMENT WIRING					
				NOTE: For cables not appearing here, see Group 0606B.					
		WO-A-5080	FM-GPW-14416	CABLE, fuel gauge to circuit breaker (issue until stock exhausted)	1	% 4	4	—	—
		WO-A-5070	FM-GPW-14406	CABLE, fuel gauge (instrument board) to fuel gauge tank unit (issue until stock exhausted)	1	% 4	4	—	—
06-5		WO-A-5072	FM-GPW-14458	CABLE, ignition switch to ammeter to blackout switch	1	% 4	—	—	—

0605C—PANEL LIGHT

0606—SWITCHES

Ref	WO No.	FM No.	Description				%	Units
06-5	WO-A-5981	FM-GPW-14432	HARNESS, filter wiring (includes the following (4) cables, w/CABLE, ammeter to horn circuit breaker (black—2 red tracings) CABLE, filter (battery ammeter) to ammeter (red—3 black tracings) CABLE, filter (coil ignition switch) to ignition switch to gasoline gauge circuit breaker (black—2 white tracings), CABLE, filter (regulator battery ammeter) to ammeter (black—2 red tracings))	7	7	2	2	1
			0605C—PANEL LIGHT					
	WO-A-1748	FM-GPW-13713	ADAPTER, instrument light (DM-29392) (issue until stock exhausted)			4		2
	WO-A-1334	FM-GPW-13704	SHIELD, instrument, lamp assembly		6		5	2
	WO-A-1411	FM-GPW-13710	SOCKET, instrument lamp, w/CABLE, assembly (includes two SOCKETS)		8	2	2	1
			0606—SWITCHES					
	WO-A-1347		CONTROL, lockout, blackout lighting switch assembly (DM-5943) (Willys only)					1
	WO-A-6152		KNOB, w/set SCREW, blackout driving lamp switch assembly (Willys only)					1
	WO-A-1348		KNOB, w/set SCREW, blackout lighting switch assembly (DM-5944) (Willys only)					1
	WO-A-1352		KNOB, w/set SCREW, instrument panel light switch assembly (DM-5999) (Willys only)	30	40	10	10	1
	WO-111062	FM-33798-S7	NUT, starting switch terminal (hex., S., thin, $\frac{5}{16}$-24NF-2)					2
	WO-52131	FM-26457-S7	SCREW, rd-hd., S., No. 8 (.164)-32NC x $\frac{1}{4}$ (instrument light switch, blackout driving light switch terminal) (DM-50033)					3
18-1	WO-A-5197	FM-31037-S7	SCREW, fill-hd., S., No. 8 (.164)-32NC x $\frac{5}{16}$ (lighting switch terminal)					7
	WO-A-6149	FM-GTB-13739	SWITCH, blackout driving light, assembly (DM-D-398) (Includes: KNOB, w/set SCREW ATTACHING PARTS)		8	2	4	1
	WO-A-1351		SWITCH, blackout driving light, assembly (after serial 163750) (DM-6000) (Willys only) (Less: KNOB, w/set SCREW)					1
	WO-A-1353		NUT, mounting, hex., S., $\frac{3}{8}$-24NF-2 (DM-53414) (Willys only)					1
	WO-52332		WASHER, lock, shakeproof, $\frac{3}{8}$ (DM-52947) (Willys only)		24	5	5	1
18-1	WO-638979	FM-GPW-13532	SWITCH, dimmer, headlight foot assembly (CL-9654) (use until exhausted then use No. A-12056)				10	1
	A-12056		SWITCH, dimmer headlight, foot, assembly (was WO-638979)					1
	WO-51492	FM-26483-S	SCREW, rd-hd., S., $\frac{1}{4}$-20NC-2 x $\frac{1}{2}$ (GM-113955)					3
	WO-52706	FM-34805-S	WASHER, lock, S., $\frac{1}{4}$ in. (BECX1G)					3
18-1	WO-A-1345		SWITCH, lighting (DM-5969) (Willys only)					1
	WO-A-1332	FM-GPW-11649	SWITCH, lighting, assembly (DM-5970) (Includes: BREAKER, circuit / CONTROL, lockout / KNOB, w/set SCREW ATTACHING PARTS)		15	3	6	1
18-1	WO-A-7225	FM-9N-11450-A	SWITCH, starting (AL-SW-4015)		36		29	1
	WO-51732	FM-20309-S7	BOLT, hex-hd., S., $\frac{1}{4}$-28NF-2 x $\frac{1}{2}$ (switch to floor pan) (BAOK4AA)					2
	WO-5914	FM-33796-S2	NUT, hex., S., $\frac{1}{4}$-28NF-2 (GM-103024) (BBBX1AA)					2

GROUP 06—ELECTRICAL (Cont'd)

Figure Number Col. 1	Official Stockage Number Col. 2	Part Number Willys Col. 3	Part Number Ford Col. 4	ITEM Col. 5	Quantity Reqd. per Unit Assy. Col. 6	12 Mos. Field Maintenance Col. 7	Major Overhaul (5th Ech) Col. 8	Total First Year Procurement Col. 9	Estimated Reqmts. per 100 Rebuilds Col. 10
				0606—SWITCHES (Cont'd)					
		WO-52031	FM-34805-S2	WASHER, lock, S., 1/4 in. (BECX1G)	2	—	—	—	—
		WO-A-1346		NUT, mounting, special, hex, S., 3/8-24NF-2 (DM-53175) (Willys only)	1	—	—	—	—
18-1		WO-A-1333	FM-GPW-13740	SWITCH, panel light (DM-5995)	1	% 2	2	6	—
		WO-A-1353		NUT, mounting, hex, S., 3/8-24NF-2 (DM-53414) (Willys only)	1	—	—	—	—
		WO-52332		WASHER, lock, S., 3/8 in. (DM-52947) (Willys only)	1	—	—	—	—
12-3		WO-A-1271	FM-11A-13480	SWITCH, stop light	1	% 4	2	12	—
		WO-A-1350	FM-356229-S	WASHER, lock, keyed, shakeproof, S., pk. zd., 17/16 in. (2 used blackout driving lamp switch terminal screw) (7 used blackout lighting switch) (DM-51104)	10	% 10	10	100	—
		WO-52424	FM-34806-S7	WASHER, lock, shakeproof, 5/16 in. (starting switch terminal)	2	—	—	—	—
				0606B—CHASSIS WIRING					
06-5		WO-A-6153	FM-GPW-13181	CABLE, blackout driving light connector to switch (black—2 white tracings) (after serial Willys 163750)	1	—	—	—	—
		WO-A-5078	FM-GPW-14457	CABLE, filter (batt.) to starting switch (before Willys serial 288835)	1	—	—	—	—
		WO-A-5079	FM-GPW-14456	CABLE, filter (coil pri.) to junction block (before Willys serial 288835)	1	—	—	—	—
		WO-A-5073	FM-GPW-14459	CABLE, filter (reg. batt.) to junction block (before Willys serial 288835)	1	—	—	—	—
		WO-A-5041	FM-GPW-18846	CABLE, voltage regulator to generator ("G" cable bond No. 6)	1	% 3	4	4	—
		WO-A-5082	FM-GPW-14465	CABLE, voltage regulator to junction block (before Willys serial 000000)	1	% 3	5	5	—
		WO-A-5598	FM-GPW-14561	CLIP, S., open 1/4 in, bolt hole 7/8 in.	8	—	—	—	—
		WO-A-5449	FM-GPW-2281	CLIP, S., open 5/16 in, bolt hole 7/32 in. (on body floor)	13	—	—	—	—
		WO-6352	FM-355836-S7	NUT, reg., hex, S., No. 10 (.190)-24NC-2 (GM-110633) (BBKX2BC)	11	—	—	—	—
		WO-5064	FM-355131-S	SCREW, machine, rd-hd., S., No. 10 (.190)-24NC-2 x 5/8 (BCNX2AC)	11	—	—	—	—
		WO-52889	FM-32866-S2	SCREW, sheet metal, pkzd., No. 10 (.190) x 5/8 clip to body floor	1	% 10	10	100	—
		WO-5580	FM-31866-S	SCREW, rd-hd., wood, No. 10 (.190) x 5/8 clip to floor	1	—	—	—	—
		WO-52221	FM-34803-S2	WASHER, lock, S., No. 10 (.190) (BECX1E)	11	—	—	—	—
		WO-A-5450	FM-GPW-9295	CLIP, S., open, 7/16 in. bolt hole 7/8 in.	6	—	—	—	—
		WO-6352	FM-355836-S7	NUT, hex, S., No. 10 (.190)-24NC-2 (GM-110633) (BBKX2BC)	6	—	—	—	—
		WO-5064	FM-355131-S	SCREW, rd-hd., S., No. 10 (.190)-24NC-2 x 5/8 (BCNX2BC)	6	—	—	—	—
		WO-52221	FM-34803-S2	WASHER, lock, S., No. 10 (.190) (BECX1E)	6	—	—	—	—
		WO-A-1289	FM-GPW-14589	CLIP, S., closed 7/16, bolt hole 13/32 (anchored at foot rest to floor pan bolt)	1	—	—	—	—
		WO-52768	FM-356305-S	WASHER, S., plain, 5/16 in.	1	—	—	—	—
		WO-78932	FM-GPW-14566	CLIP, S., z-pltd., 5/8, bolt hole 1/4 (1 req'd before 24,308 Willys trucks)	2	—	—	—	—
		WO-6352	FM-355836-S7	NUT, hex, S., No. 10 (.190)-24NC-2	2	—	—	—	—
		WO-5064	FM-355131-S	SCREW, machine, rd-hd., S., No. 10 (.190)-24NC x 5/8	2	—	—	—	—
		WO-52221	FM-34803-S2	WASHER, lock, S., No. 10 (.190) (BECX1E)	2	—	—	—	—
06-5		WO-635985	FM-GPW-14487-A	CONNECTOR (3 wire)	2	% 5	5	12	—
06-5		WO-635981	FM-GPW-14487-B	CONNECTOR (2 wire)	1	% 5	5	12	—

	Stock No.	Part No.	Description					
06-5	WO-662276	FM-GPW-13437-A	GROMMET, ⅜ in. (wiring harness through dash) (issue until stock exhausted)	1	% 4	8	—	—
	WO-345961	FM-GPW-13434-A	GROMMET, 1¼ in. (wiring harness through dash) (trailer coupling socket cable through floor pan) (issue until stock exhausted)	4	% 16	24	—	—
	WO-A-7824	FM-GPW-14305	HARNESS, wiring generator to voltage regulator and filter	1	% 3	5	4	—
	WO-A-5048	FM-GPW-14401-D	HARNESS, wiring body, long (issue until stock exhausted) Then use WO-A-9220 for maintenance before Willys serial 288835	1	% —	3	—	—
06-5	WO-9220		HARNESS, wiring, body, long (after Willys serial 288835)	1	% —	2	5	—
	WO-A-6154	FM-GPW-14402-B	HARNESS, wiring, body, left side, short (was WO-A-5048)	1	% 2	2	—	—
	WO-A-7823	FM-GPW-14432-B	HARNESS, wiring, body, right side (after Willys serial 288835)	1	% —	2	5	—
06-5	WO-A-5061	FM-GPW-14446	HARNESS, wiring, chassis, left side (after Willys serial 163750)	1	% 2	2	—	—
	WO-A-7824	FM-GPW-14305-C	HARNESS, wiring, generator to voltage regulator (after Willys serial 238835) .....(Willys only)	1				
06-5	WO-A-1665	FM-GPW-14425	HARNESS, wiring, head light (before Willys serial 238835 (issue until stock exhausted)	1	% 4	8	—	—
	WO-A-7845	FM-GPW-18361-B	KIT, repair, wiring harness	as req.				

(Includes:

	Stock No.	Part No.	Description
2	WO-A-5078	FM-GPW-14457	CABLE (filter to starter switch)
1	WO-A-5079	FM-GPW-14456	CABLE (filter to junction block)
1	WO-A-5073	FM-GPW-14459	CABLE (filter to junction block)
1	WO-A-5070	FM-GPW-14406	CABLE (gas gauge to tank)
1	WO-A-5080	FM-GPW-14416	CABLE (gas gauge to circuit breaker)
1	WO-A-5081	FM-GPW-14409	CABLE (horn to junction block)
1	WO-A-5072	FM-GPW-14458	CABLE (ignition switch to ammeter)
1	WO-A-5082	FM-GPW-14465	CABLE (voltage regulator to junction block)
1	WO-A-5074	FM-GPW-14305	HARNESS, wiring (filter)
1	WO-A-5048	FM-GPW-14401-C	HARNESS, wiring (body, long)
1	WO-A-5061	FM-GPW-14446	HARNESS, wiring (chassis)
1	WO-A-1665	FM-GPW-14425	HARNESS, wiring (head light)

0606D—CIRCUIT BREAKERS

	Stock No.	Part No.	Description					
	WO-A-1733	FM-GPW-12250A	BREAKER, circuit, 5 amp. (between ignition switch and gas gauge on instrument board)	1	% 4	2	10	—
	WO-A-1734	FM-GPW-12250-B	BREAKER, circuit, 15 amp. (between ammeter and horn)	1	% 4	2	10	—
	WO-A-1349	FM-GPW-12250-C	BREAKER, thermal, circuit, 30 amp. (included w/lighting SWITCH, assembly (DM-53097-A)	1	% 4	2	9	—
	WO-52131	FM-26496-S2	SCREW, S., rd-hd., mach., cd. or z. pltd., No. 8 (.164)-32NC-2 x ¼ (breaker terminal) (GM-122159) (BCNX1FC-15)	4				
	WO-52652	FM-34802-S2	WASHER, lock, S., cd-pltd, No. 8 (.164) (breaker terminal screw) (BCEXID-15)	4				

0606E—JUNCTION BLOCKS

	Stock No.	Part No.	Description					
	WO-639599	FM-GPW-14448-C	BLOCK, junction, 2 post (headlight wires) (before Willys serial)	2	% 2	2	6	—
	WO-A-1490	FM-GPW-14448-D	BLOCK, junction, 6 post (body wiring harness)	1	% 2	2	6	—
	WO-6352	FM-355836-S7	NUT, hex, S., No. 10 (.190)-24NC-2 (GM-110633) (BBKX2BC)	6				
	WO-5064	FM-355131-S	SCREW, rd-hd., S., No. 10 (.190)-24NC-2 x ⅜ (BCNX2BC) (block to fender splasher)	6				

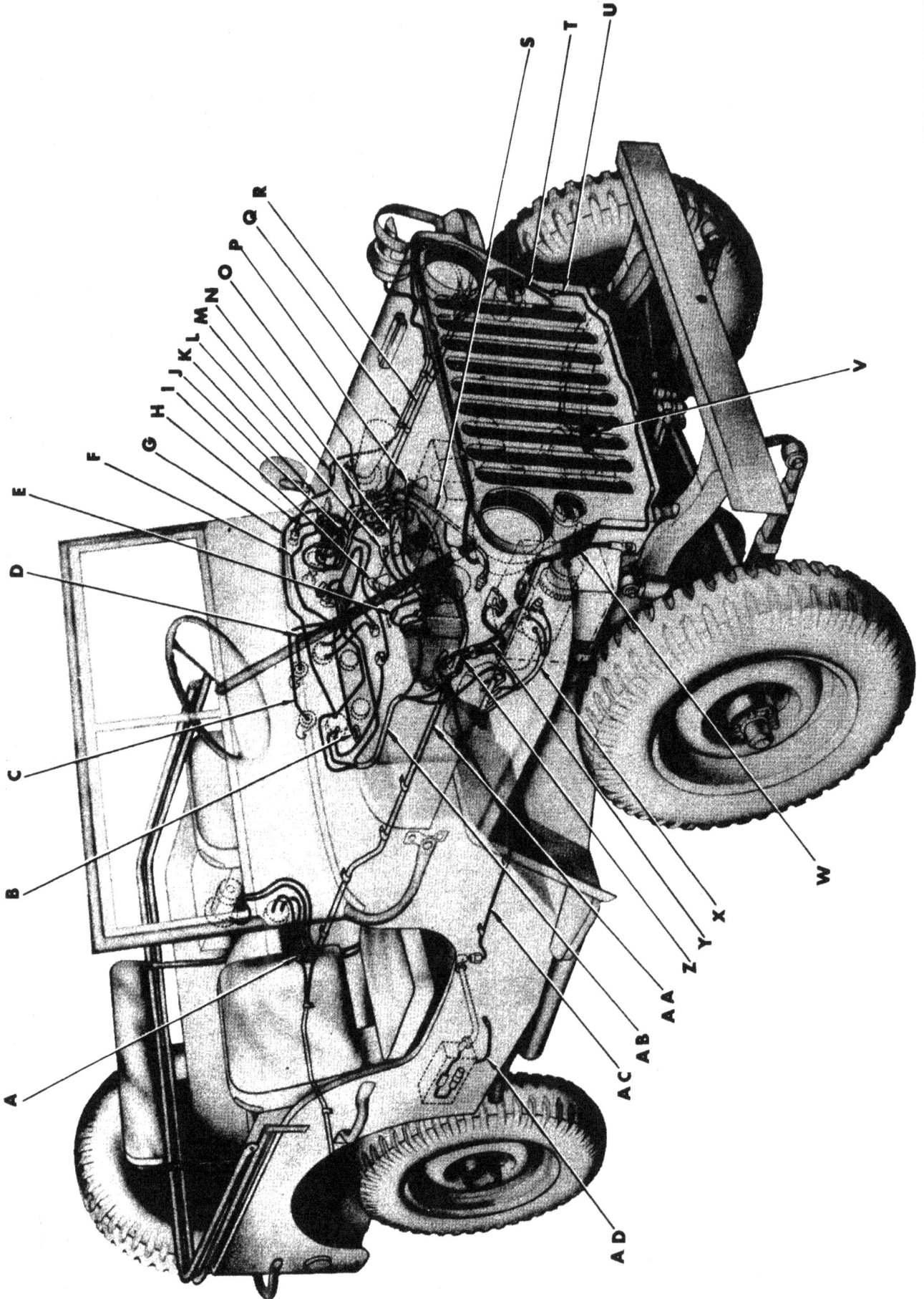

RA PD 305078

A B C D E F G H I J K L M N O P Q R S T U V W X Y Z AA AB AC AD

FIGURE 06-5—WIRING INSTALLATION

Key	Item	Willys Part No.	Ford Part No.	Gov't Group No.
A	CONNECTOR	WO-635985	FM-GPW-14487-A	0606B
B	GAUGE	WO-A-1292	FM-GPW-9276	0300A
C	SOCKET w/CABLE	WO-A-1411	FM-GPW-13710	0605C
D	CABLE	WO-A-1420	FM-GPW-12298-B	0604A
E	CABLE	WO-A-5054	(included w/WO-A-5048)	
F	CABLE	See	FM-GPW-14401-D	0606B
G	CABLE	WO-A-5072	FM-GPW-14458	0605B
H	CABLE	WO-A-6154	FM-GPW-14402-B	0606B
I	CABLE	WO-A-1418	FM-GPW-12286	0604A
J	HARNESS	WO-A-6154	FM-GPW-14402B	0606B
K	CABLE	WO-A-6153	FM-GPW-13181	0606B
L	CABLE	WO-635981	FM-GPW-12283	0604A
M	CONNECTOR	WO-A-1414	FM-GPW-14487-B	0606B
N	CABLE	WO-A-5081	FM-GPW-12284	0604A
O	CABLE		FM-GPW-14409	0609B
P	CABLE	WO-A-1412	FM-GPW-12287	0604A
Q	CABLE	WO-A-6146	FM-GPW-13175	0607D
R	HARNESS	WO-A-1665	FM-GPW-14425	0606B
S	HARNESS	WO-A-5061	FM-GPW-14446	0606B
T	CABLE	See WO-A-1436	FM-GPW-13201	0607D
U	CABLE	See WO-A-1437	FM-GPW-13200	0607D
V	STRAP	WO-A-1098	FM-GPW-14303	2602
W	STRAP	WO-635883	FM-GPW-14301	0610A
X	HARNESS	WO-A-5074	FM-GPW-14305	0606B
Y	CABLE	WO-A-1454	FM-GPW-14431	0610A
Z	CABLE	WO-A-1452	FM-GPW-14300	0610A
AA	HARNESS	WO-A-5048	FM-GPW-14401-D	0606B
AB	HARNESS	WO-A-5981	FM-GPW-14432	0605B
AC	CABLE	WO-A-8113	FM-GPW-14480-B	2601
AD	CABLE	WO-A-7640	FM-GPW-14513	2601

RA PD 305078-A

FIGURE 06-6—TRAILER ELECTRIC COUPLING SOCKET

Key	Item	Willys Part No.	Ford Part No.	Gov't Group No.
A	NUT	WO-53061	FM-34079	0606F
B	WASHER	WO-53024	FM-351279-S7	0606F
C	WASHER	WO-53313	FM-34805-S8	0606F
D	BOLT	WO-53069	(Willys only)	0606F
E	BOLT	WO-52921	(Willys only)	0606F
F	COVER assembly	WO-A-6587	FM-11YS-18149-B	0606F
G	RETAINER	WO-A-6588	FM-11YS-18198-B	0606F
H	SHIELD	WO-A-6589	FM-11YS-18193-B	0606F
I	SOCKET assembly	WO-A-6586	FM-11YS-18151-B	0606F

RA PD 305079-A

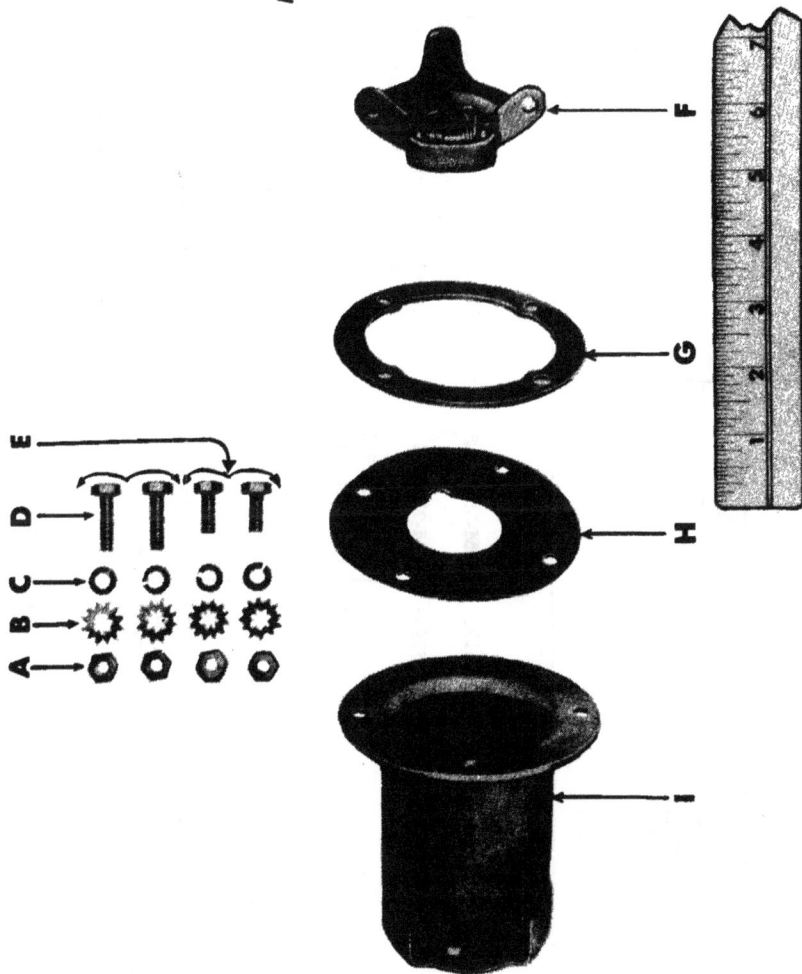

RA PD 305079

GROUP 06—ELECTRICAL (Cont'd)

Figure Number (Col. 1)	Official Stockage Number (Col. 2)	Part Number — Willys (Col. 3)	Part Number — Ford (Col. 4)	ITEM (Col. 5)	Quantity Reqd. per Unit Assy. (Col. 6)	Per 100 Major Items — 12 Mos. Field Maintenance (Col. 7)	Per 100 Major Items — Major Overhaul (5th Ech) (Col. 8)	Per 100 Major Items — Total First Year Procurement (Col. 9)	Estimated Reqmts. per 100 Rebuilds (Col. 10)
				0606E—JUNCTION BLOCKS (Cont'd)					
		WO-5272	FM-355132-S	SCREW, rd-hd., S., No. 10 (.190)-24NC-2 x ¾ (block to dash) (GM-110502) (BCNX2AG) (before utility serial 28835)	2	—	—	—	—
		WO-52994	FM-27698-S	SCREW, machine, rd-hd., S., No. 10 (.190)-32NF-2 x 5⁄16 (GM-117627) (BCCX1.1AC) (terminal to block)	10	% 10	10	100	—
		WO-A-1089	FM-356299-S	WASHER (terminal to block screw, special, S., ¼)	10	—	—	—	—
		WO-52221	FM-34803-S2	WASHER, lock, S., No. 10 (.190) (BECX1E)	16	—	—	—	—
				0606F—TRAILER ELECTRIC COUPLING					
06-6		WO-A-6356	FM-09B-14362	CABLE, trailer coupling socket to body ground	1	—	—	—	—
06-6		WO-A-6587	FM-11YS-18149-B	COVER assembly (WEB-11935-B)	1	—	—	—	—
		WO-A-8088		COVER, terminal (WEB-110242) (Willys only)	3	—	—	—	—
		WO-53061	FM-34079-S7	NUT, hex, S., cd-pltd., No. 10 (.190)-32NF-2 (socket terminal) (WEB-110477) (GM-120614)	8	—	—	—	—
06-6		WO-A-6588	FM-11YS-18198-B	RETAINER, dust shield (WEB-20099)	—	—	—	—	—
06-6		WO-A-6589	FM-11YS-18193-B	SHIELD, dust (WEB-20098)	1	—	—	—	—
06-6		WO-A-6586	FM-11YS-18151-B	SOCKET, trailer electric coupling, assembly (WEB-3529) (less COVER)	1	—	—	—	—
		WO-A-6019	FM-11YS-18142-B	SOCKET, trailer electric coupling, assembly (to accommodate trailer) (WEB-3604) (Includes: COVER / INTERNAL PARTS / ATTACHING PARTS)	1	—	—	—	—
06-6		WO-53069		BOLT, hex-hd., S., cd-pltd., ¼-28NF-2 x 1 (socket to body) (120364) (Willys only)	2	—	—	—	—
		WO-52921		BOLT, hex-hd., S., cd-pltd., ¼-28NF-2 x ¾ (socket to body) (GM-123450) (Willys only)	2	—	—	—	—
		WO-52847	FM-20364-S7	BOLT, hex-hd., S., cd-pltd., ¼-28NF-2 x ⅞ (Ford only)	4	—	—	—	—
		WO-53313	FM-33795-S7	NUT, hex., S., cd-pltd., ¼-28NF-2 (GM-120367)	4	—	—	—	—
		WO-53024	FM-34805-S8	WASHER, lock, S., z-pltd., ¼ in. S.A.E. (15⁄32 O.D. x .254 I.D. x 1⁄16 thk.)	4	—	—	—	—
		WO-A-8087	FM-351279-S7	WASHER, lock, S., cd-pltd., ¼ in. (internal, external tooth) (GM-174916)	4	—	—	—	—
		WO-53194	FM-34753-S	WASHER, terminal screw br, No. 10 (.190) (WEB-110110)	4	—	—	—	—
			FM-34903-S2	WASHER, lock, S., pkzd, shakeproof, No. 10 (.190) (GM-138534)	4	—	—	—	—

0607—HEAD LIGHTS

Fig.	Stock No.	Part No.	Name	Unit	%		
06-7	WO-A-1361	FM-GPW-13022	BOLT, mounting headlight (CB-CB-6012)	2	—	—	—
06-7	WO-A-2873	FM-GPW-13020	BOLT, hold down, headlight, w/NUT and WASHER, assembly (S., 5/16-18NC)	2	—	—	—
06-7	WO-A-1036	FM-GPW-13043	DOOR, headlight assembly (CB-CB-8523)	2	% 3	4	—
	WO-A-1731	FM-GPW-14436	CABLE, ground headlight, (issue until stock exhausted)	2	% 4	10	6
	WO-52993	FM-32852-S	SCREW, rd-hd., sheet metal, No. 10 (.190) x 3/8 (cable to bracket)	1	% 10	2	100
06-7	WO-A-1304	FM-GPW-13006-A	LIGHT, service head, left, assembly (CB-215142)	1	% 4	2	8
06-7	WO-A-1305	FM-GPW-13005-A	LIGHT, service head, right, assembly (CB-215242)	1	% 4	2	8
	WO-5916	FM-33846-S2	NUT, hex, S., 1/2-20NF-2 (to support bolt) (GM-103028) (BBBX1EA)	2	—	—	—
	WO-5009	FM-34809-S	WASHER, lock, S., 1/2 in. (BECX1M)	2	—	—	—
06-7	WO-A-2878	FM-GPW-13032	HINGE, head light support (issue until stock exhausted)	2	% 4	4	—
	WO-51485	FM-20326-S2	BOLT, hex-hd., S., 5/16-18NC-2 x 5/8 (to radiator guard) (GM-106324)	4	—	—	—
	WO-52350	FM-33797-S2	NUT, hex, S., 5/16-18NC-2	4	—	—	—
	WO-51840	FM-34806-S2	WASHER, lock, S., 5/16 in.	4	—	—	—
06-7	WO-A-5586	FM-GPW-13012	HOUSING, headlight, assembly (CB-CB-10436)	2	—	—	—
	WO-A-2466	FM-33896-S2	NUT, wing, S., 5/16-18NC-2 (headlight mounting bolt) (included w/WO-A-2873)	2	% 4	12	12
	WO-A-2875	FM-357419-S2	PIN, clevis, S., 25/64 lgth. x .248 diam., dld. f/c-pin (head light support bolt)	2	—	—	—
06-7	WO-A-1031	FM-GPW-13015	RETAINER, headlight mounting bolt (CB-CB-8010) (issue until stock exhausted)	2	—	—	—
06-7	WO-A-1032	FM-37206-S	SCREW, rd-hd., S. (CB-CB-4057)	2	% 10	4	—
	WO-52221	FM-34803-S	WASHER, lock, S. (CB-CB-300) (BCEX1E)	2	—	4	—
06-7	WO-A-1037	FM-38095-S	SCREW, headlight door (CB-CB-5099)	2	—	—	—
	WO-A-2870	FM-GPW-13071	SUPPORT, left, headlight, assembly (issue until stock exhausted)	1	—	—	—
	WO-A-2871	FM-GPW-13070	SUPPORT, right, headlight, assembly (issue until stock exhausted)	1	—	—	—
	WO-52768	FM-34706-S2	WASHER, S. plain 5/16 in. (BECX1G)	2	—	—	—
06-7	WO-A-1362	FM-GPW-13076	WIRE, left, headlight assembly	1	% 2	2	6
	WO-A-1363	FM-GPW-13075	WIRE, right, headlight assembly	1	% 2	2	6

NOTE: For support Bracket, see Group 1701.

0607A—SEALED BEAM

Fig.	Stock No.	Part No.	Name	Unit	%		
06-8	WO-A-6145	FM-GPW-13152	LAMP-UNIT, blackout, driving headlight, 6-8V., sealed, assembly (CB-CB-11267)	1	—	3	17
06-7	WO-A-1033	FM-GPW-13007	LAMP-UNIT, headlight, sealed 6V., assembly (CB-CB-8494) (Seelight unit)	2	% 10	10	46
06-10	WO-A-1074	FM-GPW-13494-A	LAMP-UNIT, service tail and stop, sealed, 21-3C.P., 6V., assembly (CB-CB-9218)	1	% 29	—	30
06-10	WO-A-1078	FM-GPW-13485-A	LAMP-UNIT, blackout stop, sealed 3 C.P. 6V., assembly (CB-CB-9234)	1	% 21	1	25
06-10	WO-A-1075	FM-GPW-13491-A	LAMP-UNIT, blackout tail, sealed 4 opening, 6V., assembly (CB-CB-9225)	2	% 19	—	30

0607B—LAMPS

Fig.	Stock No.	Part No.	Name	Unit	%		
06-9	WO-51804	FM-B-13466	LAMP, elec. incand., 6-8V., 3 C.P. single tung. fil. (MZ-63)	2	% 32	8	60
	WO-52837	FM-48-15021	LAMP, elec. incand., 6-8V., single tung. fil. (MZ-51)	2	% 16	8	60

0607D—BLACKOUT LIGHTS

NOTE: For blackout tail light see Group 0608.

Fig.	Stock No.	Part No.	Name	Unit	%		
06-8	WO-A-6146		CABLE, blackout driving light, assembly (before Willys serial 288835) (CB-CB-11503)	1	% 2	2	4

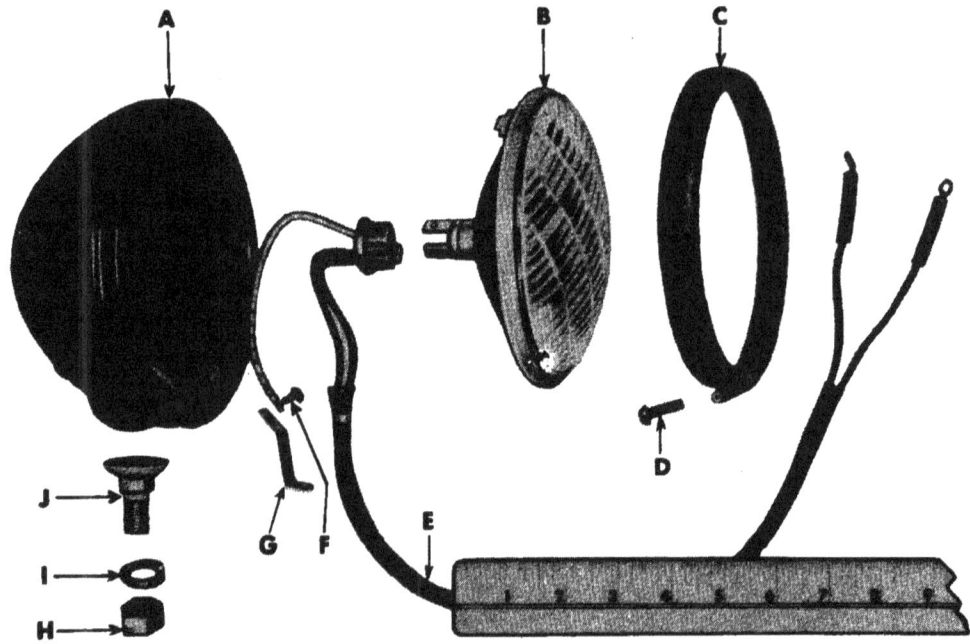

FIGURE 06-7—HEADLIGHT ASSEMBLY

Key	Item	Willys Part No.	Ford Part No.	Gov't Group No.	Key	Item	Willys Part No.	Ford Part No.	Gov't Group No.
A	HOUSING SUB assembly	WO-A-5586	FM-GPW-13012	0607	E	WIRE assembly	WO-A-1362	FM-GPW-13076	0607
B	SEELITE-UNIT	WO-A-1033	FM-GPW-13007	0607A	F	SCREW	WO-A-1032	FM-32706-S	0607
C	DOOR assembly	WO-A-1036	FM-GPW-13043	0607	G	RETAINER	WO-A-1031	FM-GPW-13022	0607
D	SCREW	WO-A-1037	FM-38095-S	0607	H	NUT	WO-5916	FM-33846-S2	0607
					I	WASHER	WO-5009	FM-34809-S	0607
					J	BOLT	WO-A-1361	FM-GPW-13022	0607

RA PD 305082

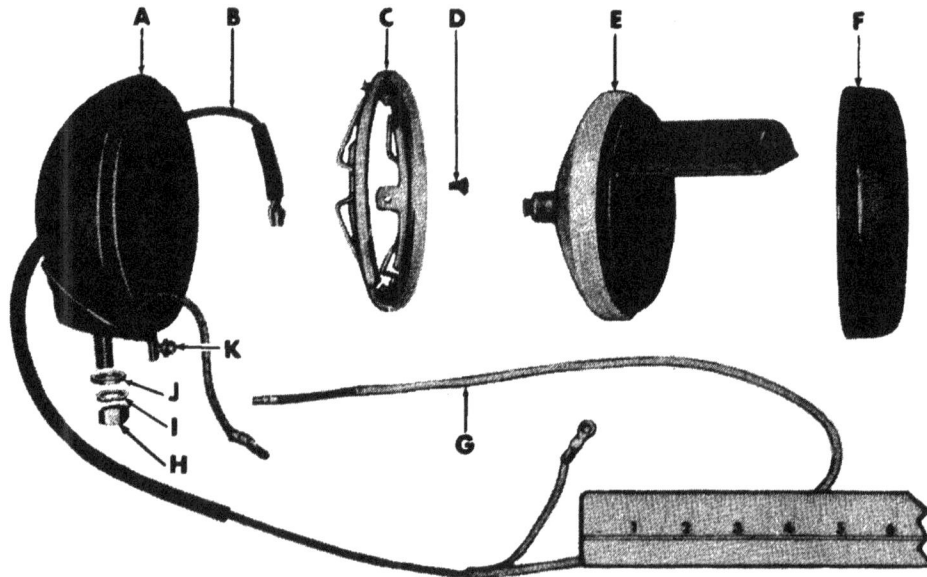

FIGURE 06-8—BLACKOUT DRIVING LIGHT ASSEMBLY

Key	Item	Willys Part No.	Ford Part No.	Gov't Group No.	Key	Item	Willys Part No.	Ford Part No.	Gov't Group No.
A	HOUSING assembly	WO-A-6143	FM-GPW-13170	0607D	F	DOOR assembly	WO-A-6144	FM-GPW-13162	0607
B	WIRE	WO-A-6147	FM-GPW-13174	0607D	G	CABLE assembly	WO-A-6146	FM-GPW-13175	0607
C	RING assembly	WO-A-6783	FM-GPW-13166	0607D	H	NUT	WO-52033	FM-33799-S2	0607
D	SCREW	WO-53071	FM-40631-S7	0607D	I	WASHER	WO-52046		0607
E	SEALED-UNIT assembly	WO-A-6145	FM-GPW-13153	0607A	J	WASHER	WO-A-6313	FM-GPW-13180	0607
					K	SCREW w/WASHER	WO-A-6148	FM-131045-S2	0607

RA PD 305080

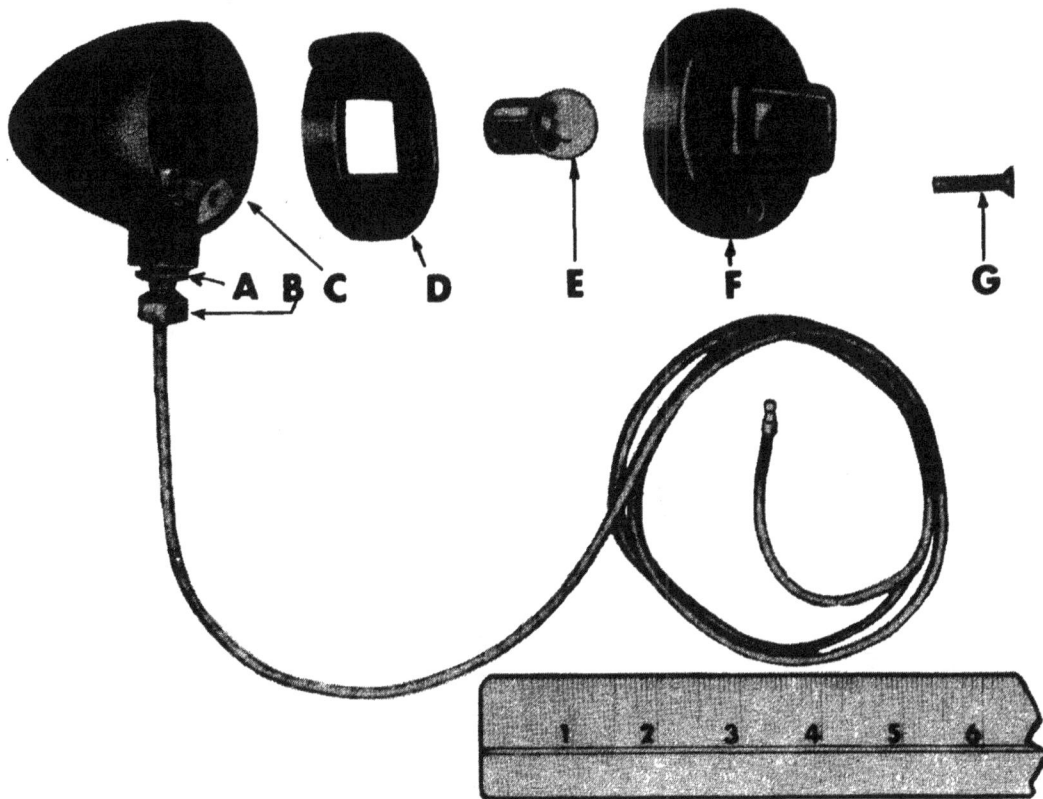

RA PD 305081

FIGURE 06-9—MARKER LIGHT

Key	Item	Willys Part No.	Ford Part No.	Gov't Group No.	Key	Item	Willys Part No.	Ford Part No.	Gov't Group No.
A	NUT	WO-5910	FM-33798-S	0607D	D	GASKET	WO-A-1071	FM-GPW-13209-B2	0607D
B	WASHER	WO-52510	FM-34941-S	0607D	E	LAMP	WO-51804	FM-B-13466	0607B
C	HOUSING assembly	WO-A-1439	FM-GPW-13217	0607D	F	DOOR	WO-A-1070	FM-GPW-13210-B	0607D
					G	SCREW	WO-A-1072	FM-28378-S2	0607D

RA PD 305081-A

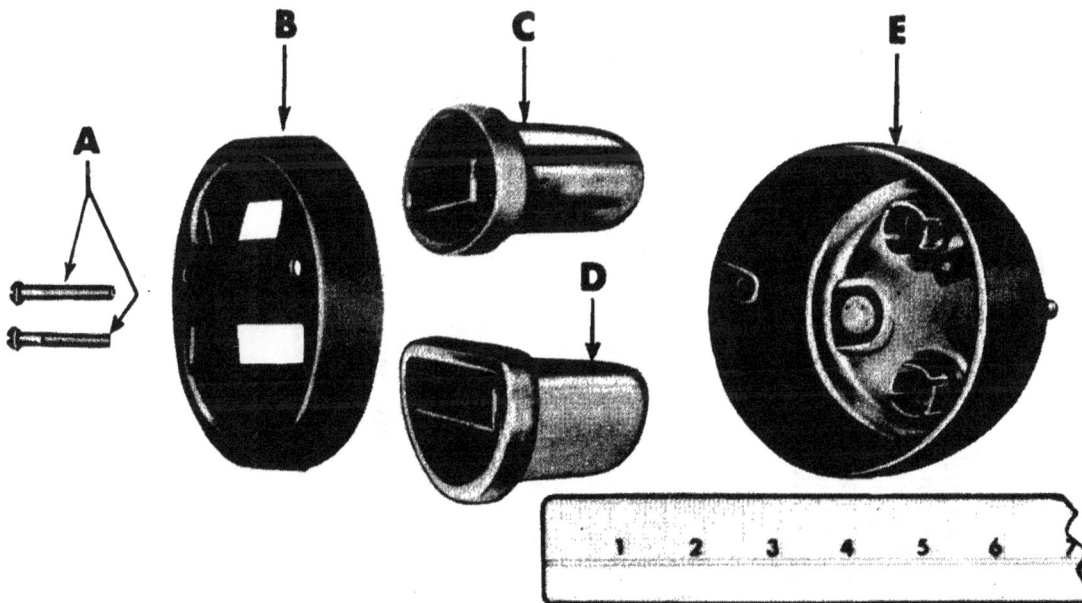

RA PD 305083

FIGURE 06-10—TAIL AND STOP LIGHT ASSEMBLY

Key	Item	Willys Part No.	Ford Part No.	Gov't Group No.	Key	Item	Willys Part No.	Ford Part No.	Gov't Group No.
A	SCREW	WO-A-1077	FM-36931-S	0608	D	LAMP-UNIT	WO-A-1075	FM-GPW-13491-A	0607A
B	DOOR	WO-A-1076	FM-GPW-13448-B	0608	E	HOUSING SUB assembly	WO-A-1073	FM-GPW-13408-B	0608
C	SERVICE UNIT	WO-A-1074	FM-GPW-13494	0607A					

RA PD 305083-A

SNL G-503

GROUP 06—ELECTRICAL (Cont'd)

| Figure Number | Official Stockage Number | Part Number | | ITEM | Quantity Reqd. per Unit Assy. | Per 100 Major Items | | | Estimated Reqmts. per 100 Rebuilds |
| | | Willys | Ford | | | 12 Mos. Field Maintenance | Major Overhaul (5th Ech) | Total FirstYear Procurement | |
Col. 1	Col. 2	Col. 3	Col. 4	Col. 5	Col. 6	Col. 7	Col. 8	Col. 9	Col. 10
				0607D—BLACKOUT LIGHTS (Cont'd)					
06-9		WO-A-11770	FM-GPW-13175-C	CABLE, blackout driving light, assembly (after Willys serial 288835)....	1	—	—	—	—
06-8		WO-A-1070	FM-GP-13210-B	DOOR, marker light, assembly (CB-9204) (issue until stock exhausted)....	2	—	4	—	—
06-9		WO-A-6144	FM-GPW-13162	DOOR, blackout driving light, assembly (CB-CB-11193)....	1	⅓ 2	2	4	—
06-9		WO-A-1071	FM-GPW-13209	GASKET, marker light door (CB-CB-9281) (issue until stock exhausted)....	2	⅓ 4	4	—	—
		WO-A-1439	FM-GPW-13217	HOUSING, marker light, w/WIRE, left, assembly (CB-CB-10461)....	1	—	—	—	—
		WO-A-1440	FM-GPW-13216	HOUSING, marker light, w/WIRE, right, assembly (CB-CB-10462)....	1	—	—	—	—
06-8		WO-A-6143	FM-GPW-13170	HOUSING, blackout driving light, assembly (CB-CB-11500)....	1	—	—	—	—
06-9		WO-A-1436	FM-GPW-13201	LIGHT, marker, left, assembly (CB-415342)....	1	⅓ 2	2	4	—
06-9		WO-A-1437	FM-GPW-13200	LIGHT, marker, right, assembly (CB-415442)....	1	⅓ 2	2	4	—
		WO-5910	FM-33798-S	NUT, hex, S, 5/16-24NF-2 (to support) (GM-103025) (BBBX1BA)....	2	—	—	—	—
		WO-52510	FM-34941-S	WASHER, lock, S, 5/16 in. (after 8,430 Willys trucks)....	2	—	—	—	—
06-8		WO-A-6142	FM-GPW-13150	LIGHT, driving, blackout assembly (CB-11495) (after Willys serial 163750) (before Willys serial 288835)....	1	⅓ 2	2	7	—
		WO-A-11768	FM-GPW-13150-C	LIGHT, driving, blackout assembly (after Willys serial 288835)....	1	—	—	—	—
		WO-52033	FM-33799-S7	NUT, hex, S, 3/8-16NC-2 (GM-123228)....	2	—	—	—	—
		WO-52046	FM-39807-S7	WASHER, lock, S, 3/8 in. (GM-115093)....	2	—	—	—	—
		WO-A-5806		NIPPLE, rubber marker (Willys only)....	2	—	—	—	—
				PAD, mounting, marker (CB-B-9276) (Willys only)....	2	—	—	—	—
06-8		WO-A-6783	FM-GPW-13166	RING, mounting, blackout driving light unit mounting (CB-CB-11199)....	1	⅓ 2	2	8	—
06-9		WO-A-1072	FM-28378-S2	SCREW, marker light door (CB-7861)....	2	⅓ 6	—	12	—
06-8		WO-53071	FM-40631-S7	SCREW, blackout driving light ground (CB-CB-7783) (GM-125760)....	1	—	—	12	—
06-8		WO-A-6148	FM-131045-S2	SCREW, marker, light door, w/WASHER, assembly (CB-CB-11263)....	1	⅓ 3	3	—	—
06-8		WO-A-6313	FM-GPW-13180	WASHER, adjusting, blackout driving light....	1	—	—	—	—
		WO-53070		WASHER, cd-pltd, S, internal teeth #8, ground screw (GM-121752) (Willys only)....	1	—	—	—	—
06-8		WO-A-6147	FM-GPW-13174	WIRE, ground, blackout driving light, assembly (CB-CB-11200)....	1	—	—	—	—
				0608—TAIL LIGHT					
06-10		WO-A-719	FM-GPW-13410	CABLE, blackout tail lamp to connector....	1	⅓ 2	2	4	—
		WO-A-1076	FM-GP-13448-B2	DOOR, blackout, tail and service tail and stop light, assembly (CB-CB-9231) (issue until stock exhausted)....	1	—	—	—	—
		WO-A-1079	FM-GP-13449-A	DOOR, blackout tail and service tail and stop light (CB-CB-9232) (issue until stock exhausted)....	1	⅓	4	—	—
06-10		WO-A-1073	FM-GPW-13408-B	HOUSING, tail and stop light, assembly (CB-CB-9212)....	1	—	4	—	—
		WO-A-1064	FM-GP-13405-B	LIGHT, blackout tail and service tail and stop, 6V, 7/8 std. lgh., assembly (CB-60142)....	2	⅓ 4	2	8	—

					DESCRIPTION		
—	8	2	%4	1	LIGHT, blackout tail and blackout stop, 6V., ⅞ std. lgh, assembly (CB-60242)	WO-A-1065	FM-GP-13404-B
	—	—	—	4	NUT, hex, S., ¼-20NC-2 (to bracket) (GM-109084) (BBAX1A)	WO-5790	FM-33795-S7
	—	—	—	4	WASHER, lock, S., ¼ in. (BECX1G)	WO-52706	FM-34805-S7
	—	—	—	4	WASHER, lock, shakeproof, S., ¼ in.	WO-352760	FM-34905-S
	—	—	—	2	SCREW, rd-hd., S. (door screw) (CB-CB-9233)	WO-A-1077	FM-36931-S2

0609—HORNS

					DESCRIPTION		
—		8	%4	1	BRACKET, horn (issue until stock exhausted)	WO-A-1389	FM-GPW-13831
		—	—	2	BOLT, hex-hd. S., 5/16-24NF-2 x ¾ (bracket to dash) (GM-100013)	WO-51396	FM-24347-S2
		—	—	2	NUT, hex, S., 5/16-24NF-2 (GM-114493) (BBDX1BA)	WO-52954	FM-33909-S7
		—	—	2	WASHER, S., 5/16 in. (plain) (GM-106262)	WO-5437	FM-34706-S7
		—	—	2	WASHER, lock, S., 5/16 in. (BECX1H)	WO-51833	FM-34806-S2
		4	%4	1	HORN assembly (SPW-B-9427) (SZE-61400)	WO-A-1312	FM-GPW-13802
		—	—	2	BOLT, hex-hd., S., ¼-28NF-2 x ⅝ (horn to bracket) (GM-106274)	WO-5920	FM-20325-S7
		—	—	2	WASHER, lock, S., ¼ in. (BECX1G)	WO-52706	FM-34805-S2
		—	—	2	NUT, hex, S., No. 8 (.164)-32NC-2 (adjusting)	WO-52182	FM-34052-S2
		—	—	2	WASHER, lock, S., No. 8 (.164) (adjusting nut) (BECX1D)	WO-52705	FM-34802-S2

0609B—HORN BUTTON AND WIRING

NOTE: For Wiring not appearing here see Group 0606B.

					DESCRIPTION		
	4	2	%2	1	BRUSH, horn wire contact (RG-032087) (GM-263549)	WO-A-302	FM-GPW-13836
	—	—	—	2	SCREW, rd-hd., mach. S., No. 8 (.164)-32NC-2 x ¼ (brush to column) (GM-100749)	WO-51040	FM-26457-S7
	—	—	—	2	WASHER, lock, S., No. 8 (.164)	WO-52754	FM-34902-S
	—	8	%4	1	BUTTON, horn (RG-450054)	WO-A-634	FM-GPW-3627
	6	2	%2	1	CABLE, horn to junction block (issue until stock exhausted) (RG-8287-32)	WO-A-5081	FM-GPW-14409
	—	—	—	1	CABLE, horn, assembly (RG-029046)	WO-A-752	FM-GPW-14171
	7	2	%2	1	CUP, horn button spring (RG-051035)	WO-A-752	FM-GPW-3646
	—	—	—	1	FERRULE, insulating (under horn button) (RG-051035)	WO-A-751	
	—	2	%2	as req.	KIT, repair, horn button (Includes: 1 WO-A-634 FM-GPW-3627 BUTTON 1 WO-638885 FM-GPW-3646 CUP, spring 1 WO-A-751 FM-GPW-3653-B FERRULE 1 WO-638884 FM-GPW-3626 SPRING 1 WO-A-750 FM-GPW-3631 WASHER, contact)	WO-A-6742	FM-GPW-18382
	—	—	—	1	NUT, horn button (also steering wheel nut. See Group 1404)		
	6	2	%2	1	RING, horn contact, assembly (RG-063991)	WO-A-747	FM-GPW-3652
	—	8	%4	1	SPRING, horn button (RG-401107) (issue until stock exhausted)	WO-638884	FM-GPW-3626
	—	—	—	1	WASHER, horn contact (RG-029049)	WO-A-750	FM-GPW-3631

0610—BATTERY

					DESCRIPTION		
—	103	40	%40	1	BATTERY, 6V., 15 plate, 116 amp. hr. (wet, charged) (U.S.L. type TS-2-15) (WB-SW-2-119) (Willys only)	WO-A-1238	
				1	BATTERY, 6V., 15 plate, 116 amp. hr. (dry, charged) (AL-TSR-2-15) (WB-SR-2-119)	WO-A-1767	FM-11AS-10658

06-10

18-1

GROUP 06—ELECTRICAL (Cont'd)

| Figure Number | Official Stockage Number | Part Number | | ITEM | Quantity Reqd. per Unit Assy. | Per 100 Major Items | | | Estimated Reqmts. per 100 Rebuilds |
| | | Willys | Ford | | | 12 Mos. Field Maintenance | Major Overhaul (5th Ech) | Total First Year Procurement | |
Col. 1	Col. 2	Col. 3	Col. 4	Col. 5	Col. 6	Col. 7	Col. 8	Col. 9	Col. 10
				0610—BATTERY (Cont'd)					
		WO-A-5433	FM-11AS-10657	BATTERY, 6V., 15 plate, 116 amp. hr. (wet dump) (unfilled and uncharged) (AL-TS-2-15)	1	—	—	—	—
		WO-A-11760		ELECTROLYTE, battery, and CONTAINER, assembly (use with dry charged battery, WO-A-1767, FM-11AS-10658) (PL-226027)	as req.	% 48	51	111	—
				0610A—BATTERY HANGERS					
				NOTE: For battery support see Group 1500.					
		WO-A-1291	FM-GPW-5165	FRAME, hold down, battery	1	—	—	—	—
		WO-A-1164	FM-GPW-5175	BOLT, S., 5/16-18NC-2 x 9 (battery clamp) (issue until stock exhausted)	2	% 2	2	6	—
		WO-300329	FM-33797-S2	NUT, hex., S., 5/16-18NC-2 (Ford only)	2	%	4	—	—
		WO-52768	FM-356305-S	NUT, wing, S., 5/16-18NC-2 (Willys only)	2	—	—	—	—
		WO-51833	FM-34806-S2	WASHER, S., 21/64 in. (plain)	2	—	—	—	—
		WO-A-1757	FM-GPW-5168	WASHER, lock, S., 5/16 in. (BECX1H)	1	—	—	—	—
		WO-A-3728	FM-33896-S2	STRAP, battery to front fender	1	—	—	—	—
		WO-52768	FM-356305-S	NUT, strap to fender bolt (wing, S., 5/16-18NC)	1	—	—	—	—
		WO-51833	FM-34806-S2	WASHER, S., 5/16 in. plain	1	—	—	—	—
		WO-323397	FM-23017-S16	WASHER, lock, S., 5/16 in. (BECX1H)	1	—	—	—	—
		WO-A-9093	FM-350343-S16	BOLT, terminal clamp (hex-hd., S., ld-pltd., 5/16-18NC x 1¼) (issued until stock exhausted)	2	% 8	16	—	—
		WO-335912	FM-350343-S16	BOLT, terminal clamp, w/NUT, assembly	2	—	—	—	—
				BOLT, terminal clamp w/NUT, assembly, (use until stock exhausted, then use WO-A-9093) (was WO-A-355912)					
		WO-A-1452	FM-GPW-14300	CABLE, battery to starting switch, positive	2	% 96	24	200	—
		WO-A-1454	FM-GPW-14431	CABLE, starting switch to cranking motor, (issue until stock exhausted)	1	% 3	3	9	—
		WO-A-9094	FM-34056-S16	NUT, hex., S., ld-pltd., 5/16-18NC (terminal clamp bolt)	2	% 4	4	—	—
		WO-A-1320		STRAP, battery ground (before Willys serial 120700) (Willys only)	1	—	—	—	—
		WO-635883	FM-GPW-14301	STRAP, battery ground (after Willys serial 120700) (all Ford trucks)	1	% 3	3	8	—
		WO-50151	FM-24369-S	BOLT, hex-hd., S., 3/8-24NF-2 x 7/8 (strap to frame) (GM-106285)	1	—	—	—	—
		WO-5901	FM-33800-S	NUT, hex., S., 3/8-24NF-2 (GM-103026) (BBBX1CA)	1	—	—	—	—
		WO-A-1680	FM-34747-S7	WASHER, S., cd-pltd. (plain) 3/8 in.	1	—	—	—	—
		WO-5010	FM-34807-S2	WASHER, lock, S., 3/8 in. (BECX1K)	1	—	—	—	—
		WO-371400	FM-11A-14452	TERMINAL, battery cable (11/32 in.)	1	% 5	5	12	—
		WO-372668	FM-19B-14451	TERMINAL, battery, positive cable clamp	1	% 5	5	12	—

GROUP 07—TRANSMISSION

0700—TRANSMISSION ASSEMBLY

0701—CASE

0703—MAIN DRIVE GEAR AND BEARINGS

Code	Stock No.	Part No.	Description	Units				
G	WO-A-7596		TRANSMISSION, w/gear shift LEVER, assembly (Willys only) (Includes: clutch release BEARING, clutch bearing carrier CARRIER, clutch bearing carrier SPRING, clutch control LEVER, and clutch control lever CABLE).	1	6	—	8	—
	WO-A-1145	FM-GPW-7000	TRANSMISSION, w/gearshift LEVER, assembly (WG-ASI-T-84-J) (issue until stock exhausted)	1	4	8	—	—
07-1	WO-A-1148	FM-GPW-7005	CASE, transmission, assembly (WG-T-84J-1A) (Includes: CUP, oil retaining PIN, countershaft washer)	1	—	3	4	10
	WO-637503	FM-GPW-7056	CUP, oil retaining, transmission (high and intermediate shift rail)	1	3	3	10	10
07-1	WO-637495	FM-GPW-7051-B	GASKET, transmission case to flywheel housing (Victorite, .015 thk.) (WG-T84D-145C)	1	38	—	40	—
	WO-A-1542	FM-GPW-13356	GASKET, SET, transmission (issue until stock is exhausted, then use WO-A-7832)	as req.	50	100	—	100
	WO-A-7832	FM-GPW-18356-B	GASKET, SET transmission (was WO-A-1542) (Includes: 1 WO-637495 FM-GPW-7051-B GASKET; 1 WO-635861 FM-GPW-7223 GASKET; 1 WO-640018 FM-GPW-7052 OIL SEAL; 1 WO-635862 FM-GPW-7207 PACKING)	as req.	18	26	48	—
	WO-5140	FM-353064-S	PLUG, pipe, I., sq-hd., ½ in. (drain and reqll)	2	—	—	—	—
07-1	WO-636885	FM-GPW-7025	BEARING, ball, annular, single row, single oil shield, snap ring groove (bore 2.834 O.D. x 1.378 I.D.) (transmission drive gear) (FB-1207-MGF) (WG-X-3204-ML)	1	3	10	15	30
	WO-639422	FM-GPW-7120	BEARING, roller, S., spherical end type .527 lgth. x .1875 diam. (transmission drive gear pilot) (BN-C-1110-Q) (WG-T-84G-26)	13	21	229	300	705
07-1	WO-A-5554	FM-GPW-7017	GEAR, main drive, S., 17 teeth (2.048 O.D.) (WG-T-84J-16A)	1	—	6	7	20
	WO-A-5553	FM-GPW-7015	GEAR, main drive, w/BEARINGS, assembly (WG-AT-84J-16A) (Includes: SNAP RINGS)	1	—	—	—	—
07-1	WO-640018	FM-GPW-7052	OIL SEAL, transmission front bearing retainer (WG-T-84J-54)	1	3	6	15	20
07-1	WO-640017	FM-GPW-7050	RETAINER, main drive gear bearing (WG-T-84J-6)	1	—	3	4	10
07-1	WO-639689	FM-20366-S	BOLT, hex-hd., S., 5/16-18NC-2 x 7/8 (WG-X-802)	3	—	—	—	—
07-1	WO-52510	FM-34941-S2	WASHER, lock, S., 5/16 in.	3	—	—	—	—
07-1	WO-635844	FM-GPW-7064	SNAP RING, main drive gear, 1-9/32 in. I.D. (WG-T-84-17)	1	3	22	30	70
07-1	WO-635846	FM-B-7070	SNAP RING, main drive gear bearing, 2-11/16 in. I.D. (WG-B-7070)	1	10	26	39	80
07-1	WO-639423	FM-GPW-7063	SNAP RING, main drive gear pilot roller bearing, 1.062 O.D. (WG-T-84G-25)	1	3	26	33	80

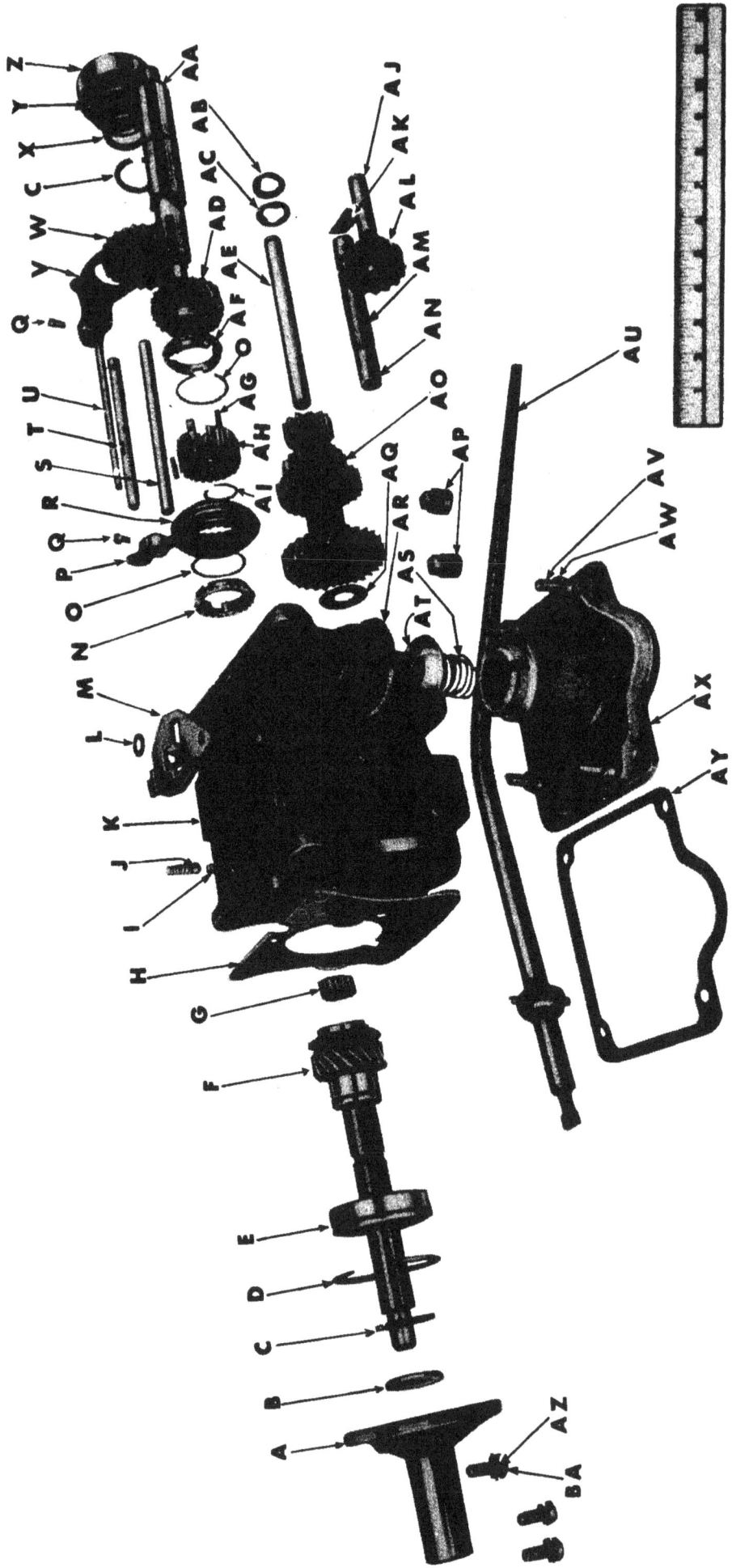

RA PD 305103

FIGURE 07-1—TRANSMISSION ASSEMBLY

Key	Item	Willys Part No.	Ford Part No.	Gov't Group No.
A	RETAINER	WO-640017	FM-GPW-7050	0703
B	OIL seal	WO-640018	FM-GPW-7052	0703
C	SNAP ring	WO-635844	FM-GPW-7064	0703
D	SNAP ring	WO-635846	FM-B-7070	0703
E	BEARING	WO-636885	FM-GPW-7025	0703
F	GEAR	WO-A-5554	FM-GPW-7017	0703
G	BEARING	WO-639422	FM-GPW-7120	0703
H	GASKET	WO-637495	FM-GPW-7051-B	0701
I	SPRING	WO-635837	FM-GPW-7234	0706B
J	BALL	WO-635838	FM-35081-S	0706B
K	CASE	WO-A-1148	FM-GPW-7005	0701
L	SPRING	WO-635839	FM-GPW-7208	0706C
M	PLATE assembly	WO-A-7260	FM-GPW-7211-B	0706C
N	RING	WO-637834	FM-GPW-7107	0704B
O	SPRING	WO-637831	FM-GPW-7109	0704B
P	FORK	WO-636196	FM-GPW-7230	0706C
Q	SCREW	WO-636200	FM-GPW-7245	0706B
R	SLEEVE	WO-637833	FM-GPW-7106	0704B
S	RAIL	WO-A-1155	FM-GPW-7241	0706B
T	RAIL	WO-A-1156	FM-GPW-7240	0706B
U	PIN	WO-635836	FM-GPW-7206	0706C
V	FORK	WO-636197	FM-GPW-7231	0706C
W	GEAR	WO-636879	FM-GPW-7100	0704B
X	SPACER	WO-A-738	FM-GPW-7062	0704
Y	WASHER	WO-A-410	FM-GPW-7080	0704
Z	BEARING	WO-A-916	FM-GP-7065	0704
AA	SHAFT	WO-A-519	FM-GPW-7061	0704
AB	WASHER	WO-A-879	FM-GPW-7126	0704C
AC	WASHER	WO-635811	FM-GPW-7129-A	0704C
AD	GEAR assembly	WO-638798	FM-GPW-7102	0704B
AE	COUNTERSHAFT	WO-638948	FM-GPW-7111	0704C
AF	RING	WO-637834	FM-GPW-7107	0704B
AG	PLATE	WO-637832	FM-GPW-7116	0704B
AH	HUB	WO-A-6319	FM-GPW-7105	0704B
AI	SNAP ring	WO-637835	FM-GPW-7059	0704B
AJ	SHAFT	WO-638952	FM-GPW-7140	0704C
AK	PLATE	WO-638949	FM-GPW-7155	0704B
AL	GEAR assembly	WO-636882	FM-GPW-7141	0704B
AM	SPACER	WO-A-880	FM-GPW-7115	0704C
AN	BUSHING	WO-A-878	FM-GPW-7121	0704C
AO	GEAR assembly	WO-A-739	FM-GPW-7113	0704B
AP	PLUG	WO-5140	FM-353064	0701
AQ	WASHER	WO-635812	FM-GPW-7119	0704C
AR	CAP	WO-A-1379	FM-BB-7220	0706A
AS	SPRING	WO-392328	FM-GPW-7227	0706A
AT	WASHER	WO-635863	FM-BB-7228	0706A
AU	LEVER assembly	WO-A-1380	FM-GPW-7210-A	0706A
AV	BOLT	WO-639689	FM-20366-S	0706A
AW	WASHER	WO-52045	FM-34806-S2	0706A
AX	HOUSING assembly	WO-635857	FM-GPW-7204	0706A
AY	GASKET	WO-635861	FM-GPW-7223	0706A
AZ	WASHER	WO-52510	FM-34941-S2	0703
BA	BOLT	WO-639689	FM-20366-S	0703

RA PD 305103-A

SNL G-503

GROUP 07—TRANSMISSION (Cont'd)

Figure Number (Col. 1)	Official Stockage Number (Col. 2)	Part Number — Willys (Col. 3)	Part Number — Ford (Col. 4)	ITEM (Col. 5)	Quantity Reqd. per Unit Assy. (Col. 6)	Per 100 Major Items — 12 Mos. Field Maintenance (Col. 7)	Per 100 Major Items — Major Overhaul (5th Ech) (Col. 8)	Per 100 Major Items — Total First Year Procurement (Col. 9)	Estimated Reqmts. per 100 Rebuilds (Col. 10)
				0704—MAIN SHAFT AND BEARINGS					
08-2		WO-A-916	FM-GP-7065	BEARING, ball annular, single row, bore 1.378 x .8268 (main shaft rear end) (MRC-307) (SKF-6307-Z) (ND-7607)	1	3	10	15	30
07-1		WO-640006	FM-GPW-7104-A2	BUSHING, main shaft second speed gear (WG-T-84C-19)	1	—	6	6	20
07-1		WO-A-519	FM-GPW-7061	SHAFT, main (dld. f/c-pin) (WG-T-84H-2)	1	2	5	9	15
07-1		WO-A-6317	FM-GPW-7060	SHAFT, main, w/GEARS, assembly (WG-1AT-84H-2)	1	—	—	—	10
07-1		WO-A-738	FM-GPW-7062	SPACER, main shaft bearing (WG-T-84J-28)	1	—	3	3	10
07-1		WO-A-410	FM-GPW-7080	WASHER, retaining, main shaft, oil (WG-T-184H-137)	2	3	10	15	30
				0704B—GEARS					
07-1		WO-A-739	FM-GPW-7113	GEARS, cluster, countershaft, w/BUSHING and SPACER, assembly (WG-AT-84J-8)	1	—	6	7	20
07-1		WO-636879	FM-GPW-7100	GEAR, sliding, mainshaft, low and reverse (O.D. 3.265 x 1¼ thk., 25 teeth) (WG-T-84F-12A)	1	—	6	7	20
07-1		WO-638798	FM-GPW-7102	GEAR, main shaft, second speed, w/BUSHING, assembly (O.D. 2.090, 24 teeth) (WG-T-84F-11A)	1	—	6	7	20
07-1		WO-636882	FM-GPW-7141	GEAR, reverse idler, w/BUSHING, assembly (O.D. 2.167 x .996 thk, 16 teeth) (WG-AT-84F-10A)	1	—	6	7	20
07-1		WO-A-6319	FM-GPW-7105	HUB, intermediate and high clutch (WG-T-84J-2½) (issue until stock exhausted)	1	4	8	—	—
07-1		WO-637832	FM-GPW-7116	PLATE, synchronizer shifting (WG-T-84F-13)	3	—	19	21	60
07-1		WO-637834	FM-GPW-7107	RING, synchronizer blocking (WG-T-84F-14)	2	—	10	11	30
07-1		WO-637833	FM-GPW-7106	SLEEVE, second and direct speed clutch (WG-T-84F-15) (issue until stock exhausted)	1	—	8	—	—
07-1		WO-637835	FM-GPW-7059	SNAP RING, high and intermediate clutch hub, I.D. 1½ (WG-4686)	1	4	22	27	70
07-1		WO-637831	FM-GPW-7109	SPRING, synchronizer (WG-4682-K)	2	3	10	10	30
07-1		WO-A-6318	FM-GPW-7124	SYNCHRONIZER UNIT, w/blocking RING, assembly (WG-1AT-84J-2½)	1	—	14	15	45
				0704C—COUNTERSHAFT AND REVERSE IDLER SHAFT					
07-1		WO-A-878	FM-GPW-7121	BUSHING, countershaft gear (WG-T-84J-167)	2	—	19	24	60
07-1		WO-635804	FM-GPW-7143	BUSHING, reverse idler gear (WG-T-84-85A)	1	—	6	6	20
07-1		WO-638948	FM-GPW-7111	COUNTERSHAFT (WG-T-84G-3)	1	—	3	4	10
07-1		WO-A-524	FM-357417-S	PIN, countershaft washer (WG-T-84J-31) (issue until stock exhausted)	1	8	16	—	—
07-1		WO-638949	FM-GPW-7155	PLATE, lock, countershaft and idler shaft (WG-T-84C-48)	1	—	3	3	10
07-1		WO-638952	FM-GPW-7140	SHAFT, reverse idler gear (WG-T-84G-35)	1	—	3	3	10
07-1		WO-A-880	FM-GPW-7115	SPACER, countershaft bushing (WG-T-84J-28) (issue until stock exhausted)	1	4	12	—	—
07-1		WO-635812	FM-GPW-7119	WASHER, thrust, countershaft, front (WG-T-84B-30A)	1	—	19	20	60

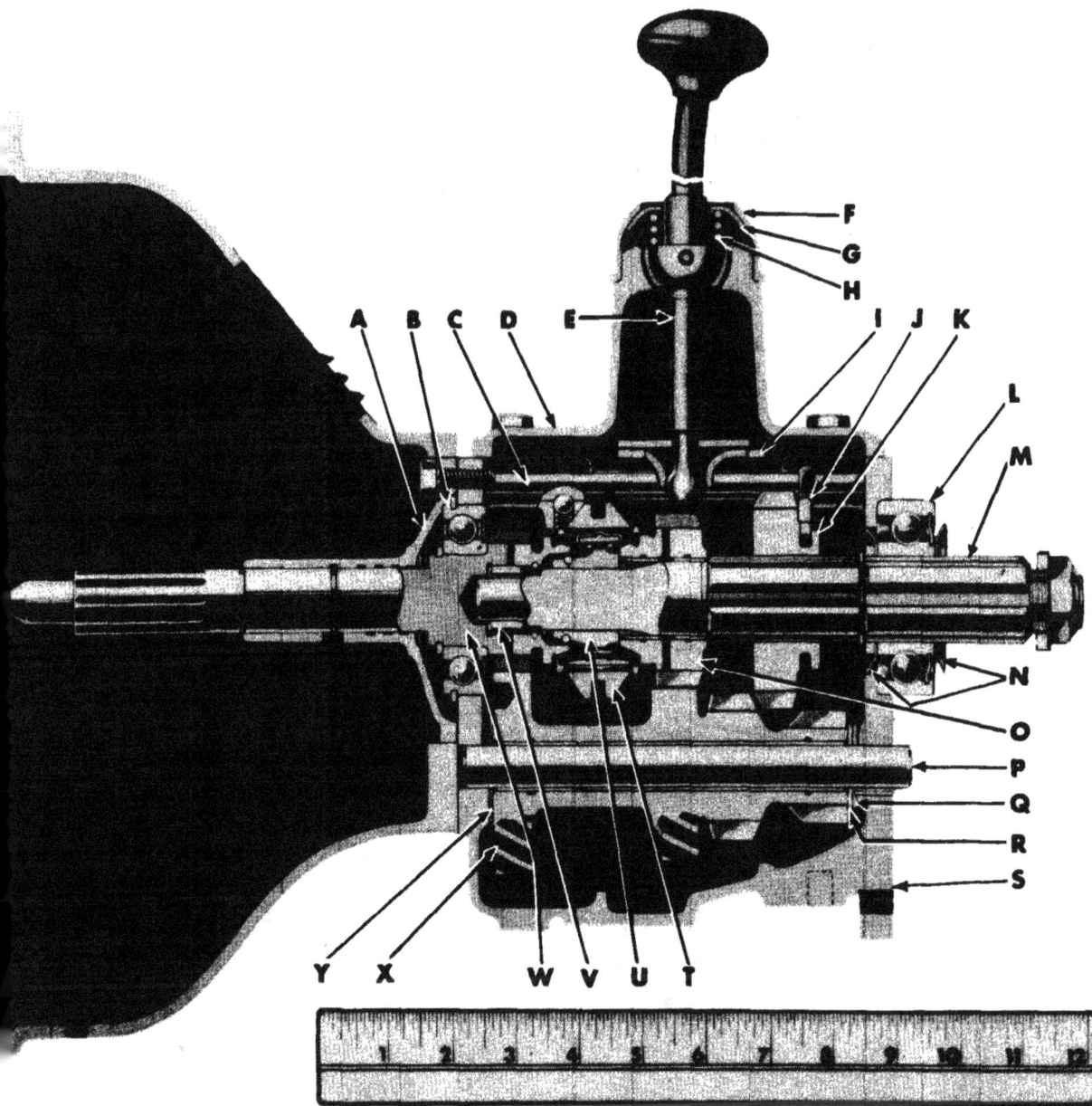

RA PD 305155

FIGURE 07-2—TRANSMISSION ASSEMBLY—CROSS-SECTIONAL VIEW

Key	Item	Willys Part No.	Ford Part No.	Gov't Group No.
A	RETAINER	WO-640017	FM-GPW-7050	0703
B	BEARING	WO-636885	FM-GPW-7025	0703
C	RAIL	WO-A-1156	FM-GPW-7240	0706B
D	HOUSING	WO-635857	FM-GPW-7204	0706A
E	LEVER	WO-A-1380	FM-GPW-7210-A	0706A
F	CAP	WO-A-1379	FM-BB-7220	0706A
G	WASHER	WO-635863	FM-BB-7228	0706A
H	SPRING	WO-392328	FM-GPW-7227	0706A
I	PLATE	WO-A-7260	FM-GPW-7214	0706A
J	FORK	WO-636197	FM-GPW-7231	0706C
K	GEAR	WO-636879	FM-GPW-7100	0704B
L	BEARING	WO-A-916	FM-GP-7065	0704
M	SHAFT	WO-A-519	FM-GPW-7061	0704
N	WASHER	WO-A-410	FM-GPW-7080	0704
O	GEAR	WO-638798	FM-GPW-7102	0704B
P	COUNTERSHAFT	WO-638948	FM-GPW-7111	0704C
Q	WASHER	WO-A-879	FM-GPW-7126	0704C
R	WASHER	WO-635811	FM-GPW-7129A	0704C
S	CASE	WO-A-1148	FM-GPW-7005	0701
T	SLEEVE	WO-637833	FM-GPW-7106	0704B
U	HUB	WO-A-6319	FM-GPW-7105	0704B
V	BEARING	WO-639422	FM-GPW-7120	0703
W	GEAR	WO-A-5554	FM-GPW-7017	0703
X	GEAR	WO-A-739	FM-GPW-7113	0704B
Y	WASHER	WO-635812	FM-GPW-7119	0704C

RA PD 305155-A

SNL G-503

GROUP 07—TRANSMISSION (Cont'd)

| | | Part Number | | | Quantity Reqd. per Unit Assy. | Per 100 Major Items | | | Estimated Reqmts. per 100 Rebuilds |
| Figure Number | Official Stockage Number | Willys | Ford | ITEM | | 12 Mos. Field Maintenance | Major Overhaul (5th Ech) | Total First Year Procurement | |
Col. 1	Col. 2	Col. 3	Col. 4	Col. 5	Col. 6	Col. 7	Col. 8	Col. 9	Col. 10
				0704C—COUNTERSHAFT AND REVERSE IDLER SHAFT (Cont'd)					
07-1		WO-635811	FM-GPW-7129-A	WAHSER, thrust, countershaft, rear (bz.) (WG-T-84-29)	1	—	19	20	60
07-1		WO-A-879	FM-GPW-7126	WASHER, thrust, countershaft, rear (S) (WG-T-84J-29)	1	—	10	12	30
				0706A—GEAR SHIFT LEVER AND PARTS					
07-1		WO-A-971	FM-GPW-7213	BALL, gearshift lever (rubber) (issue until stock exhausted)	1	16	24	—	—
		WO-A-1381	FM-GP-W-7217	BALL, gearshift lever fulcrum (WG-C-8-2½)	1	—	3	4	10
07-1		WO-A-1379	FM-BB-7220	CAP, gearshift lever housing (WG-4496-K)	1	2	2	6	5
		WO-A-3783	FM-GPW-1111286-A3	COVER, gearshift lever housing (rubber)	1	—	—	—	—
		WO-635868		BOLT, hex.hd., S., 5/16-18NC-2 x 3/4 (issue until stock is exhausted, then use WO-639689)	4	32	68	—	—
07-1		WO-639689	FM-20366-S	BOLT, hex.-hd., S., 5/16-18NC-2 x 3/4 (was WO-635868)	4	—	—	—	—
07-1		WO-52045	FM-34806-S2	WASHER, lock, S., 5/16 in. (housing cover bolt)	4	—	—	—	—
07-1		WO-635861	FM-GPW-7223	GASKET, gearshift lever housing (WG-T-84-115)	1	19	10	30	—
07-1		WO-635857	FM-GPW-7204	HOUSING, gearshift lever, assembly (Includes: PLATE RIVET)	1	—	—	—	—
07-1		WO-A-1380	FM-GPW-7210-A	LEVER, gearshift, assembly (WG-AC-84J-2A)	1	—	3	3	10
		WO-635862	FM-GPW-7207	PACKING, gearshift lever housing (issue until stock exhausted)	1	32	74	—	—
07-1		WO-A-1382	FM-GPW-7221	PIN, gearshift lever fulcrum (WG-C-84B-12A)	1	—	10	12	30
07-2		WO-A-7260	FM-GPW-7214	PLATE, guide, gearshift lever (WG-T-84B-32)	1	—	3	4	10
		WO-635860	FM-62216-S	RIVET, S., fl-hd., 3/16 x 7/16 (WG-X-3089BH)	2	—	10	25	30
07-1		WO-392328	FM-GPW-7227	SPRING, support, gearshift lever (WG-4498-K)	1	—	3	10	10
07-1		WO-635863	FM-BB-7228	WASHER, gearshift lever housing cap (WG-T-84H-137)	1	—	3	3	10
				0706B—GEARSHIFTER RAILS AND PARTS					
07-1		WO-635838	FM-353081-S	BALL, poppet, shift rail (S., 5/16 O.D.) (WG-X-2136)	2	—	10	10	30
07-1		WO-A-1385	FM-GPW-7233	PLUNGER, gearshift interlock (WG-T-84J-86A)	1	19	19	42	60
07-1		WO-A-1155	FM-GPW-7241	RAIL, shift, high and intermediate (WG-T-84J-20A)	1	—	6	7	20
07-1		WO-A-1156	FM-GPW-7240	RAIL, shift, low and reverse (WG-T-84J-21)	1	—	6	7	60
07-1		WO-636200	FM-GPW-7245	SCREW, lock, gearshift fork (WG-4418-M)	2	10	19	30	60
07-1		WO-635837	FM-GPW-7234	SPRING, shift rail, poppet ball (WG-T-84-42)	2	—	10	10	30

0706C—GEARSHIFTER YOKES OR FORKS

Fig.	Part No.	Part No.	Description	Unit				
07-1	WO-636196	FM-GPW-7230	FORK, shift, high and intermediate (WG-T-84C-23A)	1	3	3	10	10
07-1	WO-636197	FM-GPW-7231	FORK, shift, low and reverse (WG-T-84C-24A)	1	3	3	10	10
07-1	WO-A-7260	FM-GPW-7211-B	PLATE, shifter, assembly (WG-AT-84J-25)	1	—	6	5	20
07-1	WO-635836	FM-GPW-7206	PIN, shift fork guide (WG-T-84-22)	1	—	3	5	10
07-1	WO-635840	FM-357418-S	PIN, shifter plate fulcrum (WG-T-84-30)	1	3	3	10	10
07-1	WO-635839	FM-GPW-7208	SPRING, shifter plate (WG-T-84-31)	1	10	19	30	60

GROUP 08—TRANSMISSION TRANSFER

0800—TRANSMISSION TRANSFER ASSEMBLY

Fig.	Part No.	Part No.	Description	Unit				
F	WO-A-1195	FM-GPW-7700	CASE, transfer, assembly	1	6	—	8	—
	WO-A-1543	FM-GPW-18355	GASKET SET, transfer case overhaul (issue until stock is exhausted, then use WO-A-1543)	as req.	40	80	—	—
	WO-A-7443	FM-GPW-18355-B	GASKET SET, transfer case, overhaul (was WO-A-1543)	as req.	—	26	28	100

(Includes:
  1 WO-A-957 FM-GPW-7773 GASKET
  1 WO-A-1509 FM-GPW-7707 GASKET, cover, rear
  1 WO-A-954 FM-GP-7709 GASKET, cover
  2 WO-A-1134 FM-GPW-7746 GASKET, oil seal
  1 WO-A-1435 FM-GPW-7756 GASKET, transfer case to transmission)

Fig.	Part No.	Part No.	Description	Unit				
	WO-A-7445	FM-GPW-18317-B	OIL SEAL SET, transfer case	as req.	—	26	28	100

(Includes:
  2 WO-A-958 FM-GP-7770-A OIL SEAL, output shaft
  2 WO-A-974 FM-GP-7798-A OIL SEAL, shift rod)

0801—CASE

Fig.	Part No.	Part No.	Description	Unit				
08-1	WO-A-934	FM-GP-7754	BREATHER, transfer case, assembly (SP-5951X)	1	2	2	5	5
08-2	WO-A-956	FM-GPW-7774	CAP, output clutch shaft bearing, front (SP-18-19-8)	1	2	2	6	5
	WO-A-1136	FM-355554-S	BOLT, hex-hd, S., head dld. f/c-pin, 3/8-16NC-3 x 1 (SP-122-D) (5 used front cap to case, 1 used rear cap to case)	6	6	29	40	90
	WO-A-1137	FM-355533	BOLT, hex-hd, S. (head, dld. f/c-pin) 3/8-16NC-3 x 2 (rear cap to case) (SP-634-D)	3	—	10	40	30
	WO-A-1135	FM-356205-S	WASHER, cap., 5/8 O.D. x 25/64 I.D. x 1/16 thk. (SP-477-W)	9	29	230	280	720
08-1	WO-A-1507	FM-GPW-7768	CAP, output shaft bearing, w/BUSHING, rear (SP-6452X)	1	2	2	8	5
08-2	WO-A-1503	FM-GPW-7705	CASE, transfer (SP-18-15-9)	1	—	3	4	10
08-1	WO-953	FM-GP-7708	COVER, transfer case, bottom (SP-18-16-3)	1	—	3	4	10
	WO-51485	FM-20326-S7	BOLT, hex-hd, S., 5/16-18NC-2 x 5/8 (cover to case)	10	—	—	—	—
	WO-51833	FM-34806-S2	WASHER, lock, S., 5/16 in.	10	—	—	—	—
08-1	WO-A-1508	FM-GPW-7706	COVER, transfer case (rear) (SP-18-267-4)	1	—	3	4	10
	WO-6412	FM-24348-S	BOLT, hex-hd, S., 3/8-16NC-2 x 3/4 (cover to case)	5	—	—	—	—
	WO-5010	FM-34807-S	WASHER, lock, S., 3/8 in.	5	—	—	—	—
08-1	WO-A-957	FM-GPW-7773	GASKET, output clutch shaft bearing cap (SP-18-223-2)	1	—	3	15	—
08-1	WO-A-1509	FM-GPW-7707	GASKET, rear cover (SP-18-324-2)	1	13	—	15	—
	WO-A-954	FM-GP-7709	GASKET, transfer case, bottom cover (SP-18-155-2) (issue until stock exhausted)	1	50	100	—	—

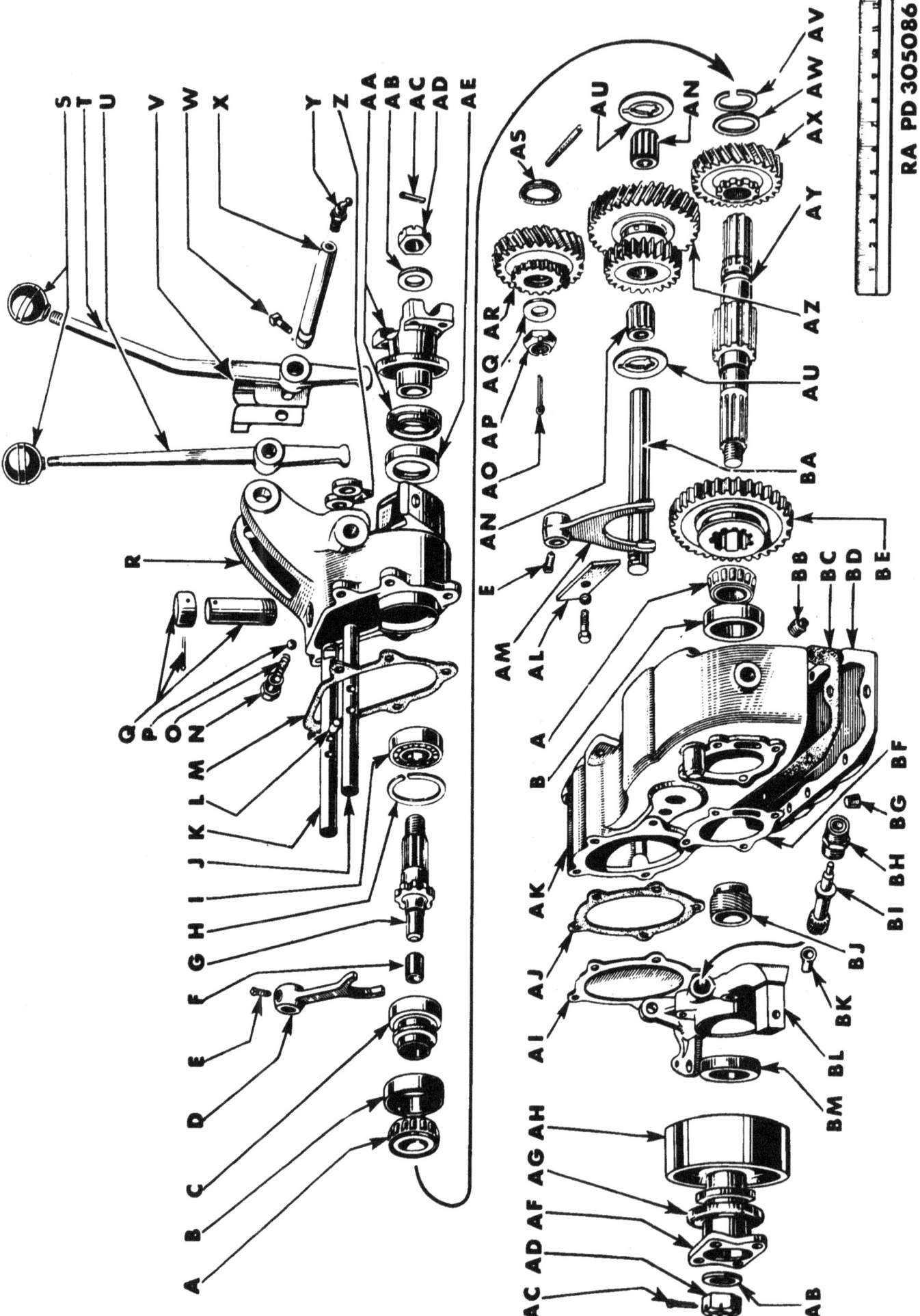

RA PD 305086

A
B
C
D
E
F
G
H
I
J
K
L
M
N
O
P
Q
R
S
T
U
V
W
X
Y
Z
AA
AB
AC
AD
AE

AF
AG
AH
AI
AJ
AK
AL
AM
AN
AO
AP
AQ
AR
AS
AT
AU
AV
AW
AX
AY
AZ
BA
BB
BC
BD
BE
BF
BG
BH
BI
BJ
BK
BL
BM

AB
AC
AD
AF
AG
AH

FIGURE 08-1—TRANSFER CASE ASSEMBLY

Key	Item	Willys Part No.	Ford Part No.	Gov't Group No.
A	ROLLERS	WO-51575	FM-GP-7723	0803
B	CUP	WO-52883	FM-OIY-1202	0803
C	GEAR	WO-A-992	FM-GP-7762	0803
D	FORK	WO-A-960	FM-GP-7711	0805A
E	SCREW	WO-A-963	FM-355550	0805A
F	BUSHING	WO-A-987	FM-GP-7777-A	0803
G	SHAFT	WO-A-975	FM-GP-7761	0803
H	RING	WO-A-976	FM-GP-7783	0803
I	BEARING	WO-A-1007	FM-GP-7719	0802
J	ROD	WO-A-1504	FM-GPW-7786	0805
K	ROD	WO-A-962	FM-GP-7787	0805
L	INTERLOCK	WO-A-965	FM-GP-7789	0805
M	GASKET	WO-A-957	FM-GPW-7773	0801
N	PLUG	WO-A-967	FM-355698-S	0805
O	SPRING	WO-A-966	FM-GP-7788	0805
P	BALL	WO-5599	FM-353075-S	0805
Q	BREATHER assembly	WO-A-934	FM-GP-7754	0801
R	CAP	WO-A-956	FM-GPW-7774	0801
S	HANDLE	WO-A-971	FM-GP-7213	0805B
T	LEVER	WO-A-1505	FM-GPW-7795	0805B
U	LEVER	WO-A-1506	FM-GPW-7710	0805
V	SPRING	WO-A-970	FM-GP-7796	0805B
W	SCREW	WO-A-973	FM-355378-S	0805B
X	PIN	WO-A-972	FM-GP-7796	0805B
Y	FITTING	WO-638224	FM-353035-A-S7	0805B
Z	SEAL	WO-A-974	(Willys only)	0805B
AA	YOKE	WO-A-1106	FM-GP-7729	0803
AB	WASHER	WO-A-1028	FM-356504-S	0803
AC	PIN	WO-5108	FM-72053-S	0803
AD	NUT	WO-A-980	FM-356125-S	0803
AE	OIL seal	WO-A-958	(Willys only)	0803
AF	FLANGE	WO-A-1105	FM-GP-4863	0803
AG	SHIELD	WO-A-1111	FM-GP-7776	0803
AH	DRUM	WO-A-1002	FM-GP-2614	1201
AI	COVER	WO-A-1508	FM-GPW-7706	0801
AJ	GASKET	WO-A-1509	FM-GPW-7707	0801
AK	CASE	WO-A-1503	FM-GPW-7768	0801
AL	PLATE	WO-A-1001	FM-GP-7767	0804
AM	FORK	WO-A-959	FM-GP-7712	0805A
AN	BEARING	WO-A-924	FM-GP-7718-A	0804
AO	PIN	WO-5397	FM-72071	0802
AP	NUT	WO-A-520	FM-356134-S-18	0802
AQ	WASHER	WO-A-1410	FM-356519-S	0802
AR	GEAR	WO-A-1510	FM-GP-7722	0802
AS	WASHER	WO-A-410	FM-356519-S	0802
AU	WASHER	WO-A-1000	FM-GP-7744	0804
AV	RING	WO-A-991	FM-GP-7784	0803
AW	WASHER	WO-A-990	FM-GP-7771	0803
AX	GEAR	WO-A-989	FM-GP-7776	0803
AY	SHAFT	WO-A-1764	FM-GP-7763	0803
AZ	GEAR	WO-A-999	FM-GP-7742	0804
BA	SHAFT	WO-A-998	FM-GP-7743	0804
BB	PLUG	WO-5140	FM-353064-S	0801
BC	GASKET	WO-A-954	FM-GP-7709	0801
BD	COVER	WO-A-953	FM-GP-7708	0801
BE	GEAR	WO-A-988	FM-GP-7765	0803
BF	SHIM	WO-A-982	FM-GP-7782-A	0801
BG	PLUG	WO-A-1104	FM-353053	0801
BH	SLEEVE	WO-636396	FM-GP-17333	0810
BI	GEAR	WO-A-1512	FM-GPW-17271	0810
BJ	GEAR	WO-A-1511	FM-GP-17285	0810
BK	BUSHING	WO-A-985	FM-GP-17277	0810
BL	CUP	WO-A-1507	FM-GPW-7768	0801
BM	OIL seal	WO-A-958	FM-GP-7776	0803

RA PD 305086-A

SNL G-503

GROUP 08—TRANSMISSION TRANSFER (Cont'd)

		Part Number				Per 100 Major Items			
Figure Number	Official Stockage Number	Willys	Ford	ITEM	Quantity Reqd. per Unit Assy.	12 Mos. Field Maintenance	Major Overhaul (5th Ech)	Total First Year Procurement	Estimated Reqmts. per 100 Rebuilds
Col. 1	Col. 2	Col. 3	Col. 4	Col. 5	Col. 6	Col. 7	Col. 8	Col. 9	Col. 10
				0801—CASE (Cont'd)					
08-1		WO-A-1435	FM-GPW-7756	GASKET, transfer case to transmission (SP-18-352-5)	1	26	—	30	—
08-1		WO-A-1104	FM-353053	PLUG, pipe, I., ½ in. (drain) (SP-18-39-2) (issue until stock exhausted)	1	4 %	8	—	—
08-1		WO-5140	FM-353064-S	PLUG, pipe, I., ½ in. (filler)	1	—	—	—	—
		WO-A-982	FM-GP-7782-A	SHIM, output shaft, rear bearing cap (.003 thk.) (SP-18-228-7)	as req.	—	—	—	—
		WO-A-983	FM-GP-7782-B	SHIM, output shaft, rear bearing cap (.010 thk.) (SP-18-228-8)	as req.	—	—	—	—
		WO-A-984	FM-GP-7782-C	SHIM, output shaft, rear bearing cap (.031 thk.) (SP-18-228-9)	as req.	—	—	—	—
		WO-A-6753	FM-GPW-18360	SHIM SET, output shaft (Includes: 4 WO-A-982 FM-GP-7782A SHIM; 2 WO-A-983 FM-GP-7782B SHIM; 2 WO-A-984 FM-GP-7782C SHIM)	as req.	—	6	5	20
				0802—DRIVE GEAR					
08-1		WO-A-1510	FM-GP-7722	GEAR, drive (O.D. 4.104 in., 27 teeth) (SP-18-8-7)	1	—	3	4	10
08-1		WO-A-520	FM-356134-S18	NUT, mainshaft (S., hex, slotted, 7/8-16NC-2) (WG-T-9-50P)	1	6	13	20	40
08-1		WO-5397	FM-72071	PIN, cotter, S., 1/8 x ½ (main shaft nut)	1	—	—	—	—
08-1		WO-A-1410	FM-356519-S	WASHER, main shaft nut (S., plain, 1⅛ O.D. x 2½2 I.D. x 5/32 thk.) (WG-T-84J-50½A)	1	6	3	10	10
				0803—DRIVEN GEARS, SHAFTS, BEARINGS					
08-1		WO-A-1007	FM-GP-7719	BEARING, ball, annular, single row, bore 1.1811 x .6299 (output clutch shaft) (MRC-206) (SKF-6206) (ND-3206)	1	—	6	7	30
08-1		WO-A-987	FM-GP-7777-A	BUSHING, output clutch shaft pilot (SP-18-24-1)	1	—	3	3	10
08-1		WO-51575	FM-GP-7723	CONE and ROLLERS, output shaft bearing (1.3125 diam. x .740 wide) (TIM-14131) (BT-BT-14131)	2	—	—	14	40
08-1		WO-52883	FM-O1Y-1202	CUP, output shaft bearing (TIM-14276) (BT-BT-14276)	2	—	13	14	40
08-1		WO-A-1105	FM-GP-4863	FLANGE, companion, rear (SP-K2-1-28)	1	3	13	8	10
08-1		WO-A-980	FM-356125-B	NUT, hex, S., slotted, ¾-20NF-3 (companion flange) (SP-124-J)	2	6	3	24	10
08-1		WO-5108	FM-72053-S	PIN, cotter, 1/8 x 1¼ (flange nut)	2	—	3	—	—
08-1		WO-A-1134	FM-GP-7746	GASKET, output clutch shaft bearing cap oil seal (SP-18-223-3)	2	13	—	22	—
08-1		WO-A-989	FM-GP-7766	GEAR, output shaft, O.D. 4.101 bore 1.250 x 1.713, 27 teeth (SP-18-8-1)	1	—	3	4	10
08-1		WO-A-988	FM-GP-7765	GEAR, sliding, output shaft O.D. 5.571 x 1.192 wide, 27 teeth (SP-18-8-2)	1	—	3	4	10
08-1		WO-A-992	FM-GP-7762	GEAR, clutch, output shaft (10 internal teeth, bore 1.088 diam. x 1.755 wide, face 2.411 diam.) (SP-18-466-1)	1	—	3	4	10
08-1		WO-A-958		OIL SEAL, output shaft (front and rear bearing cap) (SP-62-463-2) (Willys only)	2	13	—	20	—

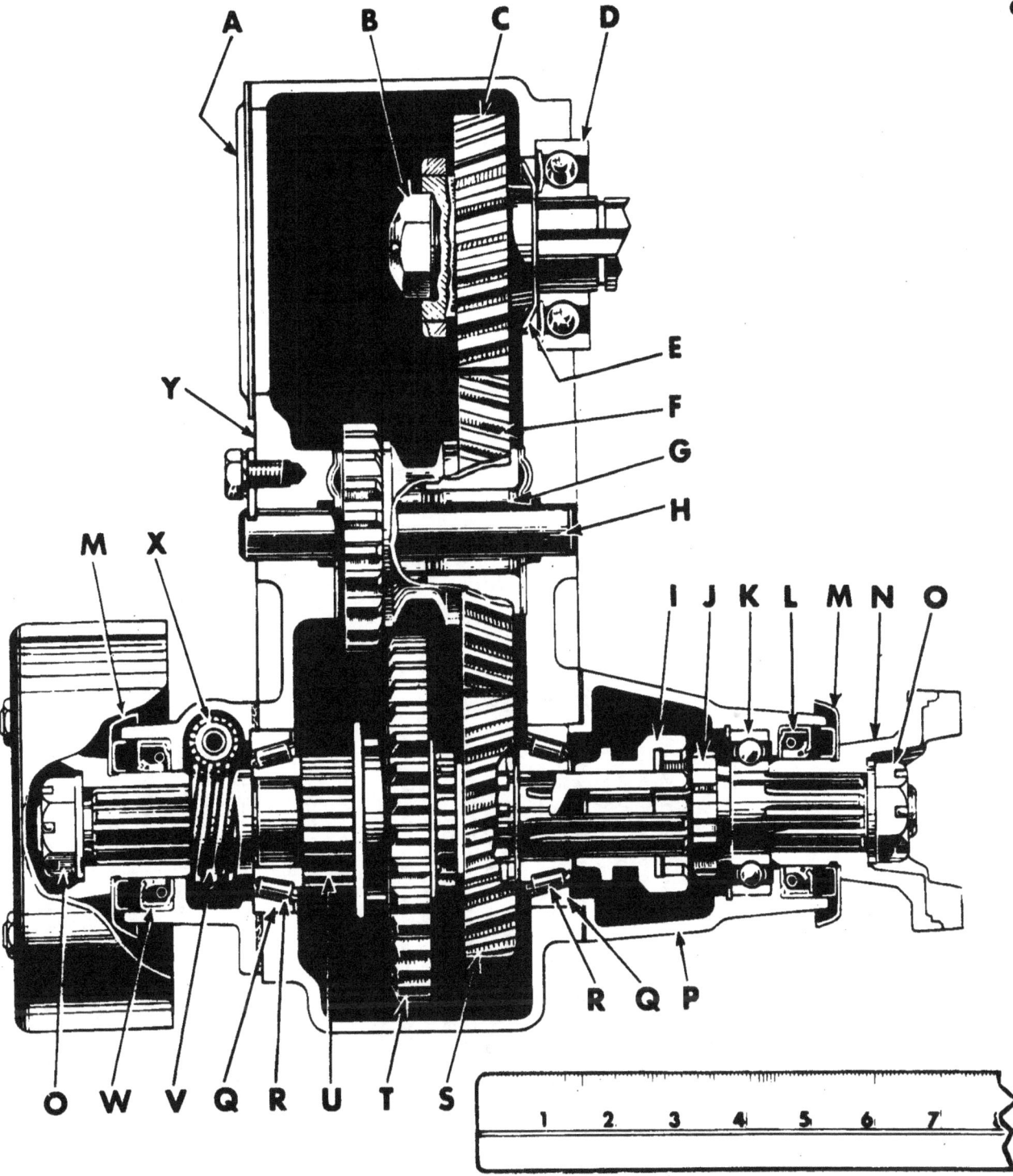

RA PD 305157

FIGURE 08-2—TRANSFER CASE—CROSS-SECTIONAL VIEW

Item	Willys Part No.	Ford Part No.	Gov't Group No.	Key	Item	Willys Part No.	Ford Part No.	Gov't Group No.
COVER	WO-A-1508	FM-GPW-7706	0801	N	YOKE	WO-A-1106	FM-GP-7729	0803
NUT	WO-A-520	FM-356134S-18	0802	O	NUT	WO-A-980	FM-356125-S	0803
GEAR	WO-A-1510	FM-GP-7722	0802	P	CAP	WO-A-956	FM-GPW-7774	0801
BEARING	WO-A-916	FM-GP-7065	0704	Q	CUP	WO-52883	FM-0IY-1202	0803
SHIELD	WO-A-1111	FM-GP-7776	0803	R	CONE and ROLLERS	WO-51575	FM-GP-7723	0803
GEAR	WO-A-999	FM-GP-7742	0804	S	GEAR	WO-A-989	FM-GP-7766	0803
BEARING	WO-A-924	FM-GP-7718-A	0804	T	GEAR	WO-A-988	FM-GP-7765	0803
SHAFT	WO-A-998	FM-GP-7742	0804	U	SHAFT	WO-A-1764	FM-GP-7763	0803
GEAR	WO-A-992	FM-GP-7762	0803	V	GEAR	WO-A-1511	FM-GP-17285	0810
SHAFT	WO-A-975	FM-GP-7761	0803	W	OIL SEAL	WO-A-958	(Willys only)	0803
BEARING	WO-A-1007	FM-GP-7719	0803	X	GEAR	WO-A-1512	FM-GPW-17271	0810
SEAL	WO-A-958	(Willys only)	0803	Y	CASE	WO-A-1503	FM-GPW-7705	0801
SHIELD	WO-A-1111	FM-GP-7776	0803					

RA PD 305157-A

SNL G-503

GROUP 08—TRANSMISSION TRANSFER (Cont'd)

| Figure Number | Official Stockage Number | Part Number | | ITEM | Quantity Reqd. per Unit Assy. | Per 100 Major Items | | | Estimated Reqmts. per 100 Rebuilds |
| | | Willys | Ford | | | 12 Mos. Field Maintenance | Major Overhaul (5th Ech) | Total First Year Procurement | |
Col. 1	Col. 2	Col. 3	Col. 4	Col. 5	Col. 6	Col. 7	Col. 8	Col. 9	Col. 10
				0803—DRIVEN GEARS, SHAFTS, BEARINGS (Cont'd)					
08-1		WO-A-975	FM-GP-7761	SHAFT, output clutch (SP-18-362-3) (issue until stock exhausted. Then use WO-A-8835 for maintenance)	1	4	8	—	—
08-1		WO-A-8835		SHAFT, output clutch, w/NUT and WASHER (was WO-A-975)	1		3	4	10
08-1		WO-A-1764	FM-GP-7763	SHAFT, output, w/BUSHING, assembly (SP-18-362-4-5866-X) (issue until stock exhausted)	1	4	4	—	10
08-1		WO-A-8841	FM-GPW-7736	SHAFT, output, w/NUT and WASHER, assembly	1		3	4	10
08-1		WO-A-1111	FM-GP-7776	SHIELD, dust (SP-5962-X) (1 used companion flange, 1 used universal joint end yoke)	2	4	—	9	5
08-1		WO-A-991	FM-GP-7784	SNAP RING, output shaft (SP-18-381-2)	1		26	27	80
08-1		WO-A-976	FM-GP-7783	SNAP RING, output clutch shaft bearing (SP-22-381-19)	1		26	28	80
08-1		WO-A-1028	FM-355504-S	WASHER, companion flange nut (SP-145-W)	2		10	20	30
08-1		WO-A-990	FM-GP-7771	WASHER, thrust, output shaft (SP-710-W)	1		3	5	10
08-1		WO-A-1106	FM-GP-7729	YOKE, end, universal joint, front (SP-K2-4-88X)	1	2	2	5	5
				0804—IDLER GEAR, SHAFT, BEARINGS					
08-1		WO-A-924	FM-GP-7718-A	BEARING, intermediate gear (HY-94322)	2		13	14	40
08-1		WO-A-999	FM-GP-7742	GEAR, intermediate (O.D. 4.962, 33 teeth) (SP-18-5-1)	1		3	4	10
08-1		WO-A-1001	FM-GP-7767	PLATE, lock, intermediate (SP-31-246-2)	1		3	3	10
08-1		WO-52189	FM-20328-S2	BOLT, hex-hd., S., 3/8-16NC-2 x 5/8	1				—
08-1		WO-5010	FM-34807-S2	WASHER, lock, S., 3/8 in.	1				—
08-1		WO-A-998	FM-GP-7743	SHAFT, intermediate (SP-18-187-1)	1		3	4	10
08-1		WO-A-1000	FM-GP-7744	WASHER, thrust, intermediate gear (SP-690-W)	1		13	12	40
				0805—SHIFTER RODS					
08-1		WO-5599	FM-353075-S	BALL, poppet, shift rod (SP-50-80-1)	2		6	36	20
08-1		WO-A-974	FM-GP-7798-A	OIL SEAL, shift rod (SP-13-463-1) (Willys only)	2	13	—	20	—
08-1		WO-A-967	FM-355698-S	PLUG, shift rod poppet (SP-854-D)	2		6	24	60
08-1		WO-A-962	FM-GP-7787	ROD, shift, front wheel drive (SP-18-67-2)	1		3	4	10
08-1		WO-A-1504	FM-GPW-7786	ROD, shift, underdrive and direct (SP-18-67-5)	1		3	4	10
08-1		WO-A-966	FM-GP-7788	SPRING, shift rod poppet (SP-22-72-5)	2		6	15	20
08-1		WO-A-964		WIRE, lock, No. 16 ga. (shift fork) (SP-LW-1) (Willys only)	2		—	—	—
				0805A—YOKES					
08-1		WO-A-960	FM-GP-7711	FORK, front wheel drive shift (SP-18-66-7)	1		3	4	10
08-1		WO-A-959	FM-GP-7712	FORK, underdrive and direct shift (SP-18-66-1)	1		3	4	10
08-1		WO-A-965	FM-GP-7789	INTERLOCK, shift rod (SP-18-21-1)	1		3	5	10
08-1		WO-A-963	FM-355550	SCREW, set, shift fork (S., 3/8-24NF, head dld. f/c pin, 3/4 in. diam.) (SP-745D)	2		13	36	40

0805B—SHIFT LEVER AND PARTS

1	2	3	4	Qty	Description	Part No.	Stock No.	Fig.
—	—	24	16	2	BALL, shift lever handle (SP-39-Q-20) (issue until stock exhausted)	FM-GP-7213	WO-A-971	08-1
—	—	74	36	1	FITTING, grease, 45° (shift lever pivot pin) (AD-1636) (issue until stock exhausted)	FM-353035-S7	WO-638224	08-1
—	6	+	6	1	GROMMET, transfer shift levers (leather)	FM-GPW-1101735-A	WO-A-3784	08-1
10	4	3	—	1	LEVER, shift, front wheel drive (SP-25-Q-1056)	FM-GPW-7710	WO-A-1506	08-1
10	4	3	—	1	LEVER, shift, underdrive and direct (SP-25-Q-1055)	FM-GPW-7793	WO-A-1505	08-1
10	3	3	—	1	PIN, pivot, shift lever (SP-363-SP)	FM-GP-7796	WO-A-972	08-1
40	20	13	—	1	SCREW, set, shift lever pivot pin (SP-433-D)	FM-355378-S	WO-A-973	08-1
10	6	3	—	2	SPRING, shift lever (SP-97-Q-13)	FM-GP-7799	WO-A-970	08-1

0809—MOUNTINGS

1	2	3	4	Qty	Description	Part No.	Stock No.	Fig.
—	—	—	—	5	BOLT, hex-hd, S., ⅜-16NC-2 x 1⅛ (transfer case to transmission)	FM-24408-S7	WO-52911	
—	—	—	—	1	BOLT, hex-hd, S., ⅜-16NC-2 x 1 (transfer case to transmission lower left) (Willys only)		WO-51405	
—	—	6	—	1	BOLT, support, transfer case (hex-hd, S., ½-20NF x 3)	FM-355741-S	WO-A-147	
—	24	3	6	1	INSULATOR, transfer case support	FM-74-6038	WO-634758	
—	5	—	3	1	NUT, hex, S., ½-20NF-2 (support bolt)	FM-33846-S	WO-5916	
—	—	16	16	1	SNUBBER, transfer case support insulator (issue until stock exhausted)	FM-GPW-7781-A	WO-634759	
—	10	—	6	1	WASHER, S., 1½ O.D. x 21/32 I.D. x 1/16 thk. (support bolt)	FM-356524-S	WO-634762	
—	—	—	—	1	WASHER, lock, S., ½ (⅞ O.D. x 17/32 I.D. x ⅛ thk.) (support bolt) (Willys only)	FM-34809-S2	WO-5009	
—	—	—	—	4	WASHER, lock, S., ⅜ S.A.E. (21/32 O.D. x 13/32 I.D. x 3/32 thk.) (case to transmission)	FM-34807-S7		

0810—SPEEDOMETER DRIVE GEARS

1	2	3	4	Qty	Description	Part No.	Stock No.	Fig.
—	4	4	4	1	BUSHING, speedometer drive gear (SP-475-2) (issue until stock exhausted)	FM-GP-17277	WO-A-985	08-1
—	4	4	4	1	GEAR, drive, speedometer (SP-18-452-2) (issue until stock exhausted)	FM-GP-17285	WO-A-1511	08-1
—	4	4	4	1	GEAR, driven, speedometer (SP-18-453-3) (issue until stock exhausted)	FM-GPW-17271	WO-A-1512	08-1
—	—	—	—	as req.	KIT, repair, speedometer drive gear (Includes: 1 WO-A-985 FM-GP-17277 BUSHING; 1 WO-A-1511 FM-GP-17285 GEAR, drive; 1 WO-A-1512 FM-GPW-17271 GEAR, driven)	FM-GPW-18314	WO-A-7837	
—	—	4	4	1	SLEEVE, speedometer driven gear (SP-18-454-4) (issue until stock exhausted)	FM-GP-17333	WO-636396	08-1

GROUP 09—PROPELLER SHAFT AND UNIVERSAL JOINT

0901—PROPELLER SHAFT ASSEMBLIES

1	2	3	4	Qty	Description	Part No.	Stock No.	Fig.
—	7	2	2	1	SHAFT, propeller, front, assembly (SP-8996-SF)	FM-GPW-3365	WO-A-1326	
—	7	2	2	1	SHAFT, propeller, rear, assembly (SP-8997-SF)	FM-GPW-4602	WO-A-1327	

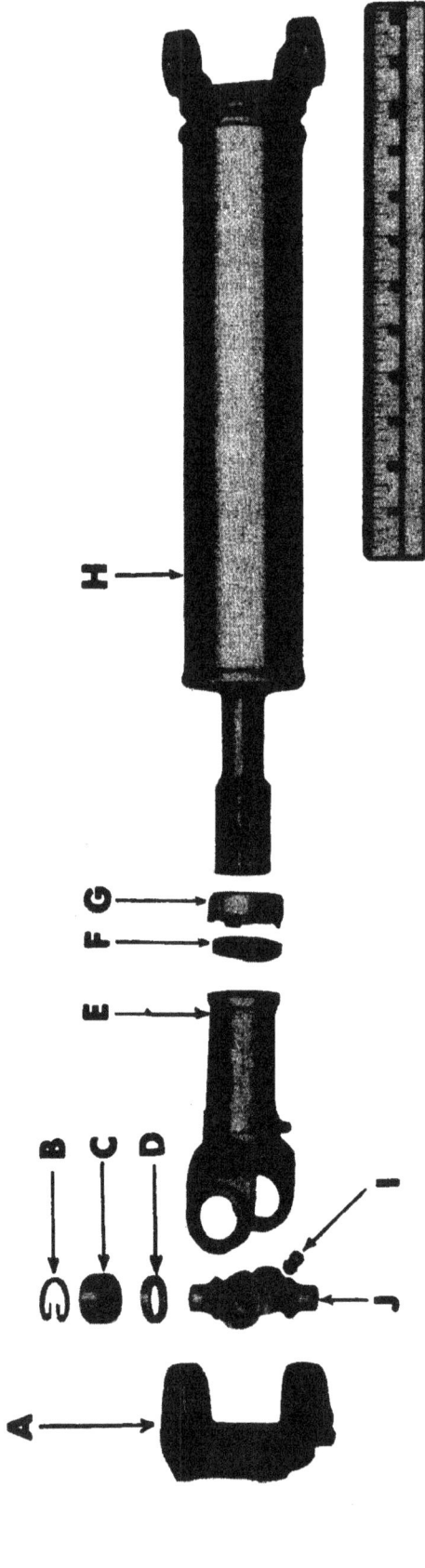

RA PD 305087

FIGURE 09-1—REAR PROPELLER SHAFT

Key	Item	Willys Part No.	Ford Part No.	Gov't Group No.
A	YOKE	WO-A-950	FM-GP-4866	0902
B	SNAP ring	WO-A-945	FM-01Y-7096	0902
C	BEARING	WO-A-1434	FM-GPW-7074	0902
D	RETAINER	WO-A-940	FM-01Y-7083	0902

Key	Item	Willys Part No.	Ford Part No.	Gov't Group No.
E	YOKE	WO-A-935	FM-GP-4841	0902
F	WASHER	WO-A-943	FM-GP-7097	0902
G	CAP	WO-A-942	FM-GP-7077	0902
H	TUBE	WO-A-1429	FM-GPW-4605	0902

RA PD 305087-A

GROUP 09—PROPELLER SHAFT AND UNIVERSAL JOINT (Cont'd)

Figure Number	Official Stockage Number	Part Number		ITEM	Quantity Reqd. per Unit Assy.	Per 100 Major Items			Estimated Reqmts. per 100 Rebuilds
		Willys	Ford			12 Mos. Field Maintenance	Major Overhaul (5th Ech)	Total First Year Procurement	
Col. 1	Col. 2	Col. 3	Col. 4	Col. 5	Col. 6	Col. 7	Col. 8	Col. 9	Col. 10
				0902—UNIVERSAL JOINTS					
09-1		WO-A-1434	FM-GPW-7074	BEARING, universal joint, assembly	4	—	—	—	—
		WO-A-490	FM-01Y-4529	BOLT, "U", S., $5/16$-24NF-3 (universal joint bearing) (SP-K2-94-29)	6	13	13	20	—
		WO-A-491	FM-33784-S2	NUT, hex., S., $5/16$-24NF-3 (SP-5-74-11) (issue until stock exhausted)	12	28	52	—	—
		WO-51833	FM-34806-S2	WASHER, lock, S., $5/16$ (SP-5-75-19)	12	—	—	—	—
09-1		WO-A-942	FM-GP-7077	CAP, dust, universal joint yoke sleeve (SP-K2-14-69)	2	3	3	6	10
09-1		WO-638792	FM-353043-S7	FITTING, grease $1/4$-28NF (AD-1641) (GM-110347)	6	—	—	—	—
		WO-A-941	FM-01T-7078-A	GASKET, universal joint trunnion (SP-K3-86-89)	16	—	—	—	—

WO No.	FM No.	Description	4	120	40	187	125	Fig.
WO-A-1433	FM-21C-18397-B	KIT, repair, universal joint journal (SP-K5-21-X)........ (Includes: 4 WO-A-1434 FM-GPW-7074 BEARING assembly; 1 WO-A-1426 FM-GPW-7084 JOURNAL assembly; 4 WO-A-945 FM-O1Y-7096 SNAP RING)	4	—	—	—	—	
WO-A-940	FM-O1Y-7083	RETAINER, trunnion gasket (SP-K2-76-17)........	16	—	—	—	—	09-1
WO-A-636568	FM-GP-4666	SHIELD, dust, front universal end yoke (SP-15009)........	1	3	3	8	10	09-1
WO-A-945	FM-O1Y-7096	SNAP RING, trunnion bearing (SP-K2-7-29)........	16	19	19	40	60	09-1
WO-A-1428	FM-GPW-3370	TUBE, front, propeller shaft, assembly (SP-K2-62-210-212-1907)........	1	—	—	—	—	09-1
WO-A-1429	FM-GPW-4605	TUBE, rear, propeller shaft, assembly (SP-K2-62-210-212-1718) (issue until stock exhausted)........	1	—	3	—	—	09-1
WO-A-943	FM-GP-7097	WASHER, universal joint yoke sleeve (cork, split, 1 37/64 O.D. x 1 I.D. x 5/16) (SP-K2-16-53)........	2	6	6	20	20	09-1
WO-A-950	FM-GP-4866	YOKE, propeller shaft flange (SP-K2-2-329)........	1	2	2	6	5	09-1
WO-A-935	FM-GP-4841	YOKE, universal joint sleeve, w/PLUG, assembly (SP-K2-3-198X)........	2	—	3	8	10	09-1

GROUP 10—FRONT AXLE

1000—FRONT AXLE ASSEMBLY

WO No.	FM No.	Description	4	120	40	187	125	Fig.
WO-A-1212		AXLE, front, assembly (Bendix Weiss joints) (SP-2058-1) (Willys only)........	1					
WO-A-1387		AXLE, front, assembly (Rzeppa joints) (SP-2058-2) (Willys only)........	1					
		NOTE: Front axles to be tagged at joints for identification.						
WO-A-1765		AXLE, front, assembly (Spicer joints) (SP-SKA-34172) (Willys only)........	1					
		NOTE: Identified by yellow paint on lower side of differential assembly.						
WO-A-6442		AXLE, front, assembly (Tracta joints) (Willys only)........	1					
		NOTE: The above (4) axles are interchangeable.						
WO-A-6029	FM-GPW-3001	AXLE, front, w/BRAKES, DRUMS and HUBS, assembly (Willys only) (This part number to be used when either Tracta or Bendix universal joints are to be supplied with this assembly)........	1	3	3	4		
WO-A-5498		AXLE, front, w/BRAKES, DRUMS and HUBS, assembly (Ford only)........	1					
WO-A-5499		AXLE, front, w/BRAKES, DRUMS and HUBS, assembly (Bendix joints) (Willys only)........	1					
		AXLE, front, w/BRAKES, DRUMS and HUBS, assembly (Rzeppa joints) (Willys only)........	1					
WO-A-6470		AXLE, front, w/BRAKES, DRUMS and HUBS, assembly (Tracta joints) (Willys only)........	1					
		NOTE: The above (3) Willys parts are interchangeable.						

1001—HOUSING

WO No.	FM No.	Description	4	120	40	187	125	Fig.
WO-A-636527	FM-355699-S2	BOLT, differential bearing cap (hex-hd., S., 1/2-13NC x 2 1/4)........	4		6	8	20	10-3
WO-A-1486	FM-GPW-2084	BRACKET, front brake hose, left (issue until stock exhausted)........	1		4			10-3
WO-A-1487	FM-GPW-2082	BRACKET, front brake hose, right (issue until stock exhausted)........	1		4			10-3
WO-A-781	FM-GP-4016	COVER, axle housing (SP-16976)........	1	1	1	4		
WO-A-51523	FM-20046-S2	BOLT, hex-hd., S., 5/16-18NC-2 x 7/8 (cover to axle housing)........	10			4	2	
WO-A-52510	FM-34941-S2	WASHER, lock, S., 5/16 in.........	10					
WO-A-1703		HOUSING, axle, w/TUBE, assembly (SP-17310-X) (Willys only)........	1		3	4		10-1
	FM-GPW-3074	HOUSING, axle, w/TUBE, assembly (Ford only)........	1				10	

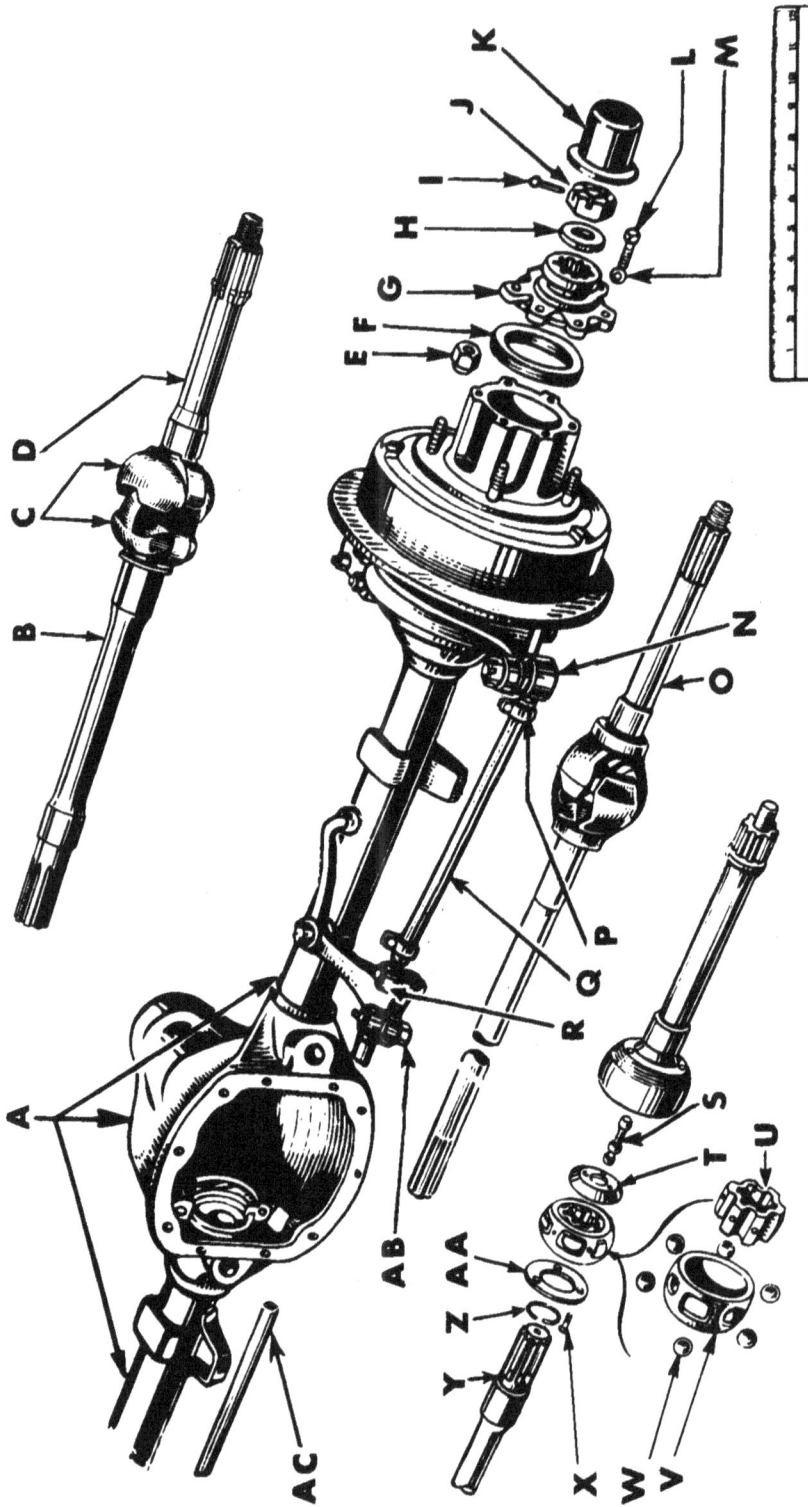

FIGURE 10-1—FRONT AXLE

RA PD 305089

Key	Item	Willys Part No.	Ford Part No.	Gov't Group No.
A	HOUSING	WO-A-1703	(Willys only)	1001
B	SHAFT	WO-A-6383	(Willys only)	1007
C	UNIVERSAL joint	WO-A-6361	(Willys only)	1007
D	SHAFT	WO-A-6382	FM-GP-1013	1302A
E	NUT	WO-A-476	FM-GP-3208-A	1007
F	SHIMS	WO-A-862	FM-GP-3204	1007
G	FLANGE	WO-A-868	FM-356504-S	1003
H	WASHER	WO-636570	FM-72071	1301C
I	PIN	WO-5397	FM-356126-S	1301C
J	NUT	WO-636569	FM-GP-1139	1301C
K	CAP	WO-A-869	FM-GP-1110	1007
L	BOLT	WO-A-760	FM-34807-S	1402
M	WASHER	WO-5010	FM-GP-3292	1007
N	END assembly	WO-A-847	(Willys only)	1402
O	SHAFT assembly	WO-A-809		1007

Key	Item	Willys Part No.	Ford Part No.	Gov't Group No.
P	CLAMP	WO-A-1706	FM-51-3287	1402
Q	TUBE	WO-A-1709	FM-GPW-3282	1402
R	CRANK	WO-A-8249	(Willys only)	1401
S	PIN	WO-A-1723	FM-GP-3218	1007
T	PILOT	WO-A-1722	FM-GP-3219	1007
U	RACE	WO-A-1720	FM-GP-3221-A	1007
V	CAGE	WO-A-1719	FM-3215-A	1007
W	BALL	WO-A-1721	FM-358074-S	1007
X	SCREW	WO-A-1725	FM-24622-S	1007
Y	SHAFT	WO-A-1727	FM-3106-A	1007
Z	SNAP ring	WO-A-1726	FM-GP-3216	1007
AA	RETAINER	WO-A-1724	FM-GP-3217	1007
AB	END assembly	WO-A-838	FM-GPW-3291	1402
AC	TUBE	WO-A-1705	FM-GPW-3281	1402

RA PD 305089-A

GROUP 10—FRONT AXLE (Cont'd)

Col. 1 Figure Number	Col. 2 Official Stockage Number	Col. 3 Willys	Col. 4 Ford	Col. 5 ITEM	Col. 6 Quantity Reqd. per Unit Assy.	Col. 7 12 Mos. Field Maintenance	Col. 8 Major Overhaul (5th Ech)	Col. 9 Total First Year Procurement	Col. 10 Estimated Reqmts. per 100 Rebuilds
				1001—HOUSING (Cont'd)					
10-3		WO-A-782	FM-GP-4035	GASKET, axle housing cover (SP-16409)	1	40	—	50	—
		WO-A-7830	FM-GPW-18365-B	GASKET SET, front axle	as req.	6	13	20	40
				(Includes:					
		WO-A-820	FM-GP-1092	2 GASKET, oil seal, assembly					
		WO-A-819	FM-GP-3135	4 SEAL, pivot					
		WO-A-818	FM-GP-3139	4 STRIP, pressure					
		WO-A-858	FM-GPW-3167	2 SEAL, steering					
		WO-A-782	FM-GP-4035	1 GASKET, housing					
		WO-636565	FM-GP-4061	1 GASKET, oil seal					
12-5		WO-A-1457	FM-GPW-2096	GUARD, front wheel brake hose, assembly (issue until stock exhausted)	2		4	—	
		WO-636577	FM-358048-S	PLUG, drain, axle housing (SP-S-780) (issue until stock exhausted)	1	8%	16	—	
		WO-636538	FM-353051-S	PLUG, filler, axle housing (SP-50-39-4) (issue until stock exhausted)	1	8%	16	—	
		WO-A-870	FM-GP-4022	PLUG, vent, axle housing (SP-16979)	1	5%		5	
		WO-636528	FM-34922-S	WASHER, lock, internal, S., 1/2 in. (differential bearing cap bolt) (SP-11586) (issue until stock exhausted)	4	68	132	—	
		WO-5010	FM-34807-S7	WASHER, lock, S., 3/8 in.	12			—	
				1002—DIFFERENTIAL ASSEMBLY					
10-3		WO-A-793	FM-GP-4206	CASE, differential (SP-16383)	1		3	4	10
		WO-A-788		DIFFERENTIAL, front assembly (SP-16968-X) (Willys only) (includes CASE, GEARS, and PINIONS)	1		3	4	10
			FM-GPW-4212	DIFFERENTIAL, front assembly (Ford only) (includes CASE, GEARS, PINIONS)	1			—	
				1003—DIFFERENTIAL GEARS, PINION AND BEARINGS					
10-3		WO-52880	FM-GP-4221	CONE and ROLLERS, differential bearing (bore 1.625 x 1.000) (TIM-24780)	2		13	14	40
10-3		WO-52876	FM-86H-4621	CONE and ROLLERS, drive pinion bearing, inner (bore 1.375 x 1.125) (TIM-31593)	1		3	4	10
10-3		WO-52878	FM-GP-4630	CONE and ROLLERS, drive pinion bearing, outer (bore 1.125 x .875) (TIM-02872)	1		3	4	10
10-3		WO-52881	FM-GP-4222	CUP, differential bearing (3.000 O.D. x .8125 wide) (TIM-24721)	1		13	14	40
10-3		WO-52877	FM-86H-4616	CUP, drive pinion bearing, inner (3.00 O.D. x .9375 wide) (TIM-31520)	1		3	4	10
10-3		WO-52879	FM-GP-4628	CUP, drive pinion bearing, outer (2.875 O.D. x .6875 wide) (TIM-02820)	1		3	4	10
		WO-636565	FM-GP-4661	GASKET, drive pinion oil seal (SP-S-171)	1	26	6	30	20
10-3		WO-A-789	FM-GPW-4209	GEAR, drive, differential, w/PINION (SP-16412-X) (Willys only)	1			6	
				GEAR, drive, differential, w/PINION (Ford only)	1			—	

FIGURE 10-3—DIFFERENTIAL ASSEMBLY—CROSS-SECTIONAL VIEW

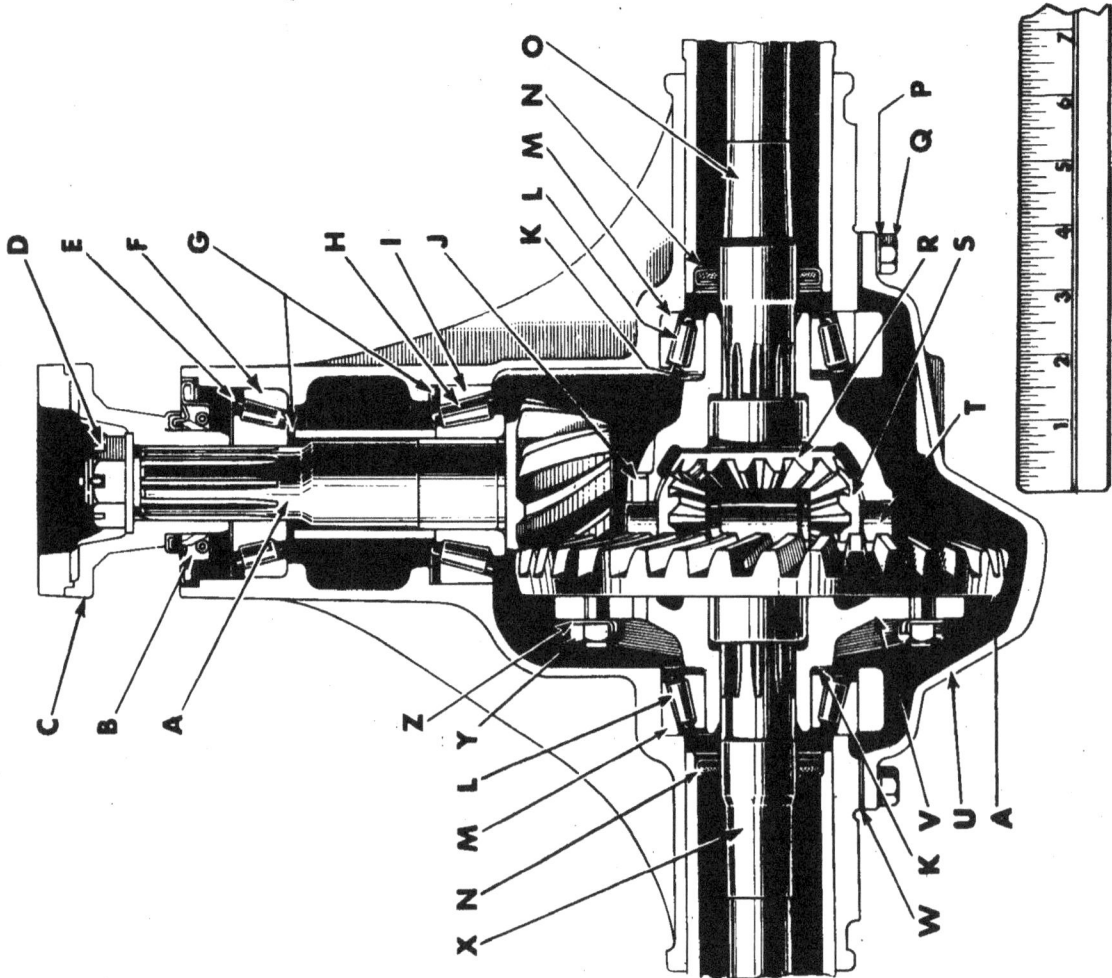

Key	Item	Willys Part No.	Ford Part No.	Gov't Group No.
A	GEARS	WO-A-789	(Willys only)	1003
B	SEAL	WO-639265	FM-GP-4676-A	1003
C	YOKE	WO-A-1445	FM-GP-4842	1104
D	NUT	WO-639569	FM-356125-S	1003
E	CONE	WO-52878	FM-GP-4630	1003
F	CUP	WO-52879	FM-GP-4628	1003
G	SHIMS	WO-A-803	FM-GP-4659-A	1003
H	CONE	WO-52876	FM-86H-4621	1003
I	CUP	WO-52877	FM-86H-4616	1003
J	PIN	WO-636360	FM-GP-4241	1003
K	SHIMS	WO-A-784	FM-GP-4229-A	1003
L	CONE	WO-52880	FM-GP-4221	1003
M	CUP	WO-52881	FM-GP-4222	1003
N	SEAL	WO-A-779	FM-GP-3034	1006
O	SHAFT	WO-A-901	FM-GP-4234	1101
P	WASHER	WO-52510	FM-34941-S2	1001
Q	BOLT	WO-51523	FM-20046-S2	1001
R	GEAR	WO-A-794	FM-GPW-4236	1003
S	GEAR	WO-A-796	FM-GPW-4215	1003
T	SHAFT	WO-A-798	FM-GP-4211	1103
U	COVER	WO-A-781	FM-GP-4016	1001
V	CASE	WO-A-793	FM-GP-4206	1103
W	GASKET	WO-A-782	FM-GP-4085	1001
X	SHAFT	WO-A-902	FM-GP-4235	1102
Y	SCREW	WO-A-871	FM-355511-S	1103
Z	STRAP	WO-A-792	FM-GP-4281	1103

RA PD 305153-A

RA PD 305153

GROUP 10—FRONT AXLE (Cont'd)

1003—DIFFERENTIAL, GEARS, PINION AND BEARINGS (Cont'd)

Col. 1 Figure Number	Col. 2 Official Stockage Number	Col. 3 Willys	Col. 4 Ford	Col. 5 ITEM	Col. 6 Quantity Reqd. per Unit Assy.	Col. 7 12 Mos. Field Maintenance	Col. 8 Major Overhaul (5th Ech)	Col. 9 Total First Year Procurement	Col. 10 Estimated Reqmts. per 100 Rebuilds
10-3		WO-A-794	FM-GPW-4236	GEAR, side, differential, 2.758 O.D. x 1⅜ in. wide, 16 teeth (SP-16385)	2	—	—	—	—
		WO-A-6743	FM-GPW-18389	KIT, repair, front axle differential gear	as req.	—	3	4	10
				(Includes:					
				2 WO-A-794 FM-GP-4236 GEAR, side					
				2 WO-A-796 FM-GPW-4215 PINION					
				1 WO-A-798 FM-GP-4211 SHAFT, pinion					
				2 WO-A-797 FM-GP-4230 WASHER, thrust					
				2 WO-A-795 FM-GP-4228 WASHER, thrust)					
		WO-A-6816	FM-GPW-18384	KIT, repair, drive gear and pinion	as req.	—	3	4	10
				(Includes:					
				1 WO-636569 FM-356126-S NUT, drive pinion					
				8 WO-A-871 FM-355511-S SCREW, drive gear					
				1 WO-A-789 FM-GPW-4209 SET, gear and pinion					
				4 WO-A-792 FM-GP-4281 STRAP, lock)					
13-1		WO-636569	FM-356126-S	NUT, hex, S., slotted, ¾-16NF-3 (drive pinion) (issue until stock exhausted)	1	20	40	—	—
10-3		WO-639265	FM-GP-4676-A	OIL SEAL, drive pinion (leather) (SP-14223)	1	13	—	20	—
10-3		WO-636571	FM-357202-S	PIN, cotter (S.) drive pinion nut (SP-7-72-39)	1	—	—	—	—
10-3		WO-636360	FM-GP-4241	PIN, lock, differential pinion shaft (SP-13449)	1	—	—	—	—
10-3		WO-A-796	FM-GPW-4215	PINION, differential (1.140 O.D., 10 teeth) (SP-15926)	2	—	—	—	—
		WO-A-871	FM-355511-S	SCREW, differential, drive gear (SP-6-73-414)	8	—	13	20	40
		WO-A-798	FM-GP-4211	SHAFT, differential pinion (SP-16075)	1	—	—	—	—
		WO-636568	FM-GP-4666	SHIELD, dust, universal joint end yoke (SP-15099)	as req.	—	—	—	—
10-3		WO-A-784	FM-GP-4229-A	SHIM, adjusting, differential bearing (.003 thk.) (SP-S-58)	as req.	—	—	—	—
		WO-A-785	FM-GP-4229-B	SHIM, adjusting, differential bearing (.005 thk.) (SP-S-59)	as req.	—	—	—	—
		WO-A-786	FM-GP-4229-C	SHIM, adjusting, differential bearing (.010 thk.) (SP-S-74)	as req.	—	—	—	—
			FM-GP-4229-E	SHIM, adjusting, differential bearing (.015 thk.) (Ford only) (SP-S-75)	as req.	—	—	—	—
		WO-A-787	FM-GP-4229-D	SHIM, adjusting, differential bearing (.030 thk.) (SP-112)	as req.	—	—	—	—
		WO-A-800	FM-GP-4660-A	SHIM, adjusting, drive pinion, large (.003 thk.) (SP-113)	as req.	—	—	—	—
		WO-A-801	FM-GP-4660-B	SHIM, adjusting, drive pinion, large (.005 thk.) (SP-114)	as req.	—	—	—	—
		WO-A-802	FM-GP-4660-C	SHIM, adjusting, drive pinion, large (.010 thk.) (SP-S-114)	as req.	—	—	—	—
		WO-A-803	FM-GP-4659-A	SHIM, adjusting, drive pinion, bearing, small (.003 thk.) (SPS-638)	as req.	—	—	—	—
		WO-A-804	FM-GP-4659-B	SHIM, adjusting, drive pinion, bearing, small (.005 thk.) (SPS-638-1)	as req.	—	—	—	—
		WO-A-805	FM-GP-4659-C	SHIM, adjusting, drive pinion, bearing, small (.010 thk.) (SPS-638-2)	as req.	—	—	—	—
		WO-A-806	FM-GP-4659-D	SHIM, adjusting, drive pinion, bearing, small (.030 thk.) (SPS-638-3)	as req.	—	—	—	—
		WO-636566	FM-GP-4619	SLINGER, oil, drive pinion bearing (SP-13575)	1	—	3	4	10
		WO-A-799	FM-GP-4668	SPACER, drive pinion bearing (SP-15367)	1	—	3	3	10
		WO-A-792	FM-GP-4281	STRAP, lock, differential, drive gear screw (SP-16866)	4	—	—	—	—
10-1		WO-636570	FM-356504-S	WASHER, drive pinion nut (SP-S-1056) (issue until stock exhausted)	1	8	12	—	—

GROUP 10—FRONT AXLE (Cont'd)

Figure Number	Official Stockage Number	Part Number Willys	Part Number Ford	ITEM	Quantity Reqd. per Unit Assy.	Per 100 Major Items 12 Mos. Field Maintenance	Per 100 Major Items Major Overhaul (5th Ech)	Per 100 Major Items Total First Year Procurement	Estimated Reqmts. per 100 Rebuilds
Col. 1	Col. 2	Col. 3	Col. 4	Col. 5	Col. 6	Col. 7	Col. 8	Col. 9	Col. 10
				1003—DIFFERENTIAL GEARS, PINION AND BEARINGS (Cont'd)					
		WO-A-795	FM-GP-4228	WASHER, thrust, differential side gear (SP-16323-2)	2	—	—	—	—
		WO-A-797	FM-GP-4230	WASHER, thrust, differential pinion (SP-16322-2)	2	—	—	—	—
		WO-A-1445	FM-GP-4842	YOKE, end, universal joint, assembly (SP-K2-4-108-X)	1	—	3	4	10
				1006—STEERING KNUCKLE, FLANGE AND ARM					
10-2		WO-A-1712	FM-GPW-3113	ARM, knuckle steering, upper left (SP-17302-X) (Includes PIN, king; PIN, lock)	1	2	2	8	5
		WO-A-1710	FM-GPW-3112	ARM, knuckle steering, upper right (SP-17301-X) (Includes PIN, king and PIN, lock)	1	2	2	8	5
		WO-A-857	FM-GPW-3171	BEARING, bell crank (SP-17212) (TR-B-1210) (issue until stock exhausted)	2	22	148	—	—
		WO-A-8249	FM-GPW-3131-B	BELL CRANK, drag link front (SP-17307) (was WO-A-1211) (issue until stock exhausted) (FM-GPW-3131)	1	8	16	—	—
10-2		WO-A-872	FM-24327-S2	BOLT, hex-hd., S., $5/16$-24NF-3 x $5/8$ (oil seal) (SP-5-73-310) (issue until stock exhausted)	16	116	234	—	—
		WO-A-821	FM-355526-S	BOLT, hex-hd., S., $3/8$-24NF-3 x $1 1/16$ (stop screw) (SP-6-73-1117)	2	6	6	50	20
10-2		WO-A-853		BUSHING, wheel bearing spindle bz, 1.376 O.D. x 1.251 I.D. x $7/8$ wide	2	6	6	50	20
		WO-A-828	FM-GP-3140	CAP, king pin bearing, lower assembly (SP-17048-X) (Includes: PIN / PIN, king / PIN, lock)	2	—	6	6	20
10-2		WO-52940	FM-GP-3161	CONE and ROLLERS, king pin bearing (bore .625 O.D. x $9/16$ wide (TIM-11590)	4	6	6	15	20
10-2		WO-A-814	FM-GP-1089	CONTAINER, steering knuckle, oil seal, assembly (half) (SP-17133-X)	4	—	—	—	—
		WO-52941	FM-GP-3162	CUP, king pin bearing, 1.687 O.D. x $3/8$ wide (TIM-11520)	4	6	6	15	20
		WO-A-820	FM-GP-1092	GASKET, steering knuckle, oil seal, assembly (SP-17041)	4	80	—	100	—
		WO-A-6882	FM-GPW-18388	SHIM SET, king pin bearing (issue until stock exhausted) (Includes:	as req.	4	12	—	—
		WO-A-830	FM-GP-3117-A	4 SHIM, .003 thk.	4				
		WO-A-831	FM-GP-3117-B	4 SHIM, .005 thk.	4				
		WO-A-832	FM-GP-3117-C	8 SHIM, .010 thk.	8				
		WO-A-833	FM-GP-3117-D	6 SHIM, .030 thk.)	6				
10-2		WO-A-812	FM-GP-3149-A	KNUCKLE, steering, left (SP-17222)	1	2	2	7	5
		WO-A-811	FM-GP-3148-A	KNUCKLE, steering, right (SP-17221)	1	2	2	7	5
		WO-A-873	FM-33911-S2	NUT, hex., S., $3/8$-24NF-3 (stop screw lock) (SP-6-74-101)	2	6	6	50	20
10-2		WO-630598	FM-33786-S2	NUT, hex., S., $3/8$-24NF-2 (steering arm stud)	16	51	51	120	160
10-3		WO-A-779	FM-GP-3034	OIL SEAL, carrier end (SP-17036)	2	3	26	40	80

Fig.	Ord. No.	Mfr. No.	Description	Units per assy.	Qty A	Qty B	Qty C	Qty D
10-2	WO-A-819	FM-GP-3135	OIL SEAL, felt, steering knuckle (half) (SP-17019)	4	—	220	—	%160
10-2	WO-A-813	FM-GP-1088	OIL SEAL, steering knuckle, assembly (half) (SP-17135-X)	4	20	20	6	%6
13-1	WO-A-824	FM-GP-3115	PIN, king (SP-16991) (issue until stock exhausted)	4	—	—	52	28
13-1	WO-A-825	FM-GP-3122	PIN, lock, king pin (S., 1/8 O.D. x 3/8 lgth.) (SP-S-957) (issue until stock exhausted)	4			292	148
10-2	WO-5140	FM-353064-S	PLUG, filler, steering arm knuckle (SP-50-39-2)	2	—	—	—	—
10-2	WO-A-830	FM-GP-3117-A	SHIM, adjusting, king pin bearing (.003 thk.) (SP-16992-1)	as req.	—	—	—	—
	WO-A-831	FM-GP-3117-B	SHIM, adjusting, king pin bearing (.005 thk.) (SP-16992-2)	as req.	—	—	—	—
	WO-A-832	FM-GP-3117-C	SHIM, adjusting, king pin bearing (.010 thk.) (SP-16992-3)	as req.	—	—	—	—
	WO-A-833	FM-GP-3117-D	SHIM, adjusting, king pin bearing (.030 thk.) (SP-16994-4)	as req.	—	—	—	—
10-2	WO-A-851	FM-GP-3105	SPINDLE, wheel bearing, w/BUSHING, assembly (SP-17202-X)	2	10	9	3	3
10-2	WO-A-818	FM-GP-3139	STRIP, pressure, felt, steering knuckle, oil seal (SP-16983)	4	—	220	—	160
10-2	WO-A-1714	FM-357703-S	STUD, S., 3/8-24NF-3 x 2 (steering arm) (SP-S-962)	12	60	40	19	19
10-2	WO-A-5504	FM-GPW-3325	STUD, S., 3/8-24NF-3 x 2 (steering arm, dowel)	4	20	35	6	6
10-2	WO-5010	FM-34807-S	WASHER, lock, S., 2 1/2 O.D. x 1 3/32 I.D. x 3/32 thk. (steering arm stud)	16	—	—	—	—
10-2	WO-52510	FM-34941-S2	WASHER, lock, oil seal bolt (SP-529-W)	16	—	—	—	—

1007—AXLE SHAFTS, UNIVERSAL JOINTS

Fig.	Ord. No.	Mfr. No.	Description	Units per assy.	Qty A	Qty B	Qty C	Qty D
10-2	WO-A-1721	FM-358074-S	BALL, S., 11/16, universal joint (SP-17232) (with Rzeppa joints WO-A-1715, FM-GPW-3206-A1 and WO-A-1728, FM-GPW-3207-A1)	12	45	50	14	14
10-2	WO-A-1725	FM-24622-S2	BOLT, axle shaft retainer (hex-hd., S., No. 8) (.164-32NC-2) (included w/Rzeppa joints) (SP-17217) WO-A-1715, FM-GPW-3206-A1 and WO-A-1728, FM-GPW-3207-A1)	6	5.	8	2	2
10-2	WO-A-6362		BUSHING, axle shaft (SP-17361) (NP-38493) (required when replacing Bendix or Rzeppa joints with Tracta type) (Willys only)	2	—	—	—	—
10-1	WO-A-1719	FM-GP-3215-A	CAGE, universal joint (SP-17230) (w/Rzeppa joints WO-A-1715, FM-GPW-3206-A and WO-A-1728, FM-GPW-3207-A1)	2	10	9	3	3
10-1	WO-A-868	FM-GP-3204	FLANGE, drive, axle shaft (SP-17153)	2	—	—	—	—
10-1	WO-A-760	FM-GP-1110	BOLT, hex-hd., s-fin, alloy S., 3/8-16NC-2 x 1 1/2 in.	12	60	50	19	19
10-2	WO-5010		WASHER, lock, S., 3/8 in.	12	—	—	—	—
10-2	WO-A-6361		JOINT, universal, axle shaft, assembly (SP-17356-X) (NP-38487) (Willys only)	2	—	—	—	—
	WO-A-6472		KIT, repair, axle shaft and universal joint, left (Tracta joint) (Willys only) (Includes: BUSHING; JOINT, universal)	as req.	—	—	—	—
10-1	WO-A-1722	FM-GP-3219	PILOT, universal joint (SP-17233) (included w/Rzeppa joints A-1715, and A-1728)	2	10	8	3	3
10-1	WO-A-1723	FM-GP-3218	PIN, universal joint pilot (SP-17224) (included w/Rzeppa joints A-1715 and A-1728)	2	10	8	3	3
10-1	WO-A-1720	FM-GP-3221-A	RACE, universal joint, inner (SP-17231) (included w/Rzeppa joints A-1715 and A-1728)	2	—	—	—	—
10-1	WO-A-1724	FM-GP-3217	RETAINER, axle shaft (SP-17216) (included w/Rzeppa joints A-1715, and A-1728)	2	—	—	—	—
10-1	WO-A-1725	FM-24622-S2	SCREW, fl-hd., S., No. 8 (.164)-32NC-2 x 11/32 (SP-17217) (included with Rzeppa joints A-1715 and A-1728)	6	—	—	—	—
10-1	WO-A-1729	FM-GP-3017-A	SHAFT, axle, left (SP-17122-4) (included w/Rzeppa joints A-1715, and A-1728)	1	5	7	2	2
10-1	WO-A-1727	FM-GP-3016-A	SHAFT, axle, right (SP-17122-3) (included w/Rzeppa joints A-1715 and A-1728)	1	5	7	2	2
10-1, 2	WO-A-6383		SHAFT, axle, inner left (SP-17360-2) (NP-38491) (Willys only)	1	—	—	—	—

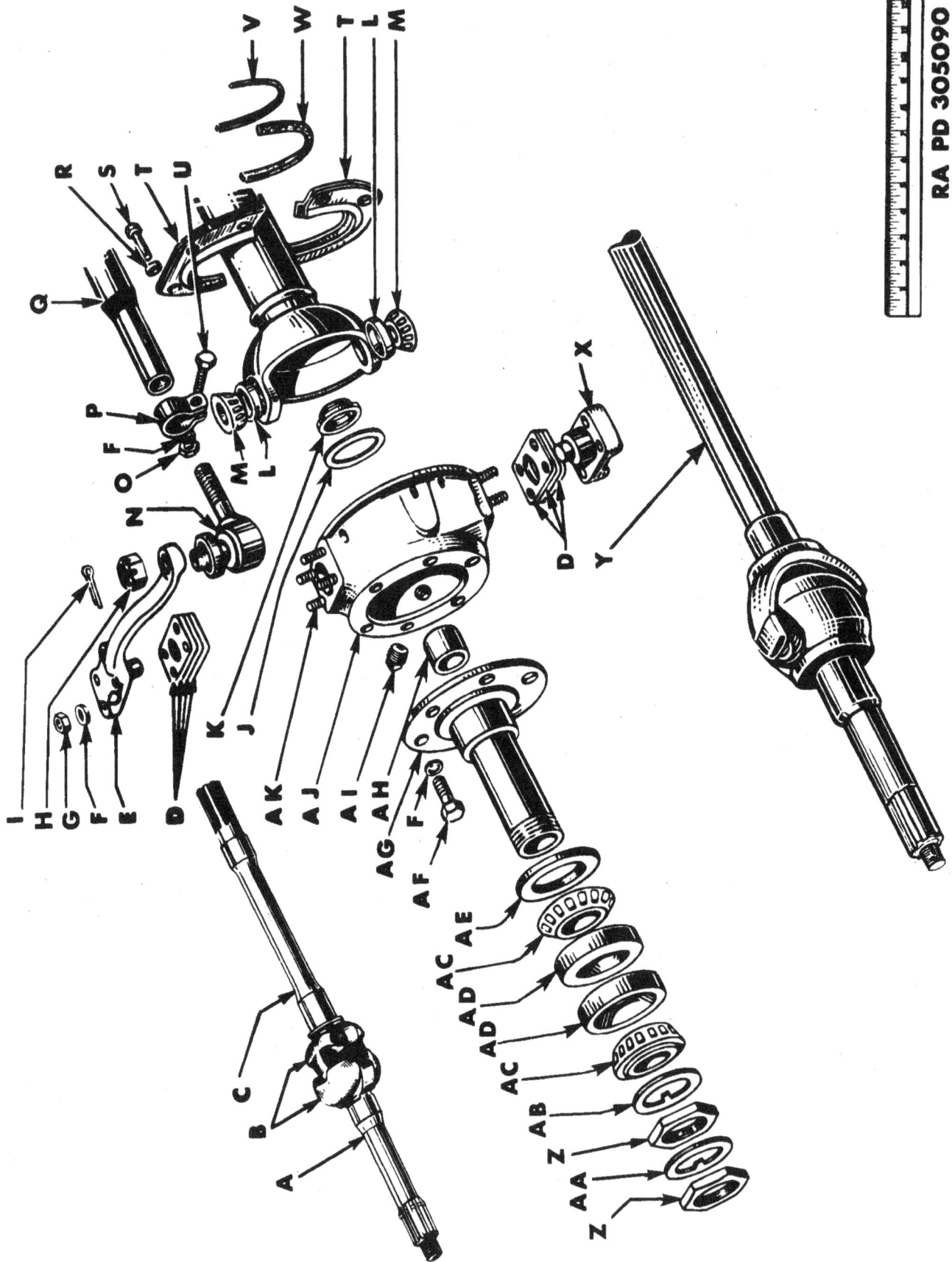

RA PD 305090

FIGURE 10-2—FRONT AXLE, STEERING KNUCKLE AND WHEEL BEARING

Key	Item	Willys Part No.	Ford Part No.	Gov't Group No.
A	SHAFT	WO-A-6382	(Willys only)	1007
B	UNIVERSAL joint assembly	WO-A-6361	(Willys only)	1007
C	SHAFT	WO-A-6383	(Willys only)	1007
D	SHIMS	WO-A-830	FM-GP-3117-A	1006
E	STEERING arm assembly	WO-A-1712	FM-GPW-3113	1006
F	WASHER	WO-5010	FM-34807-S	1006
G	NUT	WO-630598	FM-33786-S2	100
H	NUT	WO-10558	FM-351059-S7	1402
I	COTTER pin	WO-5152	FM-72025-S	1402
J	WASHER	WO-A-780	FM-CP-3374	1007
K	BUSHING	WO-A-6362	(Willys only)	1007
L	CUP	WO-52941	FM-GP-3162	1006
M	CONE and ROLLERS	WO-52940	FM-GP-3161	1006
N	END assembly	WO-A-847	FM-GP-3292	1402
O	NUT	WO-636575	FM-34083-S2	1402
P	CLAMP	WO-A-1706	FM-51-3287	1402
Q	TUBE—right	WO-A-1705	FM-GPW-3281	1402
R	WASHER	WO-52510	FM-34941-S2	1006
S	BOLT	WO-A-872	FM-24327-S2	1006
T	OIL seal assembly	WO-A-813	FM-GP-1088	1006
U	SCREW	WO-A-1707	FM-24916-S2	1402
V	FELT	WO-A-818	FM-GP-3139	1006
W	OIL seal	WO-A-819	FM-GP-3135	1006
X	CAP assembly	WO-A-828	FM-GP-3140	1006
Y	SHAFT w/JOINT assembly	WO-A-809	(Willys only)	1007
Z	NUT	WO-A-866	FM-GP-4252	1301C
AA	WASHER	WO-A-867	FM-GP-1124	1301C
AB	WASHER	WO-A-865	FM-GP-1218	1301C
AC	CONE and ROLLERS	WO-52942	FM-GP-1201	1301A
AD	CUP	WO-52943	FM-GP-1202	1301A
AE	OIL seal assembly	WO-A-864	FM-GP-1177	1301B
AF	SCREW	WO-A-877	FM-355552-S	1203
AG	SPINDLE and BUSHING assembly	WO-A-851	FM-GP-3105	1006
AH	BUSHING	WO-A-853	....	1006
AI	PLUG	WO-5140	FM-353064-S	1006
AJ	STEERING knuckle—right	WO-A-811	FM-GP-3148-A	1006
AK	STUD	WO-A-1714	FM-357703	1006

RA PD 305090-A

GROUP 10—FRONT AXLE (Cont'd)

		Part Number				Per 100 Major Items			
Figure Number	Official Stockage Number	Willys	Ford	ITEM	Quantity Reqd. per Unit Assy.	12 Mos. Field Maintenance	Major Overhaul (5th Ech)	Total First Year Procurement	Estimated Reqmts. per 100 Rebuilds
Col. 1	Col. 2	Col. 3	Col. 4	Col. 5	Col. 6	Col. 7	Col. 8	Col. 9	Col. 10
				1007—AXLE SHAFTS, UNIVERSAL JOINTS (Cont'd)					
10-2		WO-A-6384		SHAFT, axle, inner right (SP-17360-1) (NP-38492) (Willys only)	1	—	—	—	—
		WO-A-6382		SHAFT, axle, outer (SP-17359) (NP-38373) (Willys only)	2	—	—	—	—
		WO-A-810		SHAFT, axle, left, universal JOINT, assembly (Bendix) (SP-17128-2X) (Willys only)	1	—	—	—	—
10-2		WO-A-809		SHAFT, right, axle, w/universal JOINT, assembly (Bendix) (SP-17128-3X) (Willys only)	1	—	—	—	—
		WO-A-6030		SHAFT, axle, left, w/universal JOINT, assembly (Bendix or Tracta) (Willys only) (issue until stock exhausted)	1	4	8	7	—
		WO-A-6031		SHAFT, axle, right, w/universal JOINT, assembly (Bendix or Tracta) (Willys only) (issue until stock exhausted)	1	4	8	7	—
		WO-A-1728	FM-GPW-3207-A	SHAFT, axle, left, w/universal JOINT, assembly (Rzeppa) (SP-17120-4X)	1	2	2	—	5
		WO-A-1715	FM-GPW-3206-A1	SHAFT, axle, right, w/universal JOINT, assembly (Rzeppa) (SP-17120-3X)	1	2	2	—	5
		WO-A-1716	FM-GP-3200-A	SHAFT, axle, outer, w/universal JOINT, assembly (included w/Rzeppa JOINTS A-1715, and A-1728) (SP-17121-X)	2	3	3	8	10
10-1		WO-A-862	FM-GP-3208-A	SHIM, adjusting, universal joint (.010 thk.) (SP-17155-1)	as req.	—	—	—	—
		WO-A-863	FM-GP-3208-B	SHIM, adjusting, universal joint (.030 thk.) (SP-17155-2)	as req.	—	—	—	—
		WO-A-6881	FM-GPW-18336	SHIM SET, universal joint adjusting (Includes: 6 WO-A-862 FM-GP-3208-A SHIM, .010 thk. 3 WO-A-863 FM-GP-3208-B SHIM, .030 thk.)	as req.	6	6	18	20
10-2		WO-A-1726	FM-GP-3216	SNAP RING, axle shaft retainer (SP-17218)	2	5	5	12	15
		WO-A-780	FM-GP-3374	WASHER, thrust, axle shaft (SP-S-953)	2	3	3	12	10
				GROUP 11—REAR AXLE					
				1100—REAR AXLE ASSEMBLY					
10-2		WO-A-445	FM-GPW-4001	AXLE, rear, assembly (4.88 ratio) (Willys only)	1	—	—	—	—
		WO-A-5500		AXLE, rear, w/BRAKES, DRUMS and HUBS, assembly (issue until stock exhausted)	1	—	—	—	—
		WO-A-575	FM-GPW-5705	CLIP, rear spring (axle to spring)	1	3	8	4	—
		WO-339372	FM-GPW-5456	NUT, hex., S., 7/16-20NF-2 (clip nut) (issue until stock exhausted)	4	200	400	—	—
		WO-5938	FM-34838-S	WASHER, lock, S., 7/16 in.	8	—	—	—	—
		WO-A-1545	FM-GPW-18366	GASKET SET, rear axle (Use until exhausted, then use WO-A-1545)	8	—	—	—	—
		WO-A-7831	FM-GPW-18366-B	GASKET SET, rear axle (was WO-A-1545) (Includes: 2 WO-A-904 FM-GP-4032 GASKET, axle shaft 1 WO-A-782 FM-GP-4035 GASKET, housing cover 1 WO-636565 FM-GP-4661 GASKET, oil seal)	as req.	6	13	30	40

1101—HOUSING ASSEMBLY

Fig.	WO No.	FM No.	Name and description	Qty				
10-3	WO-636527	FM-355699-S2	BOLT, hex-hd., S., ½-13NC-2 x 2¼ (differential bearing cap)	4	—	6	8	20
10-3	WO-A-781	FM-GP-4016	COVER, axle housing (SP-16976)	1	1	1	4	2
10-3	WO-51523	FM-20046-S2	BOLT, hex-hd., S., 5/16-18NC-2 x ¾ (cover to housing)	8	—	—	—	—
10-3	WO-52510	FM-34941-S2	WASHER, lock, S., 5/16 in.	10	%40	—	50	—
10-3	WO-A-782	FM-GP-4035	GASKET, axle housing cover (SP-16409)	1	—	2	3	5
11-1	WO-A-888	FM-GPW-4004	HOUSING, axle, w/TUBE, assembly (SP-17226-X)	1	%8	16	—	10
11-1	WO-636577	FM-358048-S	PLUG, drain, axle housing (SP-S-780) (issue until stock exhausted)	1	—	—	5	—
11-1	WO-636538	FM-353051-S	PLUG, filler, axle housing (SP-50-39-4)	1	%5	—	—	—
	WO-A-870	FM-GP-4022	PLUG, vent, axle housing (SP-16979)	1	—	—	—	—
	WO-636528	FM-34922-S	WASHER, lock, S., internal, ½ in. (differential bearing cap screw) (issue until stock exhausted)	4	68	132	—	—

1102—AXLE DRIVE SHAFTS

Fig.	WO No.	FM No.	Name and description	Qty				
11-1	WO-A-760	FM-GP-1110	BOLT, hex-hd., S., 3/8-16NC-3 x 1½ (drive shaft bolt) (SP-6-73-124)	12	19	19	50	60
11-1	WO-A-904	FM-GP-4032	GASKET, axle shaft (SP-17146)	2	%160	8	200	35
	WO-A-6439	FM-GPW-4259	GUIDE, axle shaft (SP-17377) (issue until stock exhausted)	2	%7	26	40	—
11-1	WO-A-779	FM-GP-3034	OIL SEAL, axle inboard (SP-17036)	2	%3	2	9	5
10-3	WO-A-902	FM-GP-4235	SHAFT, rear axle, left (SP-17144-4)	1	%2	2	9	5
11-1	WO-A-901	FM-GPW-4234	SHAFT, rear axle, right (SP-17144-3)	1	%2	2	—	—
11-1	WO-5010	FM-34807-S7	WASHER, lock, S., 3/8 in. (axle shaft to hub bolt)	12	—	—	—	—

1103—DIFFERENTIAL AND CARRIER ASSEMBLY

Fig.	WO No.	FM No.	Name and description	Qty				
11-1	WO-A-793	FM-GP-4206	CASE, differential (SP-16383)	1	—	3	4	10
11-1	WO-52880	FM-GP-4221	CONE, and ROLLERS, differential bearing, bore 1.625 x 1.000 (TIM-24780)	2	—	13	14	40
11-1	WO-52881	FM-GP-4222	CUP, differential bearing, .8125 wide x 3.000 O.D. (TIM-24721)	2	—	13	14	40
11-1	WO-A-788	FM-GPW-4212	DIFFERENTIAL assembly (Ford only)	1	—	—	—	—
11-1	WO-A-794	FM-GPW-4236	DIFFERENTIAL assembly (SP-16968-X) (Willys only)	2	—	3	4	10
	WO-A-6816	FM-GPW-18384	GEAR, side, O.D. 2.758 x 1⅛ wide, 16 teeth (SP-16385)	2	—	—	—	—
			KIT, repair, drive gear and pinion	as req.	—	—	—	—
			(Includes:					
	WO-636569	FM-356126-S	1 NUT, drive pinion	1				
	WO-A-871	FM-355511-S	8 SCREW, drive gear	8				
	WO-A-789	FM-GPW-4209	1 SET, gear and pinion	1				
	WO-A-792	FM-GP-4281	4 STRAP, lock	4				
11-1	WO-A-6743	FM-GPW-18389	KIT, repair, rear axle differential gear	as req.	—	3	4	10
			(Includes:					
	WO-A-794	FM-GP-4236	2 GEAR, side	2				
	WO-A-796	FM-GPW-4215	2 MATE, pinion	2				
	WO-A-798	FM-GP-4211	1 SHAFT, pinion	1				
	WO-A-797	FM-GP-4230	2 WASHER, thrust, pinion	2				
	WO-A-795	FM-GP-4228	2 WASHER, thrust side gear	2				
11-1	WO-636360	FM-GP-4241	PIN, lock, pinion shaft (SP-13449)	1	—	6	—	60
11-1	WO-A-796	FM-GPW-4215	PINION, differential (O.D. 1.140, 10 teeth) (SP-15926)	2	—	—	—	—
11-1	WO-A-871	FM-355511-S	SCREW, drive gear (SP-73-414)	8	—	13	20	40
	WO-A-789	FM-GPW-4209	SET, drive gear and pinion (matched gear w/PINION) (SP-16412-X)	1	—	—	—	—

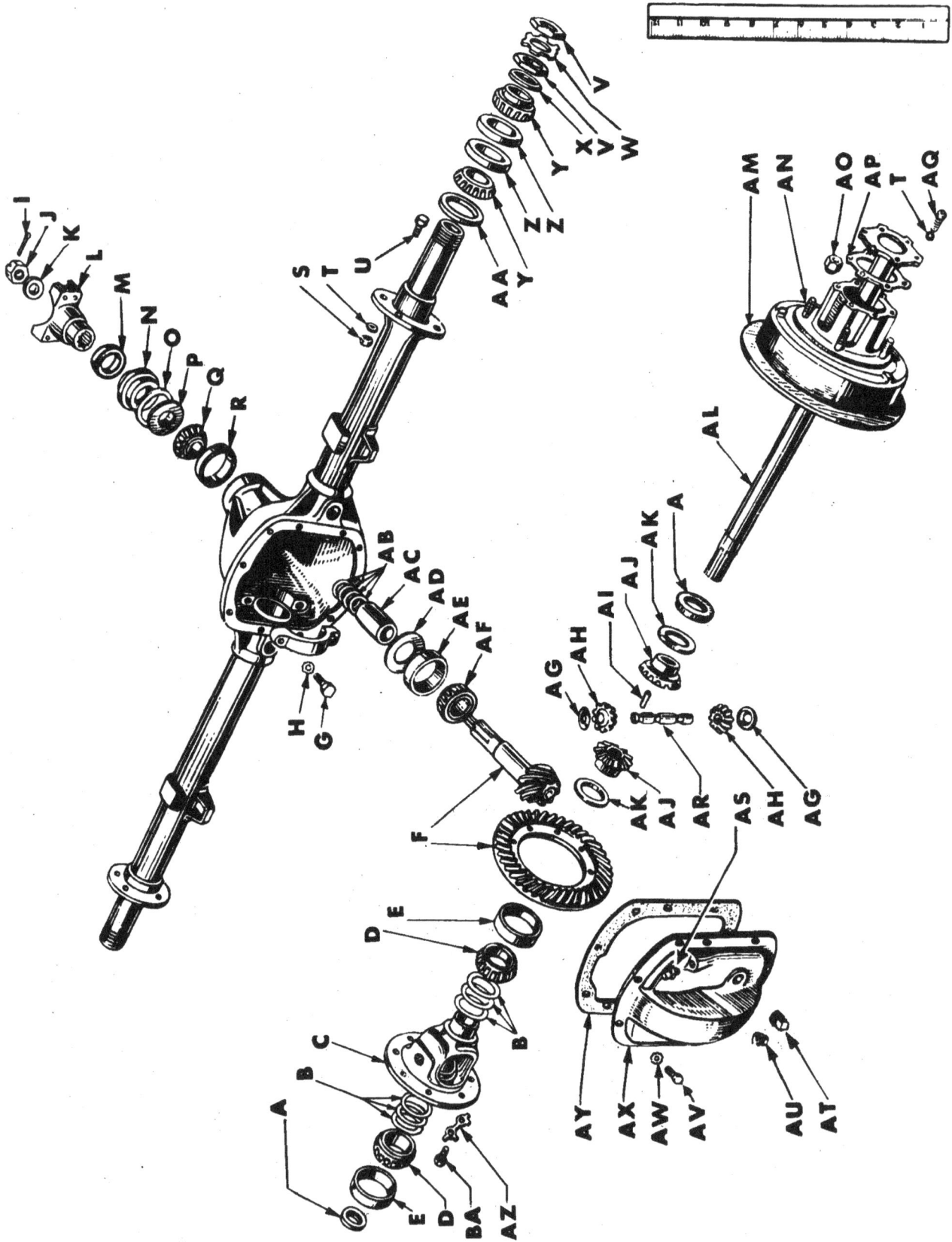

RA PD 305091

FIGURE 11-1—REAR AXLE ASSEMBLY

Key	Item	Willys Part No.	Ford Part No.	Gov't Group No.
A	SEAL	WO-A-779	FM-GP-3034	1102
B	SHIMS	WO-A-784	FM-GP-4229-A	1103
C	CASE	WO-A-793	FM-GP-4206	1103
D	CONE and ROLLERS	WO-52880	FM-GP-4221	1103
E	CUP	WO-52881	FM-GP-4222	1103
F	GEAR and PINION	WO-A-789	FM-GPW-4209	1103
G	BOLT	WO-636527	FM-355699-S2	1101
H	WASHER	WO-636528	FM-34922-S	1101
I	PIN	WO-636571	FM-357202-S	1104
J	NUT	WO-636569	FM-356126	1104
K	WASHER	WO-636570	FM-356504-S	1104
L	YOKE assembly	WO-A-1445	FM-GP-4842	1104
M	SHIELD	WO-636568	FM-GP-4666	1104
N	SEAL	WO-639265	FM-GP-4676-A	1104
O	GASKET	WO-636565	FM-GP-4661	1104
P	SLINGER	WO-636566	FM-GP-4619	1104
Q	CONE and ROLLERS	WO-52878	FM-GP-4630	1104
R	CUP	WO-52879	FM-GP-4628	1104
S	NUT	WO-636575	FM-33786-S2	1203
T	WASHER	WO-5010	FM-34807-S7	1102
U	SCREW	WO-A-903	FM-355578-S	1203
V	NUT	WO-A-866	FM-GP-4252	1301C
W	WASHER	WO-A-867	FM-GP-1124	1301C
X	WASHER	WO-A-865	FM-GP-1218	1301C
Y	CONE and ROLLERS	WO-52942	FM-GP-1201	1301A
Z	CUP	WO-52943	FM-GP-1202	1301A
AA	RETAINER	WO-A-864	FM-GP-1177	1301B
AB	SHIM	WO-A-803	FM-GP-4659-A	1104
AC	SPACER	WO-A-799	FM-GP-4668	1104
AD	SHIM	WO-A-800	FM-GP-4660-A	1104
AE	CUP	WO-52877	FM-86H-4621	1104
AF	CONE and ROLLERS	WO-52876	FM-86H-4621	1104
AG	WASHER	WO-A-797	FM-GP-4230	1103
AH	PINION	WO-A-796	FM-GPW-4215	1103
AI	PIN	WO-636360	FM-GP-4241	1103
AJ	GEAR	WO-A-794	FM-GPW-4236	1103
AK	WASHER	WO-A-795	FM-GP-4228	1103
AL	SHAFT	WO-A-901	FM-GPW-4234	1102
AM	DRUM	WO-472	(Willys only)	1302
AN	BOLT	WO-A-474	FM-GP-1107	1302A
AO	NUT	WO-A-476	FM-GP-1012	1302A
AP	GASKET	WO-A-904	FM-GP-4032	1102
AQ	BOLT	WO-A-760	FM-GP-1110	1102
AR	SHAFT	WO-A-798	FM-GPW-4209	1103
AS	PLUG	WO-A-870	FM-GP-4022	1101
AT	PLUG	WO-636538	FM-353051-S	1101
AU	PLUG	WO-636577	FM-358048-S	1101
AV	SCREW	WO-51523	FM-20046-S2	1101
AW	WASHER	WO-52510	FM-34941-S2	1101
AX	COVER	WO-A-781	FM-GP-4016	1101
AY	GASKET	WO-A-782	FM-GP-403511	1101
AZ	STRAP	WO-A-792	FM-GP-4281	1103
BA	SCREW	WO-A-871	FM-355511-S	1103

RA PD 305091-A

SNL G-503

GROUP 11—REAR AXLE (Cont'd)

Figure Number	Official Stockage Number	Willys	Ford	ITEM	Quantity Reqd. per Unit Assy.	12 Mos. Field Maintenance	Major Overhaul (5th Ech)	Total First Year Procurement	Estimated Reqmts. per 100 Rebuilds
Col. 1	Col. 2	Col. 3	Col. 4	Col. 5	Col. 6	Col. 7	Col. 8	Col. 9	Col. 10
				1103—DIFFERENTIAL AND CARRIER ASSEMBLY (Cont'd)					
11-1		WO-A-798	FM-GP-4211	SHAFT, pinion (SP-16075)	1	—	—	—	—
10-3		WO-A-784	FM-GP-4229-A	SHIM, adjusting (.003 thk.) (SP-S-58)	as req.	—	—	—	—
		WO-A-785	FM-GP-4229-B	SHIM, adjusting (.005 thk.) (SP-S-59)	as req.	—	—	—	—
		WO-A-786	FM-GP-4229-C	SHIM, adjusting (.010 thk.) (SP-S-74)	as req.	—	—	—	—
		WO-A-787	FM-GP-4229-E	SHIM, adjusting (.015 thk.) (Ford only)	as req.	—	—	—	—
			FM-GP-4229-D	SHIM, adjusting (.030 thk.) (SP-S-75)	as req.	—	—	—	—
		WO-A-6744	FM-GPW-18388	SHIM SET, rear axle, differential bearing. (Includes: 2 WO-A-784 FM-GP-4229-A SHIM, .003 — 2 WO-A-785 FM-GP-4229-B SHIM, .005 — 2 WO-A-786 FM-GP-4229-C SHIM, .010 — 2 WO-A-787 FM-GP-4229-D SHIM, .030)	as req.	—	3	5	10
11-1		WO-A-792	FM-GP-4281	STRAP, lock, drive gear screw (SP-16866)	4	—	—	—	—
11-1		WO-A-797	FM-GP-4230	WASHER, thrust, pinion (SP-16322-2)	2	—	—	—	—
11-1		WO-A-795	FM-GP-4228	WASHER, thrust, side gear (SP-16323-2)	2	—	—	—	—
				1104—DIFFERENTIAL PINION BEARING					
11-1		WO-52876	FM-86H-4621	CONE and ROLLERS, bearing (bore 1.375 x 1.125) drive pinion, inner (TM-31593)	1	—	3	4	10
11-1		WO-52878	FM-GP-4630	CONE and ROLLERS, bearing (bore 1.125 x .875) drive pinion, outer (TM-02872)	1	—	3	4	10
11-1		WO-52877	FM-86H-4616	CUP, drive pinion bearing, inner (3.00 O.D. x .9375 wide) (TM-31520)	1	—	3	4	10
11-1		WO-52879	FM-GP-4628	CUP, drive pinion bearing, outer (2.875 O.D. x .6875 wide) (TM-02820)	1	—	—	—	—
11-1		WO-636565	FM-GP-4661	GASKET, pinion oil seal, leather (SP-S-171)	1	26	—	30	—
11-1		WO-636569	FM-356126-S	NUT, hex., S., slotted, ¾-16NF-3, drive pinion (SP-S-1135) (issue until stock exhausted)	1	20	40	20	—
11-1		WO-639265	FM-GP-4676-A	OIL SEAL, pinion, leather (SP-14223)	1	13	—	—	—
11-1		WO-636571	FM-357202-S	PIN, cotter, S., 1.540 x .109 (SP-7-72-39)	1	—	—	—	—
11-1		WO-636568	FM-GP-4666	SHIELD, dust, universal joint end yoke (SP-15099)	1	—	—	—	—
11-1		WO-A-800	FM-GP-4660-A	SHIM, adjusting, bearing (large, .003) (SP-S-112)	as req.	—	—	—	—
		WO-A-801	FM-GP-4660-B	SHIM, adjusting, bearing (large, .005) (SP-S-113)	as req.	—	—	—	—
		WO-A-802	FM-GP-4660-C	SHIM, adjusting, bearing (large, .010) (SP-S-114)	as req.	—	—	—	—
		WO-A-803	FM-GP-4659-A	SHIM, adjusting, bearing (small, .003) (SP-S-638)	as req.	—	—	—	—
		WO-A-804	FM-GP-4659-B	SHIM, adjusting, bearing (small, .005) (SP-S-638-1)	as req.	—	—	—	—
		WO-A-805	FM-GP-4659-C	SHIM, adjusting, bearing (small, .010) (SP-S-638-2)	as req.	—	—	—	—
11-1		WO-A-806	FM-GP-4659-D	SHIM, adjusting, bearing (small, .030) (SP-S-638-3)	as req.	—	—	—	—

A	B	C	D	Qty	Description	Ord. No.	Mfr. No.	Fig.
10	5	3	—	as req.	SHIM SET, rear axle drive pinion (Includes:	WO-A-6745	FM-GPW-18386	
10	4	3	—	1	4 WO-A-800 FM-GP-4660-A SHIM, large, .003			
10	3	3	—	1	4 WO-A-801 FM-GP-4660-B SHIM, large, .005			
—	—	12	8	1	4 WO-A-802 FM-GP-4660-C SHIM, large, .010			
10	4	3	—	1	4 WO-A-803 FM-GP-4659-A SHIM, small, .003			
				1	4 WO-A-804 FM-GP-4659-B SHIM, small, .005			
				1	4 WO-A-805 FM-GP-4659-C SHIM, small, .010			
				1	4 WO-A-806 FM-GP-4659-D SHIM, small, .030)			
				1	SLINGER, oil drive pinion bearing (SP-13575)	WO-636566	FM-GP-4619	11-1
				1	SPACER, drive pinion bearing (SP-15367)	WO-A-799	FM-GP-4668	11-1
				1	WASHER, S, 1½ O.D. x 49/64 I.D. x ⅛ thk., drive pinion nut (SP-S-1056) (issue until stock exhausted)	WO-636570	FM-355504-S	11-1
				1	YOKE, universal joint end, w/SHIELD assembly (SP-K2-4-108X)	WO-A-1445	FM-GP-4842	11-1

GROUP 12—BRAKE

1200—BRAKE ASSEMBLY

A	B	C	D	Qty	Description	Ord. No.	Mfr. No.	Fig.
		4	%	1	BRAKE, front, left, assembly (BX-47047) (issue until stock exhausted)	WO-A-8894	FM-GPW-2011	12-2
		4	%	1	BRAKE, front, right, assembly (BX-47048) (issue until stock exhausted)	WO-A-8895	FM-GPW-2010	
		—	—	1	BRAKE, rear, left, assembly, ⅞ in. (before Willys serial 134356) (BX-452322)	WO-A-927	FM-GPW-2211	
		4	%	1	BRAKE, rear, left, assembly, ¾ in. (after Willys serial 134356) (BX-47050) (issue until stock exhausted)	WO-A-8896	FM-GPW-2211-B	
		—	—	1	BRAKE, rear, right, assembly, ⅞ in. (before Willys serial 134356) (BX-45323)	WO-A-928	FM-GPW-2210	
		4	%	1	BRAKE, rear, right, assembly, ¾ in. (after Willys serial 134356) (BX-47051) (issue until stock exhausted)	WO-A-8897	FM-GPW-2210-B	

NOTE: WO-A-8894, FM-GPW-2011 and WO-A-8895, FM-GPW-2010 may be used for service in place of WO-A-927, FM-GPW-2211 and WO-A-928, FM-GPW-2210, providing both LEFT and RIGHT brakes are changed.

1201—HAND BRAKE EMERGENCY BRAKE PARTS

A	B	C	D	Qty	Description	Ord. No.	Mfr. No.	Fig.
5	6	2	2	1	BAND, hand brake, assembly (SP-5900-X)	WO-A-1009	FM-GP-2648	12-1
	40	5	5	1	BOLT, brake band brkt. (hex-hd., S., ¼-20NC-2 x 2¾) (SP-856D) (WG-X-2428-A)	WO-A-1019	FM-355352-S7	12-1
				1	NUT, hex, S., ¼-20NC-2	WO-5790	FM-33795-S7	12-1
				1	WASHER, lock, S., ¼ in.	WO-52706	FM-34805-S2	12-1
		4		1	BRACKET, hand brake ratchet tube, w/guide SPRING and SCREW, assembly (issue until stock exhausted)	WO-639010	FM-GPW-2848	12-5
				1	BOLT, hex-hd., S., 5/16-24NF-2 x ¾ (brkt. to support)	WO-51396	FM-24347-S2	12-5
				2	NUT, hex-hd., S., 5/16-24NF-2	WO-5910	FM-33798-S2	
				2	WASHER, S. (plain) 5/16 in.	WO-5437	FM-34706-S2	
				2	WASHER, lock, S., 5/16 in.	WO-51833	FM-34806-S2	
5	8	2	2	1	BOLT, adjusting, hand brake (SP-10-B-18)	WO-A-1016	FM-O1T-2642	12-1
30	25	10	10	1	BOLT, anchor clip (hex-hd., S., 5/16-18NC-2 x 1⅝) (SP-887-D)	WO-A-1020	FM-O1T-2616-	12-1
		4	4	1	BRAKE, hand, assembly (SP-5815-X) (issue until stock exhausted)	WO-A-1008	FM-GPW-2598	F
		8	4	1	CABLE, hand brake ratchet, w/TUBE, assembly (issue until stock exhausted)	WO-A-1241	FM-GPW-2853	

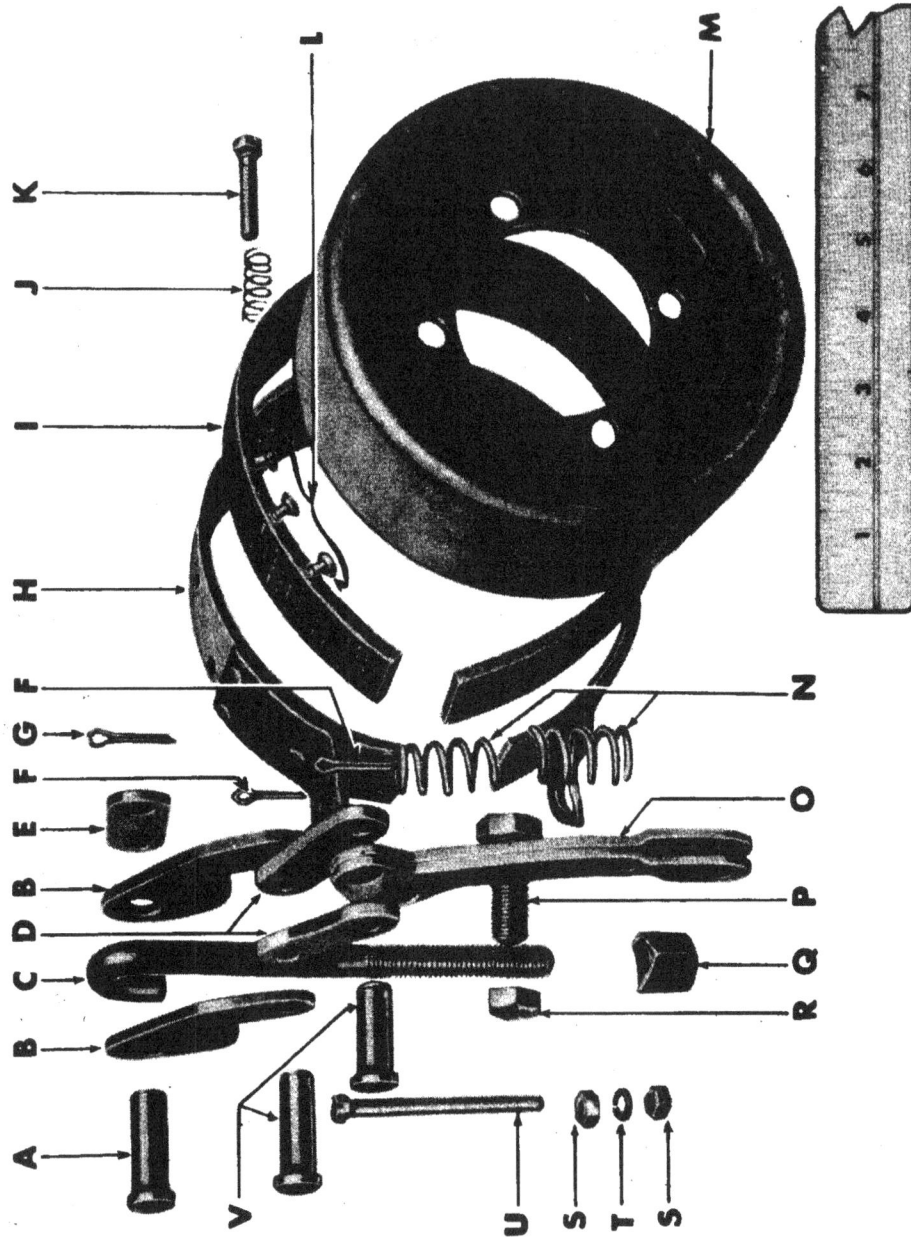

RA PD 305088

RA PD 305088-A

FIGURE 12-1—HAND BRAKE ASSEMBLY

Key	Item	Willys Part No.	Ford Part No.	Gov't Group No.
A	PIN	WO-A-1004	FM-73928-S7	1201
B	CAM	WO-A-1003	FM-GPW-2632	1201
C	BOLT	WO-A-1016	FM-01T-2642	1201
D	LINK	WO-A-1228	FM-GPW-2659	1201
E	QUADRANT	WO-A-1005	FM-GPW-2630	1201
F	PIN	WO-5354	FM-72004-S	1201
G	PIN	WO-50020	FM-72017-S	1201
H	BAND assembly	WO-A-1009	FM-GP-2648	1201
I	LINING	WO-A-1014	FM-GP-2620	1201
J	SPRING	WO-A-1021	FM-01T-2140	1201
K	SCREW	WO-A-1020	FM-01T-2616	1201

Key	Item	Willys Part No.	For I Part No.	Gov't Group No.
L	RIVET	WO-A-1015	FM-64647-S	1201
M	DRUM	WO-A-1002	FM-GP-2614	1201
N	SPRING	WO-A-1017	FM-01T-2634	1201
O	CRANK assembly	WO-A-1226	FM-GPW-2656	1201
P	CAP screw	WO-A-1227	FM-355752-S7	1201
Q	NUT	WO-A-1018	FM-01T-2805	1201
R	NUT	WO-52925	FM-33927-S7	1201
S	NUT	WO-5790	FM-33795-S7	1201
T	WASHER	WO-52706	FM-34805-S2	1201
U	BOLT	WO-A-1019	FM-355352-S7	1201
V	PIN	WO-311003	FM-73904-S7	1201

GROUP 12—BRAKE (Cont'd)

1201—HAND BRAKE EMERGENCY BRAKE PARTS (Cont'd)

Col. 1 Figure Number	Col. 2 Official Stockage Number	Col. 3 Willys	Col. 4 Ford	Col. 5 ITEM	Col. 6 Quantity Reqd. per Unit Assy.	Col. 7 12 Mos. Field Maintenance	Col. 8 Major Overhaul (5th Ech)	Col. 9 Total First Year Procurement	Col. 10 Estimated Reqmts. per 100 Rebuilds
12-1		WO-A-1003	FM-GPW-2632	CAM, hand brake (SP-5-B-15) (issue until stock exhausted)	2	4	4	—	—
		WO-A-1735	FM-GPW-2272	CLAMP, hand brake cable tube (at air cleaner brkt.) (issue until stock exhausted)	1				—
		WO-A-1533	FM-34746-S7	WASHER, S, 5/16 in. (issue until stock exhausted)	1	16	24		—
		WO-638780	FM-GPW-2279	CLAMP, hand brake cable tube (at transfer case bearing cap) (issue until stock exhausted)	1		4		—
		WO-50929	FM-20367-S7	BOLT, hex-hd, S, 5/16-24NF-2 x 7/8	2				—
		WO-5910	FM-33798-S2	NUT, hex, S, 5/16NF-2	1				—
		WO-51833	FM-34806-S7	WASHER, lock, S, 5/16 in.	2				—
		WO-A-5393	FM-GPW-2270-B	CLIP, 5/16 in., S, closed type (9/32 bolt hole) (cable tube to bell housing) (after engine Willys 114550) (All Ford trucks) (issue until stock exhausted)	1		8		—
		WO-51763	FM-20308-S2	BOLT, hex-hd, S, 1/4-20NC-2 x 1/2 (to bell housing) (after Willys engine 114550)	1	4	8		—
		WO-52706	FM-34805-S2	WASHER, lock, S, 1/4 in. (after Willys engine 114550)	1				—
		WO-A-1795		CLIP, support, hand brake cable (cable to rear engine insulator) (before Willys engine 114550) (Willys only)	1				—
12-5		WO-A-1226	FM-GPW-2656	CRANK, hand brake cable, assembly (SP-5-B-16)	1	2	2	8	5
12-1		WO-52925	FM-33927-S7	NUT, hex, S, 7/16-20NF (crank to transfer bearing cap) (issue until stock exhausted)	1	4	8		—
		WO-A-1227	FM-355752-S	SCREW, hex-hd, S, 7/16-20NF x 1 1/4	1	3	3	10	—
12-1		WO-A-1002	FM-GP-2614	DRUM, hand brake (SP-20-B-62)	1	2	2	7	5
		WO-A-997	FM-355551-S	BOLT, hex-hd, S, 3/8-20NF-3 x 1 1/8 (SP-6-73-218)	4	29	29	100	90
12-5		WO-636575	FM-33786-S2	NUT, hex, S, 3/8-24NF-2	4				—
		WO-5010	FM-34807-S2	WASHER, lock, S, 3/8 in.	4				—
12-1		WO-A-1014	FM-GP-2620	LINING, hand brake (SP-4-B-26)	1				—
12-1		WO-A-1015	FM-64647-S	RIVET, tubular, ck-hd, br. (lining to band) (SP-238-R)	14				—
		WO-A-6759	FM-GPW-18377	LINING SET, hand brake, w/RIVETS. (Includes: 1 WO-A-1014 FM-GP-2620 LINING / 15 WO-A-1015 FM-64647-S RIVET)	as req.	6	6	15	20
12-5		WO-639244		HANDLE, hand brake (issue until stock exhausted)	1		4		30
		WO-51904	FM-92047-S	SCREW, rd-hd, S, br-plted, type "U", No. 6 (.138) x 5/8 (handle to tube)	2	10	10	50	—
12-5		WO-A-1242	FM-GPW-2780	HANDLE, hand brake, w/TUBE and CABLE, assembly (issue until stock exhausted)	1				—
12-5		WO-392468	FM-357553-S18	PIN, clevis, S, 1/4 O.D. x 2 1/2 lgth. (dld. f/c-pin)	1	4	12	24	—
		WO-5067	FM-72003-S	PIN, cotter 1/16 x 1/2	1	10	10		—
12-1		WO-A-1228	FM-GPW-2659	LINK, hand brake cable crank (issue until stock exhausted)	2	4	4		—

GROUP 12—BRAKE (Cont'd)

| | | Part Number | | | | Per 100 Major Items | | | |
Figure Number	Official Stockage Number	Willys	Ford	ITEM	Quantity Reqd. per Unit Assy.	12 Mos. Field Maintenance	Major Overhaul (5th Ech)	Total First Year Procurement	Estimated Reqnts. per 100 Rebuilds
Col. 1	Col. 2	Col. 3	Col. 4	Col. 5	Col. 6	Col. 7	Col. 8	Col. 9	Col. 10
				1201—HAND BRAKE EMERGENCY BRAKE PARTS (Cont'd)					
12-1		WO-311003	FM-73904-S7	PIN, clevis, S., .3115 diam. x 1½ lgth. (crank to link and link to cam) dld. f/c-pin (SP-3-Q-76)	2	% 10	10	40	30
12-1		WO-5354	FM-72004-S	PIN, cotter, ³⁄₃₂ x ½ (SP-42-G)	2	—	16	—	—
12-1		WO-A-1018	FM-01T-2805	NUT, adjusting, hand brake band (SP-373-J) (issue until stock exhausted)	1	% 8	16	—	—
12-5		WO-A-1004	FM-73928-S7	PIN, hand brake cam (dld. f/c-pin) (SP-232-SP) (issue until stock exhausted)	1	8	3	24	10
12-1		WO-A-1006	FM-73889-S7	PIN, hand brake support quadrant (dld. f/c-pin) (SP-153-SP)	1	3	—	—	—
12-1		WO-5020	FM-72017-S	PIN, cotter, ³⁄₃₂ x ¾ (cam pin)	1	—	8	—	—
12-1		WO-52967	FM-72016-S	PIN, cotter, ³⁄₃₂ x ⅝ (quadrant pin)	1	—	3	9	10
12-1		WO-A-1005	FM-GPW-2630	QUADRANT, support, hand brake (SP-3-0-74) (issue until stock exhausted)	1	% 4	2	6	—
12-1		WO-A-1021	FM-01T-2640	SPRING, anchor clip bolt (SP-14-B-6)	1	% 3	2	6	—
12-1		WO-A-1017	FM-01T-2634	SPRING, releasing, hand brake (SP-12-B-5)	2	% 2	5	12	—
12-5		WO-635681	FM-GPW-7291	SPRING, hand brake ratchet tube	1	% 2	4	—	—
12-5		WO-A-5335	FM-GPW-2635	SPRING, retracting, hand brake (after Willys serial 102731)	1	% 5	—	12	—
12-5		WO-A-2892	FM-GPW-2852	SUPPORT, hand brake ratchet tube bracket (issue until stock exhausted)	1	—	5	—	—
		WO-52207	FM-60371-2	RIVET, rd-hd., ¼ x ⁹⁄₁₆	2	% 5	—	12	—
12-5		WO-A-1241	FM-GPW-2853	TUBE, hand brake ratchet, w/CABLE and CONDUIT, assembly	1	—	—	—	—
		WO-5437	FM-34706-S2	WASHER, plain, S, SAE std., ⁵⁄₁₆ in. (ratchet tube support)	2	—	—	—	—
				1202—SHOES AND FACINGS					
12-2		WO-116551	FM-GP-2021	LINING, brake shoe, forward, ground (BX-1067-S-3)	4	—	—	—	—
12-2		WO-116552	FM-GP-2022	LINING, brake shoe, reverse, ground (BX-1067-S-4)	4	—	—	—	—
12-2		WO-116600	FM-GPW-18367	LINING SET, brake shoe, w/RIVETS (BX-44517). (Includes: 2 WO-116551 FM-GP-2021 LINING, forward / 2 WO-116552 FM-GP-2022 LINING, reverse / 42 WO-374586 FM-351915-S RIVETS.)	as req.	72	8	93	25
				NOTE: This lining set for two (2) wheels.					
12-2		WO-374586	FM-351915-S	RIVET, tubular, br. (BX-179-S-6)	80	—	—	—	—
12-2		WO-116549	FM-GP-2018	SHOE, brake, forward, w/LINING, assembly (BX-1141-S-1)	4	% 13	—	16	4
12-2		WO-116550	FM-GP-2019	SHOE, brake, reverse, w/LINING, assembly (BX-1141-S-2)	4	% 13	—	16	4
				1203—BRAKE SHOE SUPPORT					
		WO-A-450		PLATE, backing, front and rear, brake, assembly (issue until stock exhausted then use WO-A-8898 for maintenance)	4	—	3	4	10
12-2		WO-A-8898	FM-GP-2013	PLATE, backing, front and rear brake, assembly (BX-47054) (was WO-A-450)	4	—	—	—	—

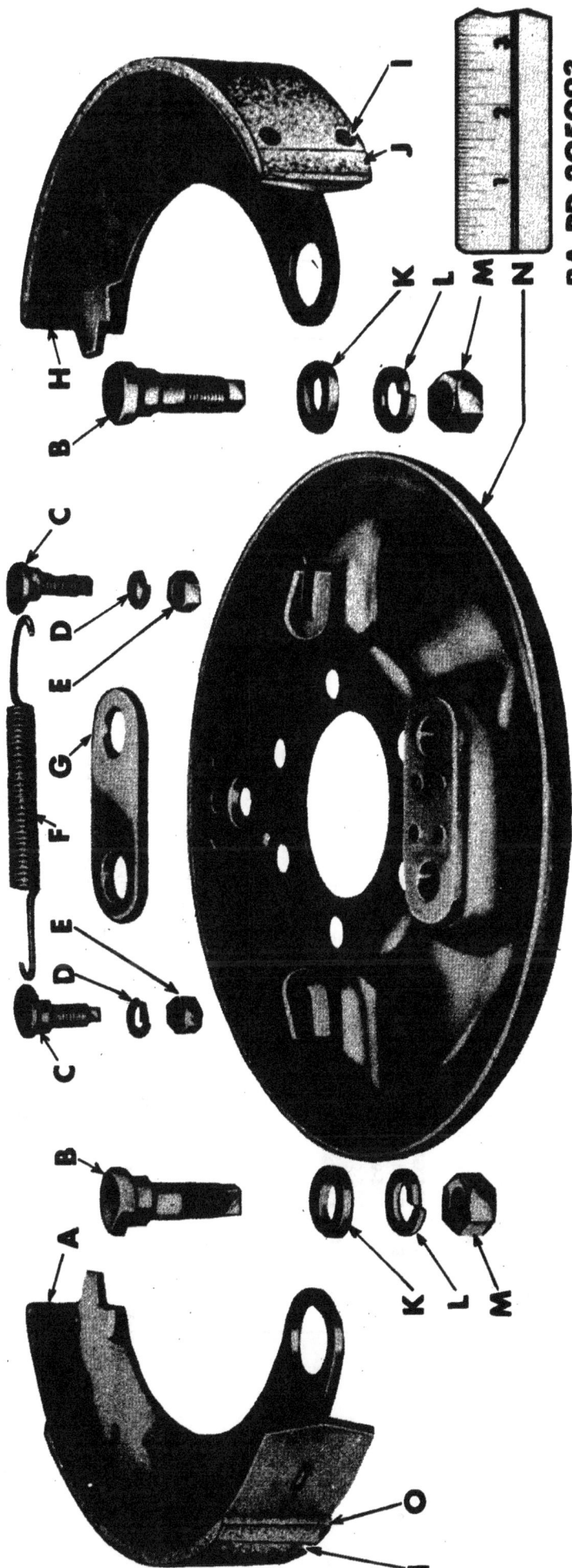

RA PD 305093

FIGURE 12-2—BRAKE ASSEMBLY

Key	Item	Willys Part No.	Ford Part No.	Gov't Group No.
I	RIVET	WO-374586	FM-351915-S	1202
J	LINING	WO-116551	FM-GP-2021	1202
K	CAM	WO-647900	FM-GP-2028	1203B
L	WASHER	WO-637923	FM-351466-S24	1203B
M	NUT	WO-637924	FM-33846-S2	1203B
N	PLATE assembly	WO-A-8898	FM-GP-2013	1203
O	LINING	WO-116552	FM-GP-2022	1202

RA PD 305093-A

Key	Item	Willys Part No.	Ford Part No.	Gov't Group No.
A	SHOE assembly	WO-116550	FM-GP-2019	1202
B	PIN	WO-637899	FM-91A-2027	1203B
C	ECCENTRIC	WO-A-754	FM-GP-2038	1203A
D	WASHER	WO-5010	FM-34807-S2	1203A
E	NUT	WO-A-755	FM-33800-S7	1203A
F	SPRING	WO-637905	FM-GP-2035	1203A
G	PLATE	WO-637901	FM-91A-2030	1203B
H	SHOE assembly	WO-116459	FM-GP-2018	1202

SNL G-503

GROUP 12—BRAKE (Cont'd)

Col. 1 Figure Number	Col. 2 Official Stockage Number	Col. 3 Willys	Col. 4 Ford	Col. 5 ITEM	Col. 6 Quantity Reqd. per Unit Assy.	Col. 7 Per 100 Major Items 12 Mos. Field Maintenance	Col. 8 Per 100 Major Items Major Overhaul (5th Ech)	Col. 9 Total First Year Procurement	Col. 10 Estimated Reqmts. per 100 Rebuilds
				1203—BRAKE SHOE SUPPORT (Cont'd)					
12-5		WO-A-903	FM-355578-S	BOLT, hex-hd., S., 3/8-24NF x 1 5/16 (rear brake plate to housing)(SP-6-73-513)	12	19	19	66	60
10-2		WO-A-377	FM-353352-S	BOLT, hex-hd., 3/8-24NF-2 x 3/4 (front brake plate to steering knuckle) (SP-IS-1310)	12	19	19	70	60
11-1		WO-636575	FM-33786-S2	NUT, hex, 3/8-24NF-2 (rear brake plates only) (SP-6-74-11)	12	—	—	—	—
12-5		WO-5010	FM-34807-S7	WASHER, lock, S., 3/8 in. (front brake plates only)	12	—	—	—	—
				1203A—GUIDE SPRINGS					
12-2		WO-A-754	FM-GP-2038	ECCENTRIC, brake shoe (BX-45771)	8	6	6	20	20
12-2		WO-A-755	FM-33800-S7	NUT, hex, S., 3/8-24NF-2 eccentric (BX-46752)	8	—	—	—	—
12-2		WO-5010	FM-34807-S2	WASHER, lock, S., 3/8 in. (BX-40-S-33)	8	—	—	—	—
12-2		WO-637905	FM-GP-2035	SPRING, return, brake shoe (BX-41545)	4	% 4	5	45	15
				1203B—ADJUSTING PIN AND ANCHOR PLATE					
12-2		WO-637900	FM-GP-2028	CAM, anchor pin (BX-41876)	8	13	13	28	40
12-2		WO-637924	FM-33846-S2	NUT, anchor pin (hex, S., 1/2-20NF-2) (BX-41708)	8	—	5	14	15
12-2		WO-637899	FM-91A-2027	PIN, anchor (BX-39953)	8	5	5	—	—
12-2		WO-637901	FM-91A-2030	PLATE, anchor pin (BX-39956) (issue until stock exhausted)	8	20	40	80	30
12-2		WO-637923	FM-351466-S24	WASHER, anchor pin (lock, S., 1/2 in.)	8	10	10	—	—
				1204—PEDAL AND SPRING					
				NOTE: For parts not listed here see Groups 0204 and 0204A.					
12-5		WO-640038	FM-358006-S8	FITTING, grease, straight (AD-1980)	1	—	—	—	—
12-5		WO-A-1359	FM-GPW-2454	PAD, brake pedal, assembly (issue until stock exhausted)	1	—	4	—	—
		WO-A-1386	FM-GPW-2452-A	PEDAL, brake, assembly (issue until stock is exhausted, then use WO-A-8253 for maintenance)	1	—	—	—	—
12-5		WO-A-8253	FM-GPW-2452-B	PEDAL, brake, assembly (was WO-A-1386)	1	—	4	—	—
		WO-A-495	FM-GPW-2473	SHAFT, brake pedal, assembly (issue until stock exhausted)	2	—	4	—	—
				1205—MASTER CYLINDER					
12-5		WO-A-1354	FM-GPW-2138	BAR, tie, master cylinder	1	—	—	—	—
		WO-51523	FM-20046-S	BOLT, hex-hd., S., 5/16-18NC-2 x 3/4 (tie bar and shield to master cylinder)	1	—	—	—	—
		WO-52836	FM-355444-S	BOLT, hex-hd., S., 5/16-24NF-2 x 3 (tie bar and shield to cylinder)	1	% 10	10	24	30
		WO-5910	FM-33798-S	NUT, hex, S., 5/16-24NF-2	1	—	—	—	—
		WO-51833	FM-34806-S2	WASHER, lock, S., 5/16 in.	2	—	—	—	—

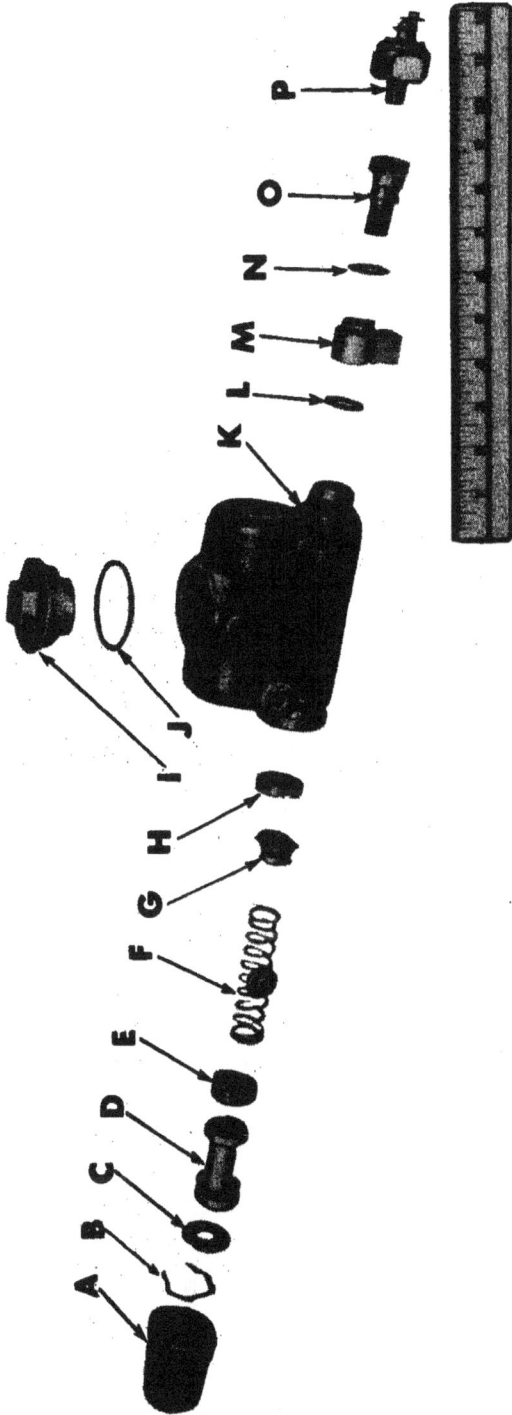

RA PD 305094

FIGURE 12-3—BRAKE MASTER CYLINDER

Key	Item	Willys Part No.	Ford Part No.	Gov't Group No.
I	CAP	WO-637608	FM-GP-2162	1205
J	GASKET	WO-637612	FM-GP-2167	1205
K	TANK	WO-637582	FM-GP-2155	1205
L	GASKET	WO-637604	FM-91A-2152	1205
M	FITTING	WO-A-557	FM-GP-2076	1205
N	GASKET	WO-637606	FM-91A-2151	1205
O	BOLT	WO-637605	FM-GP-2077	1205
P	SWITCH	WO-A-1271	FM-11A-1348	0606

RA PD 305094-A

Key	Item	Willys Part No.	Ford Part No.	Gov't Group No.
A	BOOT	WO-637602	FM-GP-2180	1205
B	WIRE	WO-637598	FM-GP-2174	1205
C	PLATE	WO-637597	FM-GP-2188	1205
D	PISTON	WO-637591	FM-GP-2169	1205
E	CUP	WO-637590	FM-GP-2173	1205
F	SPRING	WO-637587	FM-GP-2145	1205
G	VALVE	WO-637584	FM-GP-2175	1205
H	SEAT	WO-637583	FM-GP-2160	1205

FIGURE 12-4—WHEEL BRAKE CYLINDER ASSEMBLY

Key	Item	Willys Part No.	Ford Part No.	Gov't Group No.
A	BOOT	WO-637546	FM-GP-2206-A	1207
B	GUIDE	WO-637577	FM-GP-2194	1207
C	WASHER	WO-52483	FM-34905-S2	1207
D	BOLT	WO-51738	FM-20300-S7	1207
E	CYLINDER	WO-A-1502	FM-GPW-2063	1207
F	CUP	WO-637579	FM-91A-2201	1207
G	SCREW	WO-637540	FM-GP-2208	1207
H	SPRING	WO-637580	FM-GP-2205	1207

RA PD 305095-A

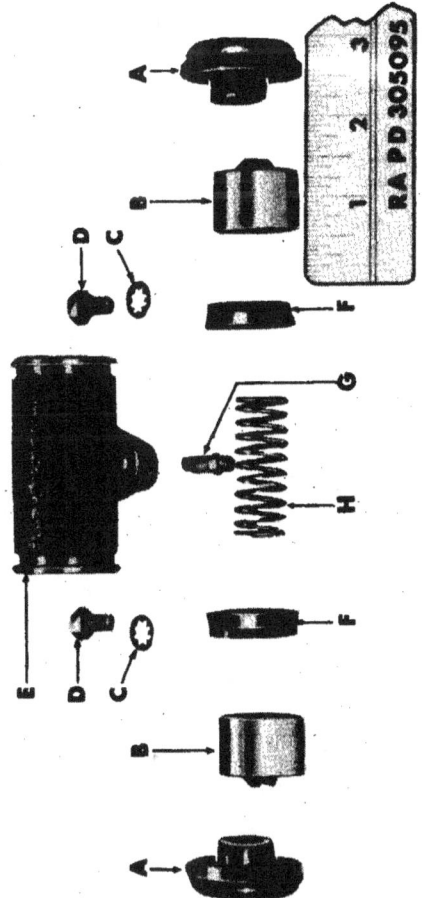

RA PD 305095

GROUP 12—BRAKE (Cont'd)

1205—MASTER CYLINDER (Cont'd)

Figure Number	Official Stockage Number	Willys	Ford	ITEM	Quantity Reqd. per Unit Assy.	12 Mos. Field Maintenance	Major Overhaul (5th Ech)	Total First Year Procurement	Estimated Reqmts. per 100 Rebuilds
		Col. 3	Col. 4	Col. 5	Col. 6	Col. 7	Col. 8	Col. 9	Col. 10
12-5		WO-A-183	FM-GPW-2462	BOLT, master cylinder to pedal (eye, S., 7/16-20NF-2) (issue until stock exhausted)	1	—	—	—	—
12-5		WO-5939	FM-33802-S2	NUT, hex., S., 7/16-20NF-2 (lock nut)	1	—	—	—	—
12-5		WO-5020	FM-72016-S	PIN, cotter, 3/32 x 3/4	1	—	—	—	—
		WO-52835	FM-356394-S2	WASHER, S., (plain) 7/16 in.	1	4 %	4	8	—
12-3		WO-637605	FM-GP-2077	BOLT, master cylinder outlet fitting	1	—	—	—	—
12-3		WO-637602	FM-GP-2180	BOOT, master cylinder (LO-S-FC-6011)	1	—	—	—	10
12-3		WO-637608	FM-GP-2162	CAP, filler, master cylinder, assembly (LO-S-FC-6018-E)	1	3 %	3	10	10
		WO-637586	FM-GP-2183	CUP, master cylinder, check valve (rubber) 1⅞ in. diam. (LO-S-FC-2919-B)	1	3 %	3	13	—
12-3		WO-637590	FM-GP-2173	CUP, master cylinder primary (rubber 1 in. bore) (LO-S-FD-2108-F)	1	—	—	—	—
		WO-637595	FM-GP-2170	CUP, master cylinder, secondary (rubber 1 in. piston) (LO-S-FD-2109-B)	1	—	—	—	—
		WO-A-556	FM-GP-2140	CYLINDER, master cylinder, assembly (LO-FE-1444)	1	10 %	10	24	—
12-3		WO-51798	FM-355398-S2	BOLT, hex-hd., S., 5/16-18NC-2 x 3 (cylinder to brkt.)	1	10 %	10	25	15
		WO-52274	FM-34706-S2	WASHER, S. (plain) 5/16 in.	1	—	—	—	15
		WO-51833	FM-34806-S2	WASHER, lock, S., 5/16 in.	1	—	—	—	20
12-3		WO-A-557	FM-GP-2076	FITTING, outlet, master cylinder (LO-S-FC-5727-A)	1	10 %	5	12	20
12-3		WO-637612	FM-GP-2167	GASKET, master cylinder filler cap (LO-S-FC-6019)	1	5 %	5	20	—
12-3		WO-637604	FM-91A-2152	GASKET, master cylinder outlet fitting (LO-S-FC-602)	1	5 %	6	20	—
		WO-637606	FM-91A-2151	GASKET, master cylinder outlet fitting bolt (LO-S-FC-603)	1	6 %	6	20	—
		WO-A-6836	FM-GPW-18370	KIT, repair, brake master cylinder (issue until stock is exhausted, then use WO-A-7838 for maintenance)	1	38	72	—	—
		WO-A-7838	FM-GPW-18370-B	KIT, repair, brake master cylinder (was WO-A-6836). (Includes:	as req.	10	3	16	10
				1 WO-637602 FM-GP-2180 BOOT					
				1 WO-637590 FM-GP-2173 CUP, primary					
				1 WO-637604 FM-91A-2152 GASKET, outlet					
				1 WO-637606 FM-91A-2151 GASKET, outlet bolt					
				1 WO-637591 FM-GP-2169 PISTON					
				1 WO-637583 FM-GP-2160 SEAT					
				1 WO-637584 FM-GP-2175 VALVE					
				1 WO-637598 FM-GP-2174 WIRE, lock)					
12-3		WO-637591	FM-GP-2169	PISTON, master cylinder, assembly (LO-S-FC-6007)	1	6 %	—	8	—
12-3		WO-637597	FM-GP-2188	PLATE, stop, master cylinder piston (LO-S-FC-2926)	1	3 %	3	8	10
		WO-637585	FM-GP-2176	RETAINER, master cylinder, check valve cup (one in. diam. cylinder) (LO-S-FC-2918-A) (issue until stock exhausted)	1	—	—	4	—
12-5		WO-637599	FM-GP-2143-A	ROD, push, master cylinder, assembly (LO-S-FC-6014) (issue until stock exhausted)	1	—	—	4	—
12-3		WO-637583	FM-GP-2160	SEAT, master cylinder valve (LO-S-FC-6010)	1	—	—	—	—

1207—WHEEL CYLINDERS

WO No.	FM No.	Description	Qty						Fig.
WO-A-647	FM-GPW-5118	SHIELD, master cylinder, assembly (issue until stock exhausted)	1	—	—	—	4	—	12-3
WO-637587	FM-GP-2145	SPRING, return, master cylinder piston, w/RETAINER, assembly (LO-S-FC-6009)	1	—	3	3	8	10	12-3
WO-637582	FM-GP-2155	TANK, supply, master cylinder (LO-S-FD-4564)	1	—	—	—	—	—	12-3
WO-637584	FM-GP-2175	VALVE, check, master cylinder, assembly (LO-S-FC-2917)	1	—	—	—	—	—	12-3
WO-637598	FM-GP-2174	WIRE, lock, piston stop, master cylinder (LO-S-FC-2927)	1	—	3	3	10	10	12-3
WO-637546	FM-GP-2206-A2	BOOT, front and rear wheel cylinder, (⅞ in. cylinder) (up to serial 137915 rear wheel only) (LO-FC-5994)	8	—	—	—	—	—	12-4
WO-A-6117	FM-GPW-2206	BOOT, rear wheel cylinder, ¾ in., (after Willys serial 137915) (LO-FC-8779)	4	—	—	—	—	—	
WO-637579	FM-91A-2201	CUP, front wheel cylinder, (1 in. diam.) (LO-S-FC-1499)	4	—	—	—	—	—	12-3
WO-637544	FM-GP-2201	CUP, rear wheel cylinder, (⅞ in. diam.) (up to Willys serial 137915) (LO-S-FC-3023)	4	—	—	—	—	—	
WO-A-6116	FM-GPW-2201	CUP, rear wheel cylinder, (¾ in. diam.)(after Willys serial 137915) (LO-FC-4158)	4	—	—	—	—	—	
WO-A-1502	FM-GP-2063	CYLINDER, front wheel brake, (LO-FD-8547)	2	—	—	—	—	—	12-4
WO-637789	FM-GP-2192	CYLINDER, rear wheel brake, (⅞ in. diam.) (before Willys serial 137915) (LO-FD-4664)	2	—	—	—	—	—	
WO-A-6111	FM-GPW-2135	CYLINDER, rear wheel brake, (¾ in. diam.) (after Willys serial 137915) (LO-FC-8782)	2	—	—	—	—	—	12-5
WO-A-1484	FM-GP-2061	CYLINDER, front wheel brake, assembly (1 in. diam.) (LO-FD-7379) (BX-45908)	2	%	5	5	14	15	13-1
WO-637787	FM-GP-2261	CYLINDER, rear wheel brake, assembly (⅞ in. diam.) (before Willys serial 137915) (LO-FD-4665) (BX-41887)	2	—	—	—	—	—	
WO-A-6110	FM-GPW-2261	CYLINDER, rear wheel brake, assembly (¾ in. diam.) (after Willys serial 137915) (LO-FD-7568-A) (BX-46491)	2	%	5	5	14	15	
WO-51738	FM-20300-S7	BOLT, hex-hd., S., ¼-20NC-2 x ⅜ (to backing plate) (BX-62-S-176) (issue until stock exhausted)	8	—	—	—	—	—	12-4
WO-52483	FM-34905-S2	WASHER, lock, S., internal, flat, ¼ in. (BX-76-S-25) (issue until stock exhausted)	8	—	—	—	—	—	12-4
WO-637577	FM-GP-2194	GUIDE, front wheel cylinder, piston and shoe, assembly (LO-FC-5997)	4	—	6	6	16	20	12-4
WO-637541	FM-GP-2196	GUIDE, rear wheel cylinder, piston and shoe, assembly (before Willys serial 137915) (LO-S-FC-5998)	4	—	—	—	—	—	
WO-A-6113	FM-GPW-2196	GUIDE, rear wheel cylinder, piston and shoe, assembly (after Willys serial 137915) (LO-FC-8778)	4	—	—	6	8	20	
WO-115962	FM-GPW-18371	KIT, repair, front wheel cylinder (LO-FC-5381). (Includes: 2 WO-637546 FM-GP-2206-A2 BOOT 2 WO-637579 FM-91A-2201 CUP)	as req.	—	19	6	28	20	
WO-115963	FM-GPW-18372	KIT, repair, rear wheel cylinder (before Willys serial 137915). (Includes: 2 WO-637546 FM-GP-2206-A2 BOOT 2 WO-637544 FM-GP-2201 CUP)	as req.	—	—	—	—	—	
WO-A-6133	FM-GPW-18368	KIT, repair, rear wheel cylinder (after Willys serial 137915). (Includes: 2 WO-A-6117 FM-GPW-2206 BOOT 2 WO-A-6116 FM-GPW-2201 CUP)	as req.	—	19	6	28	20	
WO-637540	FM-GP-2208	SCREW, bleeder, wheel cylinder (LO-S-FC-5993)	4	%	5	6	20	15	12-4

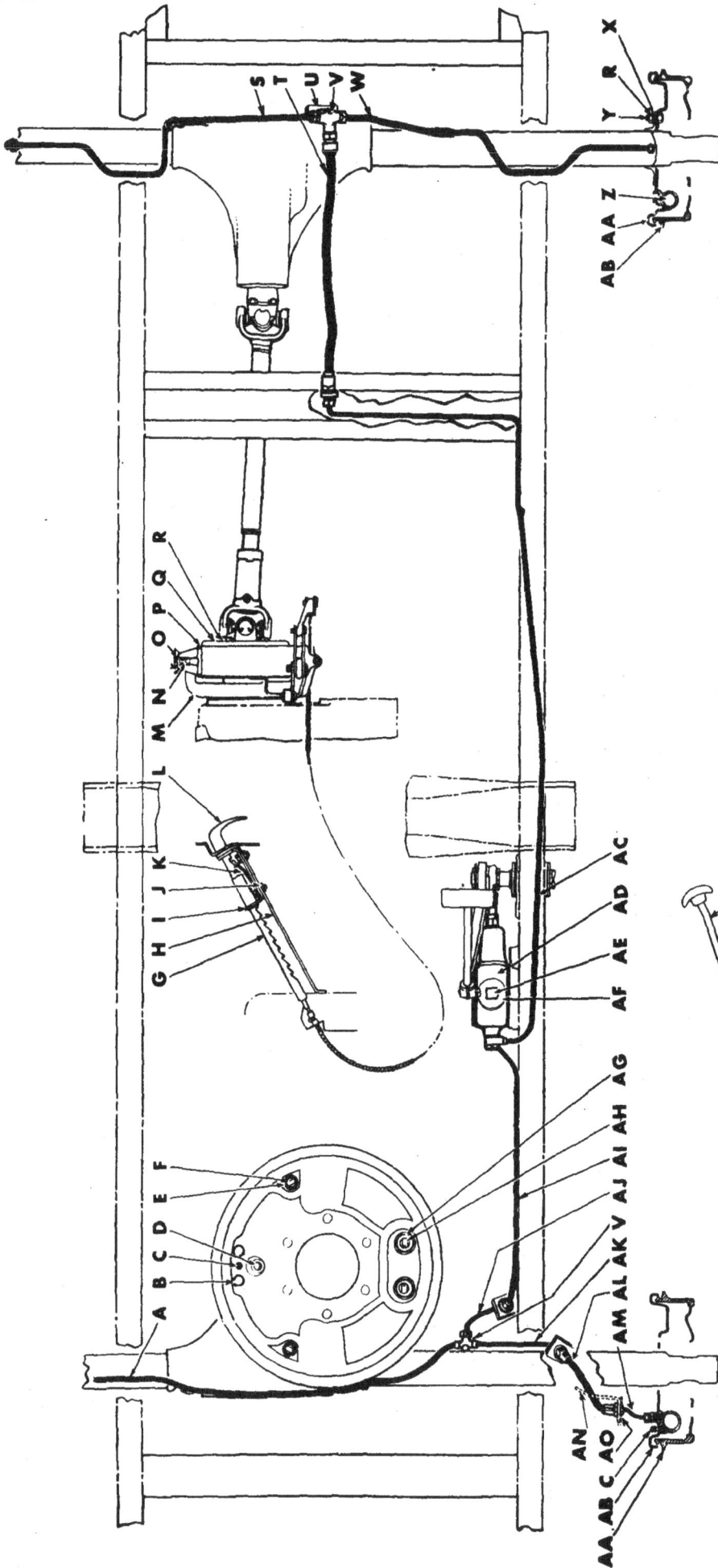

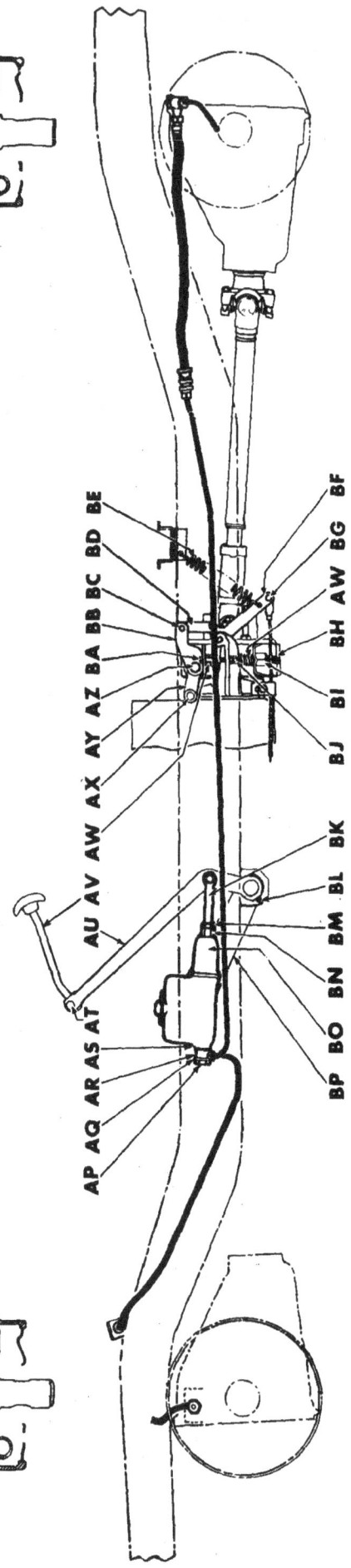

RA PD 305092

Labels (left figure): S T U V W · Y R X · AA Z · AB · L M N O P Q R · G H I J K · A B C D E F · AF AE AD AC · AM AL AK V AJ AI AH AG · AN · AA AB C AO

Labels (right figure): AU AV AW AX AY AZ BA BB BC BD BE · BJ BI BH AW BG BF · BP BO BN BM BL BK · AP AQ AR AS AT

FIGURE 12-5—BRAKE SYSTEM

Key	Item	Willys Part No.	Ford Part No.	Gov't Group No.	Key	Item	Willys Part No.	Ford Part No.	Gov't Group No.
A	TUBE	WO-A-1376	FM-GPW-2266	1209C	AI	TUBE assembly	WO-A-1377	FM-GPW-2264	1209C
B	SCREW	WO-51738	FM-20300-S7	1207	AJ	HOSE	WO-A-1373	FM-GPW-2078	1209B
C	SCREW	WO-637540	FM-GP-2208	1207	AK	TUBE assembly	WO-A-1501	FM-GPW-2263	1209C
D	CYLINDER	WO-A-1502	FM-GPW-2063	1207	AL	HOSE	WO-A-1460	FM-GPW-2079	1209B
E	NUT	WO-A-755	FM-33800-S7	1203A	AM	TUBE assembly	WO-A-1488	FM-GPW-2298	1209C
F	ECCENTRIC	WO-A-754	FM-GP-2038	1203A	AN	GUARD	WO-A-1457	FM-GPW-2096	1001
G	HANDLE w/TUBE and CABLE	WO-A-1242	FM-GPW-2780	1201	AO	CLIP	WO-637427	FM-78-2814-A	1209C
H	SUPPORT	WO-A-2892	FM-GPW-2852	1201	AP	BOLT	WO-637605	FM-GP-2077	1205
I	BRACKET assembly	WO-639010	FM-GPW-2848	1201	AQ	GASKET	WO-637606	FM-91A-2151	1205
J	NUT	WO-51396	FM-24347-S2	1201	AR	FITTING	WO-A-557	FM-GP-2076	1205
K	SPRING	WO-635681	FM-GPW-7291	1201	AS	GASKET	WO-637604	FM-91A-2152	1205
L	HANDLE	WO-639244	FM-GPW-7282	1201	AT	SCREW	WO-6157	FM-24426	0204
M	CAP	WO-A-1507	FM-GPW-7768	0801	AU	PEDAL assembly	WO-A-8253	FM-GPW-2452-B	1204
N	SPRING	WO-A-1021	FM-01T-2634	1201	AV	PAD assembly	WO-A-1359	FM-GPW-2454	1204
O	SCREW	WO-A-1020	FM-01T-2616	1201	AW	SPRING	WO-A-1017	FM-01T-2634	1201
P	BAND assembly	WO-A-1009	FM-GP-2648	1201	AX	PIN	WO-A-1006	FM-73889-S7	1201
Q	DRUM	WO-A-1002	FM-GP-2614	1201	AY	QUADRANT	WO-A-1005	FM-GPW-2630	1201
R	NUT	WO-636575	FM-33786-S2	1201	AZ	PIN	WO-A-1004	FM-73928-S7	1201
S	TUBE assembly	WO-A-5226	FM-GPW-2267	1209C	BA	SCREW	WO-A-1019	FM-355352-S7	1201
T	HOSE	WO-637424	FM-GPW-2078	1209B	BB	CAM	WO-A-1003	FM-GPW-2632	1201
U	BRACKET	WO-A-5227	FM-GPW-2274	1209C	BC	PIN	WO-311003	FM-73904-S7	1201
V	TEE	WO-637432	FM-GP-2074	1209C	BD	LINK	WO-A-1228	FM-GPW-2659	1201
W	TUBE assembly	WO-A-5225	FM-GPW-2268	1209C	BE	SPRING	WO-A-5335	FM-GPW-2635	1201
X	WASHER	WO-5010	FM-34807-S7	1203	BF	CRANK assembly	WO-A-1226	FM-GPW-2656	1201
Y	SCREW	WO-A-903	FM-355578-S	1203	BG	PIN	WO-392468	FM-357553-S18	1201
Z	CYLINDER	WO-A-6111	FM-GPW-2135	1207	BH	BOLT	WO-A-1016	FM-01T-2642	1201
AA	PLATE assembly	WO-A-8898	FM-GP-2013	1203	BI	NUT	WO-A-1018	FM-01T-2805	1201
AB	DRUM	WO-A-472	(Willys only)	1302	BJ	NUT	WO-5790	FM-33795-S7	1201
AC	TUBE assembly	WO-A-5224	FM-GPW-2265	1209C	BK	BOLT	WO-A-183	FM-GPW-2462	1205
AD	TANK	WO-637582	FM-GP-2155	1205	BL	FITTING	WO-640038	FM-358006-S8	1204
AE	CAP assembly	WO-637608	FM-GP-2162	1205	BM	NUT	WO-5939	FM-33802-S2	1205
AF	GASKET	WO-637612	FM-GP-2167	1205	BN	ROD assembly	WO-637599	FM-GP-2143-A	1205
AG	NUT	WO-637924	FM-33846-S2	1203B	BO	BOOT	WO-637602	FM-GP-2180	1205
AH	PIN	WO-637899	FM-91A-2027	1203B	BP	BAR	WO-A-1354	FM-GPW-2138	1205

RA PD 305092-A

SNL G-503

GROUP 12—BRAKE (Cont'd)

Figure Number	Official Stockage Number	Part Number Willys	Part Number Ford	ITEM	Quantity Reqd. per Unit Assy.	12 Mos. Field Maintenance	Major Overhaul (5th Ech)	Total First Year Procurement	Estimated Reqmts. per 100 Rebuilds
Col. 1	Col. 2	Col. 3	Col. 4	Col. 5	Col. 6	Col. 7	Col. 8	Col. 9	Col. 10
				1207—WHEEL CYLINDERS (Cont'd)					
12-4		WO-637580	FM-GP-2205	SPRING, front wheel cylinder cup (LO-FC-5992)................	2	3	3	8	10
		WO-637545	FM-GP-2204	SPRING, rear wheel cylinder cup (LO-FC-6003).............	2	3	3	8	10
				1209B—HOSES					
		WO-637426	FM-GP-2087	GASKET, fitting brake hose (after Willys serial 138841)............	2	%38	19	110	60
12-5		WO-A-1373	FM-GPW-2078	HOSE, front brake, assembly (11 in.) (LO-FC-8502)........	1	%6	3	20	
12-5		WO-A-1460	FM-GPW-2079	HOSE, brake, front axle, assembly (6 in.) (LO-FC-8553)........	2	%13	6	40	
12-5		WO-637424	FM-GP-2078	HOSE, rear brake, assembly (15 in.) (LO-FC-5784)........	1	%6	3	20	
				1209C—TUBES AND CLIPS					
12-5		WO-A-5227	FM-GPW-2274	BRACKET, axle brake tube tee (issue until stock exhausted)........	1	—	4		
		WO-6428	FM-20366-S	BOLT, hex-hd., 5/16-18NC-2 x 7/8 (brkt. and gear cover to housing)......	2	—	—		
		WO-A-1515	FM-GPW-2250	CLAMP, axle brake tube (after Willys serial 106763) (issue until stock exhausted)............	3	8	12		
		WO-6273	FM-34141-S2	NUT, sq. S., No. 10 (.190)-32NF-2..............	3				
		WO-6383	FM-27145-S2	SCREW, rd-hd. S., No. 10 (.190)-32NF-2..............	3				
		WO-A-1378	FM-GPW-2244	CLIP, brake pipe (closed, S., 5/16 in. bolt hole 21/64) (issue until stock exhausted)..........	2	%8	12		
		WO-52839		CLIP, tube (tube to side rail) (Carr fastener) (GM-127753) (Willys only) (issue until stock exhausted)..........	2	%20	40		
				CLIP, tube (tube to side rail) (Ford only)............	2		—		
12-5		WO-637439	FM-GPW-2223	CLIP, tube, 1/4 in. (use under head of rear axle cover bolt) (first 6000 Willys trucks) (Willys only) (issue until stock exhausted)...........	1	%4	8		
		WO-637427	FM-78-2814-A	CLIP, lock, spring (LO-FC-3052)............	6	%48	10	75	
12-5		WO-A-5449		CLIP, tube, S. (open) 5/16 in. bolt hole, 1/8 in. (Willys only) (issue until stock exhausted), (1 used tube to side rail reinforcement after Willys serial 106763, 1 used tube to frame cross member after Willys serial 106763, 1 used tube to master cylinder) (after 20677 Willys trucks)............	3	%4	4		
		WO-52840		BOLT, hex-hd., S., cd-pltd., No. 10 (.190) x 1/2 type "B" thd. (clip to master cylinder)............	1				
		WO-6352		NUT, hex, S., No. 10 (.190)-24NC-2.............	2				
		WO-5064		SCREW, rd-hd. S., No. 10 (.190)-24NC-2 x 5/8..........	2		108		
		WO-52221		WASHER, lock, S., No. 10 (.190)............	2	%52			
12-5		WO-384710	FM-GP-2133	NUT, hex. (3/16 in. inverted flared tube)............	16				
		WO-637432	FM-GP-2074	TEE, axle brake tube (LO-S-FC-5778)............	2	%6	3	12	10
		WO-6188	FM-20384-S2	BOLT, hex-hd., S., 1/4-20NC-2 x 1/2 (tee to rear axle tube) (Willys)........	2				
				(Ford)............	1				

Ref.	WO No.	FM No.	Description	Unit				
	WO-52706	FM-33795-S2	NUT, hex, S, reg, 1/4-20NC-2 (Ford only)	1	—	—	—	—
	WO-A-1377	FM-34805-S2	WASHER, lock, S, 1/4 in.	2	—	—	—	—
12-5	WO-A-1501	FM-GPW-2264	TUBE, brake, assembly, 3/16 in. (21.81 in. lgth.) (master cylinder to front hose) (issue until stock exhausted)	1	%12	24	—	—
12-5	WO-A-1376	FM-GPW-2263	TUBE, brake, assembly, 3/16 in. (6.5 in. lgth.) (tee to front brake hose, left)	1	%2	2	4	—
12-5	WO-A-1488	FM-GPW-2266	TUBE, brake, assembly, 3/16 in. (33.12 in. lgth.) (tee to front brake hose, right)	1	%2	2	8	—
12-5	WO-A-683	FM-GPW-2298	TUBE, brake, assembly, 3/16 in. (5.72 in. lgth.) (wheel cylinder to axle hose)	1	%13	—	16	—
	WO-A-630		TUBE, brake, assembly, 3/16 in. (38 1/16 in. lgth.) (master cylinder to rear hose) (before Willys serial 106763) (Willys only)	1	—	—	—	—
	WO-A-631		TUBE, brake, assembly, 3/16 in. (13 25/32 in. lgth.) (rear axle tee to rear brake, left) (before Willys serial 106763) (Willys only)	1	—	—	—	—
			TUBE, brake, assembly, 3/16 in. (30 9/32 in. lgth.) (rear axle tee to rear brake, right) (before Willys serial 106763) (Willys only)	1	—	—	—	—
12-5	WO-A-5224	FM-GPW-2265	TUBE, brake, assembly, 3/16 in. (62.94 in. lgth.) (master cylinder to rear hose) (after Willys serial 106763)	1	%2	2	4	5
12-5	WO-A-5225	FM-GPW-2268	TUBE, brake, assembly, 3/16 in. (27.38 in. lgth.) (tee to left rear brake) (after Willys serial 106763)	1	%2	2	4	5
12-5	WO-A-5226	FM-GPW-2267	TUBE, brake, assembly, 3/16 in. (28.81 in. lgth.) (tee to right rear brake) (after Willys serial 106763) (issue until stock exhausted)	1	%12	24	—	—

GROUP 13—WHEELS, HUBS AND DRUMS

1301—WHEEL ASSEMBLY

Ref.	WO No.	FM No.	Description	Unit				
13-2	WO-A-5470	FM-GPW-1029	BOLT, wheel divided rim (S, 1/2-20NF-2 x 1 11/32) (after Willys serial 120700) (KWH-25695)	40	16	16	50	—
13-2	WO-A-5471	FM-GPW-1030	NUT, hex, S, 1/2-20NF-2 (wheel divided rim bolt) (after Willys serial 120700) (KWH-22779)	40	40	40	150	—
	WO-A-5472	FM-GPW-1045	PLATE, instruction, combat wheels (KWH-25696) (after Willys serial 120700)	40	—	—	5	—
13-2	WO-A-5468	FM-GPW-1016	RIM, wheel, inner half (after Willys serial 120700) (KWH-25693)	5	3	—	5	5
13-2	WO-A-5549	FM-GPW-1024	RIM, wheel, outer half (after Willys serial 120700) (KWH-25917)	5	3	—	—	5
	WO-A-465	FM-GP-1015	RIM, wheel, w/DISC, assembly (16 x 4:00) (before Willys serial 120700) (KWH-24562)	5	—	—	5	—
	WO-A-1799		RIM, wheel, w/DISC, assembly (Willys only) (16 x 4:50, to be used with 16 x 6:50 tires and on trucks when specified)	5	3	—	—	—
	WO-A-5488	FM-GPW-1025-C	RING, lock, bead (6:00 x 16 tires on combat wheels) (after Willys serial 120700) (KWH-25930)	5	—	—	5	5
	WO-A-5467	FM-GPW-1015	WHEEL assembly (16 x 4:50) (after Willys serial 120700) (KWH-25692)	5	—	—	—	—

1301A—BEARINGS

Ref.	WO No.	FM No.	Description	Unit				
10-2	WO-52942	FM-GP-1201	CONE and ROLLERS, wheel bearing, bore 1.625 x 11/16 (TIM-18590) BOW-BT-18590)	8	%19	6	29	20
10-2	WO-52943	FM-GP-1202	CUP, wheel bearing (2.875 O.D. x 1/2 thk.) (TIM-18520, BOW, BT-18520)	8	%19	6	29	20

1301B—SEALS

Ref.	WO No.	FM No.	Description	Unit				
10-2	WO-778	FM-GP-3031-A	RETAINER, grease, wheel, end (use with Bendix or Rzeppa joints)	2	—	—	—	—
10-2	WO-A-864	FM-GP-1177	RETAINER, grease, wheel, hub assembly (SP-17004)	4	%240	80	396	250

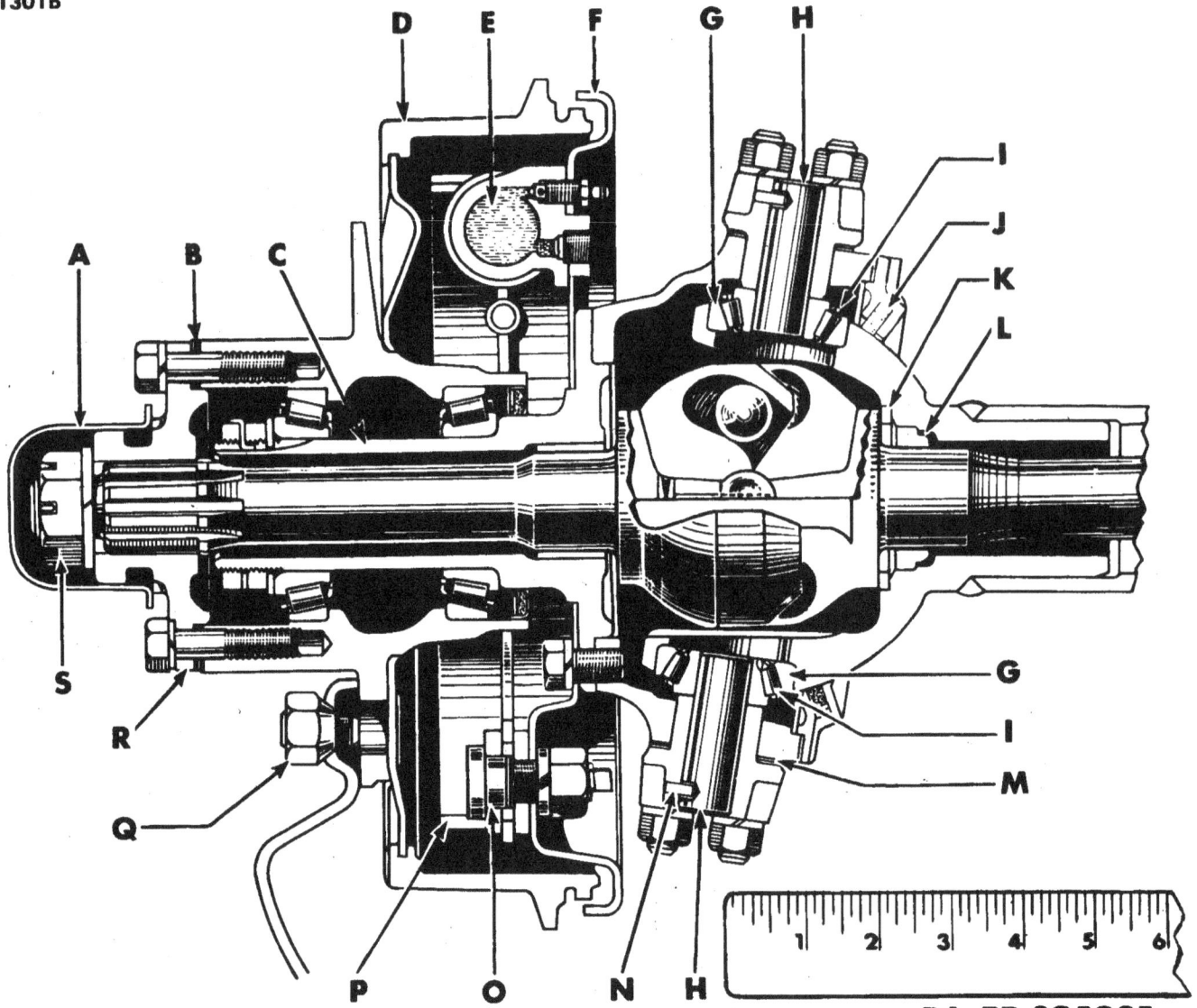

RA PD 305085

FIGURE 13-1—FRONT WHEEL AND SPINDLE—CROSS-SECTIONAL VIEW

Key	Item	Willys Part No.	Ford Part No.	Gov't Group No.
A	CAP	WO-A-869	FM-GPW-1139	1302
B	SHIM	WO-A-862	FM-GP-3208A	1007
C	SPINDLE	WO-A-851	FM-GP-3105	1006
D	DRUM	WO-A-472	Willys only	1302
E	CYLINDER assembly	WO-A-1484	FM-GPW-2061	1207
F	PLATE	WO-A-8898	FM-GP-2013	1203
G	CUP	WO-52941	FM-GP-3162	1006
H	PIN	WO-A-824	FM-GP-3115	1006
I	CONE	WO-52940	FM-GP-3161	1006
J	SEAL	WO-A-813	FM-GP-1088	1006
K	WASHER	WO-A-780	FM-GP-3374	1007
L	BUSHING	WO-A-6362	Willys only	1007
M	SHIM	WO-A-830	FM-GPW-3117-A	1006
N	PIN	WO-A-825	FM-GP-3122	1006
O	PIN	WO-637899	FM-91A-2030	1203B
P	SHOE assembly	WO-116549	FM-GP-2018	1201
Q	NUT	WO-A-475	FM-GP-1013	1302-A
R	FLANGE	WO-A-868	FM-GP-3204	1007
S	NUT	WO-636569	FM-356126-S	1003

RA PD 305085-A

FIGURE 13-2—WHEEL ASSEMBLY

Key	Item	Willys Part No.	Ford Part No.	Gov't Group No.
A	RIM, inner	WO-A-5468	FM-GPW-1016	1301
B	RIM, outer	WO-A-5549	FM-GPW-1024	1301
C	BOLT	WO-A-5470	FM-GPW-1029	1301
D	NUT	WO-A-5471	FM-GPW-1030	1301

RA PD 305159-A

RA PD 305159

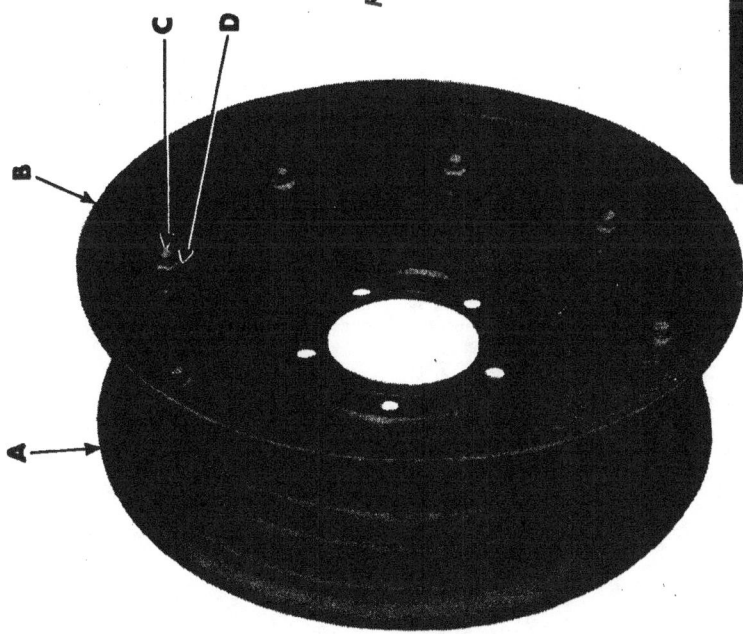

GROUP 13—WHEELS, HUBS AND DRUMS (Cont'd)

Figure Number	Official Stockage Number	Part Number		ITEM	Quantity Reqd. per Unit Assy.	Per 100 Major Items			Estimated Reqmts. per 100 Rebuilds
		Willys	Ford			12 Mos. Field Maintenance	Major Overhaul (5th Ech)	Total First Year Procurement	
Col. 1	Col. 2	Col. 3	Col. 4	Col. 5	Col. 6	Col. 7	Col. 8	Col. 9	Col. 10
				1301C—RETAINERS					
10-1		WO-636569	FM-356126-S	NUT, axle shaft (S., hex, slotted, 3/4-16NF-3) (SP-S-1135) (issue until stock exhausted)	2	% 20	40	—	—
10-2		WO-A-866	FM-GP-4252	NUT, front and rear axle wheel bearing (hex., S., 1 5/8-16 (SP-17064)	4	% 13	6	20	20
10-1		WO-5397	FM-72071	PIN, cotter, S., 1/8 x 1 1/2 (axle shaft nut)	2	—	—	—	—
10-2		WO-636570	FM-356504-S	WASHER, axle shaft nut (S., 3/4 in.) (SP-1056) (issue until stock exhausted)	2	% 8	12	—	—
10-2		WO-A-865	FM-GP-1218	WASHER, wheel bearing (S.)	2	% 6	3	10	10
10-2		WO-A-867	FM-GP-1124	WASHER, lock, wheel bearing nut (S., 1 5/8 in.) (SP-17017)	2	% 10	10	30	30

SNL G-503

GROUP 13—WHEELS, HUBS AND DRUMS (Cont'd)

Figure Number	Official Stockage Number	Part Number — Willys	Part Number — Ford	ITEM	Quantity Reqd. per Unit Assy.	Per 100 Major Items — 12 Mos. Field Maintenance	Per 100 Major Items — Major Overhaul (5th Ech)	Total First Year Procurement	Estimated Reqmts. per 100 Rebuilds
Col. 1	Col. 2	Col. 3	Col. 4	Col. 5	Col. 6	Col. 7	Col. 8	Col. 9	Col. 10
				1302—HUBS AND DRUMS					
10-1		WO-A-869	FM-GP-1139	CAP, wheel hub (SP-17071)	4	% 6	6	15	20
11-1		WO-A-472		DRUM, front and rear wheel brakes (KWH-24566) (Willys only)	4				—
		WO-A-1691		HUB, front and rear, wheel, w/BEARING, assembly (KWH-24659) (Willys only)	4				5
		WO-A-1689	FM-GP-1103	HUB, front and rear, left, wheel, w/DRUM, assembly (KWH-25647)	2	% 2	2	7	5
		WO-A-1690	FM-GP-1102	HUB, front and rear, right, wheel, w/DRUM, assembly (KWH-25646)	2	% 2	2	7	5
				1302A—STUDS AND BOLTS					
11-1		WO-A-473	FM-GP-1108	BOLT, wheel hub, S, L.H. thd., $\frac{1}{2}$-20NF-2 x 1$\frac{11}{32}$ (KWH-24566)	10	% 8	8	30	25
		WO-A-474	FM-GP-1107	BOLT, wheel hub, S, R.H. thd., $\frac{1}{2}$-20NF-2 x 1$\frac{11}{32}$ (KWH-24568)	10	% 8	8	30	—
13-1		WO-A-475	FM-GP-1013	NUT, hex, S., $\frac{1}{2}$-20NF-2 L.H. thd. (wheel bolt) (KWH-24576)	10	% 32	—	40	—
11-1		WO-A-476	FM-GP-1012	NUT, hex, S., $\frac{1}{2}$-20NF-2 R.H. thd. (wheel bolt) (KWH-24575)	10	% 32	—	40	—
				GROUP 14—STEERING					
				1401—STEERING CONNECTING ROD (DRAG LINK)					
14-1		WO-A-857	FM-GPW-3171	BEARING, steering bell crank (SP-17212)	2	26	26	60	80
		WO-A-622	FM-GPW-3332-A	COVER, dust (rubber) (CA-S-X-1109)	2	2	2	6	5
		WO-A-8249		CRANK, bell, steering drag link, front (Willys only)	1	2	2	6	5
14-1		WO-A-640038	FM-358006-S8	FITTING, grease, straight (AD-198) (issue until stock exhausted)	3	92	184	18	20
14-1		WO-A-6791	FM-GPW-18383	KIT, repair, steering connecting rod (Includes: 2 WO-A-622 FM-GPW-3332-A COVER; 1 WO-630756 FM-GPW-3323 PLUG, large; 1 WO-630757 FM-GPW-3328 PLUG, small; 2 WO-630753 FM-GPW-3326 PLUG, ball seat; 3 WO-630755 FM-GPW-3320 SEAT, ball; 2 WO-A-623 FM-GPW-3336 SHIELD; 2 WO-630754 FM-GPW-3327 SPRING)	as req.	6	6	18	20
14-1		WO-A-876	FM-356124-S	NUT, steering bell crank shaft (hex, S., $\frac{3}{4}$-20NF-2) (SP-S-1106)	1	3	3	12	8
14-1		WO-52527	FM-72062-S	PIN, cotter, S., $\frac{1}{8}$ x 1$\frac{3}{8}$ (bell crank nut)	1	—	—	—	—
		WO-A-856	FM-GP-3166	PIN, steering bell crank shaft (S, tapered, $\frac{3}{8}$ x 1$\frac{1}{8}$) (SP-17211) (issue until stock exhausted)	1	4	8	—	—
		WO-630756	FM-GPW-3323	PLUG, adjusting, steering connecting rod ball seat (slotted, large) (CA-S-5070)	1	—	—	—	—
		WO-630757	FM-GPW-3328	PLUG, adjusting, steering connecting rod ball seat (slotted, small) (CA-S-5143)	1	—	—	—	—
14-1		WO-5134	FM-72089-S	PIN, cotter $\frac{1}{8}$ x 1$\frac{1}{4}$ (ball seat plugs)	2	—	—	—	—

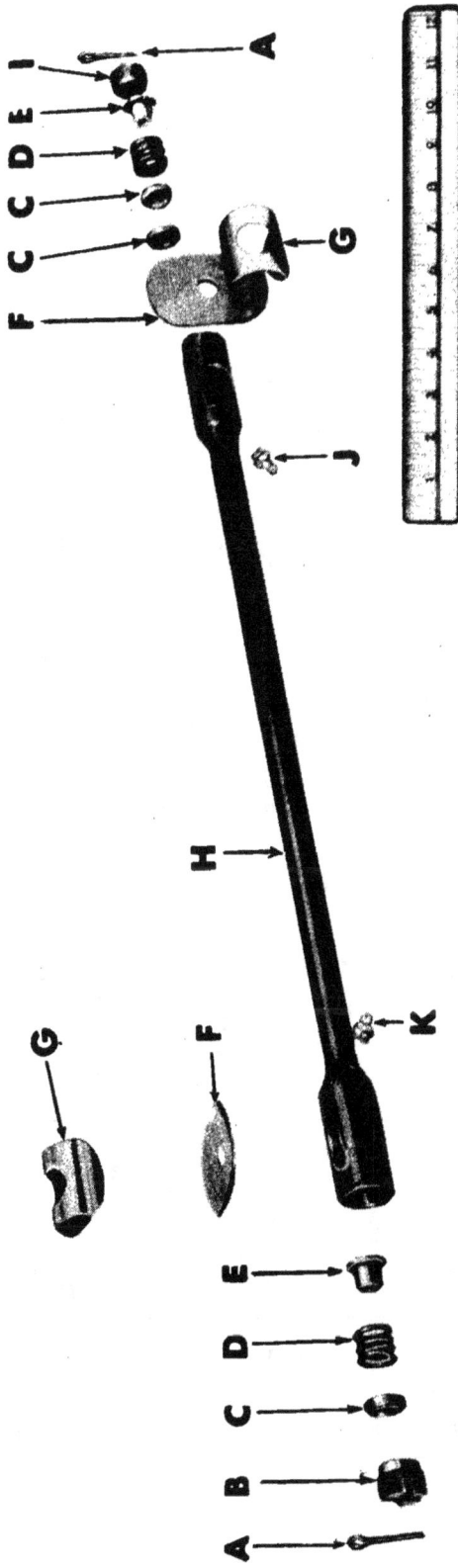

RA PD 305096

FIGURE 14-1—STEERING CONNECTING ROD

Figure 14-1 legend

Key	Item	Gov't Group No.	Ford Part No.	Willys Part No.
A	PIN	1401		WO-5134
B	PLUG	1401		WO-630756
C	BALL seat	1401		WO-630755
D	SPRING	1401		WO-630754
E	PLUG	1401		WO-630753
F	COVER	1401		WO-A-622
G	SHIELD	1401	FM-72089-S	WO-A-623
H	ROD assembly	1401	FM-GPW-3323	WO-A-8250
I	PLUG	1401	FM-GPW-3320	WO-630757
J	FITTING	1401	FM-GPW-3326	WO-640038
K	FITTING	1401	FM-GPW-3332-A	WO-640038

RA PD 305096-A

Key	Willys Part No.	Ford Part No.	Item	Qty
14-1	WO-630753	FM-GPW-3326	PLUG, steering connecting rod ball seat spring (CA-S-5144)	2
14-1	WO-A-8252	FM-GPW-3305-B	ROD, connecting, steering (CA-D-822)	1
14-1	WO-A-8250	FM-GPW-3304-B	ROD, connecting, steering, assembly	1
14-1	WO-A-858	FM-GPW-3167	SEAL, steering bell crank shaft bearing (SP-17213)	2
14-1	WO-630755	FM-GPW-3320	SEAT, ball steering connecting rod (CA-S-5069)	3
14-1	WO-A-855	FM-GPW-3165	SHAFT, steering bell crank	1
14-1	WO-A-861	FM-GPW-3170	SHIELD, dust, steering bell crank bearing (SP-17205)	1
14-1	WO-A-623	FM-GPW-3336	SHIELD, dust, steering connecting rod cover (CA-S-X-1110)	2
14-1	WO-630754	FM-GPW-3327	SPRING, steering connecting rod ball seat (CA-S-5145)	2
14-1	WO-A-859	FM-GPW-3168	WASHER, steering bell crank shaft, upper	1
14-1	WO-A-860	FM-GPW-3169	WASHER, steering bell crank shaft, lower	1

1402—TIE ROD

Key	Willys Part No.	Ford Part No.	Item	Qty
10-2	WO-A-1706	FM-51-3287	CLAMP, steering tie rod socket (SP-13439) (TP-16MN8)	4
10-2	WO-A-1707	FM-24916-S2	BOLT, hex-hd., S., 3/8-24NF x 1 5/8 (SP-6-73-326) (TP-6MN27)	4
10-2	WO-636575	FM-34083-S2	NUT, hex., S., 3/8-24NF-2 (SP-6-73-326) (issue until stock exhausted)	4
	WO-5010	FM-34807-S2	WASHER, lock, S., 3/8 in. (socket clamp screw)	4
	WO-A-844	FM-78-3336	COVER, dust, steering tie rod socket (SP-16067) (TP-14DM-43)	4

GROUP 14—STEERING (Cont'd)

Figure Number (Col. 1)	Official Stockage Number (Col. 2)	Part Number — Willys (Col. 3)	Part Number — Ford (Col. 4)	ITEM (Col. 5)	Quantity Reqd. per Unit Assy. (Col. 6)	Per 100 Major Items — 12 Mos. Field Maintenance (Col. 7)	Per 100 Major Items — Major Overhaul (5th Ech) (Col. 8)	Total First Year Procurement (Col. 9)	Estimated Reqmts. per 100 Rebuilds (Col. 10)
				1402—TIE ROD (Cont'd)					
10-2		WO-A-847	FM-GP-3292	END, steering knuckle tie rod, left, assembly (SP-17047-X) (TP-14SV90-A-7)	2	%6	6	18	20
10-1		WO-A-838	FM-GP-3291	END, steering knuckle tie rod, right, assembly (SP-17046-X) (TP-14SV89-A-7)	2	%6	6	18	20
10-2		WO-10558	FM-351059-S7	NUT, hex, S. (dld. f/c-pin) ½-20NP (steering tie rod stud nut) (SP-S-1107)	4	%3	3	25	10
10-2		WO-5152	FM-72025-S	PIN, cotter, 3/32 x 7/8 (steering tie rod stud nut)	4	—	—	—	—
		WO-A-1704	FM-GPW-3280	ROD, tie, steering, right, assembly (SP-17308-X) (issue until stock exhausted)	1	%8	12	—	—
		WO-A-1708	FM-GPW-3279	ROD, tie, steering, left, assembly (SP-17309-X) (issue until stock exhausted)	1	%8	12	—	—
			FM-78-3332-A	SEAL, dust, spindle connecting rod (Ford only)	1	—	—	—	—
		WO-A-6305		SPRING, steering tie rod socket stud (coil, S., cd-pltd.) (SP-17347) (TP-14-DS-20) (Willys only) (issue until stock exhausted)	4	%40	80	—	—
10-1		WO-A-1709	FM-GPW-3282	TUBE, steering left tie rod (14¾ in. lgth.) (SP-17295-2)	1	%3	3	10	10
10-2		WO-A-1705	FM-GPW-3281	TUBE, steering right tie rod (21.80 in. lgth.) (SP-17295-1)	1	%3	3	10	10
				1403A—GEAR ASSEMBLY					
14-2		WO-A-1116	FM-GPW-3590	ARM, steering gear	1	%2	2	5	5
14-2		WO-639104	FM-GPW-3571	BALLS, steel, steering gear cam (RG-400013)	22	—	299	400	937
14-2		WO-639190	FM-GPW-3517	BEARING, steering column, wheel end, assembly (RG-063996)	1	10	10	60	40
14-2		WO-639090	FM-GPW-3587	BUSHING, steering gear housing, inner (RG-063011)	1	—	13	16	40
14-2		WO-639091	FM-GPW-3576	BUSHING, steering gear housing, outer (RG-063012)	1	—	1	16	40
14-2		WO-A-1199	FM-GPW-3509	COLUMN, steering, w/BEARING, assembly	1	—	6	7	20
14-2		WO-639116	FM-20386-S7	COVER, steering gear housing side, assembly (RG-502540)	1	—	—	—	—
		WO-639120	FM-356303-S	BOLT, hex-hd., s-fin., alloy-S., 5/16-18NC-2 x 1 (cover to housing) (BANX1BC) (issue until stock exhausted)	4	—	16	—	—
		WO-639121		WASHER, side, cover (cop., 21/64 I.D. x 9/16 O.D. x .0375 thk.) (issue until stock exhausted)	2	12	16	—	—
14-2		WO-52045	FM-34846-S2	WASHER, lock, hv., S., 5/16 in. (19/32 O.D. x 11/32 I.D. x 3/32 thk.) (BECX3H)	2	12	16	—	—
14-2		WO-A-1760	FM-GPW-3568	COVER, steering gear housing, upper (RG-T-126000)	2	—	3	4	10
		WO-637107	FM-355426-S	BOLT, hex-hd., s-fin., alloy-S., 5/16-18NC-2 x 3/4 (cover to housing) (BANX1BA) (issue until stock exhausted)	3	—	—	—	—
14-2		WO-52045	FM-34846-S2	WASHER, lock, hv., S., 5/16 in. (19/32 O.D. x 11/32 I.D. x 3/32 thk.) (BECX3H)	3	50	150	—	—
14-2		WO-390510		COVER, steering column oil hole (Willys only)	1	—	3	6	10
14-2		WO-639102	FM-GPW-3552	CUP, steering gear cam steel balls (RG-400025)	2	—	26	30	80
14-2		WO-639119	FM-GPW-3581	GASKET, steering gear side housing cover (RG-T-129001)	1	6	13	20	—
14-2		WO-A-1239	FM-GPW-3504	GEAR, steering, w/o WHEEL, assembly (RG-T-13086)	1	3	3	9	—
14-2		WO-A-740	FM-GPW-3548	HOUSING, steering gear, assembly (RG-503284) (issue until stock exhausted)	1	—	4	—	—
		WO-51371	FM-24555-S	BOLT, hex-hd., S., 7/16-20NF-2 x 3 (gear to frame) (GM-100044)	2	—	—	—	—
		WO-51612	FM-355433-S2	BOLT, hex-hd., S., 7/16-20NF-2 x 3¾ (GM-113844)	1	—	—	—	—
		WO-5939	FM-33802-S2	NUT, hex, S., 7/16-20NF-2	3	—	—	—	—

Key	Item	Willys Part No.	Ford Part No.	Gov't Group No.
S	LEVER shaft assembly	WO-A-745	FM-GPW-3575	1403A
T	GASKET	WO-639119	FM-GPW-3581	1403A
U	SCREW	WO-639118	FM-GPW-3577	1403A
V	NUT	WO-52925	FM-33927-S7	1403A
W	COVER	WO-639116	FM-GPW-3583	1403A
X	HOUSING assembly	WO-A-740	FM-GPW-3548	1403A
Y	PLUG	WO-51091	FM-74121-S	1403A
Z	BUSHING	WO-639090	FM-GPW-3587	1403A
AA	BUSHING	WO-639091	FM-GPW-3576	1403A
AB	ARM	WO-A-1116	FM-GPW-3590	1403A
AC	WASHER	WO-5038	FM-34811-S2	1403A
AD	NUT	WO-639115	FM-356077-S8	1403A
AE	OIL seal	WO-639095	FM-GPW-3591-A	1403A
AF	BALLS	WO-639104	FM-GPW-3571	1403A
AG	COVER	WO-A-1760	FM-GPW-3568	1403A
AH	CONTACT assembly	WO-A-747	FM-GPW-3652	0609B
AI	BRUSH assembly	WO-A-302	FM-GPW-13836	0609B

RA PD 305097-A

RA PD 305097

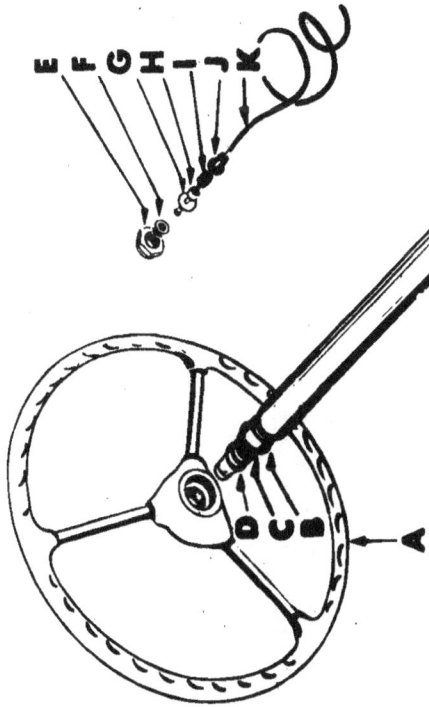

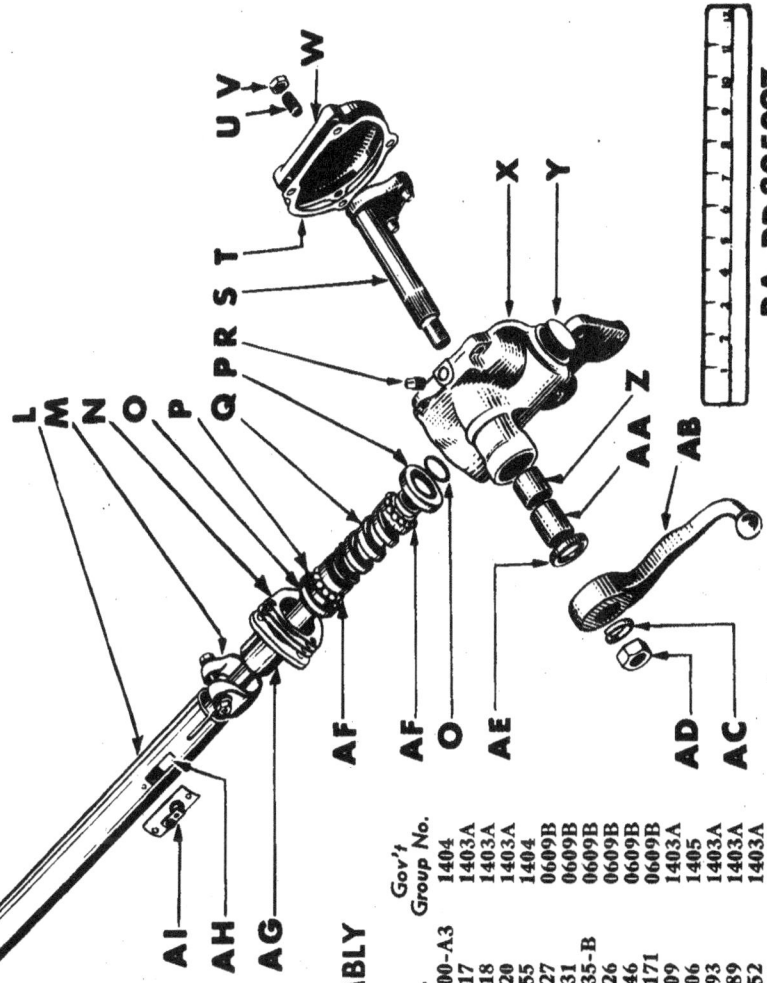

FIGURE 14-2—STEERING GEAR ASSEMBLY

Key	Item	Willys Part No.	Ford Part No.	Gov't Group No.
A	WHEEL	WO-A-6858	FM-GPW-3600-A3	1404
B	BEARING	WO-639190	FM-GPW-3517	1403A
C	SEAT	WO-639192	FM-GPW-3518	1403A
D	SPRING	WO-639191	FM-GPW-3520	1403A
E	NUT	WO-A-633	FM-GPW-3655	1404
F	BUTTON	WO-A-634	FM-GPW-3627	0609B
G	WASHER	WO-A-750	FM-GPW-3631	0609B
H	FERRULE	WO-A-751	FM-GPW-3635-B	0609B
I	SPRING	WO-638884	FM-GPW-3626	0609B
J	CUP	WO-638885	FM-GPW-3646	0609B
K	CABLE assembly	WO-A-752	FM-GPW-14171	0609B
L	COLUMN assembly	WO-A-1199	FM-GPW-3509	1403A
M	CLAMP assembly	WO-A-635	FM-GPW-3506	1405
N	SHIMS	WO-639108	FM-GPW-3593	1403A
O	RING	WO-639103	FM-GPW-3589	1403A
P	CUP	WO-639102	FM-GPW-3552	1403A
Q	TUBE assembly	WO-A-742	FM-GPW-3524	1403A
R	PLUG	WO-5085	FM-358064-S	1403A

GROUP 14—STEERING (Cont'd)

1403A—GEAR ASSEMBLY (Cont'd)

Figure Number (Col. 1)	Official Stockage Number (Col. 2)	Part Number — Willys (Col. 3)	Part Number — Ford (Col. 4)	ITEM (Col. 5)	Quantity Reqd. per Unit Assy. (Col. 6)	Per 100 Major Items — 12 Mos. Field Maintenance (Col. 7)	Per 100 Major Items — Major Overhaul (5th Ech) (Col. 8)	Per 100 Major Items — Total First Year Procurement (Col. 9)	Estimated Reqmts. per 100 Rebuilds (Col. 10)
		WO-638381	FM-356439-S	WASHER, S, ½ in. I.D. x 1¼ O.D. (use over slotted hole)	1	10	10	24	—
		WO-52874	FM-34921-S	WASHER, lock, S, shakeproof, ⁷⁄₁₆ in.	3	—	—	—	20
14-2		WO-A-745	FM-GPW-3575	LEVERSHAFT, steering gear, assembly (RG-7698-4⅝)	1	—	6	10	10
14-2		WO-639115	FM-356077-S8	NUT, steering gear, levershaft (RG-025060)	1	3	3	10	—
14-2		WO-52925	FM-33927-S7	NUT, hex, S, ⁷⁄₁₆-20 (steering gear adjusting screw jam nut)	1	—	—	—	40
14-2		WO-639095	FM-GPW-3591-A	OIL SEAL unit, steering gear housing (RG-032075)	1	3	13	20	—
14-2		WO-51091	FM-74121-S	PLUG, expansion, steering gear housing (1¼ in.)	1	—	—	—	—
14-2		WO-5085	FM-358064-S	PLUG, pipe ⅛ in. std. (filler)	1	—	—	—	—
14-2		WO-639103	FM-GPW-3589	RING, retaining, steering gear cam balls (RG-401100)	2	—	30	30	93
14-2		WO-639118	FM-GPW-3577	SCREW, adjusting, steering gear (RG-021116)	1	3	3	10	10
14-2		WO-639192	FM-GPW-3518	SEAT, steering column bearing spring (RG-028093)	1	—	6	6	20
14-2		WO-638918	FM-GPW-3563	SHIM, steering gear mounting	as req.	10	10	24	—
		WO-639108	FM-GPW-3593	SHIM, steering gear housing upper cover, S. (.002 thk.) (RG-033046)	as req.	—	—	—	—
		WO-639109	FM-GPW-3594	SHIM, steering gear housing upper cover, S. (.003 thk.) (RG-033047)	as req.	—	6	—	—
		WO-639110	FM-GPW-3595	SHIM, steering gear housing upper cover, S. (.010 thk.) (RG-033048)	as req.	—	—	—	—
		WO-A-6760	FM-GPW-18374	SHIM SET, steering gear housing upper cover	as req.	—	—	5	20
				(Includes:					
		2 WO-639108	FM-GPW-3593	SHIM, .002					
		2 WO-639109	FM-GPW-3594	SHIM, .003					
		2 WO-639110	FM-GPW-3595	SHIM, .010)					
14-2			FM-GPW-3521	SNAP RING, steering column (Ford only)	1	—	—	—	—
14-2		WO-A-1422	FM-GPW-5116	SPACER, steering gear to frame bolt	1	—	14	20	20
14-2		WO-639191	FM-GPW-3520	SPRING, steering column bearing (RG-401090)	1	—	6	6	20
14-2		WO-A-742	FM-GPW-3524	TUBE, steering gear cam and wheel, assembly with nut	1	—	6	8	—
		WO-639121	FM-356308	WASHER, steering gear housing side cover, (plain, S.) (RG-029021)	—	13	26	50	80
		WO-5038	FM-34811-S2	WASHER, lock, S, ⅝ in. (lever shaft nut)	1	—	—	—	—

1404—WHEEL ASSEMBLY

Figure Number		Willys	Ford	ITEM					
14-2		WO-A-6858	FM-GPW-3600-A3	WHEEL, steering, assembly	1	%	2	7	—
14-2		WO-633	FM-GPW-3655	NUT, hex, S. (RG-026988) (also horn button nut) (issue until stock exhausted)	1	%	8	—	—

1405—BRACKETS

Figure Number		Willys	Ford	ITEM					
14-2		WO-A-1277	FM-GPW-3682-A	BUSHING, steering column support	1	—	—	—	—
		WO-A-635	FM-GPW-3506	CLAMP, steering column, assembly (RG-502282) (issue until stock exhausted)	1	%	4	—	—

Stock No.	Piece No.	Description	Qty
WO-5910	FM-33798-S2	NUT, hex, reg., s-fin., alloy-S., 5/16-24NF-2 (BBBX1BA)	1
WO-51833	FM-34806-S2	WASHER, lock, light, S., 5/16 in. (19/32 O.D. x 11/32 I.D. x 1/16 thk.) (BECX1H)	1
WO-A-1276	FM-GPW-3511	CLAMP, steering column support (issue until stock exhausted)	1
WO-51396	FM-24347-S2	BOLT, hex-hd., s-fin., alloy-S., 5/16-24NF-2 x 3/4 (BAOX1BA) (clamp to brkt., brkt. to instrument board)	2
WO-5910	FM-33798-S2	NUT, hex, reg., s-fin., alloy-S., 5/16-24NF-2 (BBB1XBA)	2
WO-51833	FM-34806-S2	WASHER, lock, light, S., 5/16 (19/32 O.D. x 11/32 I.D. x 1/16 thk.) (BECX1H)	2
WO-638918	FM-GPW-3563	SHIM, mounting, steering column	as req.
WO-A-2859	FM-GPW-3658	SUPPORT, steering column (issue until stock exhausted)	1

GROUP 15—FRAME AND BRACKETS

1500—FRAME AND BRACKETS

Stock No.	Piece No.	Description	Qty
WO-A-1152		BRACE, machine gun mounting plate	4
WO-A-185		BRACKET, brake pedal retracting spring	1
WO-5267	FM-60416-S	RIVET, rd-hd., S., 5/16 x 3/4 (bracket to cross member)	1
WO-A-415	FM-GPW-5106	BRACKET, front cross member, intermediate	1
WO-52832	FM-60446-S	RIVET, rd-hd., S., 3/8 x 7/8 (bracket to frame)	6
WO-A-420		BRACKET, engine support, front left (lower half) (Willys only)	1
WO-A-418		BRACKET, engine support, front, left (upper half) (Willys only)	1
	FM-GPW-6030	BRACKET, engine support, front left (Ford only)	1
WO-A-421		BRACKET, engine support, front, right (lower half) (Willys only)	1
WO-A-419		BRACKET, engine support, front, right (upper half) (Willys only)	1
	FM-GPW-6029	BRACKET, engine support, front right (Ford only)	1
WO-50769	FM-60472-S	RIVET, rd-hd., S., 3/8 x 3/4 (brackets to frame)	9
WO-544	FM-GPW-5337	BRACKET, front and rear spring shackle, assembly	4
WO-5215	FM-352126-S	RIVET, rd-hd., S., 3/8 x 1 (bracket to frame) (Willys only)	8
	FM-357076-S	RIVET, huck, S., 3/8 x 7/16 (front spring shackle bracket to frame) (Ford only)	4
	FM-352127-S	RIVET, rd-hd., S., 3/8 x1 1/8 (rear spring shackle bracket to frame) (Ford only)	4
WO-A-500	FM-GPW-5341	BRACKET, front and rear spring pivot	4
WO-5215		RIVET, rd-hd., S., 3/8 x 1 (bracket to frame) (Willys only)	8
WO-A-1341	FM-GPW-5095	BRACKET, master cylinder, assembly	1
	FM-357008-S	RIVET, fl-hd., S., 1/4 x 11/16 (bracket to frame) (Ford only)	8
WO-A-1283	FM-GPW-2073	BRACKET, mounting, front brake hose (frame end)	2
	FM-357008-S	RIVET, fl-hd., S., 1/4 x 11/16 (Ford only)	2
WO-5249		RIVET, rd-hd., S. (Willys only)	2
WO-638809		BRACKET, mounting, rear brake hose (before 6,800 Willys trucks) (Willys only)	1
WO-A-1201	FM-GPW-5057	BRACKET, radiator	2
	FM-357074-S	RIVET, huck, S., 5/16 x 5/16 (bracket to frame) (Ford only)	2
	FM-352103-S	RIVET, rd-hd., S., 5/16 x 13/16 (bracket to frame) (Ford only)	2
WO-5267	FM-60416-S	RIVET, rd-hd., S., 5/16 x 3/4 (bracket to frame) (Willys only)	4
WO-A-1431		BRACKET, support, body, rear (Willys only)	2
WO-A-1202	FM-GPW-5072	BRACKET, support, radiator, left (Willys only)	1
		BRACKET, support, radiator (Ford only)	2
WO-A-508		BRACKET, support, radiator, right	1
WO-A-1514	FM-60093-S	RIVET, flat head, S., 5/16 x 11/16 (Ford only)	4
		BRACKET, support, radiator guard (Willys only)	1

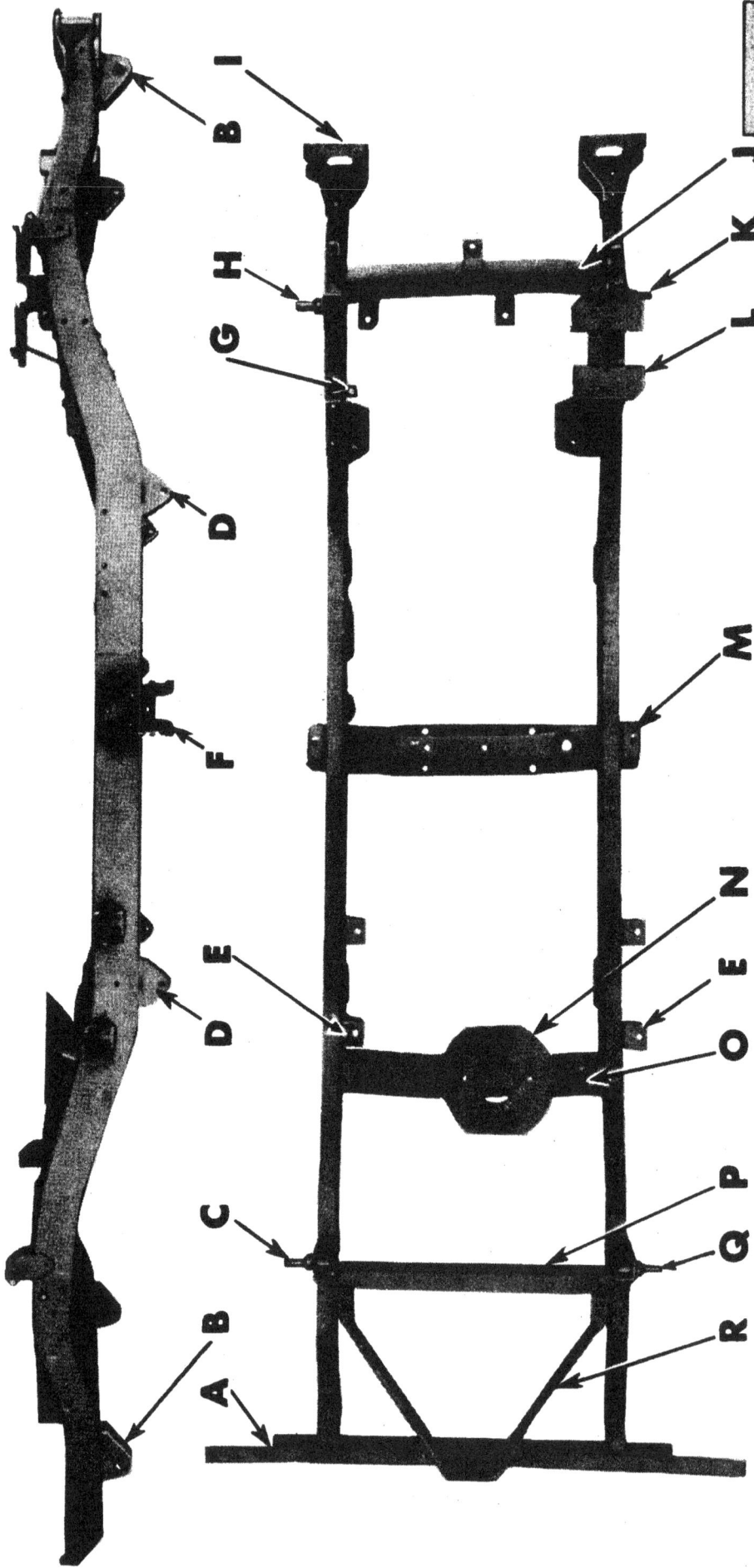

RA PD 305098

FIGURE 15-1—FRAME ASSEMBLY

Key	Item	Willys Part No.	Ford Part No.	Gov't Group No.
A	MEMBER	WO-A-547	FM-GPW-5035	1500
B	BRACKET assembly	WO-A-544	FM-GPW-5337	1500
C	BRACKET w/SHAFT	WO-A-484	(Willys only)	1500
D	BRACKET	WO-A-500	FM-GPW-5341	1500
E	OUTRIGGER	WO-A-637	FM-GPW-5077	1500
F	BRACKET	WO-A-415	FM-GPW-5106	1500
G	BRACKET	WO-A-1283	FM-GPW-2073	1500
H	BRACKET	WO-A-1204	See GPW-18075	1500
I	GUSSET	WO-A-1127	FM-GPW-17755	1500

Key	Item	Willys Part No.	Ford Part No.	Gov't Group No.
J	TUBE	WO-A-1203	FM-GPW-5019	1500
K	BRACKET	WO-A-1205	(Willys Only)	1500
L	SUPPORT assembly	WO-A-1138	(Willys only)	1500
M	MEMBER	WO-A-5127	FM-GPW-5025	1500
N	PLATE	WO-A-1151	FM-GPW-5125	1500
O	MEMBER	WO-A-1150	FM-GPW-5028	1500
P	MEMBER	WO-A-1155	(Willys only)	1500
Q	BRACKET	WO-A-485	See GPW-18040	1500

RA PD 305098-A

GROUP 15—FRAME AND BRACKETS (Cont'd)

Figure Number	Official Stockage Number	Part Number Willys	Part Number Ford	ITEM	Quantity Reqd. per Unit Assy.	12 Mos. Field Maintenance	Major Overhaul (5th Ech)	Total First Year Procurement	Estimated Reqmts. per 100 Rebuilds
Col. 1	Col. 2	Col. 3	Col. 4	Col. 5	Col. 6	Col. 7	Col. 8	Col. 9	Col. 10
				1500—FRAME AND BRACKETS (Cont'd)					
15-1			FM-99A-5090	BRACKET, support, radiator grille guard (Ford only)	1	--	--	--	--
15-1		WO-A-484		BRACKET, rear shock absorber, left, w/SHAFT, assembly (Willys only)	1	--	--	--	--
15-1		WO-A-1204		BRACKET, front shock absorber, left, w/SHAFT, assembly (frame end) (Willys only)	1	--	--	--	--
			FM-GPW-18075	BRACKET, front shock absorber, w/SHAFT, assembly (Ford only)	2	--	--	--	--
			FM-357074-S	RIVET, huck, S., 5/16 x 5/16	2	--	--	--	--
15-1		WO-A-485		BRACKET, rear shock absorber, right, w/SHAFT, assembly (Willys only)	1	--	--	--	--
15-1		WO-A-1205		BRACKET, front shock absorber, right, w/SHAFT, assembly (frame end) (Willys only)	1	--	--	--	--
15-1		WO-A-6740	FM-GPW-18040	BRACKET, rear shock absorber, w/SHAFT, assembly (Ford only)	2	--	--	--	--
				RIVET, rd-hd, S., 5/16 x 7/8	2	--	--	--	--
		WO-A-1142	FM-GPW-5084	CUP, retaining, transfer case insulator	1	--	--	--	--
			FM-352101-S	RIVET, plain, S., pointed, 5/16 x 3/4 (Ford only)	1	--	--	--	--
				FRAME assembly (Willys only) (Includes BRACES, BRACKETS, CUPS, REINFORCEMENTS, RETAINERS, TUBES, ATTACHING PARTS)	1	--	--	--	--
			FM-GPW-5005	FRAME assembly (Ford only) (Includes BRACES, BRACKETS, CUPS, REINFORCEMENTS, RETAINERS, TUBES, ATTACHING PARTS)	1	--	--	--	--
		WO-A-5415		GUARD, exhaust pipe (after Willys serial 120700) (Willys only)	1	--	--	--	--
		WO-52945		BOLT, carriage, S., 3/8-16 NC-2 x 7/8 (guard to skid plate) (Willys only)	2	--	--	--	--
		WO-5919		BOLT, hex-hd, S., 3/8-24NF-2 x 1 (guard to frame) (Willys only)	1	--	--	--	--
		WO-5544		NUT, hex, S., 3/8-16NC-2 (Willys only)	2	--	--	--	--
		WO-52101		WASHER, S., 3/8 in., S.A.E. plain (13/16 O.D. x 13/32 I.D. x 1/16 thk.) (Willys only)	3	--	--	--	--
		WO-303922		WASHER, S., 3/8 in., plain (1 1/2 O.D. x 1/16 I.D.) (Willys only)	1	--	--	--	--
		WO-5010		WASHER, lock, S., 3/8 in. (1/8 thk. x 3/32 I.D.) (Willys only)	3	--	--	--	--
15-1		WO-A-1129	FM-GPW-17753	GUSSET, front bumper, lower left	1	--	--	--	--
		WO-A-1130	FM-GPW-17752	GUSSET, front bumper, lower right	1	--	--	--	--
		WO-A-1127	FM-GPW-17755	GUSSET, front bumper, upper left	1	--	--	--	--
		WO-A-1128	FM-GPW-17754	GUSSET, front bumper, upper right	1	--	--	--	--
15-1		WO-A-1150	FM-GPW-5028	MEMBER, cross, frame, center	1	--	--	--	--
		WO-52832	FM-357100-S	RIVET, rd-hd, S., 3/8 x 7/8 (member to side rail)	2	--	--	--	--
			FM-GPW-5019	MEMBER, cross, frame, front (Ford only)	1	--	--	--	--
15-1		WO-A-5127	FM-GPW-5025	MEMBER, cross, frame, intermediate, front, assembly	1	--	--	--	--
			FM-20389-S	BOLT, hex-hd, s-fin, alloy-S, 3/8-24NF-2 x 1 (member to brace) (Ford only)	6	--	--	--	--
		WO-50151		BOLT, hex-hd, s-fin, alloy-S, 3/8-24NF-2 x 7/8 (member to brace) (Willys only)	6	--	--	--	--

GROUP 15—FRAME AND BRACKETS (Cont'd)

Figure Number Col. 1	Official Stockage Number Col. 2	Willys Col. 3	Ford Col. 4	ITEM Col. 5	Quantity Reqd. per Unit Assy. Col. 6	12 Mos. Field Maintenance Col. 7	Major Overhaul (5th Ech) Col. 8	Total First Year Procurement Col. 9	Estimated Reqmts. per 100 Rebuilds Col. 10
				1500—FRAME AND BRACKETS (Cont'd)					
		WO-5901	FM-33800-S2	NUT, hex, S., 3/8-24NF-2	6	—	—	—	—
15-1		WO-52101	FM-34747-S7	WASHER, S., S.A.E., plain (13/16 O.D. x 13/32 I.D. x 1/16 thk.)	2	—	—	—	—
		WO-5010	FM-34807-S2	WASHER, lock, S., 3/8 (3/8 I.D.)	6	—	—	—	—
		WO-A-547	FM-GPW-5035	MEMBER, cross, frame, rear	1	—	—	—	—
		WO-5326		RIVET, fl-hd, S., 3/8 x 7/8 (member to side rail) (Willys only)	4	—	—	—	—
15-1		WO-50769		RIVET, rd-hd., S., 3/8 x 3/4 (member to frame) (Willys only)	8	—	—	—	—
		WO-50769	FM-357100-S	RIVET, rd-hd., S., 3/8 x 15/16 (member to frame and side rail) (Ford only)	8	—	—	—	—
		WO-A-1153		MEMBER, cross, frame, intermediate rear (Willys only)	1	—	—	—	—
		WO-52832		RIVET, rd-hd., S., 3/8 x 7/8 (member to frame) (Willys only)	2	—	—	—	—
		WO-A-668		NUT, clinch, S., 1/4-28NF-2 (MSP-F-2005) (Willys only)	2	—	—	—	—
		WO-A-549		NUT, clinch, S., 3/8-24NF-2 (dash to frame brace fender brace to frame) (MSP-F-2009) (Willys only)	11	—	—	—	—
		WO-A-548		NUT, clinch, S., 5/16-24NF-2 (splasher apron to frame) (MSP-F-2007) (Willys only)	5	—	—	—	—
15-1		WO-A-637	FM-GPW-5077	OUTRIGGER, body rear	4	—	—	—	—
			FM-357100-S	RIVET, rd-hd., S., 3/8 x 5/16 (outrigger to frame side) (Ford only)	4	—	—	—	—
		WO-50769		RIVET, rd-hd., S., 3/8 x 3/4 (outrigger to frame side) (Willys only)	8	—	—	—	—
15-1		WO-A-1151	FM-GPW-5125	PLATE, mounting, machine gun	1	—	—	—	—
		WO-A-534	FM-GPW-5097	PLATE, reinforcement, frame, rear	1	—	—	—	—
		WO-52906		RIVET, fl-hd., S., 3/8 x 3/4 (plate to frame) (Willys only)	4	—	—	—	—
		WO-5010		WASHER, lock, S., 3/8 (3/32 I.D.) (Willys only)	1	—	—	—	—
		WO-A-493		RETAINER, pedal shaft, assembly (Willys only)	2	—	—	—	—
15-1		WO-A-416		SPACER, frame cross member, intermediate front bracket (Willys only)	2	—	—	—	—
		WO-A-1120	FM-GPW-17759	SPACER, front bumper gusset	1	—	—	—	—
		WO-A-1422	FM-GPW-5116	SPACER, steering gear to frame bolt	1	—	—	—	—
		WO-A-1138		SUPPORT, battery, assembly (before Willys serial 120700) (Willys only)	1	—	—	—	—
		WO-A-5181		SUPPORT, battery, assembly (after Willys serial 120700) (Willys only)	1	—	—	—	—
		WO-A-50769		RIVET, rd-hd., S., 3/8 x 3/4 (support to frame) (Willys only)	1	—	—	—	—
15-1		WO-1203	FM-GPW-5019	TUBE, cross, frame, front (Willys only)	1	—	—	—	—
				1502—PINTLE HOOK					
15-2		WO-A-6393	FM-GPW-5186	BOLT, eye, safety chain (after Willys serial 158372)	2	—	8	—	—
15-2		WO-6163	FM-33845-S2	NUT, hex, S., 1/2-13NC-2 (bolt to frame)	2	8	12	—	—
15-2		WO-5009	FM-34809-S	WASHER, lock, S., 1/2 in. (bolt to frame)	2	4	8	—	—
15-2		WO-A-593	FM-GPW-5182	HOOK, pintle (HLH-T-60-B)	1	—	—	—	—
15-2		WO-6923	FM-24411-S	BOLT, hex-hd., S., 7/16-20NF-2 x 1 1/8 (hook to frame)					
				Before serial 158372	4	—	—	—	—
				After serial 158372	2	—	—	—	—

FIGURE 15-2—PINTLE HOOK ASSEMBLY

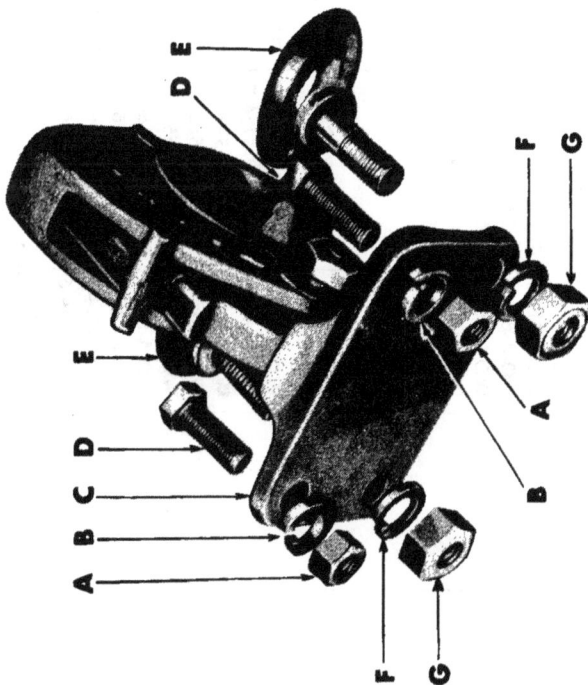

RA PD 305099

Key	Item	Willys Part No.	Ford Part No.	Gov't Group No.
A	NUT	WO-5939	FM-33802-S	1502
B	WASHER	WO-52349	FM-34808-S	1502
C	HOOK	WO-A-593	FM-GPW-5182	1502
D	BOLT	WO-6923	FM-24411-S	1502
E	EYE BOLT	WO-A-6393	FM-GPW-5186	1502
F	WASHER	WO-5009	FM-34809-S	1502
G	NUT	WO-6163	FM-N.P.N.	1502

RA PD 305099-A

Fig.	Willys	Ford	Item	Qty.
15-2	WO-5939	FM-33802-S	NUT, hex., S., $\frac{7}{16}$-20NF-2	
			Before serial 158372	4
			After serial 158372	2
15-2	WO-52349	FM-34808-S	WASHER, lock, S., external flat, $\frac{7}{16}$ in.	
			Before serial 158372	4
			After serial 158372	2
	WO-A-552		REINFORCEMENT, pintle hook frame (Willys only)	1

1505—SPARE WHEEL

Fig.	Willys	Ford	Item	Qty.
	WO-A-11701	FM-GPW-1433	BRACKET, support, spare tire, assembly	1
	WO-6299	FM-20326-S7	BOLT, hex-hd., S., $\frac{3}{8}$-16NC-2 x $\frac{7}{8}$	2
	WO-5544	FM-20406-S	NUT, hex., S., $\frac{3}{8}$-16NC-2	2
	WO-52101	FM-33797-S2	WASHER, S., $\frac{3}{8}$ in.	2
	WO-5010	FM-34806-S2	WASHER, lock, S., $\frac{3}{8}$ in.	2
	WO-A-2359	FM-GP-1012	BRACKET, support, spare wheel, assembly	1
	WO-51485	FM-GPW-1144600	BOLT, hex-hd., s-fin., alloy-S., $\frac{5}{16}$-18NC-2 x $\frac{5}{8}$ (brkt. to rear panel)	2
	WO-6660	FM-GPW-1144598	BOLT, hex-hd., s-fin., alloy-S., $\frac{5}{16}$-18NC-2 x 1$\frac{1}{4}$ (brkt. to rear panel)	2
	WO-6167		NUT, hex., s-fin., alloy-S., $\frac{5}{16}$-18NC-2	4
	WO-51840		WASHER, lock, S., $\frac{5}{16}$ in.	4
	WO-A-476		NUT, hex., S., $\frac{1}{2}$-20NF-2 (spare wheel to carrier stud)	2
	WO-A-2823		STRAINER, spare tire carrier, upper, assembly	1
	WO-A-2820		STRAINER, spare tire carrier, w/FILLER, lower, assembly	1

GROUP 16—SPRINGS AND SHOCK ABSORBERS

Figure Number	Official Stockage Number	Part Number — Willys	Part Number — Ford	ITEM	Quantity Reqd. per Unit Assy.	Per 100 Major Items — 12 Mos. Field Maintenance	Per 100 Major Items — Major Overhaul (5th Ech)	Per 100 Major Items — Total First Year Procurement	Estimated Reqmts. per 100 Rebuilds
		Col. 3	Col. 4	Col. 5	Col. 6	Col. 7	Col. 8	Col. 9	Col. 10
				1601—FRONT SPRING					
		WO-359039	FM-GPW-5781	BUSHING, front spring (.753 O.D. x .565 I.D. x 1.68 long) (spring eye)	2	—	10	12	—
		WO-116609	FM-GPW-5345-A	BOLT, front spring, center	2	48	80	150	—
		WO-116460	FM-GPW-5330-B	CLIP, front spring leaf (large for six leaves)	4	80	80	180	—
		WO-116458	FM-GPW-5330-A	CLIP, front spring leaf (small for three leaves)	4	80	80	180	—
		WO-A-612-1	FM-GPW-5313-B	LEAF, front spring, L., No. 1 w/BUSHING (36¼ in. between eye centers)	1	10	10	23	—
		WO-A-613-1	FM-GPW-5313-A	LEAF, front spring, R., No. 1 w/BUSHING (36¼ in. between eye centers)	1	10	10	23	—
		WO-A-612-2	FM-GPW-5315-B	LEAF, front spring, L., No. 2 (36¼ in. long)	1	10	10	23	—
		WO-A-613-2	FM-GPW-5315-A	LEAF, front spring, R., No. 2 (36¼ in. long)	1	10	10	23	—
		WO-A-612-3	FM-GPW-5316-B	LEAF, front spring, L., No. 3 (30 in. long) (issue until stock exhausted)	1	8	12	—	—
		WO-A-613-3	FM-GPW-5316-A	LEAF, front spring, R., No. 3 (30 in. long) (issue until stock exhausted)	1	8	12	—	—
		WO-A-612-4	FM-GPW-5317-B	LEAF, front spring, L., No. 4 (26 in. long)	1	—	—	—	—
		WO-A-613-4	FM-GPW-5317-A	LEAF, front spring, R., No. 4 (26 in. long)	1	—	—	—	—
		WO-A-612-5	FM-GPW-5318-B	LEAF, front spring, L., No. 5 (23 in. long)	1	—	—	—	—
		WO-A-613-5	FM-GPW-5318-A	LEAF, front spring, R., No. 5 (23 in. long)	1	—	—	—	—
		WO-A-612-6	FM-GPW-5319-B	LEAF, front spring, L., No. 6 (18 in. long)	1	—	—	—	—
		WO-A-613-6	FM-GPW-5319-A	LEAF, front spring, R., No. 6 (18 in. long)	1	—	—	—	—
		WO-A-612-7	FM-GPW-5320-B	LEAF, front spring, L., No. 7 (14 in. long)	1	—	—	—	—
		WO-A-613-7	FM-GPW-5320-A	LEAF, front spring, R., No. 7 (14 in. long)	1	—	—	—	—
		WO-A-612-8	FM-GPW-5321-B	LEAF, front spring, L., No. 8 (10 in. long)	1	—	—	—	—
		WO-A-613-8	FM-GPW-5321-A	LEAF, front spring, R., No. 8 (10 in. long)	1	—	—	—	—
		WO-5910	FM-33798-S	NUT, hex, s-fin, alloy-S, 5/16-24NF-2 (spring center bolt)	2	—	—	—	—
16-1		WO-A-612	FM-GPW-5311	SPRING, front, left, assembly	1	% 10	5	19	—
		WO-A-613	FM-GPW-5310	SPRING, front, right, assembly	1	% 10	5	19	—
				1601A—REAR SPRINGS					
		WO-359039	FM-GPW-5781	BUSHING, rear spring (.753 O.D. x .565 x 1.68 lgth.) (spring eye) (issue until stock exhausted)	2	160	320	—	—
		WO-116610	FM-GPW-5345-B	BOLT, rear spring center	2	80	160	—	—
		WO-116589	FM-GPW-5724-B	CLIP, rear spring leaf, large (seven leaves)	4	19	19	40	—
		WO-116459	FM-GPW-5724-A	CLIP, rear spring leaf, small (3 leaves)	4	19	19	40	—
		WO-A-614-1	FM-GPW-5563	LEAF, rear spring, No. 1, w/BUSHING (42 in. between eye centers)	2	3	3	9	—
		WO-A-614-2	FM-GPW-5565	LEAF, rear spring, No. 2 (42 in. long)	2	3	3	9	—
		WO-A-614-3	FM-GPW-5566	LEAF, rear spring, No. 3 (36 in. long) (issue until stock exhausted)	2	72	148	—	—
		WO-A-614-4	FM-GPW-5567	LEAF, rear spring, No. 4 (32 in. long)	2	—	—	—	—
		WO-A-614-5	FM-GPW-5568	LEAF, rear spring, No. 5 (28 in. long)	2	—	—	—	—
		WO-A-614-6	FM-GPW-5569	LEAF, rear spring, No. 6 (23 in. long)	2	—	—	—	—
		WO-A-614-7	FM-GPW-5570	LEAF, rear spring, No. 7 (19 in. long)	2	—	—	—	—

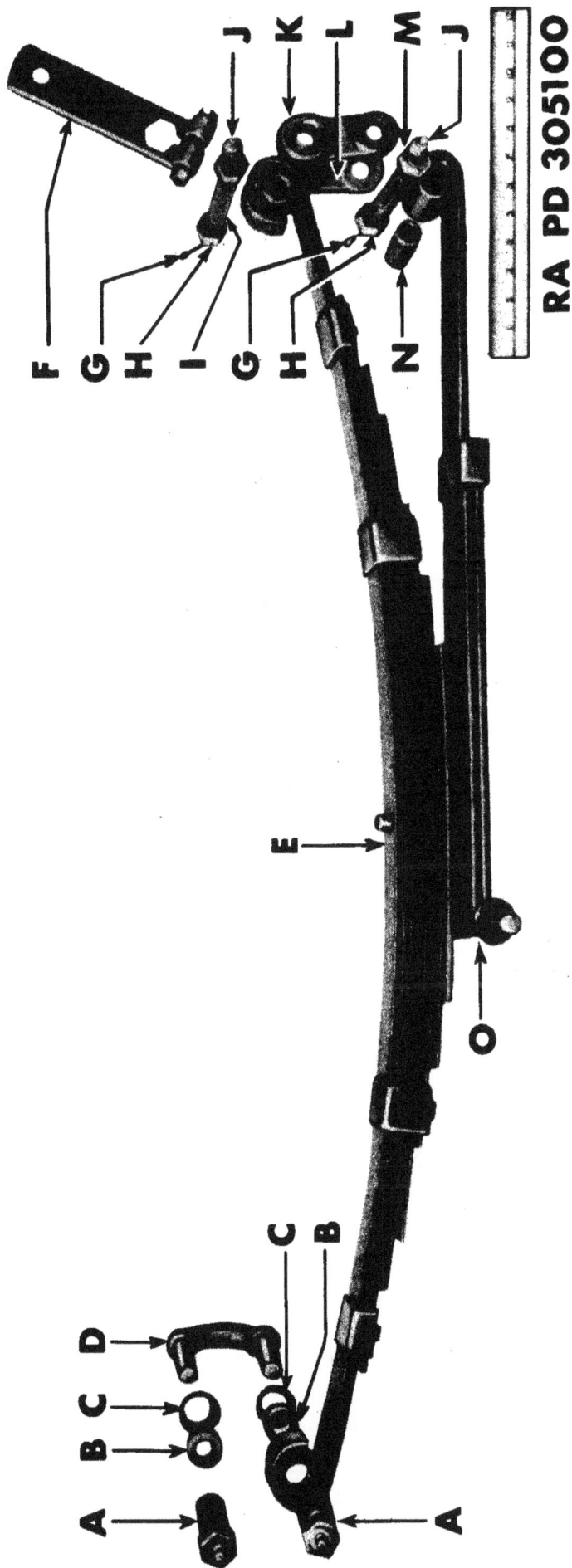

RA PD 305100

FIGURE 16-1—FRONT SPRING AND TORQUE REACTION SPRING

Key	Item	Willys Part No.	Ford Part No.	Gov't Group No.
A	BUSHING	WO-A-8255	Willys only	1602
B	GREASE seal	WO-A-515	FM-GPW-5481	1602
C	RETAINER	WO-A-1252	FM-GPW-5482	1602
D	U BOLT	WO-A-513	FM-GPW-5778	1602
E	SPRING assembly	WO-A-612	FM-GPW-5311	1601
F	BOLT lock assembly	WO-A-6326	FM-GPW-5601	1602
G	PIN	WO-5021	FM-72035-S	1602
H	NUT	WO-6436	FM-34033-S18	1602
I	BOLT	WO-384228	FM-GPW-5468	1602
J	FITTING	WO-640038	FM-358006-S8	1602
K	SPRING SHACKLE and BUSHING assembly	WO-A-6069	FM-GPW-5605	1602
L	SHACKLE and BUSHING assembly	WO-A-6068	FM-GPW-5602	1602
M	BOLT	WO-A-6067	FM-GPW-5610	1601
N	BUSHING	WO-A-6067	FM-GPW-5610	1601
O	SPRING assembly	WO-A-6066	FM-GPW-5588-A	1601

RA PD 305100-A

GROUP 16—SPRINGS AND SHOCK ABSORBERS (Cont'd)

Figure Number Col. 1	Official Stockage Number Col. 2	Part Number — Willys Col. 3	Part Number — Ford Col. 4	ITEM Col. 5	Quantity Reqd. per Unit Assy. Col. 6	12 Mos. Field Maintenance Col. 7	Major Overhaul (5th Ech) Col. 8	Total First Year Procurement Col. 9	Estimated Reqmts. per 100 Rebuilds Col. 10
				1601A—REAR SPRINGS (Cont'd)					
		WO-A-614-8	FM-GPW-5571	LEAF, rear spring, No. 8 (15 in. long)	2	—	—	—	—
		WO-A-614-9	FM-GPW-5572	LEAF, rear spring, No. 9 (11 in. long)	2	—	—	—	—
		WO-5910	FM-33798-S2	NUT, hex, S., 5/16-24NF-2	2	—	—	—	—
		WO-A-614	FM-GPW-5560	SPRING, rear, assembly (spring center bolt)	2	% 3	3	9	—
				1601B—SPRING BUMPERS					
		WO-A-481	FM-GPW-5783-A	BUMPER, front and rear axle (issue until stock exhausted)	4	8	12	12	—
		WO-51396	FM-24347-S2	BOLT, hex-hd., s-fin., alloy-S., 5/16-24NF-2 x 3/4	8	—	—	—	—
		WO-5910	FM-33798-S	NUT, hex, S., 5/16-24NF-2	4	—	—	—	—
		WO-51833	FM-34806-S2	WASHER, lock, S., 5/16 (19/32 O.D. x 11/32 I.D. x 1/16 thk.)	8	—	—	—	—
				1601C—TORQUE REACTION SPRING					
16-1		WO-A-6067	FM-GPW-5610	BUSHING, torque reaction spring (after Willys serial 146774)	1	5	5	12	—
		WO-A-6169	FM-GPW-5611	CLIP, torque reaction spring (after Willys serial 146774) (issue until stock exhausted)	1	4	4	—	—
		WO-A-6861	FM-GPW-5587	LEAF, No. 1, torque reaction spring (after Willys serial 146774) (issue until stock exhausted)	1	16	24	—	—
		WO-A-6168	FM-GPW-5590-B	LEAF, No. 2, torque reaction spring (after Willys serial 146774) (issue until stock exhausted)	1	16	24	—	—
16-1		WO-A-6066	FM-GPW-5588-A	SPRING, torque reaction, assembly (after Willys serial 146774)	1	2	2	8	—
				1602—SHACKLES AND SPRING ATTACHING PARTS					
16-1		WO-384228	FM-GPW-5468	BOLT, pivot, spring shackle (front for rear spring, rear for front spring)	3	% 10	10	28	—
16-1		WO-6436	FM-34033-S18	NUT, hex, castellated, S., 9/16-18NF-2	3	—	—	—	—
16-1		WO-5021	FM-72035-S	PIN, cotter 3/32 x 1	3	—	—	—	—
16-1		WO-A-6075	FM-GPW-5608	BOLT, torque reaction spring shackle (frt. spring rear bolt, left) (after serial 146774)	1	% 3	3	8	—
		WO-A-6074	FM-GPW-5609	BOLT, torque reaction spring shackle (shackle to torque spring) (after Willys serial 146774)	1	% 3	3	8	—
		WO-6436	FM-34033-S18	NUT, hex, castellated, S., 9/16-18NF-2	1	—	—	—	—
		WO-5021	FM-72035-S	PIN, cotter 3/32 x 1	1	—	—	—	—
16-1		WO-A-513	FM-GPW-5778	BOLT, "U", spring shackle, left hand thd., S. (front end left front spring, rear end right rear spring)	2	% 10	3	15	—
		WO-A-514	FM-GPW-5779	BOLT, "U", spring shackle, right hand thd., S. (front end right front spring, rear end left rear spring)	2	% 10	3	15	—

Fig.	Ord. part No.	Mfr. part No.	Description	Units per assembly				
	WO-A-6072	FM-GPW-5604	BUSHING, spring shackle, inner (after Willys serial 146774)	1	5	5	12	—
	WO-A-8256		BUSHING, spring shackle, assembly, right thd. (frt. end of frt. left spring, rear end right rear spring) (Willys only) (issue until stock exhausted)	2	52	108	—	—
16-1	WO-A-8255		BUSHING, spring shackle, assembly, left thd. (frt. end of frt. right spring, rear end rear left spring) (for all frame ends) (Willys only) (issue until stock exhausted)	6	132	268	12	—
	WO-A-6073	FM-GPW-5607	BUSHING, spring shackle, outer (after Willys serial 146774)	1	5	5	20	—
	WO-A-575	FM-GPW-5705-B	CLIP, front spring, left (axle to spring) (after Willys serial 146744)	2	10	10	—	—
	WO-A-1097		CLIP, front spring, right (issue until stock is exhausted, then use WO-A-6511 for maintenance)	1	%20	49	—	—
	WO-A-6511	FM-GPW-5455	CLIP, front spring, right (axle to spring) (was WO-A-1097)	1	%2	—	8	—
	WO-A-574	FM-GPW-5453	CLIP, front spring, right (axle to spring)	1	%56	2	140	—
	WO-339372	FM-GPW-5456	NUT, hex, S., special, 7/16-20NF-2 (front spring clip, right)	4	%108	56	—	—
	WO-638539	FM-33881-S2	NUT, hex, S., 7/16-20NF-2 (front spring clip, left) (issue until stock exhausted)	8	16	216	—	—
	WO-5938	FM-34848-S2	WASHER, lock, S., 7/16 in.	4	—	—	—	—
16-1	WO-638500	FM-353023-S7	FITTING, grease, 90° (SW-1911) (issue until stock exhausted)	4	92	32	—	—
	WO-640038	FM-358006-S8	FITTING, grease, straight (AD-1720) (was WO-A-392909) (issue until stock exhausted)	2	—	184	—	—
16-1	WO-A-6326	FM-GPW-5601	LOCK, spring shackle bolt, assembly (after serial 170307)	1	4	—	—	—
	WO-51308	FM-21492-S2	BOLT, hex-hd., S., 5/16-24NF x 2	1	4	4	—	—
	WO-51612	FM-350744-S2	BOLT, hex-hd., S., 7/16-20NF x 3¾ (lock to frame) (after Willys serial 170307)	1	4	3	—	—
	WO-5910	FM-33798-S	NUT, hex, S., 5/16-24 (after Willys serial 170307)	1	6	13	—	—
	WO-52510	FM-34941-S	WASHER, lock, S., 5/16 in. (after Willys serial 170307)	1	38	2	—	—
	WO-A-571	FM-GPW-5460	PLATE, spring clip, w/SHAFT, assembly, rear left (issue until stock exhausted)	2	2	2	—	—
	WO-A-572	FM-GPW-5459	PLATE, spring clip, front, w/SHAFT, assembly, left (issue until stock exhausted)	2	2	—	—	—
16-1	WO-A-568	FM-GPW-5458	PLATE, spring clip, front, w/SHAFT, assembly, right (issue until stock exhausted)	2	%10	—	—	—
16-1	WO-A-1252	FM-GPW-5482	RETAINER, spring shackle grease seal	1	%19	—	20	—
16-1	WO-A-515	FM-GPW-5481	SEAL, grease, spring shackle	8	%5	—	60	—
16-1	WO-A-6068	FM-GPW-5602	SHACKLE, spring, inner, w/BUSHING (after Willys serial 146774)	8	%10	3	6	—
	WO-A-6069	FM-GPW-5605	SHACKLE, spring, outer, w/BUSHING (after Willys serial 146774)	1	14	6	6	—

1603—SHOCK ABSORBERS AND MOUNTINGS

Fig.	Ord. part No.	Mfr. part No.	Description	Units per assembly				
	WO-A-6902		ABSORBER, shock, front, assembly (MAE-11465) (after Willys serial 197066) (Willys only)	2	%10	2	16	—
		FM-GPW-18045	ABSORBER, shock, front, assembly (Gabriel) (Ford only)	2	%19	—	30	—
16-2	WO-A-6903		ABSORBER, shock, rear, assembly (MAE-11466) (after Willys serial 197066) (Willys only)	2	%5	3	10	—
		FM-GPW-18080	ABSORBER, shock, rear, assembly (Gabriel) (Ford only)	2	%10	—	16	—
	WO-116640		BASE, front, shock absorber, assembly (MAE-12626) (Willys only)	2	—	—	—	—
	WO-116641		BASE, rear, shock absorber, assembly (MAE-12627) (Willys only)	2	—	—	—	—
16-2	WO-637936	FM-GPW-18060	BUSHING, shock absorber mounting pin, rubber (HP-779)	16	14	14	50	—
16-2	WO-116631		BUSHING, shock absorber rebound spring seat (MAE-12448) (Willys only)	4	—	—	—	—
16-2	WO-638343		DISC, shock absorber spring seat (MAE-108496) (Willys only)	4	—	—	—	—
16-2	WO-637810		GASKET, shock absorber bushing (MAE-10875) (Willys only)	4	—	—	—	—
	WO-116637		GUIDE, shock absorber piston rod, w/SEAL, assembly (MAE-12507) (Willys only)	4	—	—	—	—
16-2	WO-116642		HEAD, front, shock absorber, assembly (MAE-12628) (Willys only)	2	—	—	—	—
	WO-116643		HEAD, rear, shock absorber, assembly (MAE-12629) (Willys only)	2	—	—	—	—

GROUP 16—SPRINGS AND SHOCK ABSORBERS (Cont'd)

Col. 1 Figure Number	Col. 2 Official Stockage Number	Col. 3 Willys	Col. 4 Ford	Col. 5 ITEM	Col. 6 Quantity Reqd. per Unit Assy.	Col. 7 12 Mos. Field Maintenance	Col. 8 Major Overhaul (5th Ech)	Col. 9 Total First Year Procurement	Col. 10 Estimated Reqmts. per 100 Rebuilds
				1603—SHOCK ABSORBERS AND MOUNTINGS (Cont'd)					
		WO-A-8810		KIT, repair, rear shock absorber (Willys only) . . . . (Includes:	as req.	6	7	20	—
				1 WO-638343 DISC					
				1 WO-637810 GASKET					
				4 WO-637936 GROMMET					
				1 WO-116637 GUIDE w/SEAL					
				1 WO-116632 NUT					
				1 WO-116630 PISTON					
				1 WO-116644 PLATE, adjusting					
				1 WO-116626 PLATE, valve backing					
				1 WO-116625 SEAT					
				1 WO-116636 SLEEVE					
				1 WO-116627 SPACER, metering					
				1 WO-116631 SPACER, piston nut					
				1 WO-637803 SPRING, intake valve					
				1 WO-116633 SPRING, valve					
				1 WO-116639 TUBE					
				1 WO-116624 VALVE, compression assembly					
				1 WO-637804 VALVE, intake					
				1 WO-116629 WASHER, piston support					
				1 WO-116634 WASHER, spring seat)					
		WO-A-8809		KIT, repair, front shock absorber (Willys only) . . . . (Includes:	as req.	13	14	40	
				1 WO-638343 DISC					
				1 WO-637810 GASKET					
				4 WO-637936 GROMMET					
				1 WO-116637 GUIDE w/SEAL					
				1 WO-116632 NUT					
				1 WO-116630 PISTON					
				1 WO-116644 PLATE, adjusting					
				1 WO-116626 PLATE, valve					
				1 WO-116625 SEAT					
				1 WO-116635 SLEEVE					
				1 WO-116628 SPACER, metering					
				1 WO-116631 SPACER, piston nut					
				1 WO-637803 SPRING, intake valve					
				1 WO-116633 SPRING, valve					
				1 WO-116638 TUBE					

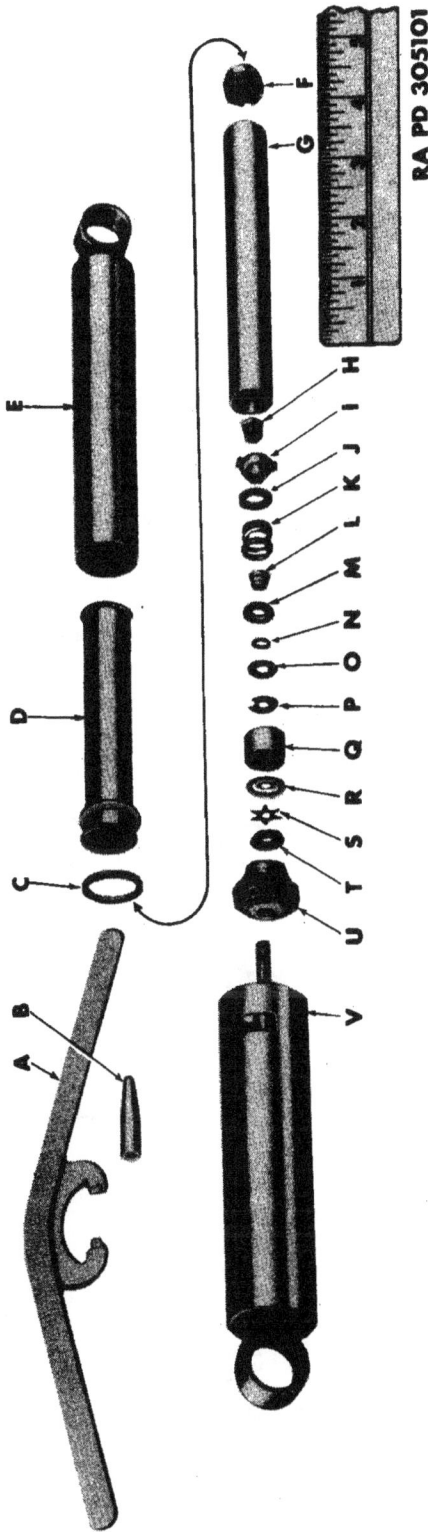

RA PD 305101

FIGURE 16-2—SHOCK ABSORBER ASSEMBLY

Key	Item	Willys Part No.	Ford Part No.	Gov't Group No.
A	WRENCH	WO-A-7778	(Willys only)	2301
B	THIMBLE	WO-A-7779	(Willys only)	2301
C	GASKET	WO-637810	(Willys only)	1603
D	SLEEVE assembly	WO-116635	(Willys only)	1603
E	BASE assembly	WO-116640	(Willys only)	1603
F	VALVE assembly	WO-116624	(Willys only)	1603
G	TUBE	WO-116638	(Willys only)	1603
H	NUT	WO-116632	(Willys only)	1603
I	PLATE	WO-116644	(Willys only)	1603
J	WASHER	WO-116634	(Willys only)	1603
K	SPRING	WO-116633	(Willys only)	1603

Key	Item	Willys Part No.	Ford Part No.	Gov't Group No.
L	BUSHING	WO-116631	(Willys only)	1603
M	WASHER	WO-116629	(Willys only)	1603
N	PLATE	WO-116626	(Willys only)	1603
O	DISC	WO-638343	(Willys only)	1603
P	SPACER	WO-116628	(Willys only)	1603
Q	PISTON	WO-116630	(Willys only)	1603
R	VALVE	WO-637804	(Willys only)	1603
S	SPRING	WO-637803	(Willys only)	1603
T	SEAT	WO-116625	(Willys only)	1603
U	GUIDE and SEAL assembly	WO-116637	(Willys only)	1603
V	HEAD assembly	WO-116643	(Willys only)	1603

RA PD 305101-A

1	WO-116624	VALVE, compression, assembly	
1	WO-637804	VALVE, intake	
1	WO-116629	WASHER, piston support	
1	WO-116634	WASHER, spring seat	
4		NUT, hex, S, shock absorber piston rod (MAE-12449) (Willys only)	WO-116632
4		PISTON, shock absorber (MAE-10966-A) (Willys only)	WO-116630
4		PLATE, adjusting, shock absorber (MAE-12631) (Willys only)	WO-116644
4		PLATE, back, shock absorber rebound valve (MAE-10856-8) (Willys only)	WO-116626
4		SEAT, shock absorber spring (MAE-10855) (Willys only)	WO-116625
2		SLEEVE, front, shock absorber, assembly (MAE-12468) (Willys only)	WO-116635
2		SLEEVE, rear, shock absorber, assembly (MAE-12469) (Willys only)	WO-116636
2		SPACER, metering, front shock absorber (MAE-10863-2) (Willys only)	WO-116628
2		SPACER, metering, rear shock absorber (MAE-10863-1) (Willys only)	WO-116627

16-2
16-2
16-2
16-2
16-2
16-2

GROUP 16—SPRINGS AND SHOCK ABSORBERS (Cont'd)

| Figure Number | Official Stockage Number | Part Number | | ITEM | Quantity Reqd. per Unit Assy. | Per 100 Major Items | | | Estimated Reqmts. per 100 Rebuilds |
| | | Willys | Ford | | | 12 Mos. Field Maintenance | Major Overhaul (5th Ech) | Total First Year Procurement | |
Col. 1	Col. 2	Col. 3	Col. 4	Col. 5	Col. 6	Col. 7	Col. 8	Col. 9	Col. 10
				1603—SHOCK ABSORBERS AND MOUNTINGS (Cont'd)					
16-2		WO-637803		SPRING, shock absorber piston intake valve (MAE-10639-B) (Willys only)	4	—	—	—	—
16-2		WO-116633		SPRING, shock absorber rebound valve (MAE-12463) (Willys only)	4	—	—	—	—
16-2		WO-116638		TUBE, pressure, front shock absorber (MAE-12620) (Willys only)	2	—	—	—	—
16-2		WO-116639		TUBE, pressure, rear shock absorber (MAE-12621) (Willys only)	2	—	—	—	—
16-2		WO-116624		VALVE, compression, shock absorber, assembly (MAE-83-C1) (Willys only)	4	—	—	—	—
16-2		WO-637804		VALVE, intake, shock absorber piston (MAE-10640-B) (Willys only)	4	—	—	—	—
16-2		WO-116634		WASHER, shock absorber adjusting plate (MAE-12464) (Willys only)	4	—	—	—	—
16-2		WO-A-227	FM-356525	WASHER, shock absorber mounting pin (S., 1¼ O.D. x 4/64 I.D. x 5/64 thk.)	8	29	10	60	—
16-2		WO-52946	FM-72037-S	PIN, cotter, 3/16 x 1	8	—	—	—	—
16-2		WO-116629		WASHER, shock absorber piston support (MAE-10906-A) (Willys only)	4	—	—	—	—
				GROUP 17—HOOD, FENDERS, APRONS					
				1701—FENDERS					
		WO-A-2662	FM-GPW-16132-	ANTI-SQUEAK, front fender to cowl side panel, lower	2	—	—	—	—
		WO-A-3158	FM-GPW-16159	ANTI-SQUEAK, front fender to cowl side panel, upper (issue until stock exhausted)	2	—	—	—	—
		WO-A-3179		BOLT, welding, S., 5/16-18NC x 5/8 (battery holder to fender)	1	% 4	8	—	—
		WO-A-2843	FM-GPW-16095	BRACE, left front fender	1	—	—	—	—
		WO-A-2844	FM-GPW-16094	BRACE, right front fender	1	—	—	—	—
		WO-52600	FM-24449-S2	BOLT, hex-hd., S., 3/8-24NF x 1⅜ (brace to chassis)	4	—	—	—	—
		WO-52819	FM-34707-S2	WASHER, S. (plain) 3/8 in.	4	—	—	—	—
		WO-51304	FM-34807-S2	WASHER, lock, S., 3/8 in.	4	—	—	—	—
		WO-A-2942	FM-GPW-16006	FENDER, front left, w/SPLASHER, assembly	1	—	—	—	—
		WO-A-2943	FN-GPW-16005	FENDER, front right, w/SPLASHER, assembly	1	—	—	—	—
		WO-51523	FM-20046-S2	BOLT, hex-hd. S., 5/16-18NC x ¾ (fender to cowl)	8	—	—	—	—
		WO-6167	FM-34130-S2	NUT, hex, S., 5/16-18NC	2	—	—	—	—
		WO-A-4116		WASHER, S. (plain) 5/16 in. (was WO-71633) (issue until stock exhausted)	8	% 28	52	—	—
		WO-53029		WASHER, lock, S., external, shakeproof, 5/16 in.	8	—	—	—	—
		WO-53025		WASHER, lock, S., cd-pltd., internal, external, 5/16 in. (was WO-A-51833)	10	—	—	—	—
		WO-A-2891	FM-GPW-16105	FILLER, front fender brace, wood	2	—	—	—	—
				1702—SPLASH APRON					
		WO-A-3159	FM-GPW-16133	ANTI-SQUEAK, apron to frame	3	—	—	—	—
		WO-A-2841	FM-GPW-16083	APRON, splash, left front fender, assembly	1	—	—	—	—
		WO-A-2842	FM-GPW-16082	APRON, splash, right front fender, assembly	1	—	—	—	—

Part No.	Part No.	Description	Qty
WO-50163	FM-20349-S2	BOLT, hex-hd., S., 3/8-24NF-2 x 3/4 (apron to fender)	4
WO-51396	FM-20347-S2	BOLT, hex-hd., S., 5/16-24NF-2 x 3/4 (apron to top of frame)	1
WO-50921		WASHER, S. (plain) 3/8 in. (was WO-5455)	4
WO-71633	FM-351370-S2	WASHER, S. (plain) 5/16 in.	1
WO-53026		WASHER, lock, S., cd-pltd., internal, external, 3/8 in. (was WO-5010)	4
WO-51840	FM-34806-S7-8	WASHER, lock, S., 5/16 in.	1
WO-A-3096	FM-GPW-16128	BRACE, mounting, voltage regulator	2
WO-A-2375	FM-350976-S2	NUT, clinch, 5/16-24NF-2 (regulator to apron)	4
WO-53135		WASHER, lock, S., cd-pltd., external, 3/8 in.	4

1704—HOOD

Part No.	Part No.	Description	Qty
WO-A-3059	FM-GPW-16684	BRACKET, hood catch	2
WO-51514	FM-20308-S7-8	BOLT, hex-hd., S., 1/4-20NC-2 x 5/8 (brkt. to hood)	4
WO-5790	FM-33795-S2	NUT, hex, S., 1/4-20NC-2	4
WO-5121	FM-34705-S2	WASHER, S., plain, 1/4 in.	4
WO-52031	FM-34805-S2	WASHER, lock, S., 1/4 in.	4
WO-A-4683	FM-GTB-16848-A	BUMPER, hood top windshield, assembly (wood)	2
WO-A-4679		NUT, S., prong type, sleeve, No. 10 (.190)-32NF-2 (Willys only)	4
WO-6627		SCREW, rd-hd., S., No. 10 (.190)-32NF-2 x 7/8 (bumper to hood) (Willys only)	4
WO-52221	FM-34803-S2	WASHER, lock, No. 10 (.190) (Willys only)	4
WO-A-2896	FM-GPW-16892	CATCH, hood, assembly	2
WO-51514	FM-20308-S7-8	BOLT, hex-hd., s-fin., alloy-S., 1/4-20NC-2 x 5/8 (catch to fender)	4
WO-5790	FM-33795-S2	NUT, reg. hex., S., 1/4-20NC-2	4
WO-5121	FM-34706-S2	WASHER, S., plain, 1/4 in.	4
WO-52031	FM-34805-S2	WASHER, lock, S., 1/4 in.	4
WO-A-2188	FM-GPW-16802	HINGE, hood, assembly	1
WO-52168	FM-20324-S2	BOLT, hex-hd., S., 1/4-20NC-2 x 5/8 (hinge to cowl top)	5
WO-52167	FM-24344	BOLT, hex-hd., S., 1/4-20NC-2 x 3/4 (hinge to cowl top)	2
WO-52702	FM-34705-S2	WASHER, S., 1/4 in.	5
WO-53204		WASHER, lock, S., cd-pltd., internal, external, 1/4 in. (Willys only)	6
WO-52031	FM-34805-S2	WASHER, lock, S., rust proof, 1/4 in.	3
WO-A-3225	FM-GPW-16610	HOOD assembly (after Willys serial 103545)	1
WO-A-2836		HOOD assembly (before Willys serial 103545)	1
WO-A-4680	FM-GTB-16847	WEBBING, hood top windshield bumper	2
WO-53103	FM-32475-S2	SCREW, wood, ov-hd., S., No. 6 (.138) x 1 (webbing to bumper)	4

GROUP 18—BODY

1800—BODY

Part No.	Part No.	Description	Qty
WO-A-12237	FM-GPW-1000000	BODY assembly	1

(Includes:
BOWS, top
BRACES
BRACKETS
CLAMPS
COMPARTMENT, glove

GROUP 18—BODY (Cont'd)

| Figure Number | Official Stockage Number | Part Number | | ITEM | Quantity Reqd. per Unit Assy. | Per 100 Major Items | | Total First Year Procurement | Estimated Reqmts. per 100 Rebuilds |
| | | Willys | Ford | | | 12 Mos. Field Maintenance | Major Overhaul (5th Ech) | | |
Col. 1	Col. 2	Col. 3	Col. 4	Col. 5	Col. 6	Col. 7	Col. 8	Col. 9	Col. 10
				1800—BODY (Cont'd)					
				COMPARTMENT, tool					
				HANDLES					
				LIGHTS, tail					
				MIRROR, rear view, outside					
				SEALS					
				SEATS					
				ATTACHING PARTS)					
		WO-A-12237	FM-GPW-1100001	BODY assembly	1	—	—	—	—
				(Includes:					
				FLOORS					
				PANELS)					
				1800A—BODY HANDLES					
		WO-A-2389	FM-GPW-1129672	HANDLE, body, outside, rear corner	2	—	—	—	—
		WO-A-2390	FM-GPW-1129670	HANDLE, body, outside front	2	—	—	—	—
		WO-51485	FM-20326-S2	BOLT, hex-hd., S., $\frac{5}{16}$-18NC-2 x $\frac{5}{8}$ (handles to body)	16	—	—	—	—
		WO-52350	FM-3379752	NUT, hex, S., $\frac{5}{16}$-18NC-2	16	—	—	—	—
		WO-51840	FM-34806-S2	WASHER, lock, S., $\frac{5}{16}$ in.	16	—	—	—	—
		WO-A-2168	FM-GPW-1154810-A	LOOP, footman (on body side panel)	5	—	—	—	—
		WO-6352	FM-GPW-355836-S7	NUT, hex, S., No. 10 (.190)-24NC-2 (loop to sides)	10	—	—	—	—
		WO-6290	FM-GPW-355162-S2	SCREW, fl-hd., S., No. 10 (.190)-24NC-2 x $\frac{1}{2}$	10	—	—	—	—
		WO-52221	FM-GPW-34803-S7	WASHER, lock, S., No. 10 (.190)	10	—	—	—	—
				1801—BODY MOUNTING PARTS					
		WO-A-4415	FM-GPW-1102511	LINER, body sill, thick	2	—	—	—	—
		WO-A-4414	FM-GPW-1102510	LINER, body sill, thin	as req.	—	—	—	—
				1802—BODY PILLARS					
		WO-A-2815		PILLAR, cowl, left	1	—	—	—	—
		WO-A-2816		PILLAR, cowl, right	1	—	—	—	—
		WO-52863		BOLT, hex-hd., S., $\frac{3}{8}$-16NC-2 x $1\frac{5}{8}$ (pillar to floor)	2	—	—	—	—
		WO-5544	FM-33799-S	NUT, hex, S., $\frac{3}{8}$-16NC-2	2	—	—	—	—

1802A—BODY PANELS

WO No.	FM No.	Description	Qty.
WO-5455	FM-34747-S	WASHER, I., plain, 3/8 in.	4
WO-5010	FM-34807-S7	WASHER, lock, S., 3/8 in. (13/16 O.D. x 13/32 I.D. x 1/16 thd.)	2
WO-A-2138	FM-GPW-1102396	BOLT, eye, safety strap (S., 3/8-16NC-2)	2
WO-5544	FM-33799-S2	NUT, hex, S., 3/8-16NC-2 (bolt to body)	2
WO-51304	FM-34807-S2	WASHER, lock, S., 3/8 in.	2
WO-A-2853	FM-GPW-1140449	BRACKET, engine crank retaining clamp, assembly	2
WO-A-2940	FM-GPW-1140441	BRACKET, retaining, engine crank	1
WO-A-2950	FM-GPW-1110895	FILLER, rear floor pan cross sill (wood)	1
WO-A-2311		GUSSET, front floor toe board to frame, left assembly	1
WO-A-2312		GUSSET, front floor toe board to frame, right assembly	1
WO-50163	FM-355498-S7	BOLT, hex-hd., S., 3/8-24NF-2 x 3/4	2
WO-50921		WASHER, S., plain, pkzd., 3/8 in. (was WO-5455)	2
WO-53135		WASHER, lock, S., external, 3/8 in.	2
WO-53026		WASHER, lock, internal, external, 3/8 in. (GM-178551)	2
WO-A-2933	FM-355909-S2	NUT, clinch, special, S. (1/4 in. pipe tap, std.) (for 1/4 in. drain plug in floor)	2
WO-A-2278		NUT, clinch, S. (1/4-20NC-2) (transmission cover)	7
WO-A-2375	FM-350976-S2	NUT, clinch, S. (5/16-24NF-2) (transmission cover and accelerator pedal)	1
WO-A-2263	FM-351015-S7	NUT, clinch, S. (3/8-16NC-2) (front floor rear cross sill assembly)	1
WO-662010	FM-353202-S	NUT, clinch, S., 5/16-18NC-2 (cowl side panel for front fender)	6
WO-A-2466	FM-33896-S2	NUT, wing, S., 5/16-18NC-2 (crank retaining brkt.)	1
WO-A-12003	FM-GPW-1111140	PAN, front floor	1
WO-51545	FM-20514-S2	BOLT, hex-hd., S., 3/8-16NC x 1 3/4 (thru left front cross sill)	2
WO-51954		BOLT, hex-hd., S., 3/8-16NC x 2 (thru front cross sills to frame brkt.)	2
WO-5544	FM-33799-S	NUT, hex, S., 3/8-16NC-2	4
WO-52893	FM-356016-S7	NUT, hex, S., stamped 3/8-16NC-2	4
WO-50921		WASHER, S., plain, 3/8 (was WO-5455)	8
WO-53135		WASHER, lock, S., external, 3/8 in.	4
WO-53026		WASHER, lock, S., internal, external, 3/8 in.	4
WO-A-12017	FM-GPW-1111218	PAN, rear floor	1
WO-51405	FM-20514-S2	BOLT, hex-hd., S., 3/8-16NC x 1 (to frame, left side)	1
WO-51954		BOLT, hex-hd., S., 3/8-16NC-2 x 2 (to frame, right side)	1
WO-52893	FM-356016-S7	NUT, hex, S., 3/8-16NC-2 (stamped)	1
WO-5544	FM-33799-S	NUT, hex, 3/8-16NC-2	1
WO-50921		WASHER, S. Pkzd, plain, 3/8 in. (was WO-5455)	2
WO-53135		WASHER, lock, S., external, 3/8 in.	3
WO-53026		WASHER, lock, S., cd-pltd. internal, external, 3/8 in.	3
WO-A-2756	FM-GPW-1127847	PANEL, body, left side	1
WO-A-2757	FM-GPW-1127846	PANEL, body, right side	1
WO-A-2758	FM-GPW-1140324	PANEL, body, rear	1
WO-52857		BOLT, hex-hd., S., 5/16-18NC-2 x 1 5/8	2
WO-52845		NUT, hex, S., 5/16-18NC-2 (stamped)	2
WO-6167		NUT, hex, S., 5/16-18NC-2	2
WO-A-4116		WASHER, S., Pkzd, plain, 5/16 in.	4
WO-53025		WASHER, lock. S., internal, external, cd-pltd, 5/16 in.	4
WO-A-12011	FM-GPW-1102039	PANEL, cowl, left side	1
WO-A-12010	FM-GPW-1102038	PANEL, cowl, right side	1
WO-A-12105	FM-GPW-1102015	PANEL, cowl, top	1

GROUP 18—BODY (Cont'd)

Figure Number Col. 1	Official Stockage Number Col. 2	Part Number — Willys Col. 3	Part Number — Ford Col. 4	ITEM Col. 5	Quantity Reqd. per Unit Assy. Col. 6	Per 100 Major Items — 12 Mos. Field Maintenance Col. 7	Per 100 Major Items — Major Overhaul (5th Ech) Col. 8	Per 100 Major Items — Total First Year Procurement Col. 9	Estimated Reqmts. per 100 Rebuilds Col. 10
				1802A—BODY PANELS (Cont'd)					
		WO-A-12025	FM-GPW-1127851	PANEL, dash, assembly	1	—	—	—	—
		WO-A-12023	FM-GPW-1127850	PANEL, rear left wheel house, inner	1	—	—	—	—
		WO-A-12024	FM-GPW-1127849	PANEL, rear right wheel house, inner	1	—	—	—	—
		WO-A-2771	FM-GPW-1127848	PANEL, left wheel house, top	1	—	—	—	—
		WO-A-2772	FM-GPW-	PANEL; right wheel house, top	1	—	—	—	—
		WO-A-2939	1103446-A	PLATE, tapping, axe clamp	1	—	—	—	—
		WO-A-2879	FM-GPW-1111238	PLATE, tapping, driver seat, front floor	1	—	—	—	—
		WO-A-2190	FM-GPW-1102130	PLATE, tapping, hood hinge (on cowl)	1	—	—	—	—
		WO-A-2940	FM-GPW-1128242	PLATE, tapping, shovel clamp	2	—	—	—	—
		WO-A-5120	FM-358019-S	PLUG, body drain hole cover (pipe, slotted, ¼ in.)	2	—	—	—	—
		WO-5247		RIVET, rd-hd, S., ¼ x ⅜	8	—	—	—	—
		WO-A-3144	FM-GPW-1162606	REST, foot, rear floor assembly	2	—	—	—	—
		WO-51485	FM-20326-S2	BOLT, hex-hd., S., 5⁄16-18NC x 5⁄8 (to wheel house panel)	3	—	—	—	—
		WO-51523	FM-20346-S2	BOLT, hex-hd., S., 5⁄16-18NC x ¾ (to floor pan)	1	—	—	—	—
		WO-6167	FM-33797-S	NUT, hex, S., 5⁄16-18NC-2	4	—	—	—	—
		WO-52274	FM-34746-S2	WASHER, S., plain, 5⁄16 in.	4	—	—	—	—
		WO-51840	FM-34806-S2	WASHER, lock, S., pltd., 5⁄16 in.	4	—	—	—	—
		WO-A-2470	FM-GPW-1140447	RETAINER, engine crank	2	—	—	—	—
		WO-A-3222	FM-GPW-1111155	RISER, front to rear floor pan, assembly	1	—	—	—	—
		WO-51406	FM-356016-S7	BOLT, hex-hd., S., ⅜-16NC-2 x 1¼	2	—	—	—	—
		WO-52893	FM-33799-S	NUT, hex., S., ⅜-16NC-2 stamped	2	—	—	—	—
		WO-5544		NUT, hex., S., ⅜-16NC-2	2	—	—	—	—
		WO-50921		WASHER, S., Plzd., plain ⅜ in. (was WO-5455)	4	—	—	—	—
		WO-53026		WASHER, lock, S., cd-pltd., internal, external, ⅜ in.	4	—	—	—	—
		WO-53135		WASHER, lock, S., shakeproof, external, ⅜ in.	4	—	—	—	—
		WO-52236	FM-27068-S	SCREW, fl-hd., S., No. 10 (.190)-24-NC-2 x ⅜ (footman loop hole plug cap)	2	—	—	—	—
		WO-A-2968	FM-GPW-1128244	SHEATH, axe	1	—	—	—	—
		WO-A-2837		SILL, cross, front floor pan, left assembly (Willys only)	1	—	—	—	—
		WO-A-2838	FM-GPW-1110610	SILL, cross, front floor pan, right assembly	1	—	—	—	—
		WO-A-2935	FM-GPW-1110625	SILL, cross, front floor pan, rear assembly	1	—	—	—	—
		WO-A-2948	FM-GPW-1110750	SILL, cross rear floor pan	1	—	—	—	—
		WO-A-2325	FM-GPW-1128286	STRAINER, body side panel to wheel house	1	—	—	—	—
		WO-A-3120	FM-GPW-1144620	SUPPORT, rear body to frame, inner	2	—	—	—	—
		WO-A-2336	FM-GPW-1144606	SUPPORT, rear body to frame, outer	2	—	—	—	—
				1803—FLOOR MATS AND SEALS					
		WO-A-2990	FM-GPW-1112115	COVER, inspection, floor, brake cylinder	1	—	—	—	—

			No. req'd
WO-52170	FM-24308-S2	BOLT, hex-hd., S., 1/4-20NC x 1/2	5
WO-52706	FM-34805-S2	WASHER, lock, S., 1/4 in.	5
WO-A-2759	FM-GPW-1111331	COVER, rear floor pan, left, shock absorber assembly	1
WO-A-2760	FM-GPW-1111330	COVER, rear floor pan, right, shock absorber assembly	1
WO-A-4592	FM-GPW-1111162	COVER, trailer electric plug housing (after Willys serial 200679)	1
WO-A-2982	FM-GPW-1112110	COVER, floor, transmission	1
WO-52170	FM-24308-S2	BOLT, hex-hd., S., 1/4-20NC x 1/2 (to floor pan)	7
WO-52706	FM-34805-S2	WASHER, lock, S., 1/4 in.	7
WO-A-3784	FM-GPW-1101735-A	GROMMET, floor, transfer shift lever (rubber)	1
WO-12055		PLATE, floor seal, dimmer switch (Willys only)	1
WO-A-2919	FM-GPW-1112119	RING, floor gearshift lever housing cover	1
WO-52167	FM-20344-S2	BOLT, hex-hd., S., 1/4-20NC-2 x 3/4	1
WO-52142	FM-32866-S2	SCREW, binding hd., S., No. 10 (.190) x 5/8	3
WO-A-2917	FM-GPW-1112158	RING, floor seal, steering column	1
WO-52809	FM-32924-S7-8	SCREW, binding hd., S., No. 10 (.190) x 1/2	4
WO-A-2918	FM-GPW-1112117	RING, floor transmission shift lever grommet	1
WO-52167	FM-20344-S2	BOLT, hex-hd., S., 1/4-20NC-2 x 3/4	2
WO-52142	FM-32866-S7	SCREW, binding hd., S., No. 10 (.190) x 1/2	3
WO-A-3782	FM-GPW-1111520-A	SEAL, floor, accelerator pedal rod	1
WO-A-2932	FM-GPW-1111137-A	SEAL, floor brake cylinder inspection cover	1
WO-A-3116	FM-GPW-1111299-A	SEAL, floor, fuel line hole floor	1
WO-A-3207	FM-GPW-9082-A	SEAL, fuel tank stone guard, front	1
WO-A-3208	FM-GPW-9085-A	SEAL, fuel tank stone guard, left	1
WO-A-3209	FM-GPW-9083-A	SEAL, fuel tank stone guard, rear	1
WO-A-3206	FM-GPW-9071	SEAL, fuel tank stone guard, right	1
WO-A-3783	FM-GPW-M11286-A	SEAL, floor gearshift lever housing cover (rubber)	1
WO-A-3943	FM-GPW-1102330-A	SEAL, glove compartment door (after serial 120697)	1
WO-A-12054		SEAL, floor, headlight foot switch (Willys only)	1
WO-A-3173		SEAL, radiator side air deflector, left (before Willys serial 125809)	1
WO-A-3575	FM-GPW-8155	SEAL, radiator side air deflector, left (after Willys serial 125809)	1
WO-A-3182		SEAL, radiator side air deflector to radiator (after Willys serial 125809) (Willys only)	1
WO-A-2979	FM-GPW-8222	SEAL, radiator side air deflector to radiator, right (before Willys serial 125809) (Willys only)	1
WO-A-3574	FM-GPW-8166	SEAL, radiator top air deflector (before Willys serial 125809)	1
WO-A-2992	FM-GPW-11514-A	SEAL, radiator top air deflector (after Willys serial 125809)	1
WO-A-2931		SEAL, floor starter switch	1
WO-A-2994	FM-GPW-1101698-A	SEAL, floor steering column	1
	FM-GPW-1112116-A	SEAL, floor, transmission cover	1

1804—SEATS

WO-A-2983	FM-355569-S2	BOLT, anchor, safety strap (3/8-16NC-2)	2

GROUP 18—BODY (Cont'd)

1804—SEATS (Cont'd)

| Figure Number | Official Stockage Number | Part Number | | ITEM | Quantity Reqd. per Unit Assy. | Per 100 Major Items | | | Estimated Reqmts. per 100 Rebuilds |
| | | Willys | Ford | | | 12 Mos. Field Maintenance | Major Overhaul (5th Ech) | Total First Year Procurement | |
Col. 1	Col. 2	Col. 3	Col. 4	Col. 5	Col. 6	Col. 7	Col. 8	Col. 9	Col. 10
		WO-6169	FM-33799-S2	NUT, hex, S., 3/8-16NC-2	2	—	—	—	—
		WO-51304	FM-34807-S2	WASHER, lock, S., 3/8 in.	2	—	—	—	—
		WO-A-2453	FM-GPW-1162552	BRACKET, front seat pivot (passenger)	2	—	—	—	—
		WO-51561	FM-60375-S	RIVET, rd-hd., S., 3/8 x 9/16	2	—	—	—	—
		WO-A-2515	FM-GPW-1160314	BRACKET, front seat, rear stop	2	—	—	—	—
		WO-A-2833	FM-GPW-1162414	BRACKET, pivot, rear seat frame	2	—	—	—	—
		WO-A-2787	FM-GPW-1162402	BRACKET, retaining, rear seat back frame	4	—	—	—	—
		WO-52217	FM-33795-S2	NUT, hex, S., 1/4-20NC-2 (brkt. to body)	4	—	—	—	—
		WO-52702	FM-34745-S	WASHER, S. (plain) 1/4 in.	4	—	—	—	—
		WO-52706	FM-34805-S2	WASHER, lock, S., 1/4 in.		—	—	—	—
		WO-A-2886	FM-GPW- 1160327-B	FRAME, front seat, assembly (driver)	1	—	—	—	—
		WO-5922	FM-24407-S	BOLT, hex-hd., S., 5/16-24NF-2 x 1 1/8 (frame to wheel house top panel)	1	—	—	—	—
		WO-51523	FM-20346-S2	BOLT, hex-hd., S., 5/16-18NC-2 x 3/4 (frame to floor)	2	—	—	—	—
		WO-6167	FM-33797-S	NUT, hex, S., 5/16-18NC-2 (frame to floor)	3	—	—	—	—
		WO-5437	FM-34706-S2	WASHER, S. (plain) 5/16 in.	1	—	—	—	—
		WO-71633	FM-351370-S2	WASHER, S. (special) 5/16 in. (frame to floor)	1	—	—	—	—
		WO-51840	FM-34806-S2	WASHER, lock, S., 5/16 in.	1	—	—	—	—
		WO-A-2925	FM-GPW-1160326	FRAME, front seat, back assembly (passenger)	1	—	—	—	—
		WO-51523	FM-20346-S2	BOLT, hex-hd., S., 5/16-18NC-2 x 3/4 (to floor pan)	2	—	—	—	—
		WO-6167	FM-33797-S	NUT, hex, S., 5/16-18NC-2	2	—	—	—	—
		WO-71633	FM-351370-S2	WASHER, S. (special) 5/16 in.	2	—	—	—	—
		WO-51840	FM-34806-S7-8	WASHER, lock, S., 5/16 in.	2	—	—	—	—
		WO-A-2782	FM-GPW-1161326	FRAME, rear seat, assembly	1	—	—	—	—
		WO-A-2783	FM-GPW-1161300	FRAME, rear seat back, assembly	1	—	—	—	—
		WO-A-2607	FM-355965-S2	NUT, tee, S., 5/16-24NF-2 (in drivers seat frame)	1	—	—	—	—
		WO-A-2466	FM-33896-S2	NUT, wing, S., 5/16-18NC-2 x 1 (on rear seat for tire pump) (after Willys serial 193040)	1	—	—	—	—
		WO-A-2910	FM-GPW-1162181	PANEL, front seat body side crash pad, left assembly	1	—	—	—	—
		WO-A-2911	FM-GPW-1162180	PANEL, front seat body side crash pad, right assembly	1	—	—	—	—
		WO-654934	FM-355900-S2	NUT, tee, S., 1/4-20NC-2 (on crash pads)	8	—	—	—	—
		WO-51539	FM-36868-S2	SCREW, rd-hd., S., 1/4-20NC-2 x 3/4 (pads to side panels)	8	—	—	—	—
		WO-52706	FM-34805-S2	WASHER, lock, S., reg., 1/4 in.	8	—	—	—	—
		WO-A-2945	FM-GPW-1162410	RETAINER, rear seat pivot tube	2	—	—	—	—
		WO-52170	FM-20308-S2	BOLT, hex-hd., S., 1/4-20NC-2 x 1/2	2	—	—	—	—
		WO-52217	FM-33799-S2	NUT, hex, S., 1/4-20NC-2	2	—	—	—	—
		WO-52706	FM-34805-S2	WASHER, lock, S., reg., 1/4 in.	2	—	—	—	—
		WO-A-1443	FM-GPW-17955	RETAINER, (see upper handle (after Willys serial 193040)	1	—	—	—	—

WO No.	FM-GPW No.	Description	Qty
WO-A-2830	FM-GPW-1161302	SPRING, retaining, rear seat frame	2
WO-52170	FM-20308-S2	BOLT, hex-hd., S., 1/4-20NC-2 x 1/2 (spring to wheel house panel)	4
WO-52217	FM-33795-S2	NUT, hex, S., 1/4-20NC-2	4
WO-52031	FM-34805-S2	WASHER, lock, S., 1/4 in.	4
WO-A-2883	FM-GPW-1131414	STRAP, safety, assembly	2
WO-A-3029		SUPPORT, rear seat frame to wheel house (Willys only)	2

1804A—CUSHIONS

WO No.	FM-GPW No.	Description	Qty
WO-A-2986	FM-GPW-1162900-B	CUSHION, front seat, assembly (driver and passenger)	2
WO-52651	FM-355836-S7	NUT, hex, S., No. 10 (.190)-24NC-2	10
WO-50814		SCREW, fl-hd., S., No. 10 (.190)-24NC-2 x 1/2	10
WO-52221	FM-34803-S2	WASHER, lock, S., No. 10 (.190)	10
WO-A-2788	FM-GPW-1160780	FRAME, rear seat cushion, assembly	1
WO-A-3114	FM-GPW-1166401	PAD, crash, front seat body side, left, assembly	1
WO-A-3115	FM-GPW-1166400	PAD, crash, front seat body side, right, assembly	1
WO-A-3108	FM-GPW-1160026	SEAT, trimmed, rear, assembly	1
WO-A-3107	FM-GPW-1160016-B	SEAT, trimmed, front, assembly	1
WO-A-3980		SPRING, front seat back assembly	2
WO-A-3984		SPRING, front seat body side crash pad, left, assembly	1
WO-A-3985		SPRING, front seat body side crash pad, right, assembly	1
WO-A-3981	FM-GPW-1163206	SPRING, front seat cushion, assembly (driver and passenger)	2
WO-A-3982	FM-GPW-1166800-B	SPRING, rear seat back, assembly	1
WO-A-3983	FM-GPW-1163846-B	SPRING, rear seat cushion, assembly	1
WO-A-11729		TRIM, front seat back cover, assembly	2
WO-A-11730		TRIM, front seat cushion cover, assembly	2
WO-A-11731		TRIM, rear seat back cover, assembly	1
WO-A-11732		TRIM, rear seat cushion cover, assembly	1
WO-52990	FM-32989-S2	SCREW, ctsk. ov-hd., S., No. 10 (.190) x 1/2	32
WO-52968	FM-34886-S	WASHER, ctsk., S., No. 10 (.190)	42

1809—INSTRUMENT PANEL AND MOUNTING PARTS

WO No.	FM-GPW No.	Description	Qty
WO-A-3155	FM-GPW-1104364	BRACE, instrument panel, circuit breaker, w/BRACKET, assembly	1
WO-5247	FM-60332-S2	RIVET, rd-hd., S., 1/4 x 3/8 (brace to dash)	2
WO-A-1765	FM-GPW-1101670	PAD, dash panel	1
WO-A-3005	FM-GPW-1101610	PANEL, dash, assembly	1
WO-A-3578	FM-GPW-1104320	PANEL, instrument assembly	1

1809A—COMPARTMENT AND PARTS

WO No.	FM-GPW No.	Description	Qty
WO-A-3536	FM-GPW-1106085	BOLT, glove compartment lock (after 20,700 Willys trucks up to 37,914 Willys trucks)	1
WO-A-3531	FM-GPW-1106084-A	CASE, glove compartment lock, assembly (after 20,700 Willys trucks up to 37,914 Willys trucks)	1

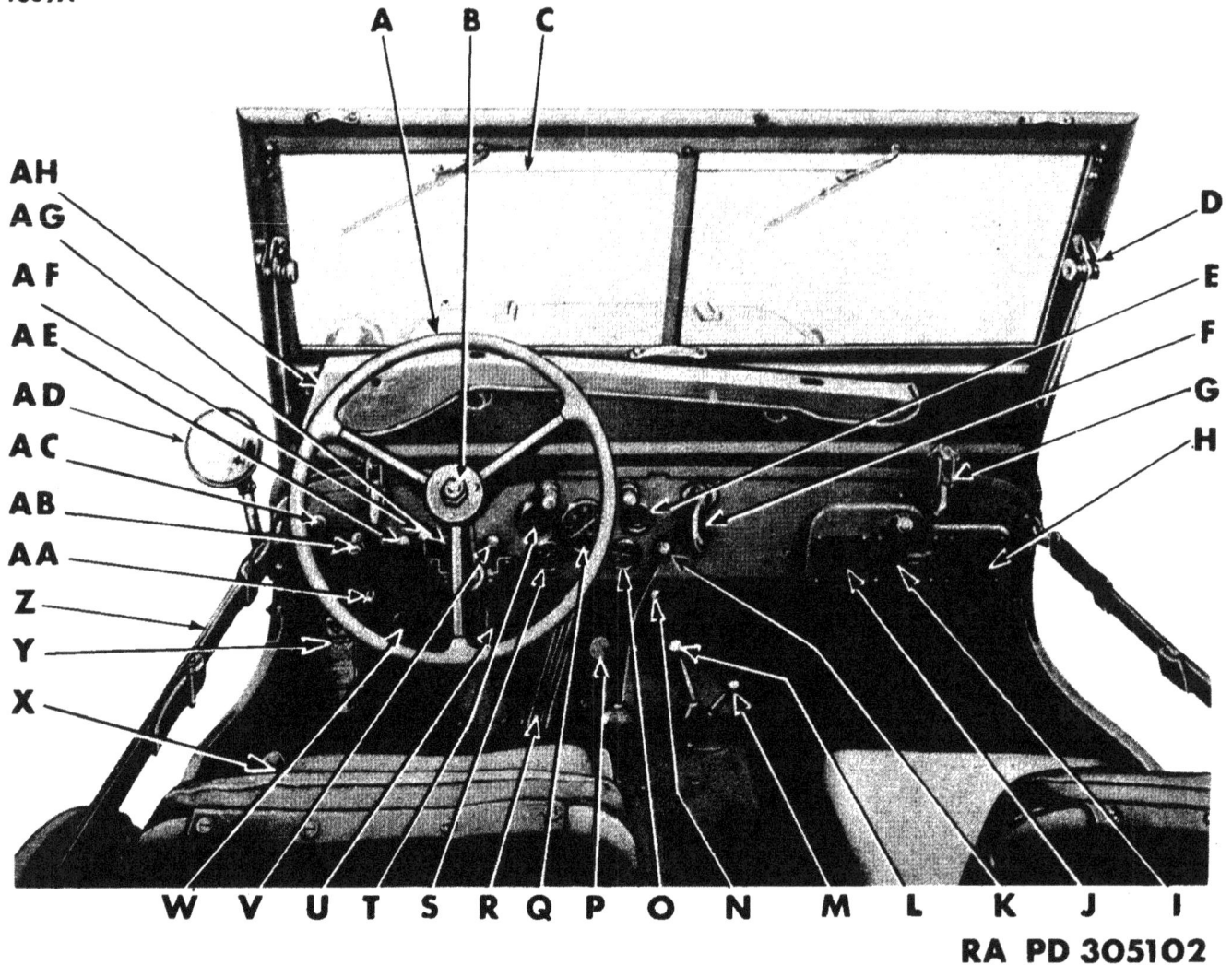

RA PD 305102

FIGURE 18-1—FRONT COMPARTMENT

Key	Item	Willys Part No.	Ford Part No.	Gov't Group No.
A	WHEEL	WO-A-6858	FM-GPW-3600-A-3	1404
B	BUTTON	WO-A-634	FM-GPW-3627	0609B
C	WIPER	WO-A-11433	FM-GPW-17500	1811D
D	ARM	WO-A-2235	FM-GP-1103302	1811
E	AMMETER	WO-A-8186	FM-GPW-10850	0605
F	HANDLE	WO-A-1242	FM-GPW-2780	1201
G	CLAMP assembly	WO-A-2227	FM-GP-1103482-A	1811
H	PLATE	WO-A-1330	FM-GPW-1101621-A	2203
I	PLATE	WO-A-11757	FM-GPW-4101629-A	2203
J	PLATE	WO-A-1331	FM-GPW-1101627-A	2203
K	LEVER	WO-A-1380	FM-GPW-7210-A	0706A
L	LEVER	WO-A-1506	FM-GPW-7710	0805B
M	LEVER	WO-A-1505	FM-GPW-7793	0805B
N	SWITCH	WO-A-7225	FM-9N-11450-A	0606
O	INDICATOR	WO-A-8188	FM-GPW-10883	0605
P	REST	WO-A-1225	FM-GPW-9716	0303
Q	SPEEDOMETER	WO-A-8180	FM-GPW-17255-A	2204
R	PEDAL	WO-A-6851	FM-GPW-9735-B	0303
S	GAUGE	WO-A-8190	FM-GPW-9273	0605
T	GAUGE	WO-A-8184	FM-GPW-9280	0605
U	PEDAL	WO-A-8253	FM-GPW-2452-B	1204
V	SWITCH	WO-A-1333	FM-GPW-13740	0606
W	PEDAL	WO-A-405	FM-GPW-7250	0204
X	TANK	WO-A-6618	FM-GPW-9002-B	0300
Y	EXTINGUISHER	WO-A-616	FM-GPW-17100	2402
Z	STRAP	WO-A-2883	FM-GPW-1131414	1804
AA	SWITCH	WO-A-12056	See GPW-13532	0606
AB	SWITCH	WO-A-1332	FM-GPW-11649	0606
AC	SWITCH	WO-A-6149	FM-GTB-13739	0606
AD	MIRROR	WO-A-2934	FM-21CS-17682-B	2202
AE	CHOKE	WO-A-1307	FM-GPW-97303	0301B
AF	SWITCH	WO-A-6811	FM-GPW-3686-B	0604C
AG	THROTTLE	WO-A-1308	FM-GPW-9778	0303A
AH	HOLDER	WO-A-11319	FM-GPW-1153100	1820

RA PD 305102-A

GROUP 18—BODY (Cont'd)

1809A—COMPARTMENT AND PARTS (Cont'd)

Figure Number Col. 1	Official Stockage Number Col. 2	Willys Col. 3	Ford Col. 4	ITEM Col. 5	Quantity Reqd. per Unit Assy. Col. 6	12 Mos. Field Maintenance Col. 7	Major Overhaul (5th Ech) Col. 8	Total First Year Procurement Col. 9	Estimated Reqmts. per 100 Rebuilds Col. 10
		WO-A-3652	FM-GPW-1106087-A	CUP, retaining, glove compartment door lock (after 20,700 Willys trucks up to 37,914 Willys trucks)	1	—	—	—	—
		WO-53020	FM-37621-S7	SCREW, S., mach., fil-hd., No. 6 (.138)-32NC-2 x 5/16 cup to lock) (after 37,914 Willys trucks)	2				
		WO-52614	FM-34801-S7	WASHER, lock, S., No. 6 (.138)	2	—	—	—	—
		WO-A-3532	FM-GPW-1106068	CYLINDER, glove compartment lock, assembly (after Willys trucks up to 37,914 Willys trucks)	1	—	—	—	—
		WO-A-3434	FM-GPW-1106024-B	DOOR, glove compartment, w/HINGE, assembly (after Willys serial 120698, up to serial Willys 137909)	1				
		WO-A-3835	FM-GPW-1106050	DOOR, glove compartment, w/HINGE, assembly (after Willys serial 137909)	1	—	—	—	—
		WO-A-3436		HINGE, glove compartment door, assembly (after Willys serial 120698)	1	—	—	—	—
		WO-6352	FM-355836-S7	NUT, hex, S., No. 10 (.190)-24NC-2	3				
		WO-5182	FM-355158-S2	SCREW, rd-hd., S., No. 10 (.190)-24NC-2 x 3/8 (hinge to panel)	3				
		WO-52221	FM-34803-S2	WASHER, lock, S., No. 10 (.190)	3	—	—	—	—
		WO-A-3823	FM-GPW-1106084-B	LOCK, glove compartment door, assembly (after 37,914 Willys trucks)	1	—	—	—	—
		WO-A-3537	FM-356123-S7	NUT, glove compartment door lock (hex., S., pkzd., thin, 3/4-28) (after 20,700 Willys trucks up to 37,914 Willys trucks)	1				
		WO-A-3538		PLATE, tapping, glove compartment hinge (after Willys serial 120,697)	1	—	—	—	—
		WO-A-3943	FM-GPW-1102330-A	SEAL, glove compartment door (after Willys serial 156083)	1	—	—	—	—
		WO-A-3818	FM-GPW-1106064	STRIKER, glove compartment door lock (after Willys serial 137909)	1				
		WO-5182	FM-355158-S2	SCREW, hr-hd., S., No. 10 (.190)-24NC-2 x 3/8	1				
		WO-52220		WASHER, S. (plain) No. 10 (.190)	1	—	—	—	—
		WO-52221	FM-34803-S2	WASHER, lock, S., No. 10 (.190)	1	—	—	—	—

1811—WINDSHIELD ASSEMBLY

Figure Number Col. 1	Official Stockage Number Col. 2	Willys Col. 3	Ford Col. 4	ITEM Col. 5	Quantity Reqd. per Unit Assy. Col. 6	12 Mos. Field Maintenance Col. 7	Major Overhaul (5th Ech) Col. 8	Total First Year Procurement Col. 9	Estimated Reqmts. per 100 Rebuilds Col. 10
18-1		WO-A-2235	FM-GP-1103302	ARM, adjusting, windshield	2	—	—	—	—
		WO-A-2485	FM-24454-S2	BOLT, windshield clamp bracket (hex.-hd., S., 1/4-20NC-2 x 1 1/16) (AN X-1754)	4				
		WO-A-2798	FM-GPW-1103050	BRACKET, windshield mounting clamp catch	2	—	—	—	—
		WO-52963	FM-36046-S2	SCREW, S., fl-hd., 1/4-20NC-2 x 3/4	4				
		WO-A-2213	FM-GPW-1103162	BRACKET, windshield pivot, assembly	2	—	—	—	—
		WO-51485	FM-355449-S2	BOLT, hex-hd., S., 5/16-18NC-2 x 5/8 (to cowl)	4				
		WO-6167	FM-33797-S2	NUT, hex., S., 5/16-18NC-2	4	—	—	—	—
		WO-51840	FM-34806-S2	WASHER, lock, S., 1/4 in.	4	—	—	—	—
		WO-A-2232	FM-GPW-1103304	BRACKET, windshield swing arm, w/STUD (5/16-18NC-2)	2	—	—	—	—

GROUP 18—BODY (Cont'd)

1811—WINDSHIELD ASSEMBLY (Cont'd)

Figure Number	Official Stockage Number	Part Number Willys	Part Number Ford	ITEM	Quantity Reqd. per Unit Assy.	12 Mos. Field Maintenance	Major Overhaul (5th Ech)	Total First Year Procurement	Estimated Reqmts. per 100 Rebuilds
	Col. 1 / Col. 2	Col. 3	Col. 4	Col. 5	Col. 6	Col. 7	Col. 8	Col. 9	Col. 10
		WO-A-4300	FM-GP-1103310	BRACKET, windshield adjusting arm, w/BUSHING, assembly	2	—	—	—	—
		WO-A-2989		CATCH, hold down, windshield, assembly (before Willys serial 103545)......(Willys only)	2	—	—	—	—
		WO-A-3197	FM-GPW-1103027	CATCH, hold down, windshield, assembly (after Willys serial 103545)	2	% 6	8	—	—
		WO-51514	FM-20324-S2	BOLT, hex-hd., S., ¼-20NC-2 x ⅝ (catch to hood top panel)	4	—	—	—	—
		WO-5790	FM-33795-S2	NUT, hex., S., ¼-20NC-2	4	—	—	—	—
		WO-5121	FM-34705-S2	WASHER, S. (plain) ¼ in.	4	—	—	—	—
		WO-52031	FM-34805-S2	WASHER, lock, S., ¼ in.	4	—	—	—	—
		WO-A-2791	FM-GPW-1103488	CATCH, windshield clamp (to instrument board)	2	—	—	—	—
		WO-A-2501	FM-GPW-1151289	CHAIN, windshield thumb screw	2	—	—	—	—
18-1		WO-A-2227	FM-GP-1103482-A	CLAMP, windshield, assembly	2	—	—	—	—
		WO-A-4120	FM-352692-S15	FASTENER, windshield curtain (top to windshield) (CNC-519)	10	—	—	—	—
		WO-A-4303	FM-GP-1103350	HANDLE, windshield frame	1	—	—	—	—
		WO-A-2238	FM-B-45482-C	KNOB, windshield adjusting arm (⅝-18NC-2)	2	—	—	—	—
		WO-A-2234	FM-GP-1103334	LOOP, hold down, windshield	2	—	—	—	—
		WO-A-2481	FM-34176-S2	NUT, windshield frame screw (hex., br., No. 10 (.190)-32NF-2 std. acorn) (AN X-1751)	3	—	—	—	—
		WO-A-2301		PANEL, windshield, outer (before Willys serial 103544)	1	—	—	—	—
		WO-A-3190		PANEL, windshield, outer (after Willys serial 103544)	1	—	—	—	—
		WO-A-2939	FM-GPW-1103446-A	PLATE, tapping, windshield clamp catch	2	—	—	—	—
		WO-A-3203	FM-GPW-1103028	RETAINER, windshield hold down catch (BA B-194) (after Willys serial 103545)	2	—	—	—	—
		WO-A-2386	FM-GPW-16892	RETAINER, windshield hold down catch (before Willys serial 103545)	2	—	—	—	—
		WO-51514	FM-20308-S7-8	BOLT, hex-hd., S., ¼-20NC-2 x ⅝ (retainer to hood)	4	—	—	—	—
		WO-5790	FM-33795-S	NUT, hex, S., ¼-20NC-2	4	—	—	—	—
		WO-5121	FM-34706-S2	WASHER, S., plain ¼ in.	4	—	—	—	—
		WO-52031	FM-34805-S2	WASHER, lock, S., ¼ in.	4	—	—	—	—
		WO-A-3189	FM-GPW-1151297	RING, windshield thumb screw chain	4	—	—	—	—
		WO-A-2239	FM-81W-811189	SCREW, pivot, windshield adjusting arm (rd-hd., S., ¼-28NF-2 x .47)(AN X-1747)	2	% 6	6	—	—
		WO-A-2487	FM-33249-S2	SCREW, windshield bracket (ov. ctsk-hd., S., No. 10 (.190) x ½, type "Z") (AN X-1755)	4	—	—	25	—
		WO-A-2480	FM-38259-S2	SCREW, windshield frame (special, rd-hd., S., No. 10 (.190)-32NF-2 x ⅞) (AN X-1750)	4	% 6	6	50	—
		WO-A-2482	FM-38192-S2	SCREW, windshield frame, corner (rd-hd., I., No. 8 (.164)-32NC x ⁷⁄₁₆)	4	—	—	—	—

Ord. No.	Mfr. Part No.	Description					
WO-A-2214	FM-355449-S2	SCREW, windshield pivot, thumb (Mail I., 3/8-24)	2	% 10	10	20	—
WO-A-2483	FM-99W-8103311	SPACER, windshield adjusting arm (S., 21/32 O.D. x 29/64 I.D.) (AN S-384)	2	% 6	6	25	—
WO-A-2490	FM-355286-S	WASHER, windshield adjusting arm (.688 O.D. x .349 I.D. x .067 thk.) (AN X-1637)	6	% 6	6	30	—
WO-A-4260	FM-356361-S	WASHER, spring, windshield adjusting arm (S., .56 O.D. x .34 I.D.) (AN S-1044)	4	% 10	10	40	—
WO-A-2220	FM-356373-S2	WASHER, windshield pivot spring (S., 1 in. O.D. x 21/32 I.D. x .015 thk., 3/32 crown)	4	% 14	14	25	—
WO-A-2486	FM-34805-S2	WASHER, lock, windshield clamp bracket screw (S., 1/4, .475 O.D. x .269 I.D. x .063 thk.) (AN X-1700)	4	—	—	—	—
WO-52960	FM-34803-S7	WASHER, lock, S., pkzd, No. 10 (.190)	3	—	—	—	—
WO-52961	FM-34802-S	WASHER, lock, S., pkzd, No. 8 (.164)	4	—	—	—	—
WO-A-2796		WINDSHIELD, assembly (before Willys serial 103545)	1	—	—	—	—
		(Includes:					
		ARMS					
		BRACKETS					
		CLAMPS					
		GLASS					
		ATTACHING PARTS)					
WO-A-3210	FM-GPW-1103010	WINDSHIELD assembly (after Willys serial 103545)	1	—	—	—	—
		(Includes:					
		ARMS					
		BRACKETS					
		CLAMPS					
		GLASS					
		ATTACHING PARTS)					

1811A—WINDSHIELD GLASS

Ord. No.	Mfr. Part No.	Description					
WO-A-2226	FM-GP-1103014	FRAME, windshield, w/GLASS, assembly	1	2	2	7	—
WO-A-2478	FM-GP-1103100	GLASS, windshield	2	—	—	—	—

1811B—WINDSHIELD FRAME MOULDING AND SEALS

Ord. No.	Mfr. Part No.	Description					
WO-A-2246	FM-GPW-1103030	FRAME, windshield, lower, assembly	1	—	—	—	—
WO-A-2255	FM-GP-1103020	FRAME, windshield, upper, assembly	1	—	—	—	—
WO-A-2776		FRAME, windshield, tubular (before Willys serial 103545) (Willys only)	1	—	—	—	—
WO-A-3204	FM-GPW-1103268	FRAME, windshield, tubular (after Willys serial 103545)	1	—	—	—	—
WO-A-2479	FM-GP-1103106	PADDING, windshield glass (Everseal, 1 1/8 in. x 73 in.)	2	10	10	24	—
WO-A-2250	FM-GP-1103110	WEATHERSTRIP, windshield	1	—	—	—	—
WO-A-2476	FM-GPW-1103080-A	WEATHERSTRIP, windshield frame to cowl	1	—	—	—	—
WO-A-2489	FM-33184-S2	SCREW, windshield to cowl weatherstrip (S., binding hd., sheet metal type "Z" No. 10 (.190) x 1/2)	15	—	—	—	—

1811D—WINDSHIELD WIPER ARMS AND BLADES

Ord. No.	Mfr. Part No.	Description					
WO-A-2512	FM-GP-17535	ARM, windshield wiper, w/BLADE assembly (includes RIVET)	2	—	—	—	—
WO-A-11519		ARM, windshield wiper, w/CLIP, assembly	2	—	—	—	—
WO-A-4687		BLADE, windshield wiper	2	—	—	—	—

GROUP 18—BODY (Cont'd)

Col. 1 Figure Number	Col. 2 Official Stockage Number	Col. 3 Willys	Col. 4 Ford	Col. 5 ITEM	Col. 6 Quantity Reqd. per Unit Assy.	Col. 7 12 Mos. Field Maintenance	Col. 8 Major Overhaul (5th Ech)	Col. 9 Total First Year Procurement	Col. 10 Estimated Reqmts. per 100 Rebuilds
				1811D—WINDSHIELD WIPER ARMS AND BLADES (Cont'd)					
		WO-A-2513	FM-GP-17531	HANDLE, wiper, inner w/KNOB, assembly	2	—	—	—	—
		WO-A-11432	FM-GPW-17534	HANDLE, windshield wiper, w/tie ROD, assembly (TRI 80540)	1	%10	10	50	—
		WO-A-2588	FM-34054-S2	NUT, hex, S., No. 12 (.216)-24NF-2	2	%2	2	8	—
		WO-53243	FM-63848-S	RIVET, S., tubular, ov-hd., type "A" (blade to arm)	2	—	—	—	—
		WO-A-2584	FM-351303-S2	WASHER, windshield wiper handle to frame (S., 1½ O.D. x $\frac{9}{32}$ I.D. x .032 thk.)	2	—	—	—	—
		WO-A-2587	FM-34824-S2	WASHER, lock, S., No. 12 (.216) (.266 I.D. x $\frac{1}{32}$ thk.) (TRI 2328)	2	%10	10	100	—
18-1		WO-A-11433	FM-GPW-17500	WIPER SET, tandem, windshield	1	—	—	—	—
				(Includes:					
				ARM w/BLADE, assembly					
				HANDLE w/ROD, assembly					
				NUT					
				WASHERS)					
				1817—TOOL BOX					
		WO-A-3198	FM-GPW-1146152	BRACKET, tool compartment lock, assembly	1	—	—	—	—
		WO-A-2774		FRAME, tool compartment, left	1	—	—	—	—
		WO-A-2775		FRAME, tool compartment, right	1	—	—	—	—
		WO-A-2811		HINGE, tool compartment lid, assembly (before Willys serial 118599)	4	—	—	—	—
		WO-A-3226	FM-GPW-1146132	HINGE, tool compartment lid, assembly (after Willys serial 118599)	4	—	—	—	—
		WO-A-2810		LID, tool compartment, w/HINGE, assembly (before Willys serial 118599)	2	—	—	—	—
		WO-A-3227	FM-GPW-1146100	LID, tool compartment, w/HINGE, assembly (after Willys serial 118599)	2	—	—	—	—
		WO-6352	FM-355836-S7-8	NUT, hex, S., No. 10 (.190)-24NC-2	6	—	—	—	—
		WO-5182		SCREW, rd-hd., S., No. 10 (.190)-24NC-2 x $\frac{3}{8}$	6	—	—	—	—
		WO-52221	FM-34803-S2	WASHER, lock, S., No. 10 (.190)	6	—	—	—	—
		WO-A-2895	FM-GPW-1143501	LOCK, tool compartment lid, assembly (TA OP-528)	2	—	—	—	—
		WO-6352	FM-355836-S7-8	NUT, hex, S., No. 10 (.190)-24NC-2	4	—	—	—	—
		WO-52236		SCREW, rd-hd., S., No. 10 (.190)-24NC-2 x $\frac{3}{4}$	4	—	—	—	—
		WO-52221	FM-34803-S2	WASHER, lock, S., No. 10 (.190)	4	—	—	—	—
		WO-A-3933		SEAL, tool compartment lid (short)	2	—	—	—	—
		WO-A-3934		SEAL, tool compartment lid (long)	2	—	—	—	—
		WO-A-2832	FM-GPW-1146126	STRIKER, tool compartment lid lock	2	—	—	—	—
				1817A—SPARE GASOLINE CARRIER					
		WO-A-4123	FM-GPW-1140330	BRACKET, spare gas line can assembly	1	—	—	—	—
		WO-53031	FM-20348-S2	BOLT, hex-hd., S., $\frac{3}{8}$-16NC-2 x $\frac{3}{4}$ (brkt. to rear panel) -GM-122122)	4	—	—	—	—

WO No.	Ford No.	Description	Qty.
WO-52046	FM-34807-S2	WASHER, lock, S., 3/8 in.	4
WO-A-4127		STRAP, spare gasoline can, w/BUCKLE, assembly	1
WO-53056		RIVET, tubular 9/64 x 11/32	8
WO-53057		WASHER, S., 7/16 O.D. x 5/32 I.D.	8
WO-A-4128		STRAP, spare gasoline can, w/TIP, assembly	1

1820—GUN HOLDERS

WO No.	Ford No.	Description	Qty.
WO-A-11581		BRACKET, support, rifle holder to windshield, left, assembly	1
WO-A-11584		BRACKET, support, rifle holder to windshield, right, assembly	1
WO-A-11945	FM-GPW-1153124	BUMPER, rifle holder retainer clamp, assembly (BUE-505285)	1
	FM-GPW-1153106	CAM, rifle holder retainer handle (Ford only)	1
	FM-62323-S2	RIVET, ov-hd., tubular, 7/64 x 13/32 (Ford only)	1
WO-A-11942	FM-GPW-1153104	HANDLE, rifle holder retainer, w/CAM, assembly (BUE-505284)	1
WO-A-11319	FM-GPW-1153100	HOLDER, rifle, assembly (BUE-505275)	1
WO-53266	FM-23908-S2	BOLT, step, S., 5/16-18NC-2 x 3/4 (holder to windshield)	2
WO-53037	FM-33797-S2	NUT, reg., hex., S., 5/16-18NC-2 (holder to windshield bolt)	2
WO-53032		WASHER, flat, S., 3/8 in. (holder to windshield bolt) (Willys only)	2
WO-53047	FM-34707-S2	WASHER, flat, S., 5/16 in. (holder to windshield bolt) (Ford only)	2
WO-A-11944	FM-34806-S2	WASHER, lock, S., 5/16 in. (holder to windshield bolt)	2
	FM-GPW-1153110	PLATE, rifle holder retainer clamp handle (BUE-505288)	1
	FM-36801-S2	SCREW, rd-hd., No. 12 (.216)-24NC-2 x 3/8 (Ford only)	4
	FM-34804-S2	WASHER, lock, S., No. 12 (.216)	4
WO-A-11946	FM-GPW-1153123	SPRING, retaining, rifle holder, assembly (BUE-505282)	1
WO-A-11940	FM-GPW-1153107	SPRING, rifle holder retainer handle cam (BUE-505287)	1

GROUP 21—BUMPERS AND GUARDS

2101—BUMPERS

WO No.	Ford No.	Description	Qty.
WO-A-1117	FM-GPW-17750	BAR, bumper, front	1
WO-51823	FM-355456-S	BOLT, hex-hd., S., 5/16-24NF-2 x 4 1/2 (bar to frame)	4
WO-5910	FM-33798-S2	NUT, hex., S., 5/16-24NF-2	4
WO-51833	FM-34806-S2	WASHER, lock, S., 5/16 in.	4
WO-A-1157	FM-GPW-17775	BUMPERETTE, rear	2
WO-50163	FM-24349-S2	BOLT, hex-hd., S., 3/8-24NF-2 x 3/4 (bumperette to frame)	8
WO-5901	FM-33800-S	NUT, hex., S., 3/8-24NF-2	8
WO-5010	FM-34807-S	WASHER, lock, S., 3/8 in.	8
WO-A-1147	FM-GPW-17751	FILLER, front bumper bar, wood	1

2103—RADIATOR GUARDS

WO No.	Ford No.	Description	Qty.
WO-A-2858		FRAME, radiator guard, assembly (before Willys serial 125809 only)	1
WO-6299		BOLT, hex-hd., S., 3/8-16NC-2 x 7/8 (frame to chassis) (Willys only)	3
WO-5544		NUT, hex., S., 3/8-16NC-2 (Willys only)	3
WO-5455		WASHER, plain S., 3/8 in. (Willys only)	3
WO-51304		WASHER, lock, S., 3/8 in. (Willys only)	3
WO-A-3615	FM-GPW-8307	GUARD, radiator, assembly (after Willys serial 125809)	1
WO-51485	FM-20326-S	BOLT, hex-hd., S., 5/16-18NC-2 x 5/8 (guard to fender)	6

21-1

FIGURE 21-1—RADIATOR GUARD

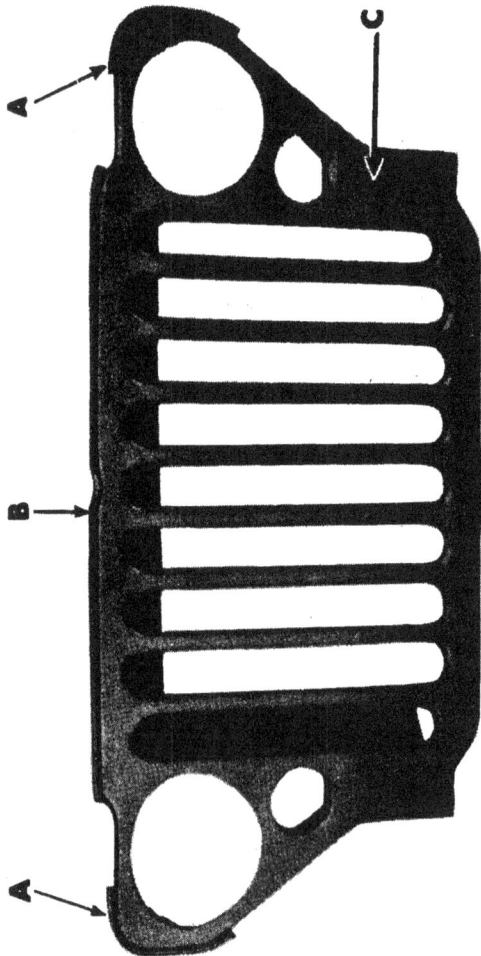

Key	Item	Willys Part No.	Ford Part No.	Gov't Group No.
A	LINER	WO-A-3095	FM-GPW-8384	2103
B	LINER	WO-A-3094	FM-GPW-8349	2103
C	GUARD	WO-A-3615	FM-GPW-8307	213

RA PD 305158-A

RA PD 305158

GROUP 18—BODY (Cont'd)

Figure Number	Official Stockage Number	Part Number Willys	Part Number Ford	ITEM	Quantity Reqd. per Unit Assy.	12 Mos. Field Maintenance	Major Overhaul (5th Ech)	Total First Year Procurement	Estimated Reqmts. per 100 Rebuilds
Col. 1	Col. 2	Col. 3	Col. 4	Col. 5	Col. 6	Col. 7	Col. 8	Col. 9	Col. 10
				2103—RADIATOR GUARDS (Cont'd)					
		WO-6167	FM-33797-S	NUT, hex, S., $5/16$-18NC-2 (before Willys serial 125809)	6	—	—	—	—
		WO-A-3550	FM-34130-S2	NUT, sp, S., $5/16$-18NC-2 (after Willys serial 125809)	6	—	—	—	—
		WO-5437	FM-34706-S7	WASHER, S., (plain) $5/16$ in.	6	—	—	—	—
		WO-51840	FM-34806-S8	WASHER, lock, S., $5/16$ in.	6	—	—	—	—
21-1		WO-A-3094	FM-GPW-8348	LINER, radiator guard hood, center	1	—	—	—	—
21-1		WO-A-3095	FM-GPW-8349	LINER, radiator guard hood, side	2	—	—	—	—
		WO-51182	FM-62627-S3	RIVET, split, $3/64$ x $3/16$ (5 used center, 6 used sides—Willys) (4 used center, 7 used sides—Ford) (Willys)	11	—	—	—	—
		WO-A-3549		RETAINER, radiator guard to fender nut (after Willys serial 125809) (Willys only)	6	—	—	—	—

2103A—LAMP GUARDS

WO No.	FM No.	Description	Qty
WO-A-4118	FM-GPW-13176	GUARD, blackout driving lamp, w/SUPPORT, assembly (after Willys serial 163750)	1
WO-53050	FM-20387-S7	BOLT, hex-hd., S., 5/16-24NF-2 x 1 (GP-112657) (after Willys serial 163750)	1
WO-53048	FM-24347-S7	BOLT, hex-hd., S., 5/16-24NF-2 x 3/4 (GM-118772)	2
WO-50802	FM-33798-S7	NUT, hex., S., 5/16-24NF-2 (GM-115729)	3
WO-A-4116		WASHER, S., cd-pltd., 13/16 O.D. x 11/32 I.D. (Willys only)	1
WO-53047	FM-34806-S2	WASHER, lock, S., 5/16 in.	1
WO-53025	FM-356309-S7	WASHER, lock, S., shakeproof, 5/16 in.	4

GROUP 22—MISCELLANEOUS BODY, CHASSIS AND ACCESSORY

2201—TARPAULINS

WO No.	FM No.	Description	Qty
WO-A-3070	FM-GPW-1102980	COVER, headlight	2
WO-A-3073		COVER, windshield, assembly (before Willys serial 103545)	1
WO-A-3211	FM-GPW-1103214	COVER, windshield, assembly (after Willys serial 103545)	1
WO-A-2909		PAULIN, deck top, assembly (before Willys serial 103545) (Willys only)	1
WO-A-3216	FM-GPW-1152700	PAULIN, deck top, assembly (after Willys serial 103545)	1

2201A—BOWS

WO No.	FM No.	Description	Qty
WO-A-2898	FM-GPW-1151272	BOW, top front, assembly	1
WO-A-2897	FM-GPW-1151266	BOW, top, assembly (includes bow, top front, assembly)	1
WO-5216	FM-60470-S	RIVET, rd-hd., 5/16 x 1 (bow to pivot)	2
WO-A-2744	FM-GPW-1129069	BRACKET, left hand rail, front, assembly	1
WO-A-2745	FM-GPW-1129068	BRACKET, right hand rail, front, assembly	1
WO-51391	FM-24310-S	BOLT, hex-hd., S., 5/16-18NC-2 x 1/2 (bracket to body)	2
WO-52350	FM-33797-S2	NUT, hex., S., 5/16-18NC-2	2
WO-51840	FM-34806-S2	WASHER, lock, S., 5/16 in.	2
WO-A-2754	FM-GPW-1153030	BRACKET, top bow pivot	2
WO-51495	FM-20326-S2	BOLT, hex-hd., S., 5/16-18NC-2 x 5/8 (bracket to body)	2
WO-52350	FM-33797-S2	NUT, hex., S., 5/16-18NC-2	6
WO-52987	FM-24650-S2	SCREW, fl-hd., S., 5/16-18NC-2 x 5/8	4
WO-51840	FM-34806-S2	WASHER, lock, S., 5/16 in.	6
WO-A-2501	FM-GPW-1151289	CHAIN, thumb screw	2
WO-A-3113	FM-348148-S	EYELET, curtain fastener (CND-0)	16
WO-A-3112	FM-95633-S	FASTENER, single post type, curtain (rivet, tubular) (CNC-12)	11
WO-A-2168	FM-GPW-1154810	LOOP, footman	10
WO-6352	FM-355836-S7	NUT, hex., S., No. 10 (.190)-24NC-2	20
WO-6290	FM-355162-S2	SCREW, fl-hd., S., No. 10 (.190)-24NC-2 x 1/2 (loop to body)	20
WO-52221	FM-34803-S7	WASHER, lock, S., No. 10 (.190)	20
WO-A-2901	FM-GPW-1151270	PIVOT, top bow, assembly	2
WO-52189	FM-20328-S2	BOLT, hex-hd., S., 3/8-16NC-2 x 5/8 (pivot to bracket)	2
WO-5455	FM-34747-S	WASHER, S. (plain) 3/8 in.	2
WO-51304	FM-34807-S2	WASHER, lock, S., 3/8 in.	2

GROUP 22—MISCELLANEOUS BODY, CHASSIS AND ACCESSORY (Cont'd)

Col. 1 Figure Number	Col. 2 Official Stockage Number	Col. 3 Willys	Col. 4 Ford	Col. 5 ITEM	Col. 6 Quantity Reqd. per Unit Assy.	Col. 7 12 Mos. Field Maintenance	Col. 8 Major Overhaul (5th Ech)	Col. 9 Total First Year Procurement	Col. 10 Estimated Reqmts. per 100 Rebuilds
				2201A—BOWS (Cont'd)					
		WO-A-2900	FM-GPW-1153142	PIVOT, top bow front	2	—	—	—	—
		WO-A-3111	FM-353104-S	PLATE, back, clincher (CNC-58)	11	—	—	—	—
		WO-A-3189	FM-GPW-1151297	RING, thumb screw chain	4	—	—	—	—
		WO-A-2214	FM-GPW-1150498	SCREW, thumb, top bow, assembly ($\frac{3}{8}$-24NF)	2	—	—	—	—
		WO-A-3051	FM-95705-S	SOCKET, curtain fastener (CNC-515)	14	—	—	—	—
		WO-A-3110	FM-GPW-1152720	STRAP, top lock, assembly	2	—	—	—	—
		WO-A-3141	FM-GPW-1152950	STRAP, top roll, assembly	2	—	—	—	—
		WO-A-3109	FM-GPW-1152730	STRAP, hold down, top, rear, assembly	6	—	—	—	—
		WO-A-2639	FM-356522-S2	WASHER, anti-rattle, top bow pivot	2	—	—	—	—
		WO-A-3054	FM-95627-S	WASHER, curtain eyelet fastener (CNC-98)	32	—	—	—	—
		WO-A-2902	FM-351370-S2	WASHER, top bow, front pivot (S, 1 in. O.D. x $\frac{21}{64}$ I.D.)	2	—	—	—	—
				2201F—SIDE CURTAINS					
		WO-A-2998	FM-GPW-1120041	CURTAIN, left, side, assembly	1	—	—	—	—
		WO-A-2999	FM-GPW-1120040	CURTAIN, right, side, assembly	1	—	—	—	—
		WO-A-3052	FM-95626-S	FASTENER, side curtain (male, flush type) (CNC-50)	14	—	—	—	—
		WO-A-3053	FM-95628-S	WASHER, side curtain fastener (CNC 60)	14	—	—	—	—
				2202—REAR VIEW MIRRORS					
18-1		WO-A-2934	FM-21CS-17682-B	MIRROR, rear view, assembly	1	% 7	2	15	—
		WO-6989	FM-34054-S2	NUT, hex, S., No. 12 (.216)-24NC-2	6	—	—	—	—
		WO-6988	FM-36845-S2	SCREW, rd-hd, S., No. 12 (.216)-24NC-2 x $\frac{5}{8}$	6	—	—	—	—
		WO-51969	FM-34804-S2	WASHER, lock, S., No. 12 (.216)	6	—	—	—	—
		WO-A-11850	FM-34804-S2	MIRROR, rear view with washer and nut, assembly	1	% 56	6	73	—
		WO-A-11862		NUT, mirror to tube, (S., hex., $\frac{1}{4}$-28NF)	1	—	—	—	—
		WO-A-11861		WASHER, lock, S., $\frac{1}{4}$ in.	1	—	—	—	—
				2203—IDENTIFICATION AND CAUTION PLATES					
18-1		WO-A-1330	FM-GPW-1101621-A	PLATE, caution	1	—	—	—	—
18-1		WO-A-11757	FM-GPW-1101629-A	PLATE, name	1	—	—	—	—
18-1		WO-A-1331	FM-GPW-1101627-A	PLATE, shifting	1	—	—	—	—
		WO-53013		NUT, hex, S., stamped, No. 6 (.138)-32NC-2 (Willys only)	12	—	—	—	—

				Qty.		Ford & Willys Part No.	Ordnance Part No.	Group
—	—	—	—	12	RIVET, ov-hd., fr., 5/64 x 3/16 (Ford only)		WO-53014	
—	—	—	—	12	SCREW, rd-hd., S., No. 6 (.138)-32NC x 5/16 in. (Willys only)	FM-357016-S19		

2203A—REFLECTORS

				Qty.		Ford & Willys Part No.	Ordnance Part No.	Group
—	32	10	% 19	4	REFLECTOR, reflex, assembly (CB-3121441)	FM-GPW-13380-A	WO-A-1306	
—	—	—	—	8	NUT, hex, S., reg., 1/4-28NF-2	FM-33796-S	WO-5914	
—	—	—	—	8	SCREW, rd-hd., S., 1/4-28NF-2 x 1/2	FM-36825-S	WO-52834	
—	—	—	—	8	WASHER, lock, light, S., 1/4 in.	FM-34805-S	WO-52706	

2204—SPEEDOMETER AND PARTS

				Qty.		Ford & Willys Part No.	Ordnance Part No.	Group
—	—	—	—	1	BRACKET, mounting, speedometer (AL-SPK-66) (AL speedometer only) (Willys only)		WO-A-8242	
—	—	—	—	1	BRACKET, mounting, speedometer (KS 40333) (KS speedometer only) (Willys only)		WO-A-8126	
—	—	—	—	1	BUSHING, rubber	FM-GPW-13434-A2	WO-345961	
—	—	—	—	2	CLIP, mounting, closed type, S., 5/16 in. (1 used speedometer cable to dash with starter to starter switch cable, 1 used speedometer cable to frame)	FM-GPW-14561	WO-A-5449	
—	—	—	—	2	NUT, hex, S., No. 10 (.190)-24NC-2	FM-355836-S7-8	WO-6352	
—	—	—	—	2	SCREW, rd-hd., S., No. 10 (.190)-24NC-2 x 5/8 in.	FM-355130-S7	WO-5064	
—	—	—	—	2	WASHER, lock, S., 1/4 in.	FM-34803-S2	WO-52221	
50	10	29	% 2	1	SHAFT, flexible, speedometer drive, assembly (53 5/32 in. long) (AL SPS-1105-X)	FM-GPW-17262	WO-A-1344	18-1
—	12	3	% 6	1	SPEEDOMETER, luminous dial, assembly (AL SPK 4003) (KS 40355)	FM-GPW-17255-A	WO-A-8180	
—	—	—	—	2	NUT, wing, S., No. 10 (.190)-32NF-2 (AL 8X-72B) (Auto-Lite speedometer to brkt. only) (Willys only)		WO-A-8127	
—	—	—	—	2	NUT, wing, S., No. 10 (.190)-32NF (KS 40350) (King Seeley Speedometer to brkt. only) (Willys only)		WO-A-8125	
—	—	—	—	2	WASHER, lock, S. (special) No. 10 (.190) (AL 12X-196) (Willys only)		WO-A-8129	
—	—	—	—	2	WASHER, lock, S., No. 10 (.190) (KS-5131) (Willys only)		WO-53194	
—	24	6	% 13	1	TUBE, speedometer drive, assembly (52 in. long) (AL SPS-L100R)	FM-GPW-17261	WO-A-1343	
—	—	—	—	1	TUBE, speedometer drive, w/SHAFT, assembly (AL SPS-2099M)	FM-GPW-17260	WO-A-1267	

GROUP 23—GENERAL USE STANDARDIZED PARTS

2301—TOOLS

				Qty.		Ford & Willys Part No.	Ordnance Part No.	Group
—	8	3	3	1	ADAPTER, grease gun, push type (req'd for universal joints) (SW-6334)	FM-GPW-17126	WO-A-11765	23-1
—	—	—	—	1	BAG, tool	FM-GPW-17005	WO-A-372	41-B-15	23-1
—	—	—	—	1	BRACKET, oil can (BA-SK-1670)	FM-GPW-17037	WO-A-313	
—	—	—	—	2	NUT, hex, S., 1/4-20NC-2	FM-33795-S2	WO-5790	
—	—	—	—	2	SCREW, rd-hd., S., 1/4-20NC x 1/2 (brkt. to body)	FM-26483-S7	WO-51492	
—	—	—	—	2	WASHER, lock, S., 1/4 in.	FM-34805-S2	WO-52706	
—	—	—	—	1	CRANK, starting	FM-GP-17036	WO-A-239	8-G-615	23-1
—	—	—	—	1	GAUGE, tire pressure		WO-A-6855	23-1
—	—	—	—	1	GUN, lubricating, hand type, 16 oz.	FM-GPW-18325	WO-A-12058	41-G-1344-40

GROUP 23—GENERAL USE STANDARDIZED PARTS (Cont'd)

2301—TOOLS (Cont'd)

Figure Number (Col. 1)	Official Stockage Number (Col. 2)	Willys (Col. 3)	Ford (Col. 4)	ITEM (Col. 5)	Quantity Reqd. per Unit Assy. (Col. 6)	12 Mos. Field Maintenance (Col. 7)	Major Overhaul (5th Ech) (Col. 8)	Total First Year Procurement (Col. 9)	Estimated Reqmts. per 100 Rebuilds (Col. 10)
23-1		WO-A-213	FM-GP-17125	GUN, lubricating, hand type, 9 oz. capacity (issue until stock is exhausted, then issue 41-G-1344-40)	—				—
23-1	41-H-523	WO-A-373	FM-GP-17042	HAMMER, machinist, ball peen, 16 oz	1	3	3	8	—
23-1		WO-306715	FM-GPW-17011-A	HANDLE, spark plug wrench	1	3	3	8	—
23-1	41-J-66	WO-A-1240	FM-GPW-17080	JACK, screw, w/handle (1½ ton)	1	—	—	—	—
23-1	13-O-1530	WO-A-379	FM-GP-17038	OILER, straight spout, spring bottom (capacity ½ pt.)	1	3	3	8	—
23-1	41-P-1650	WO-A-374	FM-GP-17028	PLIERS, combination, slip joint, 6 in.	1	3	3	8	—
23-1		WO-A-1339	FM-GPW-17090	PULLER, wheel hub.	1	—	—	—	—
23-1	8-P-5000	WO-A-7511	FM-GPW-17052	PUMP, tire, w/air CHUCK	1	—	—	—	—
23-1	41-S-1076	WO-A-375	FM-GP-17026	SCREW DRIVER, common, heavy duty, integral handle, 6 in. (TGBX1A)	1	3	3	9	—
23-1		WO-A-378		WRENCH SET, engineers	1	3	3		—
				(Consists of:					
				1 41-W-991 WRENCH, engineers, $\frac{3}{8}$ and $\frac{7}{16}$					
				1 41-W-1003 WRENCH, engineers, $\frac{1}{2}$ and $\frac{19}{32}$					
				1 41-W-1005-5 WRENCH, engineers, $\frac{9}{16}$ and $\frac{11}{16}$					
				1 41-W-1008-10 WRENCH, engineers, $\frac{5}{8}$ and $\frac{25}{32}$					
				1 41-W-1012-5 WRENCH, engineers, $\frac{3}{4}$ and $\frac{7}{8}$ set)					
		WO-A-371		TOOL, SET	1	—	—	—	—
				(Consists of:					
				BAG					
				HAMMER					
				PLIERS					
				SCREW DRIVER					
				WRENCH, adjustable					
				WRENCH, bleeder screw					
				WRENCH, crescent					
				WRENCH, set)					
		WO-A-1162	FM-GPW-17003	TOOL, SET	1	—	—	—	—
				(Consists of:					
				ADAPTER, grease gun					
				GAUGE, tire pressure					
				GUN, grease					
				HAMMER					
				HANDLE, spark plug wrench					
				JACK					
				KIT, tool					
				OILER (oil can)					
				PULLER, wheel hub					

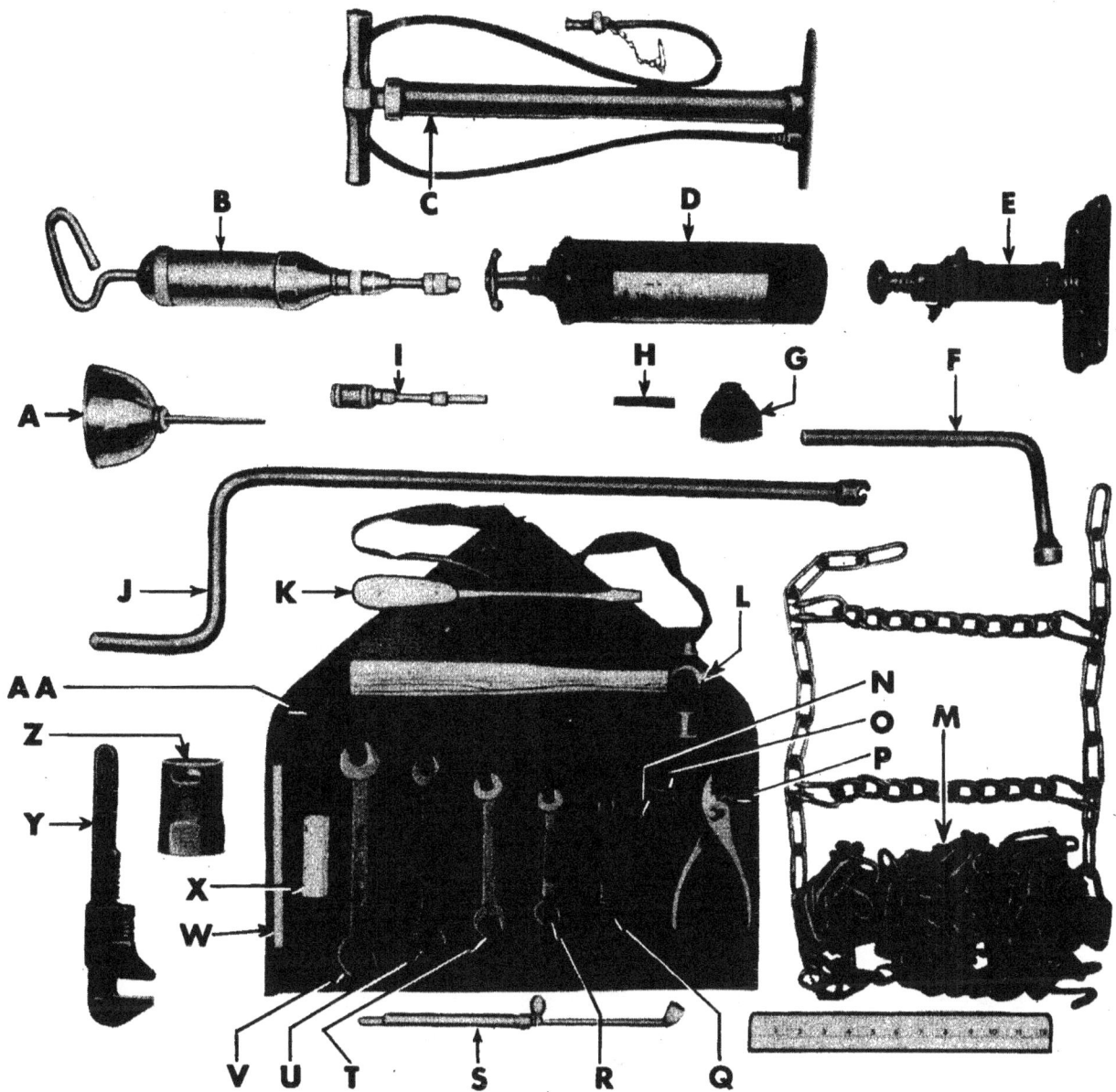

RA PD 305177

FIGURE 23-1—TOOLS

Key	Item	Willys Part No.	Ford Part No.	Gov't Group No.
A	OILER	WO-A-379	FM-GP-17038	2301
B	GUN, grease	WO-A-213	FM-GP-17125	2301
C	PUMP, tire	WO-A-7511	FM-GPW-17025	2301
D	EXTINGUISHER, fire	WO-A-616	FM-GPW-17100	2402
E	JACK	WO-A-1240	FM-GPW-17080	2301
F	WRENCH, wheel	WO-A-348	FM-GPW-17035	2301
G	PULLER, wheel	WO-A-1339	FM-GPW-17090	2301
H	WRENCH		FM-GP-17062	2301
I	ADAPTER	WO-A-11765	FM-GPW-17126	2301
J	CRANK	WO-A-239	FM-GPW-17036	2301
K	SCREWDRIVER	WO-A-375	FM-GPW-17020	2301
L	HAMMER	WO-A-373	FM-GP-17042	2301
M	CHAINS, tire	WO-A-7687	FM-GPW-18136-B	2301B
N	WRENCH	WO-A-5130	FM-GPW-17030	2301
O	WRENCH	WO-A-1492	FM-GPW-17091	2301
P	PLIERS	WO-A-734	FM-GP-17028	2301
Q	WRENCH	WO-A-596	FM-GP-17043	2301
R	WRENCH	WO-A-597	FM-GP-17044	2301
S	GAUGE, tire	WO-A-6855	FM-GPW-18325	2301
T	WRENCH	WO-A-598	FM-GP-17045	2301
U	WRENCH	WO-A-599	FM-GP-17046	2301
V	WRENCH	WO-A-600	FM-GP-17047	2301
W	HANDLE	WO-306715	FM-GPW-17011-A	2301
X	WRENCH, spark plug	WO-637635	FM-GPW-17017-A	2301
Y	WRENCH	WO-A-377	FM-GP-17021	2301
Z	WRENCH	WO-A-692	FM-17033	2301
AA	BAG, tool	WO-A-372	FM-GPW-17005	2301

RA PD 305177-A

SNL G-503

GROUP 23—GENERAL USE STANDARDIZED PARTS (Cont'd)

		Part Number				Per 100 Major Items			
Figure Number	Official Stockage Number	Willys	Ford	ITEM	Quantity Reqd. per Unit Assy.	12 Mos. Field Main-tenance	Major Overhaul (5th Ech)	Total First Year Procure-ment	Estimated Reqmts. per 100 Rebuilds
Col. 1	Col. 2	Col. 3	Col. 4	Col. 5	Col. 6	Col. 7	Col. 8	Col. 9	Col. 10

2301—TOOLS (Cont'd)

Col. 1	Col. 2	Col. 3	Col. 4	Col. 5	Col. 6	Col. 7	Col. 8	Col. 9	Col. 10
				WRENCH, 17/64					
				WRENCH, drain plug					
				WRENCH, socket (transmission)					
				WRENCH, socket					
				WRENCH, spark plug)					
23-1		WO-A-5130	FM-GPW-17030	WRENCH, hydraulic brake bleeder screw	1	—	—	—	—
23-1	41-W-449	WO-A-377	FM-GP-17021	WRENCH, screw, adjustable, auto type, 11 in. (2¾ in. capacity)	1	—	—	—	—
23-1	41-W-991	WO-A-1100	FM-GPW-17062	WRENCH, drain plug	1	3	3	8	—
23-1	41-W-1003	WO-A-596	FM-GP-17043	WRENCH, engineers double hd., alloy S., $3/8$ and $7/16$	1	—	—	—	—
23-1		WO-A-597	FM-GP-17044	WRENCH, engineers double hd., alloy S., $1/2$ and $19/32$	1	—	—	—	—
23-1	41-W-1005-5	WO-A-598	FM-GP-17045	WRENCH, engineers double hd., alloy S., $9/16$ and $11/16$	1	—	—	—	—
23-1	41-W-1008-10	WO-A-599	FM-GP-17046	WRENCH, engineers double hd., alloy S., $5/8$ and $25/32$	1	—	—	—	—
23-1	41-W-1012-5	WO-A-600	FM-GP-17047	WRENCH, engineers double hd., alloy S., $3/4$ and $7/8$	1	—	—	—	—
23-1		WO-A-1492	FM-GPW-17091	WRENCH, fluted socket head, (use on transmission) (WG-T-84J-100)	1	5	6	5	—
23-1		WO-A-348	FM-GPW-17035	WRENCH, socket (KHW 24832)	1	3	3	10	—
23-1		WO-637635	FM-GPW-17017-A	WRENCH, socket, spark plug	1	3	3	10	—
16-2		WO-A-7778		WRENCH, special (shock absorbers) (MAE T-317) (Willys only)	1	—	3	4	4
23-1		WO-A-692	FM-GP-17033	WRENCH, wheel hub	1	3	3	8	—
16-2		WO-A-7779		THIMBLE (shock absorber piston rod guide oil seal protector) (MAE T-347) (Willys only)	1	—	—	—	—

2301A—PIONEERING EQUIPMENT

Col. 1	Col. 2	Col. 3	Col. 4	Col. 5	Col. 6	Col. 7	Col. 8	Col. 9	Col. 10
		WO-A-3082	FM-GPW-1128258	BRACKET, shovel	1	—	—	—	—
		WO-52987	FM-24650-S2	NUT, hex., S., $5/16$-18NC-2 (brkt. to side)	2	—	—	—	—
		WO-52350	FM-33797-S2	SCREW, fl-hd., S., $5/16$-18NC-2 x $5/8$	2	—	—	—	—
		WO-5455	FM-34747-S	WASHER, S., plain, $5/16$ in.	2	—	—	—	—
		WO-51840	FM-34806-S2	WASHER, lock, S., $5/16$ in.	2	—	—	—	—
		WO-A-2601	FM-GPW-1128256	BUCKLE (for axe, shovel, and top straps)	12	—	—	—	—
		WO-A-2995	FM-GPW-1128256	CLAMP, axe, front	1	—	—	—	—
		WO-52350	FM-33797-S2	NUT, hex., S., $5/16$-28NF-2	4	—	—	—	—
		WO-52274	FM-34706-S2	WASHER, S., plain, $5/16$ in.	2	—	—	—	—
		WO-51840	FM-34806-S7-8	WASHER, lock, S., $5/16$ in.	4	—	—	—	—
		WO-A-2984	FM-GPW-1128254	CLAMP, axe rear, assembly	1	—	—	—	—
		WO-A-3139	FM-GPW-1128267	STRAP, axe, assembly	1	—	—	—	—
		WO-A-3137	FM-GPW-1128237	STRAP, shovel, front, assembly	1	—	—	—	—
		WO-A-3135	FM-GPW-1128247	STRAP, shovel, rear, assembly	1	—	—	—	—

2301B—TIRE CHAINS

2302—BRAKE FLUID

2304B—BOLTS

WO No.	FM No.	Description	Unit				
WO-A-7687	FM-GPW-18136-B	CHAIN, tire, heavy duty, type "TS" 600 x 16	2 pr.	—	—	—	—
WO-A-1798		CHAIN, tire, heavy duty, type "D" 6.50 x 16 (Willys only)	2 pr.	—	—	—	—
WO-A-8279		FLUID, brake (Lo 21-11)	12 oz.	% 10	10	23	—
WO-52840		BOLT, hex-hd., S, pltd., No. 10 (.190) x 1/2, type "B" thd.	—	10	—	—	—
WO-51738	FM-20300-S7	BOLT, hex-hd., S, 1/4-20NC-2 x 3/8	—	19	19	100	100
WO-52170	FM-20308-S2	BOLT, hex-hd., S, pkzd., 1/4-20NF-2 x 1/2	—	39	39	100	100
WO-51763	FM-20308-S2	BOLT, hex-hd., S, 1/4-20NC-2 x 1/2	—	20	40	100	100
WO-51732		BOLT, hex-hd., S, 1/4-28NF-2 x 1/2 (issue until stock exhausted)	—	10	10	—	—
WO-51514	FM-20324-S	BOLT, hex-hd., S, 1/4-20NC-2 x 5/8	—	—	—	—	—
WO-52168	FM-20324-S2	BOLT, hex-hd., S, pkzd., 1/4-20NC-2 x 5/8	—	—	—	—	—
WO-52167	FM-20344-S2	BOLT, hex-hd., S, pkzd., 1/4-20NC-2 x 3/8	—	20	—	—	—
WO-6188	FM-20384-S2	BOLT, hex-hd., S, 1/4-28NF-2 x 1/8	—	4	8	—	—
WO-52132		BOLT, hex-hd., S, 1/4-20NC-2 x 3/4	—	—	40	—	—
WO-51732		BOLT, hex-hd., S, 1/4-28NF-2 x 7/8 (issue until stock exhausted)	—	—	—	—	—
WO-51798		BOLT, hex-hd., S, 1/4-28NF-2 x 3 (issue until stock exhausted)	—	—	—	—	—
WO-51485	FM-20326-S7	BOLT, hex-hd., S, 5/16-18NC-2 x 5/8	—	170	73	300	300
WO-52132	FM-24327-S7	BOLT, hex-hd., S, 5/16-24NF-2 x 5/8	—	36	36	100	100
WO-51523	FM-20346-S2	BOLT, hex-hd., S, 5/16-18NF-2 x 3/4	—	98	42	200	200
WO-51396	FM-24347-S2	BOLT, hex-hd., S, 5/16-24NF-2 x 3/4	—	10	10	100	100
WO-6428	FM-20366-S	BOLT, hex-hd., S, 5/16-18NC-2 x 7/8	—	46	44	100	100
WO-50929	FM-24367-S2	BOLT, hex-hd., S, 5/16-24NF-2 x 7/8	—	70	10	200	200
WO-51486	FM-24386-S2	BOLT, hex-hd., S, 5/16-18NC-2 x 1	—	42	58	100	100
WO-5934	FM-34387-S7	BOLT, hex-hd., S, 5/16-24NF-2 x 1	—	88		100	100
WO-53050	FM-20387-S7	BOLT, hex-hd., S, pkzd., 5/16-24NF-2 x 1 (GM-11267)	—	10	10	50	50
WO-53131	FM-20407-S7	BOLT, hex-hd., S, cd-pltd., 5/16-24NF-2 x 1 1/8 (GM-123499)	—	90	10	—	—
WO-6157	FM-24426-S2	BOLT, hex-hd., S, 5/16-18NC-2 x 1 1/4	—	29	16	50	50
WO-6660	FM-20406-S	BOLT, hex-hd., S, 5/16-16NC-2 x 1 1/4 (issue until stock exhausted)	—	8	—	—	—
WO-5922	FM-24427-S7	BOLT, hex-hd., S, 5/16-18NF-2 x 1 1/8	—	10	10	50	50
WO-52857		BOLT, hex-hd., S, 5/16-18NC-2 x 1 5/8	—	10	10	50	50
WO-50992	FM-24505-S7	BOLT, hex-hd., S, 5/16-24NF-2 x 1 3/4	—	10		50	50
WO-51308	FM-21492-S2	BOLT, hex-hd., S, 5/16-24NF-2 x 2	—	10	10	50	50
WO-51858	FM-355442-S	BOLT, hex-hd., S, 5/16-18NC-2 x 2 1/2	—	—	—	—	—
WO-51798	FM-355398-S	BOLT, hex-hd., S, 5/16-18NC-2 x 3	—	—	—	—	—
WO-51823		BOLT, hex-hd., S, 5/16-24NF-2 x 4 (issue until stock exhausted)	—	—	—	—	—
WO-52189	FM-24328-S2	BOLT, hex-hd., S, 3/8-16NC-2 x 5/8	—	10	40	100	100
WO-6412	FM-24348-S	BOLT, hex-hd., S, 3/8-16NC-2 x 3/4	—	10	13	—	—
WO-52543		BOLT, hex-hd., S, cd-pltd., 3/8-16NC-2 x 3/4	—	90	10	100	100
WO-53031	FM-20348-S2	BOLT, hex-hd., S, rust proofed 3/8-16NC-2 x 3/4	—	—	—	—	—
WO-6299		BOLT, hex-hd., S, 3/8-16NC-2 x 7/8	—	—	—	—	—
WO-51405	FM-24388-S	BOLT, hex-hd., S, 3/8-16NC-2 x 1	—	20	20	100	100
WO-52911	FM-24408-S	BOLT, hex-hd., S, 3/8-16NC-2 x 1 1/8	—	31	12	100	100

23-1

GROUP 23—GENERAL USE STANDARDIZED PARTS (Cont'd)

Figure Number	Official Stockage Number	Part Number: Willys	Part Number: Ford	ITEM	Quantity Reqd. per Unit Assy.	Per 100 Major Items: 12 Mos. Field Maintenance	Per 100 Major Items: Major Overhaul (5th Ech)	Per 100 Major Items: Total First Year Procurement	Estimated Reqmts. per 100 Rebuilds
Col. 1	Col. 2	Col. 3	Col. 4	Col. 5	Col. 6	Col. 7	Col. 8	Col. 9	Col. 10
				2304B—BOLTS (Cont'd)					
		WO-51406	FM-24428	BOLT, hex-hd., S., 3/8-16NC-2 x 1 1/4	—	10	10	100	100
		WO-52863		BOLT, hex-hd., S., 3/8-16NC-2 x 1 5/8	—	—	—	—	—
		WO-51545		BOLT, hex-hd., S., 3/8-16NC-2 x 1 3/4	—	—	—	—	—
		WO-51954		BOLT, hex-hd., S., 3/8-16NC-2 x 2	—	—	—	—	—
		WO-6486	FM-24534-S	BOLT, hex-hd., S., 3/8-16NC-2 x 2 1/4	—	—	—	—	—
		WO-6184	FM-24430-S	BOLT, hex-hd., S., 7/16-14NC-2 x 1 1/4	—	10	10	100	100
		WO-52945	FM-23395-S	BOLT, carriage, sq-shank, S., 3/8-16NC-2 x 7/8	—	—	—	—	—
		WO-52983	FM-23393-S2	BOLT, carriage, sq-shank, S., 5/16-18NC-2 x 7/8	—	—	—	—	—
		WO-52372	FM-23498-S2	BOLT, carriage, sq-shank, S., 5/16-18NC-2 x 1 3/4 (issue until stock exhausted)	—	8	16	—	—
				2304C—CHAINS					
		WO-638457	FM-GPW-6260	CHAIN, drive, camshaft, 1/2 x 1 wide	—	—	—	—	—
				2304D—HOSE CLAMP					
		WO-53108	FM-GPW-6772	CLAMP, hose, S., 3/4 in. dia.	—	—	—	—	—
		WO-52226	FM-60-8287	CLAMP, hose, S., 1 13/16 in. I.D. (issue until stock exhausted)	—	100	200	—	—
		WO-635097	FM-GPW-9653	CLAMP, hose, S., 2 9/32 in., I.D.	—	—	—	—	—
				2304E—ELBOWS					
		WO-384549	FM-GPW-9268	ELBOW, inverted flared tube, br.	—	—	—	—	—
		WO-384569	FM-9N-18686	ELBOW, inverted flared tube, 1/4 br.	—	—	—	—	—
		WO-A-7180	FM-GPW-14521	ELBOW, Mall-1., z-pltd., 1/2-14 pipe thd., 15/16-18 U.S.S. thd.	—	—	—	—	—
		WO-A-6885	FM-GPW-6722	ELBOW, pipe 1/4 in.	—	—	—	—	—
				2304F—FITTINGS					
		WO-640038	FM-350006-S8	FITTING, grease (AD-1980)	—	23	11	40	40
		WO-392909	FM-353027-A-S7	FITTING, grease, straight, 1/8 inch (AD-1720)	—	32	32	80	80
		WO-638500	FM-353023-S7	FITTING, grease, 90 degrees	—	10	10	40	40
		WO-638792	FM-353043-S7	FITTING, grease, straight, 1/4-28NF-2	—	29	10	40	40
		WO-638224	FM-353035-S7	FITTING, grease, 45 degrees	—	5	5	40	40

2304G—NUTS

WO No.	FM No.	Description						
WO-53013		NUT, hex, stamped, S., cd-pltd. No. 6 (138)-32NC-2 (GM-174884)	—	—	—	—	—	—
WO-6214	FM-34052-S2	NUT, hex, reg, S., No. 8 (164)-32NC-2	—	—	—	—	—	—
WO-52182	FM-34052-S2	NUT, hex, stamped, br., No. 8 (164)-32NC-2	—	45	45	—	200	200
WO-6352	FM-355836-S7	NUT, hex, reg, S., No. 10 (190)-24NC-2	—	—	—	—	—	—
WO-52651	FM-355836-S7	NUT, hex, reg, S., cd-pltd, No. 10 (190)-24NC-2	—	10	10	—	100	100
WO-5848		NUT, hex, reg, br, No. 10 (190)-32NF-2	—	—	—	—	—	—
WO-6536		NUT, hex, reg, S., No. 10 (190)-32NF-2 (AL-2235)	—	54	—	—	—	—
WO-53061		NUT, hex, reg, S., cd-pltd, No. 10 (190)-32NF-2	—	4	8	—	—	—
WO-6273	FM-34141-S2	NUT, sq., reg, S., No. 10 (190)-32NF-2	—	132	268	—	200	200
WO-6989	FM-34054-S2	NUT, hex, reg, S., No. 12 (214)-24NC-2 (issue until stock exhausted)	—	10	10	—	—	—
WO-52217	FM-33795-S2	NUT, hex, reg, S., pkzd., 1/4-20NC-2	—	38	24	—	—	—
WO-5790	FM-33795-S2	NUT, hex, reg, S., 1/4-20NC-2 (issue until stock exhausted)	—	16	30	—	—	—
WO-53132	FM-33823-S	NUT, hex, br., 1/4-20NC-2 (GM-114379)	—	49	19	—	200	200
WO-52165	FM-34129-S2	NUT, hex, S., pkzd., 1/4-20NC-2	—	78	—	—	—	—
WO-52217	FM-33795-S2	NUT, hex, S., pkzd., 1/4-20NC-2	—	—	—	—	100	100
WO-51802		NUT, sq., S., 1/4-20NC-2 (issue until stock exhausted)	—	—	—	—	—	—
WO-5914	FM-33796-S2	NUT, hex, S., 1/4-28NF-2	—	—	—	—	200	200
WO-52804		NUT, hex, S., thin, 1/4-28NF-2	—	—	—	—	—	—
WO-6167	FM-33797-S2	NUT, hex, S., 5/16-18NC-2	—	—	128	—	200	200
WO-52845		NUT, hex, spring, S., type B, 5/16-18NC-2	—	—	5	—	—	—
WO-52350	FM-33797-S2	NUT, hex, thin, S., rust proof 5/16-18NC-2	—	—	40	—	—	—
WO-52891		NUT, hex, spring, S., stamped 5/16-24NF-2 (GM-147510)	—	—	—	—	—	—
WO-52885		NUT, hex, S., cd-pltd, 5/16-24NF-2	—	—	365	—	—	—
WO-50802	FM-33798-S7	NUT, hex, S., pkzd., 5/16-24NF-2	—	—	—	—	—	—
WO-52954	FM-33909-S7	NUT, hex, thin, S., 5/16-24NF-2 (GM-114493)	—	265	—	—	—	—
WO-5910	FM-33798-S7	NUT, hex, S., 5/16-24NF-2	—	5	16	—	500	500
WO-53033	FM-33799-S2	NUT, hex, S., rust proof, 3/8-16NC-2 (GM-123228)	—	20	20	—	200	200
WO-6196		NUT, hex, half hex, S., 3/8-16NC-2 (issue until stock exhausted)	—	—	—	—	—	—
WO-5544		NUT, hex, S., 3/8-16NC-2	—	391	16	—	—	—
WO-52893	FM-356016-S7	NUT, hex, stamped, type B, S., 3/8-16NC-2 (GM-147511)	—	—	—	—	800	800
WO-53289		NUT, hex, Sez-proof, S., 3/8-16NC-2	—	—	—	—	—	—
WO-5901	FM-33800-S7	NUT, hex, S., 3/8-24NF-2	—	—	—	—	—	—
WO-50922	FM-33800-S2	NUT, hex, S., pkzd., 3/8-24NF-2 (CAR-105A-13)	—	—	—	—	—	—
WO-52542		NUT, hex, S., cd-pltd., 3/8-24NF-2	—	—	—	—	—	—
WO-52909	FM-33800-S7	NUT, hex, stamped, S., cd-pltd., 3/8-24NF-2 (GM-107322)	—	—	—	—	—	—
WO-53288		NUT, hex, Sez-proof, S., 3/8-24NF-2	—	8	10	—	—	—
WO-53287		NUT, hex, Sez-proof, S., 3/8-24NF-2	—	20	—	—	—	—
WO-5336		NUT, hex, S., 7/16-14NC-2 (issue until stock exhausted)	—	16	20	—	200	200
WO-5939	FM-33802-S2	NUT, hex, S., 7/16-20NF-2	—	20	20	—	200	200
WO-52925	FM-33927-S7	NUT, hex, thin, S., 7/16-20NF-2	—	10	6	—	—	—
WO-52825	FM-356028-S	NUT, hex, stamped, S., 7/16-20NF-2 (GM-107322)	—	68	32	—	40	40
WO-6163	FM-33845-S2	NUT, hex, stamped, S., 1/2-13NC-2	—	10	16	—	100	100
WO-5916	FM-33846-S2	NUT, hex, stamped, S., 1/2-20NF-2	—	68	20	—	50	50
WO-52804	FM-33832-S2	NUT, hex, thin, S., 1/2-20NF-2	—	16	6	—	—	—
WO-10558	FM-351059	NUT, hex, S., dld. f/c-pin., 1/2-20NF-2 (issue until stock exhausted)	—	16	32	—	—	—
WO-6436	FM-34033-S18	NUT, hex, slotted, S., 9/16-18NF-2	—	29	29	—	100	100

GROUP 23—GENERAL USE STANDARDIZED PARTS (Cont'd)

Figure Number	Official Stockage Number	Part Number		ITEM	Quantity Reqd. per Unit Assy.	Per 100 Major Items			Estimated Reqmts. per 100 Rebuilds
		Willys	Ford			12 Mos. Field Maintenance	Major Overhaul (5th Ech)	Total First Year Procurement	
Col. 1	Col. 2	Col. 3	Col. 4	Col. 5	Col. 6	Col. 7	Col. 8	Col. 9	Col. 10
				2304Y—PLUGS					
		WO-51091		PLUG, expansion, 1¼ in. (issue until stock exhausted)	—	68	132	—	—
		WO-51460		PLUG, expansion, 1¾ in. (issue until stock exhausted)	—	68	132	—	—
		WO-52525	FM-353052-S	PLUG, pipe, slotted, ⅛ in.	—	3	3	40	40
		WO-5085	FM-358064-S	PLUG, pipe, ⅛ in.	—	3	32	40	40
		WO-5138	FM-353055-S7	PLUG, pipe, ¼ in.	—	26	12	40	40
		WO-376373	FM-358063-S	PLUG, pipe, S., ck, ⅜ in.	—	—	16	20	20
		WO-5140	FM-353064-S	PLUG, pipe, ½ in.	—	14	5	20	20
		WO-636538	FM-353051-S	PLUG, pipe, sq-hd., ¾ x 5/16	—	6	6	40	40
		WO-636577	FM-358048-S	PLUG, pipe, special, ¾ in.	—	6	6	40	40
				2304J—RIVETS					
		WO-635860	FM-62216-S	RIVET, fl-hd., S., 3/16 x 7/16	—	—	—	—	—
		WO-51977	FM-62218-S	RIVET, fl-hd., S., ¼ x 7/16	—	—	—	—	—
		WO-52207	FM-60371-S	RIVET, fl-hd., S., ¼ x 9/16 (issue until stock exhausted)	—	4	8	—	—
		WO-52906		RIVET, fl-hd., S., ⅜ x ¾	—	—	—	—	—
		WO-5326		RIVET, fl-hd., S., ⅜ x ⅞	—	—	—	—	—
		WO-5247		RIVET, rd-hd., S., ¼ x ⅜	—	—	—	—	—
		WO-5249		RIVET, rd-hd., S., ¼ x ⅝	—	—	—	—	—
		WO-5267	FM-60416-S	RIVET, rd-hd., S., 5/16 x ¾	—	—	—	—	—
		WO-5216	FM-60470-S	RIVET, rd-hd., S., 5/16 x 1	—	—	—	—	—
		WO-51561	FM-60375-S	RIVET, rd-hd., S., ⅜ x 9/16	—	—	—	—	—
		WO-50769		RIVET, rd-hd., S., ⅜ x ¾	—	—	—	—	—
		WO-52832	FM-357100-S	RIVET, rd-hd., S., ⅜ x ⅞	—	—	—	—	—
		WO-5215		RIVET, rd-hd., S., ⅜ x 1	—	—	—	—	—
		WO-51182	FM-62627-S3	RIVET, split, tubular, S., 9/64 x 9/16	—	—	—	—	—
		WO-53056		RIVET, ov-hd., tubular, br., 9/64 x 11/32	—	—	—	—	—
		WO-53243	FM-63848-S	RIVET, ov-hd., tubular, br., ⅛ x 15/32	—	—	—	—	—
		WO-51182	FM-62627-S3	RIVET, split tubular, S., 9/64 x 9/16	—	—	—	—	—
		WO-53056		RIVET, ov-hd., tubular, br., 9/64 x 11/32	—	—	—	—	—
		WO-53243	FM-63848-S	RIVET, ov-hd., tubular, br., ⅛ x 15/32	—	—	—	—	—
				2304K—SCREWS					
		WO-52990	FM-32989-S2	SCREW, ctsk., ov-hd., S., No. 10 (.190) x ½	—	—	—	—	—
		WO-50814		SCREW, fl-hd., S., No. 10 (.190)-24NC-2	—	—	—	—	—

					ITEM	PART NO.	STOCK NO.
—	—	—	—	—	SCREW, fl-hd., S., 5/16-18NC-2 x 5/8	FM-24650-S2	WO-52987
—	—	—	—	—	SCREW, fl-hd., S., 5/16-18NC-2 x 5/8	FM-33797-S2	WO-52350
—	—	—	—	—	SCREW, fl-hd., S., 1/4-20NC-2 x 3/4	FM-36046-S2	WO-52963
—	—	—	—	—	SCREW, binding head, S., No. 10 (.190) x 1/2	FM-32924-S8	WO-52809
—	—	—	—	—	SCREW, binding head, S., No. 10 (.190) x 1/2	FM-32866-S7	WO-52142
—	—	—	—	—	SCREW, ov-hd. S., No. 12 (.216)-24NC-2 x 3/8	FM-31079	WO-51819
200	200	54	75	—	SCREW, rd-hd., machine, S., No. 6 (.138)-32NC-2 x 5/16		WO-53014
100	100	16	16	—	SCREW, rd-hd., machine, S., No. 8 (.164)-32NC-2 x 1/4		WO-5253
—	—	—	—	—	SCREW, rd-hd., machine, S., No. 8 (.164)-32NC-2 x 1/4	FM-26457-S7	WO-51040
—	—	—	—	—	SCREW, rd-hd., machine, S., No. 8 (.164)-32NC-2 x 3/8	FM-355095-S7	WO-5556
—	—	—	—	—	SCREW, rd-hd., machine, S., No. 8 (.164)-32NC-2 x 1/2	FM-26480-S2	WO-6287
—	—	—	—	—	SCREW, rd-hd., machine, S., No. 10 (.190)-24NC-2 x 3/8	FM-355158-S2	WO-5182
—	—	—	—	—	SCREW, rd-hd., machine, S., No. 10 (.190)-24NC-2 x 3/8		WO-5236
200	200	10	10	—	SCREW, rd-hd., machine, S., No. 10 (.190)-24NC-2 x 1/2		WO-5113
100	100	10	10	—	SCREW, rd-hd., machine, S., No. 10 (.190)-24NC-2 x 5/8	FM-26496-S	WO-5064
200	200	16	16	—	SCREW, rd-hd., machine, S., No. 10 (.190)-24NC-2 x 5/8		WO-5780
200	200	19	19	—	SCREW, rd-hd., machine, S., No. 10 (.190)-24NC-2 x 3/4	FM-355132-S	WO-5272
—	—	—	—	—	SCREW, rd-hd., machine, S., No. 10 (.190)-32NF-2 x 5/16	FM-27698-S	WO-52994
—	—	—	—	—	SCREW, rd-hd., machine, S., No. 10 (.190)-32NF-2 x 7/16 (issue until stock exhausted)		WO-52060
100	100	—	32	—	SCREW, rd-hd., machine, S., No. 10 (.190)-32NF-2 x 7/8	FM-27145-S2	WO-6627
—	—	—	—	—	SCREW, rd-hd., machine, S., No. 10 (.190)-32NF-2 x 1	FM-26498-S2	WO-6383
—	—	—	—	—	SCREW, rd-hd., machine, S., No. 12 (.216)-24NC-2	FM-36845-S2	WO-53267
—	—	—	—	—	SCREW, rd-hd., machine, S., No. 12 (.126)-24NC-2 x 5/8	FM-26147-S7	WO-6988
100	100	—	3	—	SCREW, rd-hd., machine, S., 1/4-20NC-2 x 3/8	FM-26483-S	WO-52036
200	200	—	48	—	SCREW, rd-hd., machine, S., 1/4-20NC-2 x 1/2 (GM-113955)	FM-36868-S2	WO-51492
—	—	—	—	—	SCREW, rd-hd., machine, S., 1/4-20NC-2 x 3/4		WO-51539
—	—	—	—	—	SCREW, rd-hd., machine, S., 7/16-32NF	FM-37621-S7	WO-53020
—	—	—	—	—	SCREW, fil-hd., machine, S., No. 6 (.138)-32NC-2 x 5/16	FM-31628-S7	WO-113439
—	—	—	—	—	SCREW, fil-hd., machine, S., No. 10 (.190)-32NF-2 x 1/2	FM-32852-S	WO-52993
—	—	—	—	—	SCREW, rd-hd., sheet metal, S., No. 10 (.190) x 3/8	FM-92047-S	WO-51904
—	—	—	—	—	SCREW, rd-hd., machine, S., br-pltd., No. 6 (.138) x 5/16	FM-26457-S7	WO-52131
—	—	—	—	—	SCREW, rd-hd., machine, S., cd-pltd., No. 8 (.164)-32NC-2		WO-52781
—	—	—	—	—	SCREW, rd-hd., machine, S., cd-pltd., No. 8 (.164)-32NC-2 x 3/8	FM-355162-S	WO-51537
—	—	—	—	—	SCREW, fl-hd., S., cd-pltd., No. 10 (.190)-24NC-2 x 1/2		WO-53252
—	—	—	—	—	SCREW, rd-hd., machine, S., cd-pltd., No. 10 (.190)-32NF-2 x 3/8		WO-53220
—	—	—	—	—	SCREW, rd-hd., machine, S., cd-pltd., No. 10 (.190)-32NF-2 x 3/8		WO-53103
—	—	—	—	—	SCREW, wood, ov-hd., S., No. 6 (.138) x 1	FM-31866-S	WO-5580
—	—	—	—	—	SCREW, wood, rd-hd., S., No. 10 (.190) x 5/8		

2304L—WASHERS

					ITEM	PART NO.	STOCK NO.
200	200	10	19	—	WASHER, plain, S., No. 8 (.164)	FM-34802-S2	WO-51532
200	200	10	20	—	WASHER, plain, S., No. 10 (.164)	FM-34745-S2	WO-52702
100	100	39	159	—	WASHER, plain, S., 1/4 in.	FM-34706-S2	WO-5437
200	200	—	130	—	WASHER, plain, S., 5/16 in.	FM-34746-S2	WO-53274
200	200	30	46	—	WASHER, plain, S., 3/8 in.	FM-34747-S2	WO-52101
—	—	10	30	—	WASHER, plain, S., 3/8 in.	FM-34707-S7	WO-5455
—	—	12	8	—	WASHER, plain, S., 7/16 in.		WO-52835

GROUP 23—GENERAL USE STANDARDIZED PARTS (Cont'd)

Figure Number	Official Stockage Number	Part Number		ITEM	Quantity Reqd. per Unit Assy.	Per 100 Major Items			
		Willys	Ford			12 Mos. Field Maintenance	Major Overhaul (5th Ech)	Total First Year Procurement	Estimated Reqmts. per 100 Rebuilds
Col. 1	Col. 2	Col. 3	Col. 4	Col. 5	Col. 6	Col. 7	Col. 8	Col. 9	Col. 10
				2304L—WASHERS (Cont'd)					
		WO-52768	FM-356305	WASHER, plain, S., 21/64 in.	—	20	10	200	200
		WO-636570	FM-356504-S	WASHER, plain, S., 1½ O.D. x 49/64 I.D. x ⅛ thk.	—	4	12	40	40
		WO-52705	FM-34802-S2	WASHER, lock, regular, S., No. 8 (.160)	—	36	20	200	200
		WO-113440	FM-34803-S7	WASHER, lock, regular, S., cd-pltd., No. 10 (.190)	—	126	87	300	300
		WO-51969	FM-34804-S2	WASHER, lock, regular, S., No. 12 (.214)	—	19	—	100	100
		WO-52706	FM-34805-S2	WASHER, lock, regular, S., ¼ in.	—	130	79	300	300
		WO-51833	FM-34806-S2	WASHER, lock, regular, S., 5/16 in.	—	766	224	1200	1200
		WO-5051		WASHER, lock, regular, S., 5/16 in.	—	28	52	—	—
		WO-5010	FM-34807-S7	WASHER, lock, regular, S., ⅜ in.	—	571	421	1200	1200
		WO-5059	FM-34808-S2	WASHER, lock, regular, S., 7/16 in.	—	6	—	100	100
		WO-5938	FM-34838-S	WASHER, lock, regular, S., 7/16 in. heavy duty	—	56	56	200	200
		WO-5009	FM-34809-S	WASHER, lock, regular, S., ½ in.	—	81	231	400	400
		WO-5038	FM-34811-S2	WASHER, lock, regular, S., 5/8 in.	—	10	10	200	200
		WO-52754	FM-43902-S	WASHER, lock, internal tooth, S., No. 8 (.164)	—	5	5	200	200
		WO-52483	FM-34905-S7	WASHER, lock, internal tooth, S., No. 14	—	19	19	200	200
		WO-52428	FM-34906-S	WASHER, lock, internal tooth, S., 5/16 in.	—	32	48	200	200
		WO-52332	FM-34907-S7	WASHER, lock, internal tooth, S., ⅜ in.	—	78	59	200	200
		WO-52874	FM-34921-S	WASHER, lock, internal tooth, S., 7/16 in.	—	26	26	200	200
		WO-636528	FM-34922-S	WASHER, lock, internal tooth, S., ½ in.	—	102	—	100	100
		WO-52510	FM-34941-S7	WASHER, lock, external tooth, S., 5/16 in.	—	80	80	200	200
				2304M—WIRE LOCKING					
		WO-637598	FM-GP-2174	WIRE, lock piston stop (LO-S-FC-2927)	—	—	—	—	—
		WO-636298	FM-GPW-8576	WIRE, retaining, water pump brg.	—	—	—	—	—
				2305—MISCELLANEOUS COTTER PINS					
		WO-5067	FM-72083-S	PIN, cotter, S., 1/16 x ½	—	—	—	—	—
		WO-5354	FM-72004-S	PIN, cotter, S., 3/32 x ½	—	—	—	—	—
		WO-52967	FM-72016-S	PIN, cotter, S., 3/32 x 5/8	—	—	—	—	—
		WO-5020	FM-72-16-S	PIN, cotter, S., 3/32 x ¾	—	—	—	—	—
		WO-5152	FM-72025-S	PIN, cotter, S., 3/32 x ⅞	—	—	—	—	—
		WO-5021	FM-72035-S	PIN, cotter, S., 3/32 x 1	—	—	—	—	—
		WO-5397	FM-72071	PIN, cotter, S., ⅛ x 1½	—	—	—	—	—
		WO-5108	FM-72053-S	PIN, cotter, S., ⅛ x 1¼	—	—	—	—	—

2305A—WOODRUFF KEYS

2310—WINTERIZATION FIELD KIT

		Qty	Description	FM No.	WO No.
—	—	—	PIN, cotter, S., 1/8 x 1 3/8	FM-72062-S	WO-52527
—	—	—	PIN, cotter, S., 1/8 x 1 1/4	FM-72089-S	WO-5134
—	—	—	PIN, cotter, S., 3/16 x 1	FM-72037-S	WO-52946
8	16	—	KEY, Woodruff, S., No. 5 (AL-X-260)	FM-74144-S	WO-A-1641
8	16	—	KEY, Woodruff, S., No. 6 (AL-X-261)	FM-74175-S7	WO-5017
8	16	—	KEY, Woodruff, S., No. 9	FM-74178-S	WO-5036
—	—	—	KEY, Woodruff, S., No. 13	FM-74182-S	WO-50917
		1	CRATE, winterization field kit		WO-A-11848
		as req.	KIT, field, winterization (Willys only)		WO-A-11847
			(Consists of:		
		as req.	KIT, field, battery plug-in receptacle		WO-A-11792
			(Consists of:		
		1	CABLE, positive, assembly		WO-A-12135
		1	BOLT, hex-hd, s-fin, alloy-S., 3/8-16NC-2 x 1 3/8 (cable to cranking motor attaching bolt)		WO-52600
		1	WASHER, lock, S., internal, external tooth, 3/8 in. (cable to cranking motor attaching bolt)		WO-53026
		2	CABLE, receptacle to ground, assembly		WO-A-11787
		1	CLIP, wire, 5/8 in., (cable to gusset)		WO-78923
		2	NUT, hex, reg., S., No. 10 (.190)-24NC-2 (clip to gusset screw)		WO-6352
		1	SCREW, rd-hd, machine, S., No. 10 (.190)-24NC-2 x 5/16 (clip to gusset)		WO-52548
		1	WASHER, lock, reg, S., No. (.190) (clip to gusset screw)		WO-5168
		1	DRAWING, installation, battery plug-in receptacle field kit		WO-A-11851
		1	RECEPTACLE, battery plug-in, 2 pole type, assembly		WO-A-11772
		4	BOLT, hex-hd, s-fin, alloy-S., 1/4-20NC-2 x 1 (receptacle to fender)		WO-53199
		4	NUT, hex, reg., S., 1/4-20NC-2 (receptacle to fender bolt)		WO-52217
		4	WASHER, lock, reg., S., 1/4 in. (receptacle to fender bolt)		WO-53084
		1	REINFORCEMENT, splash apron		WO-A-11773
		4	BOLT, hex-hd, s-fin, alloy-S., 1/4-20NC-2 x 1/2 (reinforcement to fender)		WO-52170
		4	NUT, hex, reg., S., 1/4-20NC-2 (reinforcement to fender bolt)		WO-52217
		4	WASHER, plain, SAE, std., 1/4 in. (reinforcement to fender bolt)		WO-53178
		4	WASHER, lock, reg., S., 1/4 in. (reinforcement to fender bolt)		WO-53084
		as req.	KIT, field, blanket		WO-A-11815
			(Consists of:		
		1	BLANKET, brush guard, assembly		WO-A-11816
		1	BLANKET, hood, assembly		WO-A-12038
		1	BLANKET, under fender, left, assembly		WO-A-11828
		1	BLANKET, under fender, right, assembly		WO-A-11829
		1	BLANKET, under motor, assembly		WO-A-11821
		1	CARTON, blanket field kit		WO-A-11832

23-2
23-2

RA PD 305174

FIGURE 23-2—COLD STARTING BLANKET FIELD KIT

Key	Item	Willys Part No.	Ford Part No.	Gov't Group No.
A	BLANKET	WO-A-12038	(Willys only)	2310
B	BLANKET	WO-A-11816	(Willys only)	2310

RA PD 305174-A

RA PD 305173

FIGURE 23-3—COLD STARTING HEATER FIELD KIT

Key	Item	Willys Part No.	Ford Part No.	Gov't Group No.
A	STACK	WO-A-8214	(Willys only)	2310
B	CLAMP	WO-A-8208	(Willys only)	2310
C	ELBOW	WO-52805	(Willys only)	2310
D	STOVE	WO-A-7156	(Willys only)	2310
E	HOSE	WO-A-7036	(Willys only)	2310
F	COCK	WO-A-7000	(Willys only)	2310
G	LINE	WO-A-1367	FM-GPW-9282	0304
H		See WO-A-7156	(Willys only)	2310
I	LINE	WO-A-8214	(Willys only)	2310
J	COCK	WO-A-7058	(Willys only)	2310
K	TANK	WO-A-8097	(Willys only)	2310

RA PD 305173-A

GROUP 23—GENERAL USE STANDARDIZED PARTS (Cont'd)

Figure Number	Official Stockage Number	Part Number Willys	Part Number Ford	ITEM	Quantity Reqd. per Unit Assy.	12 Mos. Field Maintenance	Major Overhaul (5th Ech)	Total First Year Procurement	Estimated Reqmts. per 100 Rebuilds
Col. 1	Col. 2	Col. 3	Col. 4	Col. 5	Col. 6	Col. 7	Col. 8	Col. 9	Col. 10
				2310—WINTERIZING FIELD KIT (Cont'd)					
		WO-A-1173		CLIP, retaining spring (S., 2⅞ in., lgth., under motor blanket to flywheel housing)	1	—	—	—	—
		WO-A-11833-1		DRAWING, installation, under motor blanket	1	—	—	—	—
		WO-A-11833-2		DRAWING, installation, radiator and brush guard	1	—	—	—	—
		WO-A-2924		FASTENER, curtain, P.K. type, single stud	8	—	—	—	—
		WO-A-3625		FASTENER, curtain, single stud, No. 10 (.190)-32NF-2	21	—	—	—	—
		WO-52216		NUT, hex, reg, S., No. 10 (.190)-32NF-2	21	—	—	—	—
		WO-52221		WASHER, lock, S., light, No. 10 (.190)	21	—	—	—	—
		WO-A-11950		SPRING, retaining, under motor blanket	4	—	—	—	—
		WO-A-11826		SPRING, retaining, under motor blanket, assembly	12	—	—	—	—
		WO-A-11793		KIT, field, body enclosure	as req.	—	—	—	—
				(Consists of:					
		WO-A-11814		CARTON, body enclosure field kit	1	—	—	—	—
		WO-A-11794		CURTAIN, side, left, assembly	1	—	—	—	—
		WO-A-11795		CURTAIN, side, right, assembly	1	—	—	—	—
		WO-A-12200		INSTRUCTIONS, body enclosure field kit	1	—	—	—	—
		WO-A-11813		TOP assembly	1	—	—	—	—
		WO-A-7154		KIT, field, cold starting primer installation	as req.	—	—	—	—
				(Consists of:					
		WO-A-7108		CARTON, cold starting primer installation field kit	1	—	—	—	—
		WO-A-6950		CLIP, S., 5/16 in.	1	—	—	—	—
		WO-A-7858		EQUIPMENT, cold starting primer (furnished by Ordnance Dept. for packaging by W. O. Co.) (Consists of:	1	—	—	—	—
				1 WO-A-6879 ELBOW, primer (DV-BU-710)					
				1 WO-A-6878 ELBOW, primer, assembly (DV-PR-20-700)					
				1 WO-A-6880 NUT, primer tube, w/SLEEVE, assembly (DV-BU-010)					
				1 WO-A-6876 PRIMER assembly (DV-PR-20)					
				1 WO-A-6877 TEE, primer, assembly (DV-PR-1100))					
		WO-A-6954		TUBING, intake manifold tee to elbow (S., ⅛ in., 7⅞ in., lgth.)	1	—	—	—	—
		WO-A-6955		TUBING, intake manifold tee to primer (S., ⅛ in., 7⁹⁄₁₆ in., lgth.)	1	—	—	—	—
		WO-A-6956		TUBING, primer to gas strainer (S., ⁹⁄₁₆ in., 30³⁄₁₆ in., lgth.))	1	—	—	—	—
		WO-A-7155		KIT, field crankcase ventilation	as req.	—	—	—	—
				(Consists of:					
		WO-A-6019		BODY, crankcase ventilator assembly	1	—	—	—	—

		WO-A-7186		CARTON, crankcase ventilation field kit	1	—	—	—	—
		WO-53108		CLAMP, hose, ¾ in.	2	—	—	—	—
		WO-A-6918		CONNECTION, hose, flexible	1	—	—	—	—
		WO-384549		ELBOW, inverted flared tube, 5/16 in., ⅛ in. pipe thread	1	—	—	—	—
		WO-A-6885		ELBOW, inverted flared tube, 5/16 in., ¼ in. pipe thread	1	—	—	—	—
		WO-634811		GASKET, intake to exhaust manifold	1	—	—	—	—
		WO-630299		GASKET, ventilator to valve spring cover	1	—	—	—	—
		WO-A-6525		INDICATOR, oil filler cap and level, assembly	1	—	—	—	—
		WO-A-6969		INSTALLATION, crankcase ventilator	1	—	—	—	—
		WO-A-1166		MANIFOLD, intake, assembly	1	—	—	—	—
		WO-A-9008		PLUG, driving, oil filler tube (wood)	1	—	—	—	—
		WO-A-6922		TUBE, crankcase ventilator valve, assembly	1	—	—	—	—
		WO-A-6915		TUBE, oil, filler, assembly	1	—	—	—	—
		WO-A-6911		TUBE, air cleaner, w/BRACKET, assembly	1	—	—	—	—
		WO-A-6895		VALVE, crankcase ventilator, assembly)	1	—	—	—	—
		WO-A-11846		KIT, field heater	as req.	—	—	—	—
				(Consists of:					
		WO-A-11844		CARTON, heater field kit	1	—	—	—	—
		WO-53308		CLAMP, heater tube coupling hose	8	—	—	—	—
		WO-A-11865		DRAWING, installation, heater field kit	1	—	—	—	—
		WO-A-8558		GASKET, cylinder head	1	—	—	—	—
		WO-A-11839		HEATER assembly	1	—	—	—	—
		WO-53031		BOLT, hex-hd., s-fin., alloy-S., ⅜-16NC-2 x ¾ (heater to dash)	1	—	—	—	—
		WO-53033		NUT, hex., reg., S., ⅜-16NC-2	2	—	—	—	—
		WO-50921		WASHER, S., plain, SAE, std., ⅜ in.	3	—	—	—	—
		WO-52046		WASHER, lock, reg., S., ⅜ in.	2	—	—	—	—
		WO-A-11838		HOSE, coupling, heater tube	4	—	—	—	—
		WO-A-11845		SWITCH, heater, assembly	1	—	—	—	—
		WO-A-11837		TUBE, inlet, heater	1	—	—	—	—
		WO-A-11836		TUBE, outlet, heater	1	—	—	—	—
		WO-A-11840		VALVE, shut off, heater, assembly	2	—	—	—	—
		WO-53114		CLAMP, hose, ⅞ in.	8	—	—	—	—
		WO-A-8210		CLAMP, stove stack	1	—	—	—	—
		WO-6159		BOLT, hex-hd., s-fin., alloy-S., ¼-20NC-2 x 1 (clamp to bracket and stack)	1	—	—	—	—
		WO-5790		NUT, hex., reg., S., ¼-20NC-2 (clamp to bracket and stack bolt)	1	—	—	—	—
		WO-52706		WASHER, lock, SAE, std., ¼ in. (clamp to bracket and stack bolt)	1	—	—	—	—
23-3		WO-A-7058		COCK, shut off (WH-775)	1	—	—	—	—
		WO-387249		CONNECTOR, inverted flared tube, 5/16 in.	1	—	—	—	—
		WO-A-7043		CONNECTOR, tee-block (DV-X-697)	1	—	—	—	—
		WO-A-7001		ELBOW, street, std., 45°, ¼ in.	1	—	—	—	—
23-3		WO-52805		ELBOW, street, std., 90°, ¼ in.	1	—	—	—	—
		WO-A-6999		FITTING, hose, special, ¼ in. pipe to ½ in. hose	1	—	—	—	—
		WO-639650		GASKET, water outlet elbow	1	—	—	—	—
		WO-662420		GROMMET, 3/16 in., (rubber, split)	1	—	—	—	—
		WO-A-5080		WIRE, heater switch to ignition switch, assembly)	1	—	—	—	—
23-3		WO-A-7156		STOVE, winterization field kit	as req.	—	—	—	—
				(Consists of:					
		WO-A-8206		BRACKET, mounting, stove to frame	1	—	—	—	—
		WO-53242		BOLT, carriage, S., ⅜-16NC-2 x 2½ (bracket to frame)	1	—	—	—	—

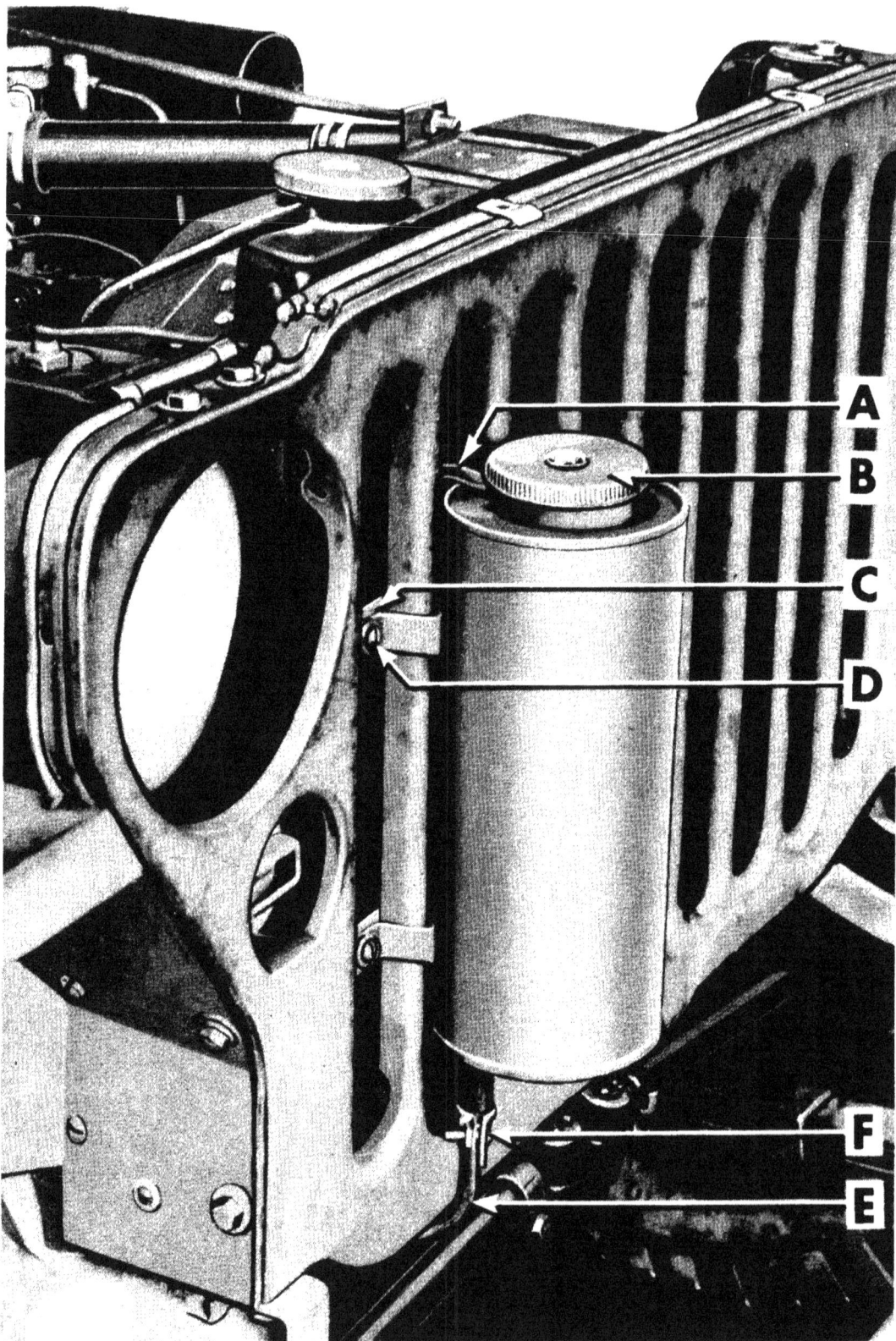

RA PD 305172

FIGURE 23-4—DESERT COOLING KIT

Key	Item	Willys Part No.	Ford Part No.	Gov't Group No.
A	TUBE	WO-A-6945		2309
B	TANK	WO-A-6933		2309
C	STRAP	WO-A-6944		2309
D	SCREW	WO-52468		2309
E	HOSE	WO-A-6946		2309
F	CLAMP	WO-A-6947		2309

RA PD 305172-A

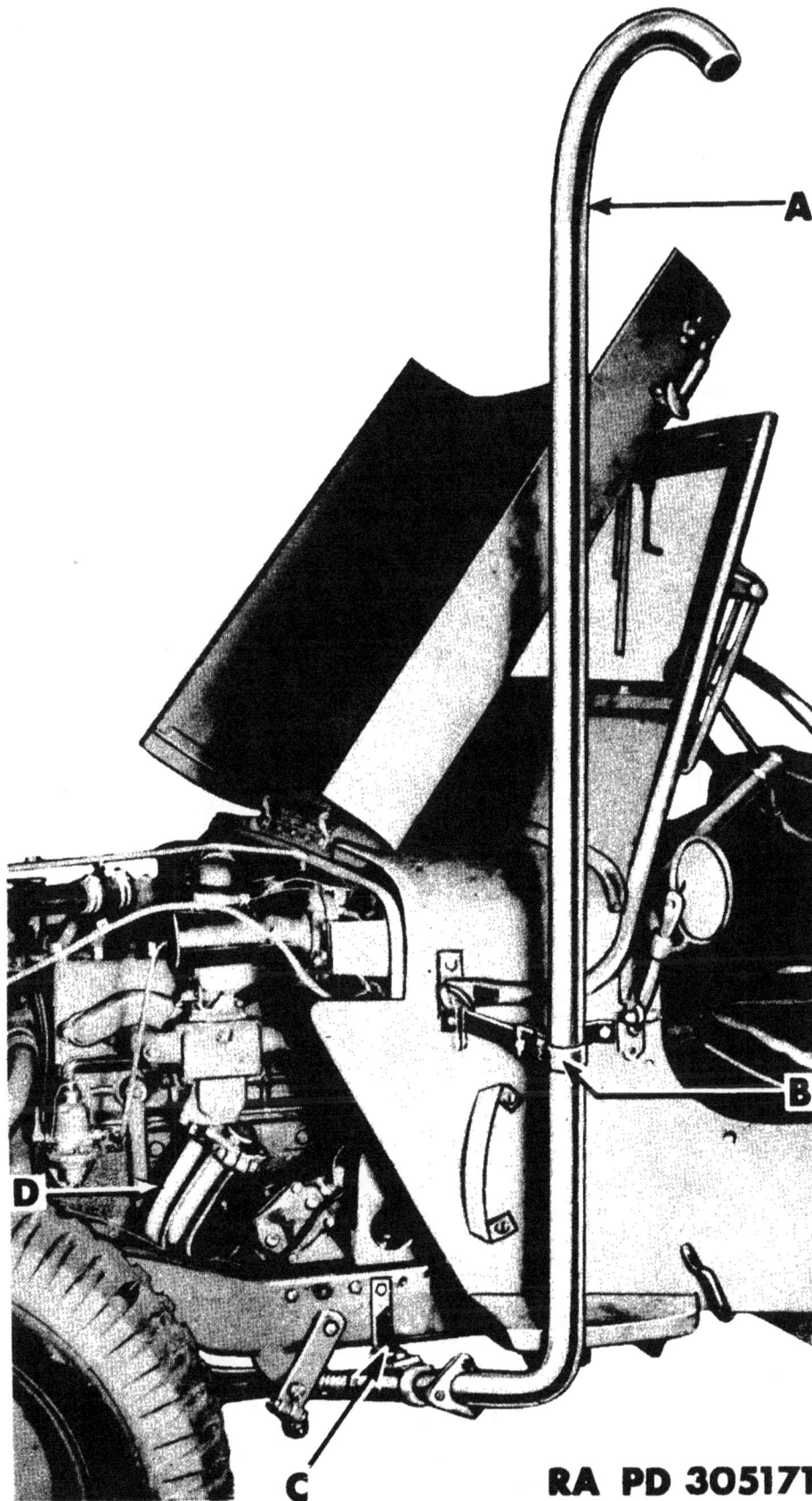

RA PD 305171

FIGURE 23-5—DEEP WATER EXHAUST

Key	Item	Willys Part No.	Ford Part No.	Gov't Group No.
A	EXTENSION	WO-A-7816		2309
B	BRACKET	WO-A-8196		2309
C	BRACKET	WO-A-8197		2309
D	PIPE	WO-A-7814		2309

RA PD 305171-A

SNL G-503

GROUP 23—GENERAL USE STANDARDIZED PARTS (Cont'd)

| Figure Number | Official Stockage Number | Part Number | | ITEM | Quantity Reqd. per Unit Assy. | Per 100 Major Items | | | Estimated Reqmts. per 100 Rebuilds |
| | | Willys | Ford | | | 12 Mos. Field Maintenance | Major Overhaul (5th Ech) | Total First Year Procurement | |
Col. 1	Col. 2	Col. 3	Col. 4	Col. 5	Col. 6	Col. 7	Col. 8	Col. 9	Col. 10
				2310—WINTERIZING FIELD KIT (Cont'd)					
		WO-50051		BOLT, hex-hd., s-fin., alloy-S., 3/8-24NF-2 x 2½ (brkt. to frame)	1	—	—	—	—
		WO-5544		NUT, hex, reg., S., 3/8-16NC-2 (bracket to frame bolt)	1	—	—	—	—
		WO-5901		NUT, hex, reg., S., 3/8-24NF-2 (bracket to frame bolt)	1	—	—	—	—
		WO-5010		WASHER, lock, SAE, std., 3/8 in. (bracket to frame bolt)	2	—	—	—	—
23-3		WO-A-8208		BRACKET, stove stack, front	1	—	—	—	—
		WO-A-8209		BRACKET, stove stack, rear	1	—	—	—	—
		WO-A-7059		BUSHING, reducing, pipe 3/8 x 1/4	1	—	—	—	—
		WO-A-7137		CAP, fuel tank	1	—	—	—	—
		WO-A-7187		CARTON, perfection stove field kit	1	—	—	—	—
		WO-A-8212		HEATER, Superfex engine (furnished by Ordnance Dept. for packaging by W. O. Co.)	1	—	—	—	—
23-3		WO-6412		BOLT, hex-hd., s-fin., alloy-S., 3/8-16NC-2 x 3/4 (to bracket)	4	—	—	—	—
		WO-5010		WASHER, lock, reg., S., 3/8 in.	4	—	—	—	—
		WO-A-7036		HOSE, stove, long	2	—	—	—	—
		WO-A-7037		HOSE, stove, short	2	—	—	—	—
23-3		WO-A-8213		INSTRUCTION SET, cold starting stove	1	—	—	—	—
		WO-A-8214		LINE, fuel, tank to stove, 5/16 x 16 3/8	1	—	—	—	—
		WO-A-7040		NIPPLE, hose, 1/4 in. pipe	3	—	—	—	—
		WO-A-7581		PLATE, instruction, cold starting stove	1	—	—	—	—
		WO-53013		NUT, stamped, cd-pltd., S., type C, No. 6 (.138)-3NC-2	4	—	—	—	—
		WO-53014		SCREW, rd-hd., machine, cd-pltd., S., No. 6 (.138)-32NC-2 x 5/16 (plate to windshield)	4	—	—	—	—
23-3		WO-A-8219		SPACER, (steel tubing 5/8 O.D. x 1 3/8 lgth. x 3/32 thk.)	1	—	—	—	—
23-3		WO-A-8211		STACK, Perfection stove	1	—	—	—	—
		WO-A-8097		TANK, fuel	1	—	—	—	—
		WO-52132		BOLT, hex-hd., s-fin., alloy-S., 5/16-24NF-2 x 5/8 (fuel tank to cowl)	3	—	—	—	—
		WO-5910		NUT, hex, reg., S., 5/16-24NF-2 (tank to cowl bolt)	3	—	—	—	—
		WO-5437		WASHER, plain, S., 5/16 in. (tank to cowl bolt)	3	—	—	—	—
		WO-51833		WASHER, lock, reg., S., 5/16 in. (tank to cowl bolt)	3	—	—	—	—
23-3		WO-A-7000		TEE, 1/4 in., std.	1	—	—	—	—
23-3		WO-A-8224		THERMOSTAT, 180°	1	—	—	—	—
		WO-A-8207		TUBE, (stove)	2	—	—	—	—
				2310A—DESERT COOLING KIT					
		WO-A-6940		KIT, desert cooling	as req.	—	—	—	—
				(Composed of:					
		WO-A-6942		1 CAP, radiator filler, assembly					

Part No.	Description	Qty
WO-A-6949	1 CARTON, desert cooling kit	
WO-A-6947	4 CLAMP, hose, 5/8 in.	
WO-A-6941	1 CORE, radiator, and SHROUD, assembly	
WO-A-6943	1 DRAWING, installation, desert cooling kit	
WO-A-6946	2 HOSE, radiator surge tank tube	
WO-A-6944	2 STRAP, mounting radiator surge tank	
	4 WO-52161 NUT, hex, S., 1/4-28NF-2	
	4 WO-52468 SCREW, rd-hd, machine, S., 1/4-28NF-2 x 5/8 (strap to guard)	
	4 WO-53058 WASHER, lock, S., 1/4 in.	
WO-A-6933	1 TANK, surge, radiator	
WO-A-6945	1 TUBE, radiator surge tank)	

2310E—DEEP WATER EXHAUST KIT

Part No.	Description	Qty
WO-A-7726	KIT, deep water exhaust pipe	as req.
	(Composed of:)	
WO-A-8204	1 BAG (for standard parts)	
WO-A-8197	1 BRACKET, support, lower, assembly	
	(Includes:	
	1 WO-A-7986 BRACKET, support, lower	
	1 WO-A-658 INSULATOR, muffler support	
	2 WO-638058 PLATE, muffler support insulator	
	1 WO-50929 BOLT, hex-hd., s-fin., alloy-S., 5/16-24NF-2 x 1 (plate and insulator to strap)	
	1 WO-5934 BOLT, hex-hd., S., 5/16-24NF-2 x 1 (plate and insulator to bracket)	
	2 WO-5910 NUT, hex, reg, S., 5/16-24NF-2	
	2 WO-51833 WASHER, lock, reg, S., 5/16 in.	
	1 WO-A-657 STRAP, support, muffler)	
WO-A-8196	1 BRACKET, support, upper, assembly	
	(Includes:	
	1 WO-A-7988 BRACKET, support, upper	
	1 WO-A-7987 CLAMP, pipe	
	1 WO-50922 BOLT, hex-hd., s-fin., alloy-S., 5/16-24NF-2 x 7/8 (clamp to plate and insulator)	
	1 WO-5910 NUT, hex, reg, S., 5/16-24NF-2	
	1 WO-51833 WASHER, lock, reg, S., 5/16 in.)	
WO-A-8203	1 CARTON (accommodates 5 kits)	
WO-A-7816	1 EXTENSION, exhaust pipe	
	3 WO-33050 BOLT, hex-hd., s-fin., alloy-S., 5/16-24NF-2 x 1 (extension to pipe)	
	4 WO-50802 NUT, hex, S., 5/16-24NF-2	
	4 WO-83047 WASHER, lock, S., 5/16 in.	
WO-634814	1 GASKET, exhaust pipe	
WO-A-7967	1 GASKET, exhaust pipe flange	
WO-A-7813	1 INSTRUCTIONS, deep water exhaust pipe kit	
WO-A-7814	1 PIPE, exhaust, assembly	

GROUP 23—GENERAL USE STANDARDIZED PARTS (Cont'd)

Figure Number	Official Stockage Number	Part Number — Willys	Part Number — Ford	ITEM	Quantity Reqd. per Unit Assy.	12 Mos. Field Maintenance	Major Overhaul (5th Ech)	Total First Year Procurement	Estimated Reqmts. per 100 Rebuilds
Col. 1	Col. 2	Col. 3	Col. 4	Col. 5	Col. 6	Col. 7	Col. 8	Col. 9	Col. 10
				2310E—DEEP WATER EXHAUST KIT (Cont'd)					
				(Includes:					
				1 WO-6486 BOLT, hex-hd., S., $\frac{3}{8}$-16NC-2 x 2¼ (flange bolt)					
				1 WO-A-7984 CLAMP					
				1 WO-5922 BOLT, hex-hd., S., $\frac{5}{16}$-24NF-2 x 1⅛ (clamp to pipe)					
				1 WO-5910 NUT, hex, S., reg., $\frac{5}{16}$-24NF-2					
				1 WO-5437 WASHER, plain, SAE. std., $\frac{5}{16}$ in.					
				1 WO-51833 WASHER, lock, reg., S., $\frac{5}{16}$ in.					
				GROUP 24—FIRE EXTINGUISHER					
				2402—PORTABLE SYSTEMS AND PARTS					
23-1		WO-A-693	FM-GPW-17097	BRACKET, mounting, fire extinguisher	3	—	—	—	—
		WO-6989	FM-34054-S	NUT, hex, S., No. 12 (.216)-24NC-2	6	—	—	—	—
		WO-53267	FM-26498-S2	SCREW, rd-hd., S., No. 12 (.216)-24NC-2 (brkt. to body)	6	—	—	—	—
		WO-51969	FM-34804-S2	WASHER, lock, S., No. 12 (.216)	6	—	—	—	—
		WO-A-616	FM-GPW-17100	EXTINGUISHER, fire, w/HOLDER (FYR-D-10-A)	1	—	—	—	—
		WO-6214	FM-34052-S2	NUT, hex, S., No. 8 (.164)-32NC-2	6	—	—	—	—
		WO-6287	FM-26480-S2	SCREW, rd-hd., S., No. 8 (.164)-32NC-2 x ½ (holder to brkt.)	6	—	—	—	—
		WO-52705	FM-34802-S2	WASHER, lock, S., No. 8 (.164)	6	—	—	—	—
				GROUP 26—RADIO					
				2601—RADIO INSTALLATION PARTS					
06-5		WO-A-8114	FM-GPA-18861-B	BOX, radio terminal, assembly	1	—	—	—	—
		WO-51514		BOLT, hex-hd., s-fin., alloy-S., ¼-28NF-2 x ½ (box to body panel) (Willys only)	4	—	—	—	—
		WO-5790	FM-33795-S7	NUT, hex, reg., S., ¼-28NF-2 (Willys only)	4	—	—	—	—
			FM-36813-S7	NUT, hex, reg., S., ¼-20NC-2 (Ford only)	4	—	—	—	—
		WO-A-7715		SCREW, rd-hd., machine, ¼-20NC-2 x $\frac{7}{16}$ (Ford only)	4	—	—	—	—
		WO-5121		WASHER, S., flat, ¼ in. (Willys only)	4	—	—	—	—
				WASHER, S., plain, ¼ in. (Willys only)	4	—	—	—	—
		WO-53024	FM-351274-S7	WASHER, lock, internal, external, cd-pltd., S., ¼ in. (GM-174916)	4	—	—	—	—
		WO-A-7640		CABLE, ground, w/TERMINALS, assembly	1	—	—	—	—
		WO-52543	FM-GPW-14513	BOLT, hex-hd., S., cd-pltd., $\frac{3}{8}$-24NF-2 x ¾ (Cable to frame) (Willys only)	1	—	—	—	—

FM No.	WO No.	Description					Qty
FM-20346-S7	WO-52542	BOLT, hex-hd., S., cd-pltd., 5/16-18NC-2 x 3/4 (cable to frame) (Ford only)	—	—	—	—	1
33797-S7	WO-53135	NUT, hex., S., cd-pltd., 3/8-24NF-2 (Willys only)	—	—	—	—	1
356309-S7	WO-53036	NUT, hex., S., cd-pltd., 5/16-18NC-2 (Ford only)	—	—	—	—	1
	WO-53135	WASHER, lock, S., cd-pltd., internal, external tooth, 5/16 in. (Ford only)	—	—	—	—	3
	WO-A-8113	WASHER, lock, S., cd-pltd., internal, external tooth, 3/8 in. (Willys only)	—	—	—	—	1
FM-GPW-14480-B	WO-A-6809	WASHER, lock, S., pkzd., internal, external tooth, 3/8 in. (Willys only)	—	—	—	—	2
		CABLE, starting switch to radio box, assembly	—	—	—	—	1
FM-GPW-14464		CLIP, closed, S., 1/2 in., bolt hole 13/64 in. (Willys only)	—	—	—	—	3
		(1 used cable to speedometer clip bolt; 2 used cable to frame side member)					
		CLIP, closed, S., 3/8 in. (Ford only)	—	—	—	—	3
		(1 used cable to speedometer clip bolt; 2 used cable to frame side member)					
FM-GPW-14481-B	WO-A-5450	CLIP, closed, S., 1/16 in. (cable to frame side member) (Willys only)	—	—	—	—	1
FM-GPW-14521	WO-6352	NUT, hex., S., No. 10 (.190)-24NC-2 (Willys only)	—	—	—	—	1
FM-GPW-14532	WO-5113	SCREW, rd-hd, machine, S., No. 10 (.190)-24NC-2 x 1/2 (Willys only)	—	—	—	—	1
FM-GPW-18841	WO-52221	WASHER, lock, sherardized, No. 10 (.190) (Willys only)	—	—	—	—	1
	WO-A-8116	CONDUIT, 1/2 in., (thin wall)	—	—	—	—	1
FM-GPW-14536	WO-A-7180	ELBOW, mall-I, z-pltd., 1/2-14 pipe thd., 15/16-18 U.S.S. thd. (AE-9212)	—	—	—	—	2
FM-GPW-14531	WO-A-7182	NUT, hex., thin, S., 1/2 in. pipe thd. (TB-141)	—	—	—	—	2
	WO-A-7600	FILTER, radio terminal box (.5 mfd.) (DT-CHC-201) (SPR-JX-131) (MLL-A-205244) (SOL-EV-128)	6	6	2	% 2	1
	WO-A-7181	GLAND, conduit	—	—	—	—	2
FM-34914-S7	WO-A-7636	GLAND, radio cable conduit nipple	—	—	—	—	1
FM-GPW-14537	WO-7718	LOOM, split, starting switch cable, 7/16 x 1 1/2 in. NIPPLE, radio cable conduit, S., z-pltd., internal thds. 1/2-14 pipe, external thds. 13/16-18 (AE-9821) (Willys only)	—	—	—	—	3
FM-GPW-14534	WO-53054	WASHER, flat, S., cd-pltd., internal tooth 7/8 in. (GM-138566)	—	—	—	—	2
FM-GPW-18876	WO-A-7947	RING, sealing, gland to conduit	—	—	—	—	2
FM-20407-S7	WO-A-7638	SEAL, gland to radio cable conduit	—	—	—	—	1
FM-33798-S7	WC-A-5038	STRAP, bond, radio box negative terminal to floor ground connection	—	—	—	—	1
FM-34706-S7	WO-53131	BOLT, hex-hd., S., cd-pltd., 5/16-24NF-2 x 1 1/8 (strap to floor) (GM-213499)	—	—	—	—	1
FM-356309-S7	WO-A-1532	NUT, hex., S., cd-pltd., 5/16-24NF-2	—	—	—	—	2
	WO-52725	WASHER, S., flat, cd-pltd., 21/64 I.D. x 5/8 O.D. x 1/2 thk. (GM-138485)	—	—	—	—	1
FM-91BS-14463	WO-53025	WASHER, lock, S., internal, external tooth cd-pltd., 5/16 in. (GM-178532) (Willys)	—	100	10	% 10	3
		(Ford)	—	—	—	—	4
FM-GPW-18877-B	WO-A-7466	TERMINAL, S., 21/64 in. hole (Willys)	—	—	—	—	2
FM-GPW-14320		TERMINAL, S., 21/64 in. hole (Ford only)	—	—	—	—	1
FM-11A-14452	WO-A-7645	TERMINAL, S., 13/32 in. hole	—	—	—	—	1
FM-33823-S	WO-371400	TERMINAL, S., 11/32 in., (battery cable)	—	—	—	—	1
FM-34805-S7	WO-33132	NUT, hex., br., 1/4-20NC-2 (negative terminal post (GM-114379)	—	—	—	—	3
	WO-52667	WASHER, lock, light, S., cd-pltd., 1/4 in. (GM-174916)	—	—	—	—	3

2603—RADIO SUPPRESSION SYSTEM PARTS
(Before Willys serial 289,001)

FM No.	WO No.	Description					Qty
FM-20346-S7	WO-A-5919	BOLT, hex-hd., cd-pltd., S., 3/8-24NF-2 x 1 (bond No. 10 to engine plate) (GM 100026)	—	—	—	—	1

GROUP 26—RADIO (Cont'd)

2603—RADIO SUPPRESSION SYSTEM PARTS (Cont'd)
(Before Willys Serial 289,001)

Figure Number (Col 1)	Official Stockage Number (Col 2)	Part Number — Willys (Col 3)	Part Number — Ford (Col 4)	ITEM (Col 5)	Quantity Reqd. per Unit Assy. (Col 6)	12 Mos. Field Maintenance (Col 7)	Major Overhaul (5th Ech) (Col 8)	Total First Year Procurement (Col 9)	Estimated Reqmts. per 100 Rebuilds (Col 10)
		WO-A-1648	FM-20311-S7	BOLT, hex-hd., cd-pltd., S., $5/16$-24NF-3 x $1/2$ (1 used Bond No. 1 to hood; 1 used Bond No. 2 to hood)	2	% 10	—	50	50
		WO-A-1694	FM-GPW-14561	CLIP, closed, S., tinned, $1/4$, $7/32$ bolt hole (1 used Bond No. 4 to speedometer; 1 used Bond No. 5 to gasoline (Willys); 4 used Bond No. 23 (Willys); 1 used blackout driving light cable (Ford); 1 used head light wiring harness (Ford))	6	—	—	—	—
		WO-A-1693	FM-GPW-146211	CLIP, closed, S., tinned, $3/8$, $13/64$ bolt hole (Willys) (1 used Bond No. 4 to heat indicator cable; 1 used Bond No. 6 to choke control (Ford); 1 used Bond No. 6 to throttle control; 1 used Bond No. 6 to oil gauge line)	4	—	—	—	—
		WO-A-5980	FM-GPW-18960-B	FILTER GROUP, w/COVER, assembly (SOL type Ev. 103, DTC 1107DE, SPR type JX-17, MLL 134739) (after Willys serial 137916)	1	% 3	3	9	9
		WO-A-1517		FILTER GROUP, w/COVER, assembly (before Willys serial 137916) (ER type L-7)	1	—	—	—	—
		WO-A-1287	FM-GPW-18936-A	FILTER-UNIT, generator regulator (mounts on regulator) (issue until stock exhausted)	1	% 4	1	—	—
		WO-A-5337	FM-GPW-18935-A	FILTER-UNIT, generator to ground to armature circuit (after Willys serial 112925)	1	% 5	5	12	12
		WO-662276	FM-GPW-13437-A2	GROMMET, $5/8$ in. (harness to filter group cover) (issue until stock exhausted)	1	% 4	8	—	—
		WO-51396	FM-24347-S2	BOLT, hex-hd., s-fin., alloy-S., $5/16$-24NF-2 x $3/4$ (filter to dash)	4	—	—	—	—
		WO-5910	FM-33798-S7	NUT, hex-hd., S., $5/16$-24NF-2	4	—	—	—	—
		WO-51833	FM-34806-S2	WASHER, lock, S., $5/16$ in.	4	—	—	—	—
		WO-A-1550	FM-351025-S7	NUT, hex., S., cd-pltd., $7/16$-20NF-2 (Bond No. 3 to cylinder head stud)	2	—	—	—	—
		WO-A-1532	FM-33798-S7	NUT, hex., S., cd-pltd., $5/16$-24NF-2 (1 used Bond No. 3 and 4 to dash stud; 1 used Bond No. 6 to dash stud; 2 used Bond No. 13 to frame front cross tube stud; 2 used Bond No. 14 to frame front cross tube stud; 2 used Bond No. 15 to frame front cross member stud; 2 used Bond No. 16 to body floor pan stud)	10	% 10	10	100	—

Fed. Stock No.	Mfr. Part No.	Description	Qty	%			
WO-A-1546	FM-34084-S	NUT, hex, S, cd-pltd., 7/16-20NF (issue until stock exhausted) (2 used Bond No. 13 radiator to support brkt. and bond strap to stud, right; 2 used Bond No. 14 radiator to support brkt. and bond strap to stud, left)	4	% 28	52	—	—
WO-A-1701	FM-355835-S	NUT, hex, S, cd-pltd., No. 10 (.190)-24NC-2 (for SCREW WO-A-1700)	8	% 10	—	100	—
WO-A-1700	FM-355130-S7	SCREW, rd-hd., S, cd-pltd., No. 10 (.190)-24NC-2 x 1/2 (5 used clip to bond strap)	8	—	—	—	—
WO-53220		SCREW, rd-hd., S, cd-pltd., No. 10 (.190)-32NF-2 x 3/8, terminal screw (3 used head lamp harness clip to fender splasher)	6	—	—	—	—
WO-53252		SCREW, rd-hd., S, cd-pltd., w/WASHER, No. 10 (.190)-32NF-2 x 3/8 (for use on SOL, SPR and MLL filter group only)	6	—	—	—	—
WO-A-5038	FM-GPW-18876	STRAP, bond No. 1 (cowl to hood, right, bond No. 2, cowl to hood, left) (issue until stock exhausted) (for use with DTC filter group only)	2	% 16	32	—	—
WO-A-5033	FM-GPW-18858	STRAP, bond No. 3 (cylinder head to dash)	1	% 3	—	6	—
WO-A-5035	FM-GPW-18849	STRAP, bond No. 4 (hand brake cable, speedometer cable, heat indicator cable to dash)	1	% 3	16	—	—
WO-A-5039	FM-GPW-18853	STRAP, bond No. 5 (gas line to dash) (issue until stock exhausted)	1	% 8	—	6	—
WO-A-5040	FM-GPW-18850	STRAP, bond No. 6 (choke and throttle controls and oil gauge to dash)	1	% 3	—	—	—
WO-A-5037	FM-GPW-18857	STRAP, bond No. 7 (generator mounting bolt to starter motor brkt.) (issue until stock exhausted)	1	% 8	16	—	—
WO-A-5041	FM-GPW-18846	STRAP, bond No. 8 (generator voltage regulator ground) (issue until stock exhausted)	1	% 12	16	—	—
WO-A-5034	FM-GPW-18873	STRAP, bond No. 9 (ignition coil to cylinder block) (issue until stock exhausted)	1	% 8	16	—	—
WO-A-1098	FM-GPW-14303	STRAP, bond No. 10 (engine front plate to frame right. This is regular engine ground strap) (before serial 288835) (issue until stock exhausted)	1	—	4	—	—
WO-A-5036	FM-GPW-18874	STRAP, bond No. 11 (engine front plate to frame left) (issue until stock exhausted)	1	% 8	16	—	—
WO-A-5027	FM-GPW-18874	STRAP, bond No. 12 (exhaust pipe ground connection to frame) (issue until stock exhausted)	1	% 8	16	—	—
WO-A-5032	FM-GPW-18859	STRAP, bond No. 13 (radiator stud, right, to frame, and STRAP, bond No. 14 (radiator stud, left, to frame) (issue until stock exhausted)	2	% 16	32	—	—
WO-A-5031	FM-GPW-18872	STRAP, bond No. 15 (rear engine support to frame cross member stud) (issue until stock exhausted)	1	% 8	16	—	—
WO-A-5030	FM-GPW-18871	STRAP, bond No. 16 (transfer case to body floor stud)	1	% 3	—	6	—
WO-A-1699	FM-GPW-18852	STRAP, bond No. 21 (hood to brush guard, right, and STRAP, bond No. 22 (hood to brush guard, left) (issue until stock exhausted)	2	% 16	32	—	—
WO-A-6320	FM-GPW-18812-B	SUPPRESSOR, ignition	4	% 32	8	48	—
WO-A-1680		WASHER, S, cd-pltd. (plain) 3/8 in. (used with STRAPS No. 10, 11, 12) (issue until stock exhausted)	5	% 32	68	—	—
WO-5455		WASHER, S, cd-pltd. (plain) 3/8 in. (used with STRAPS No. 10, 11, 12)	3	—	—	—	—
WO-A-1702	FM-34703-S7	WASHER, S, cd-pltd., No. 10 (.190) (harness clip to fender splasher) (issue until stock exhausted)	5	—	—	—	—
WO-A-1680	FM-34707-S7	WASHER, S, cd-pltd., 3/8 in. (issue until stock exhausted)	2	% 32	68	—	—
WO-A-1547	FM-34708-S	WASHER, S. (plain) (used with STRAPS No. 13 and 14) (issue until stock exhausted)	2	% 16	24	—	—
WO-51833	FM-34806-S2	WASHER, lock, S, cd-pltd., 5/16 in.	4	—	—	—	—
WO-52859		WASHER, lock, S, cd-pltd., internal tooth No. 10 (.190) use with terminal screw on SOL, SPR and MLL filter group units only)	6	%	—	—	—

GROUP 26—RADIO (Cont'd)

2603—RADIO SUPPRESSION
(After Willys Serial 288,835)

Figure Number Col. 1	Official Stockage Number Col. 2	Part Number — Willys Col. 3	Part Number — Ford Col. 4	ITEM Col. 5	Quantity Reqd. per Unit Assy. Col. 6	Per 100 Major Items — 12 Mos. Field Maintenance Col. 7	Per 100 Major Items — Major Overhaul (5th Ech) Col. 8	Total First Year Procurement Col. 9	Estimated Reqmts. per 100 Rebuilds Col. 10
		WO-A-1694	FM-GPW-14561	CLIP, closed, S., pltd., ¼ in., bolt hole ⁷⁄₃₂ in. (bond No. 4)	1				
		WO-A-1693	FM-GPW-14621	CLIP, closed, S., pltd., ³⁄₁₆ in., bolt hole ¹³⁄₆₄ in., (bond No. 4)	1				
		WO-A-1701	FM-355836-S7	NUT, hex., S., No. 10 (.190)-24NC-2 (bond No. 4)	2				
		WO-A-1700	FM-355130-S7	SCREW, rd-hd., S., pltd., No. 10 (.190)-24NC-2 x ½	2				
		WO-A-1702	FM-34703-S7	WASHER, plain, S., pltd., No. 10 (.190) (¼ I.D. x ⁹⁄₁₆ O.D.)	2				
		WO-52221	FM-34803-S	WASHER, lock, S., std., No. 10 (.190)	2				
		WO-A-7848	FM-GPW-18938-C	FILTER, 0.1 MFD. (mounted on coil stud)	1				
		WO-A-8884	FM-GPW-18937	FILTER, 0.01 MFD. (mounted to ignition switch)	1				
		WO-6536	FM-34079-S7	NUT, hex., S., No. 10 (.190)-32NF-2	1				
		WO-52250	FM-27068-S7	SCREW, rd-hd., S., No. 10 (.190)-32NF-2 x ⅜	1				
		WO-53023	FM-356256-S7	WASHER, lock, S., cd-pltd., internal, external teeth, No. 10 (.190)	2				
		WO-A-7849	FM-GPA-18841	FILTER, 0.5 MFD. (mounted on starter motor)	1				
		WO-A-8010	FM-GPW-18938-B-1-2	FILTER, 0.25 MFD. (voltage regulator) (mounted to regulator)	1				
		WO-A-8883	FM-GPW-18937-B	FILTER UNIT, 0.01 MFD., generator regulator field (mounted to regulator)	1	% 2	2	6	6
		WO-A-5337	FM-GPW-18935-A-1-2-3	FILTER UNIT, generator ground to armature circuit	1				
		WO-52164	FM-33802-S7	NUT, hex., S., pltd., ⁷⁄₁₆-20NF-2 (radiator to support brkt.)	2				
		WO-52815	FM-356028-S7	NUT, hex., S., stamped, ⁷⁄₁₆-20NF-2 (radiator to support brkt.)	2				
		WO-A-11766	FM-GPW-18876-B	STRAP, bond No. 1 and 2, hood to cowl	2				
		WO-A-5033		STRAP, bond No. 3, cylinder head to dash	1				
		WO-A-5035	FM-GPW-18449	STRAP, bond No. 4, hand brake, heat indicator and speedometer cables to dash	1				
		WO-A-7826	FM-GPW-18840	STRAP, bond No. 7, generator mounting bolt to starter motor bracket	1				
		WO-A-5036	FM-GPW-18874	STRAP, bond No. 11, engine front plate to frame, left	1				
		WO-A-1702	FM-34703-S7	WASHER, plain, S., pltd., No. 10 (.190) (clip to heat indicator and speedometer cable strap)	2				
		WO-52221	FM-34803-S	WASHER, lock, S., std., No. 10 (.190) (clip to heat indicator and speedometer cable strap)	2				
		WO-53303	FM-34943-S7	WASHER, lock, S., cd-pltd., external, ⁷⁄₁₆ in. (radiator to support bracket)	2				
		WO-53027	FM-356314-S7	WASHER, lock, S., cd-pltd., internal, external, ⁷⁄₁₆ in. (radiator to support bracket)	2				

MAINTENANCE TOOLS FOR ORDNANCE PERSONNEL

The tools required for Ordnance organizations for the maintenance of the Truck, ¼-Ton, 4 x 4, Command Reconnaissance, are listed in SNL G-27, Volume 1, "Tools, Maintenance, for Repair of Automotive Vehicles," and are issued as required.

MAINTENANCE PARTS FOR ARMAMENT SIGHTING AND FIRE CONTROL EQUIPMENT

For Maintenance Parts see ADDENDUM for the following SNL's:

GUN, Machine, Cal. .50, Browning, M2, Heavy Barrel, Flexible.............................SNL A-39
                                        or
GUN, Machine, Cal. .30, Browning, M1919A4, Flexible.....................................SNL A-6
                                        or
RIFLE, Automatic, Cal. .30, Browning, M1918A2...SNL A-4
MOUNT, Machine Gun, Cal. .30, M48..SNL A-55, Sec. 32
                                        or
MOUNT, Truck, Pedestal, M31...SNL A-55, Sec. 18

SECTION 3

VEHICULAR SPARE PARTS
AND
EQUIPMENT
LIST

TABLE OF CONTENTS

VEHICULAR SPARE PARTS AND EQUIPMENT LIST

Figure Number Col. 1	Official Stockage Number Col. 2	Ordnance Drawing Number Col. 3	Mfg.'s Number Col. 4	Item Col. 5	Quan. Per Vehicle Col. 6	Note Symbol Col. 7
				VEHICULAR SPARE PARTS		
1-A			FM-GP-17008 / WO-A7686	BAG, spare parts (Fed. St. No. 8-B-11)	1	
			FM-GPW-8620 / WO-A1495	BELT, fan, generator and water pump (DAY-No. V-5) (Fed. St. No. 33-B-76)	1	
1-J			FM-GPW-1720 / WO-A5986	CAPS, tire valve (screwdriver type) (boxed five) (Fed. St. No. 8-C-650)	1	
1-H			FM-B1724 / WO-A5491	CORES, tire valve (boxed five) (Fed. St. No. 8-C-6750)	1	
1-F			FM-B13466 / WO-51804	LAMP, elect., incand., 6-8-v., sgle.-tung-fil., No. 63, 3 cp., (D C cand., bay. base, G 6 bulb, C-2R fil.) (Fed. St. No. 17-L-5215)	1	
1-D			FM-GPW-13494A / WO-A1078	LAMP UNIT, blackout stop, sealed, one opening, 6-8-v., 3 cp., assembly (CB-9234) (GL-5933121) (Fed. St. No. 8-L-421)	1	
1-B		C84908K / C84923K / C91706C	FM-GPW-13491A / WO-A1075	LAMP-UNIT, blackout tail, sealed, four opening, 6-8-v., 3 cp., assembly (CB-9225) (GL-5933078) (Fed. St. No. 8-L-415)	1	
1-C		C84908J	FM-GPW-13485A / WO-A1074	LAMP-UNIT, service tail and stop, sealed, 6-8-v., 21-3 cp., assembly (CB-9218) (GL-5933104) (KD-8039) (Fed. St. No. 8-L-419)	1	
1-L	42-P-5347		FM-GPW-18318 / WO-A7683	PIN, cotter, split, S., (type B) (assorted in small box)	1	
				(Consists of:		
		BFAX1BS		20 PIN, cotter, split, S., (type B), $\frac{1}{16} \times 1\frac{1}{2}$ (Fed. St. No. 42-P-5470)		
		BFAX1CT		25 PIN, cotter, split, S., (type B), $\frac{3}{32} \times 2$ (Fed. St. No. 42-P-5580)		
		BFAX1DT		20 PIN, cotter, split, S., (type B), $\frac{1}{8} \times 2$ (Fed. St. No. 42-P-5690)		
				5 PIN, cotter, split, S., (type B), $\frac{5}{32} \times 2\frac{1}{2}$		
		BFAX2AD		5 PIN, cotter, split, S., (type B), $\frac{3}{16} \times 2\frac{1}{2}$ (Fed. St. No. 42-P-5890))		
1-G			FM-GPW-12405 / WO-A538	PLUG, spark, w/GASKET (Fed. St. No. 17-P-5365)	1	
			FM-GPW-1509	TIRE, 6.00 x 16, mud and snow (Fed. St. No. 8-T-5964)	1	
			FM-GPW-1655	TUBE, 6.00 x 16, heavy duty	1	
			FM-GPW-1015	WHEEL, 16 x 400, assembly	1	
				(Consists of:		
			FM-GPW-1029	8 BOLT, divided rim		
			FM-GPW-1030	8 NUT		
			FM-GPW-1016	1 WHEEL, combat, inner half		
			FM-GPW-1027	1 WHEEL, combat, outer half		
				VEHICULAR ACCESSORIES		
2-AD			FM-GP-17126 / WO-A11765	ADAPTER, lubricating gun	1	M-3 W
2-J	41-B-15		FM-GPW-17005 / WO-A372	BAG, tool	1	M-8 W
				BOOK, Standard Nomenclature List No. G503	Note 1	

Group	Federal stock No.	Part No.	Description	Qty	Ref.
		TM-9-803	BOOK, Technical Manual, 9-803	Note 1	W
2-D	8-C-2358	FM-GPW-18136B; WO-A7687	CHAINS, tire, single pneumatic, 6.00 x 16 (type TS) (4 link spacing)	4	W
		FM-GPW-1102980; WO-A3070	COVER, head light	2	W
2-B		FM-GPW-1103214; WO-3211	COVER, windshield	1	W
		FM-GPW-17036; WO-A289	CRANK, starting, engine	1	
		FM-GPW-1120041; WO-A2998	CURTAIN, side, left	1	
		FM-GPW-1120040; WO-A2999	CURTAIN, side, right	1	
2-P	58-E-202	FM-GPW-17101A; WO-A8429	EXTINGUISHER, fire, carbon tetrachloride, 1 qt. size, w/BRACKET, assembly	1	K-2 W
2-Y	8-G-615	FM-GPW-18325; WO-A6855	GAGE, tire pressure (general service type)	1	J-9 W
			GUIDE, lubr., War Dept. No. 501	Note 1	
2-X	41-G-1344-40	FM-GP-17125; WO-A12058	GUN, lubricating, hand type	1	M-3 W
2-E	41-H-523	FM-GP-17042; WO-A373	HAMMER, machinist, ball peen, 16 oz.	1	J-6 W
2-V		FM-GPW-17011A; WO-306715	HANDLE, spark plug wrench	1	J-4 W
2-H	41-J-66	FM-GPW-17080; WO-A1240	JACK, screw, w/HANDLE, assembly (1½ Ton)	1	M-3 W
			KEY, padlock (code H-700)	2	
			GUIDE, lubr., War Dept. No. 501	Note 1	
2-G	13-O-1530	FM-GP-17038; WO-A379	OILER, S, straight spout, spring bottom (capacity ½ pt.)	1	H-8 W
	42-L-14280	FM-GP-17028; WO-A374	PADLOCK, 1¾ in. (type 1-C)	2	M-3 W
2-C	41-P-1650	FM-GPW-17090; WO-A1339	PLIERS, combination, slip joint, 6 in.	1	H-8 W
2-AB	41-P-2962-700	FM-GPW-17052; WO-A7511	PULLER, wheel hub	1	J-2 W
2-N	8-P-5000	FM-GPW-17020; WO-A375	PUMP, tire, w/air CHUCK	1	W
2-F	41-S-1076	TGBX1A; FM-GPW-17058; WO-A7684	SCREWDRIVER, common heavy duty, integral handle, 6 in.	1	M-3 W
1-K	17-T-805	FM-GPW-1152700; WO-A2909	TAPE, friction, black, grade A, ¾ in. wide (8 oz. roll)	1	J-4 W
1-E		FM-GPW-17060; WO-A7685	TARPAULIN, deck top, assembly	1	H-9 W
	22-W-650	FM-GP-17021; WO-A377	WIRE, iron spool	1	W
2-A	41-W-448	FM-GP-17062; WO-A1100	WRENCH, screw, adjustable, auto type, 11 in., (2¾ in. cap.)	1	J-4 W
2-AA			WRENCH, drain plug	1	J-4 W

Note 1. One per vehicle. Additional issue by the Adjutant General is as indicated in FM-21-6.

VEHICULAR ACCESSORIES—Cont'd.

Figure Number	Official Stockage Number	Ordnance Drawing Number	Mfg.'s Number	Item	Quan. Per Vehicle	Note Symbol
Col. 1	Col. 2	Col. 3	Col. 4	Col. 5	Col. 6	Col. 7
2-Q	41-W-991		FM-GP-17043 WO-A596	WRENCH, engrs., dble-hd., alloy-S., $\frac{3}{8}$ x $\frac{7}{16}$	1	J-4 W
2-R	41-W-1003		FM-GP-17044 WO-A596	WRENCH, engrs., dble-hd., alloy-S., $\frac{1}{2}$ x $\frac{19}{32}$	1	J-4 W
2-S	41-W-1005-5		FM-GP-17045 WO-A598	WRENCH, engrs., dble-hd., alloy-S., $\frac{9}{16}$ x $\frac{11}{16}$	1	J-4 W
2-T	41-W-1008-10		FM-GP-17046 WO-A599	WRENCH, engrs., dble-hd., alloy-S., $\frac{5}{8}$ x $\frac{25}{32}$	1	J-4 W
2-U	41-W-1012-5		FM-GP-17047 WO-A600	WRENCH, engrs., dble-hd., alloy-S., $\frac{3}{4}$ x $\frac{7}{8}$	1	J-4 W
2-K	41-W-2459-500		FM-GPW-17091 WO-A1492	WRENCH, fluted, socket hd., screw (WG-T84J-100)	1	J-4 W
2-L			FM-GPW-17030 WO-A5130	WRENCH, hydraulic brake, bleeder screw ($\frac{17}{64}$)	1	J-4 W
2-W	41-W-3335-50		FM-GPW-17017A WO-637635	WRENCH, spark plug, socket	1	J-4 W
2-AC	41-W-3825-200		FM-GP-17033 WO-A692	WRENCH, wheel bearing nut	1	J-4 W
2-Z	41-W-3837-55		FM-GP-17035 WO-A348	WRENCH, wheel nut, socket (Kelsey-Hayes 24832)	1	W

ARMAMENT

The armament and mounts listed hereunder are installed in the vehicle prior to issue of the vehicle to using troops. Caliber and model of gun, type of mount and basis of issue dependent upon pertinent T/O & E. Issue only spare parts and accessories for that gun and mount which is authorized.

Figure Number	Official Stockage Number	Ordnance Drawing Number	Mfg.'s Number	Item	Quan. Per Vehicle	Note Symbol
		51-84		GUN, machine, cal. .30, Browning, M2, M1919A4, flexible (for list of all parts see SNL A-6) or		A-1 W
		51-70		GUN, machine, cal. .50, Browning, M2, HB, flexible (for list of all parts see SNL A-39) (Note 1) or		A-1 W
		51-102		RIFLE, automatic, cal. .30, Browning, M1918A2 (for list of all parts see SNL A-4)		A-1 W
		E6266		MOUNT, machine gun, cal. .30, M48 (for list of all parts see Section 32, SNL A-55) or		
		D47980		MOUNT, truck, pedestal, M31 (for list of all parts see Section 18, SNL A-55)		A-1 W

Note 1—BMG, cal. .50, M2, HB not used when Mount, machine gun, cal. .30, M48 is authorized.

ARMAMENT SPARE PARTS; FOR:

GUN, MACHINE, CAL. .30, BROWNING, M2, M1919A4, FLEXIBLE

Figure Number	Official Stockage Number	Ordnance Drawing Number	Mfg.'s Number	Item	Quan. Per Vehicle	Note Symbol
	A005-01-00010	C64142		ACCELERATOR	1	A-19
	A006-01-00020	A170491		BAND, lock, front barrel, bearing	1	A-6
	A006-01-00050	D35233		BARREL	1	A-6
	A005-01-00180	B147299		BOLT, assembly	1	A-19
	A005-01-00210	A157375		BUSHING, belt feed lever pivot	1	A-19

Stock No.	Part No.	Description	Qty	Fig.
A006-01-00340	C9801	COVER, assembly	1	A-6
A005-01-00490	C64139	EXTENSION, barrel, assembly	1	A-19
A005-01-00541	C121076	EXTRACTOR, assembly	1	A-19
A005-01-00570	C9182	FRAME, lock, assembly	1	A-19
A005-01-00800	B131317	LEVER, cocking	2	A-19
A005-01-00820	B17503	LEVER, feed, belt	2	A-19
A005-01-00830	B147214	LOCK, breech	1	A-19
A005-01-00840	A196284	NUT, belt feed lever pivot	1	A-19
A005-01-00900	C8461	PAWL, feed, belt	2	A-19
A005-01-00910	B147216	PAWL, holding, belt	1	A-19
A005-01-00930	B131253	PIN, accelerator, assembly	3	A-19
A005-01-00960	B131255	PIN, belt feed pawl, assembly	1	A-19
A005-01-00970	B147217	PIN, belt, holding pawl, split	1	A-19
A005-01-01010	A20567	PIN, cocking lever	2	A-19
A005-01-01090	C9186	PIN, firing, assembly	2	A-19
A005-01-01170	A20503	PIN, trigger	2	A-19
A005-01-01190	A157434	PIVOT, belt feed lever	1	A-19
A005-01-05200	B131251	PLUNGER, barrel, assembly	1	A-19
A005-01-01780	B147222	ROD, driving spring, assembly	1	A-19
A005-88-02096	A196283	SCREW, belt feed lever pivot	1	A-19
A005-01-01950	C64137	SEAR	1	A-6
A005-01-02040	B131262	SLIDE, feed, belt, assembly	1	A-19
A006-01-01240	A135057	SPRING, barrel plunger	1	A-19
A005-01-02100	B147224	SPRING, belt feed pawl	1	A-19
A005-01-02140	B147225	SPRING, belt feed holding pawl	1	A-19
A005-01-02160	B17513	SPRING, cover extractor	1	A-19
A006-01-01275	B212654	SPRING, driving	1	A-6
A005-01-02240	B147230	SPRING, locking barrel	2	A-19
A005-01-02300	B131265	SPRING, sear, assembly	2	A-19
A005-01-02320	B147231	SPRING, trigger pin	2	A-19
A005-01-02470	C8476	TRIGGER	2	A-19
H001-15-19004	BEAX1D	WASHER, lock, internal teeth, reg, S, No. 5 (0.125)	1	H-1

GUN, MACHINE, CAL. .50, BROWNING, M2, HB, FLEXIBLE

Stock No.	Part No.	Description	Qty	Fig.
A037-01-00010	B8914	ARM, belt, feed pawl	1	A-19
A039-01-00008	D28253A	BARREL (spare)	1	A-39
A019-01-00731	A152835	DISK, buffer	1	A-19
A037-01-00500	B8976	EXTENSION, firing pin, assembly	1	A-19
A037-01-00520	B8959	EXTRACTOR, assembly	1	A-39
	B7918A	LEVER, cocking	1	A-19
A037-01-00940	B8961	PAWL, feed, belt, assembly	1	A-19
A037-01-01010	B8962	PIN, belt feed pawl, assembly	1	A-19
H001-08-11027	BFAX1BE	PIN, cotter, split, S., (type B) 1/16 x 3/4)	1	H-1
H001-08-11039	BFAX1CE	PIN, cotter, split, S., (type B) 3/32 x 3/4	1	H-1
	BFAX1DD	PIN, cotter, split, S., (type B) 1/8 x 5/8	1	H-1
A002-01-01100	B17171	PIN, firing	1	A-19
A037-01-01310	A13515	PLUNGER, belt feed lever	1	A-19
A037-01-01470	C64305	ROD, driving spring, w/SPRING, assembly	1	A-19
A037-01-01703	B261110	SLIDE, feed belt, assembly	1	A-19

VEHICULAR SPARE PARTS AND EQUIPMENT LIST (Cont'd.)

Figure Number	Official Stockage Number	Ordnance Drawing Number	Mfg.'s Number	Item	Quan. Per Vehicle	Note Symbol
Col. 1	Col. 2	Col. 3	Col. 4	Col. 5	Col. 6	Col. 7
				ARMAMENT SPARE PARTS; FOR:		
				GUN, MACHINE, CAL. .50, BROWNING, M2, HB, FLEXIBLE (Cont'd)		
	A037-01-01780	A351220		SLIDE, sear	1	A-39
	A037-01-01775	A13516		SPRING, belt feed lever plunger	1	A-19
	A037-01-01775	A9351		SPRING, belt feed pawl	1	A-19
	A036-01-00145	A153146		SPRING, belt holding pawl	1	A-19
	A037-01-01805	B7941		SPRING, cover extractor	1	A-19
	A037-01-01820	B8908		SPRING, locking barrel	1	A-19
	A002-01-02120	A9524		SPRING, sear	1	A-19
	A020-01-00890	A13424		STUD, bolt	1	A-19
				RIFLE, AUTOMATIC, CAL. 30, BROWNING, M1918A2		
	A004-01-00310	B19636		CONNECTOR	1	A-4
	A004-01-00370	C9090		EXTRACTOR	1	A-4
	A004-02-00650	C64076		MAGAZINE, assembly (cap, 20 rds.)	Note 1	A-4
	A004-01-00830	C64074		PIN, retaining, gas cylinder tube, assembly	1	A-4
	A004-01-00850	B19680		PIN, retaining, trigger guard, assembly	1	A-4
	A004-01-00890	A22238		PIN, trigger	2	A-4
				Note 1. Magazines are issued in a quantity sufficient for ammunition authorized.		
	A004-01-01331	B147490		SPRING, change and stop, lever, assembly	1	A-4
	A004-01-01330	B19697		SPRING, change lever stop, assembly	1	A-4
	A004-01-01370	A22202		SPRING, extractor	1	A-4
	A004-01-01410	B147134		SPRING, magazine catch	1	A-4
	A004-01-01450	B147131		SPRING, recoil	1	A-4
	A004-01-01460	B19662		SPRING, sear	1	A-4
				MOUNT, MACHINE GUN, CAL. .30, M48		
				Quantities are per twenty-five mounts.		
	A188031	A188031		BUSHING, automatic, rifle, locking pin	2	A-19
	A236999	A236999		CHAIN, automatic rifle locking pin. assembly	1	A-55
				(Composed of:		
		SDAX6B		2 CHAIN, welded, machs, short twist link, 0.105 in. (5 links)		H-8
		SCAX1D		1 HOOK, "S", steel, 0.105 x 1⁵⁄₁₆ reach		H-2
	A005-02-01290	A142460		1 SWIVEL, chain)		A-19
	A051-02-00050	A188032		CHAIN, bushing, assembly	1	A-19
				(Composed of:		
		SDAX6B		2 CHAIN, welded, machs, short twist link, 0.105 in. (5 links)		H-8
		SCAX1B		2 HOOK, "S", steel, 0.120 x 1⁷⁄₁₆ reach		H-2
	A005-02-01290	A142460		1 SWIVEL, chain)		A-19

Federal stock No.	Piece mark	Description	Qty	Group
	A336250	LOCK, pintle, assembly	1	A-55
		(Composed of:		
	A336242	1 LOCK, pintle		A-55
	A336237	1 NUT, hand, pintle lock		A-55
H001-07-25660	BBSX4AC	1 RETAINER, pintle lock, brazed assembly)		A-55
	A188028	NUT, safety, S. (Elastic Stop type), 3/8-24NF-3 ..... (Note 1)		H-1
		PIN, locking, automatic rifle, assembly	2	A-19
		(Composed of:		
	CCAX1B	2 BALL, chr-alloy-S., grade 2, 3/16 in.		H-2
	A188027	1 BODY, automatic rifle locking pin		A-55
A055-02-09302	A230490	1 HANDLE, locking pin ((Note 2)		A-55
M005-02-56170	FAAX1B	1 SPRING, compression, S., 0.024 diam. stock, 0.185 O.D., 8 coils)		M-5
	A188030	PIN, locking, machine gun, assembly .... (Note 1)	2	A-19
		(Composed of:		
	CCAX1B	2 BALL, chr-alloy-S., grade 2, 3/16 in.		H-2
	A188029	1 BODY, machine gun locking pin		A-55
A055-02-09302	A230490	1 HANDLE, locking pin (Note 2)		A-55
M005-02-56170	FAAX1B	1 SPRING, compression, S., 0.024 diam. stock, 0.185 O.D., 8 coils)		M-5
		Note 1—Pin assemblies of early manufacture had body of the same piece mark with integral handle and did not require HANDLE A230490.		
		Note 2—Identical with machine gun locking pin HANDLE A230490, and so carried and reviewed in SNL A-55, Sec. 19.		
H001-15-30009	BCBX1CA	SCREW, cap, hex-hd., S., 3/8-24NF-2 x 3/4.	2	H-1
	BEBX1K	WASHER, plain, S., SAE std., 3/8 in.	2	H-1

MOUNT, TRUCK, PEDESTAL, M31

Quantities are per twenty-five mounts.

Federal stock No.	Piece mark	Description	Qty	Group
	A188028	PIN, locking, automatic rifle, assembly .... (Note 1)	1	A-55
		(Composed of:		
	CCAX1B	2 BALL, chr-alloy-S., grade 2, 3/16 in.		H-2
	A188027	1 BODY, automatic rifle locking pin		A-55
A055-02-09302	A230490	1 HANDLE, locking pin (Note 2)		A-55
M005-02-56170	FAAX1B	1 SPRING, compression, S. 0.024 diam., stock. 0.185 O.D., 8 coils)		M-5
	A188030	PIN, locking, machine gun, assembly .... (Note 1)	1	A-55
		(Composed of:		
	CCAX1B	2 BALL, chr-alloy-S., grade 2, 3/16 in.		H-2
	A188029	1 BODY, machine gun locking pin		A-55
A055-02-09302	A230490	1 HANDLE, locking pin (Note 2)		A-55
M005-02-56170	FAAX1B	1 SPRING, compression, S. 0.024 diam. stock 0.185 O.D., 8 coils)		M-5
	A176083	PIN, locking, traveling lock, assembly	1	A-55
		(Composed of:		
	CCAX1B	2 BALL, chr-alloy-S., grade 2, 3/16 in.		H-2
	A176082	1 BODY, traveling lock, locking pin		A-55
A055-02-09302	A230490	1 HANDLE, locking pin (Note 2)		A-55
M005-02-56170	FAAX1B	1 SPRING, compression, S., 0.024 diam. stock, 0.185 O.D., 8 coils)		M-5
		Note 1—Pin assemblies of early manufacture had body of the same piece mark with integral handle and did not require HANDLE A230490.		
		Note 2—Identical with machine gun locking pin HANDLE A230490, and so carried and reviewed in SNL A55, Sec. 19.		

VEHICULAR SPARE PARTS AND EQUIPMENT LIST (Cont'd.)

ARMAMENT ACCESSORIES; FOR:

GUN, MACHINE, CAL. .30, BROWNING, M1919A4, FLEXIBLE

Figure Number	Official Stockage Number	Ordnance Drawing Number	Mfg.'s Number	Item	Quan. Per Vehicle	Note Symbol
Col. 1	Col. 2	Col. 3	Col. 4	Col. 5	Col. 6	Col. 7
	A005-06-00020	C3951		BELT, ammunition, cal. .30, M1917 (250 rd.)	5	T-5 W
		FM 23-50		BOOK, Field Manual 23-50, Browning Machine Gun Caliber .30, HB, M1919A4 (mounted in combat vehicles)	Note 1	
				Note 1—One per packing container. Additional issue by the Adjutant General is as indicated in FM 21-6.		
	A005-06-00081	D44070		BOX, ammunition, cal. .30, M1 (empty)	5	T-5
	M-003-01-01930	B108828		BRUSH, chamber, cleaning, M6	1	M-3
	M003-01-02020	C4035		BRUSH, cleaning, cal. .30, M2	3	M-3
	M003-01-02965	B147310		CAN, tubular (¾ in. diam. x 2½ in. w/screw top)	1	M-3
	M003-01-03160	C6573		CASE, cleaning rod, cal. .30, M1	1	M-3 W
	A019-01-00300	C59656		CASE, spare bolt, M2	3	A-19 W
	M008-01-00570	D28243		CHEST, steel, M5 (STAX1AB modified) (Note 1)	1	M-8 W
	A019-01-00640	D30674		COVER, spare barrel, M9	1	A-19 W
	M003-01-04590	C59696		ENVELOPE, spare parts, M1	1	A-19
	A019-01-00820	C3854		EXTRACTOR, ruptured cartridge, Mk. IV	1	A-19 W
	A019-01-01240	D1262		MACHINE, browning belt filling M1918	Note 2	A-19 W
	M003-01-04570	C59737		OILER, oval, 3 oz., w/cap and chain (Note 3)	1	M-3
	M003-01-10050	C59736		OILER, rectangular, 12 oz., w/cap and chain (Note 3)	1	M-3
	M003-01-12020	B147001		REFLECTOR, barrel, cal. .30	1	M-3 W
	M003-01-12690	D8237		ROD, cleaning, jointed, cal. .30, M1	1	M-3 W
	M008-01-00990	D7349		ROLL, spare parts, M13	1	M-8 W
				Note 1—Issue 49-I-82 CHEST, accessory and spare parts, M1917, until CHEST, steel, M5 is obtainable.		
				Note 2—One per four guns or major fraction thereof.		
				Note 3—To be issued until supply is exhausted.		
	M008-01-01110	D7389		ROLL, tool, M12	1	M-8 W
	J004-01-04360	TGAX1A		SCREWDRIVER, comm., normal duty, 3 in. blade	1	J-4 W
	A006-03-00310	C68334		WRENCH, combination, M6	1	A-19 W
	A006-03-00320	B147277		WRENCH, socket, front barrel bearing plug	1	A-6 W

GUN, MACHINE, CAL. .50, BROWNING, M2, HB, FLEXIBLE

Figure Number	Official Stockage Number	Ordnance Drawing Number	Mfg.'s Number	Item	Quan. Per Vehicle	Note Symbol
	38-B-992-27	FM.-23-65		BOOK, Field Manual 23-65, Browning Machine Gun, Caliber .50, HB, M2 (mounted on combat vehicles)	Note 1	A-19
		C4037		BRUSH, cleaning, cal. .50, M4	4	A-19 W
	A019-01-00540	C64274		CASE, cleaning rod, cal. .50, M15	1	A-19 W
		C61331		CHUTE, metallic belt link, M1	1	A-19 W
	A039-88-00260	D33912		COVER, spare barrel, M13 (for 45 in. barrel)	1	A-19
		C59696		ENVELOPE, spare parts M1	2	A-19 W
		C64392		EXTRACTOR, ruptured cartridge, cal. .50, M5	1	A-19 W
		D35441		ROD, cleaning, jointed, caf. .50, M7	1	A-19 W
	A019-01-02710	D28242		WRENCH, combination, cal. .50, M2	1	A-19 W

RIFLE, AUTOMATIC, CAL. .30, BROWNING, M1918A2

Reference No.	Part No.	Description	Qty	Symbol
A004-02-00051	FM 23-15	BOOK, Field Manual 23-15, Browning, Automatic Rifle, Cal. .30, M1918A2	Note 1	
38-B-992	D28362	BRUSH, chamber, cleaning M1	1	A-4 W
B003-02-00030	C4035	BRUSH, cleaning, cal. .30, M2	2	M-3
	C64173	BRUSH and THONG, cal. .30, complete.	1	M-3
M003-01-15411		(Consisting of:		
	C64174	1—BRUSH, thong, cal. .30		
	C64175	1—THONG)		

Note 1—One per packing container. Additional issue by the Adjutant General is as indicated in FM 21-6.

Reference No.	Part No.	Description	Qty	Symbol
24-C-534	C6573	CASE, cleaning rod, cal. .30, M1 (for carrying Rod D8237)	1	M-3 W
A004-02-00151	C64177	COVER, front sight	1	A-4 W
M003-01-04570	15-18-102B	ENVELOPE, fabric, one-button, 3 x 3⅛ in.	1	M-3
M003-01-04580	15-18-102A	ENVELOPE, fabric, two-button, 3 x 4⅞ in.	1	M-3
41-E-555	C7912	EXTRACTOR, ruptured cartridge, Mk. II	1	M-3 W
A004-02-00190	C7913	FILLER, magazine	2	A-4 W
13-0-1280	C59737	OILER, oval, 3 oz., w/cap and chain	1	M-3
41-R-2330-975	B147001	REFLECTOR, barrel, cal. .30 ....(Note 1)	1	M-3 W
41-R-2564	D8398	ROD, cleaning, cal. .30, M2A1	1	M-3
41-A-2567	D8237	ROD, cleaning, cal. .30, jointed, M1	1	M-3 W
B003-02-00151	D44058	SLING, gun, M1 (web) or		M-3 W
A004-02-00250	20-18-25	SLING, gun, M1907, modified	1	M3- W
41-T-3081-110	C64144	TOOL, cleaning, gas cylinder	1	A-4 W
41-T-3085-250	C64145	TOOL, combination	1	A-4 W

MOUNT, MACHINE GUN, CAL. .30, AMMUNITION BOX, M1

Reference No.	Part No.	Description	Qty	Symbol
	E6288	ADAPTER, machine gun, cal. .30, ammunition box, M1	1	A-55 W

Note 1—To be issued until supply is exhausted.

MOUNT, TRUCK, PEDESTAL, M31

Reference No.	Part No.	Description	Qty	Symbol
	C38571	TRAY, ammunition (for cal. .30 machine gun)	1	W A-55
	D38607	TRAY, ammunition (for cal. .50 machine gun)	1	W A-55

MAJOR ITEM

Note	Symbol	Class	Division
	W		TRUCK, ¼-Ton, 4 x 4, Command Reconnaissance

NOTES

Items listed in Section 3 of this SNL will be stored under SNL G-503 unless the note symbol indicates otherwise.

The major item indicated by (φ) is stored, issued, and reviewed under "Major Items," SNL G-1.

Note		
φ		
W	(a)	Items indicated by (W) are non-expendable parts and the issue of such items must be in accordance with par. 3b (2), AR 35-6540.
	(b)	Items not marked by (W) are expendable in accordance with AR 35-6620.

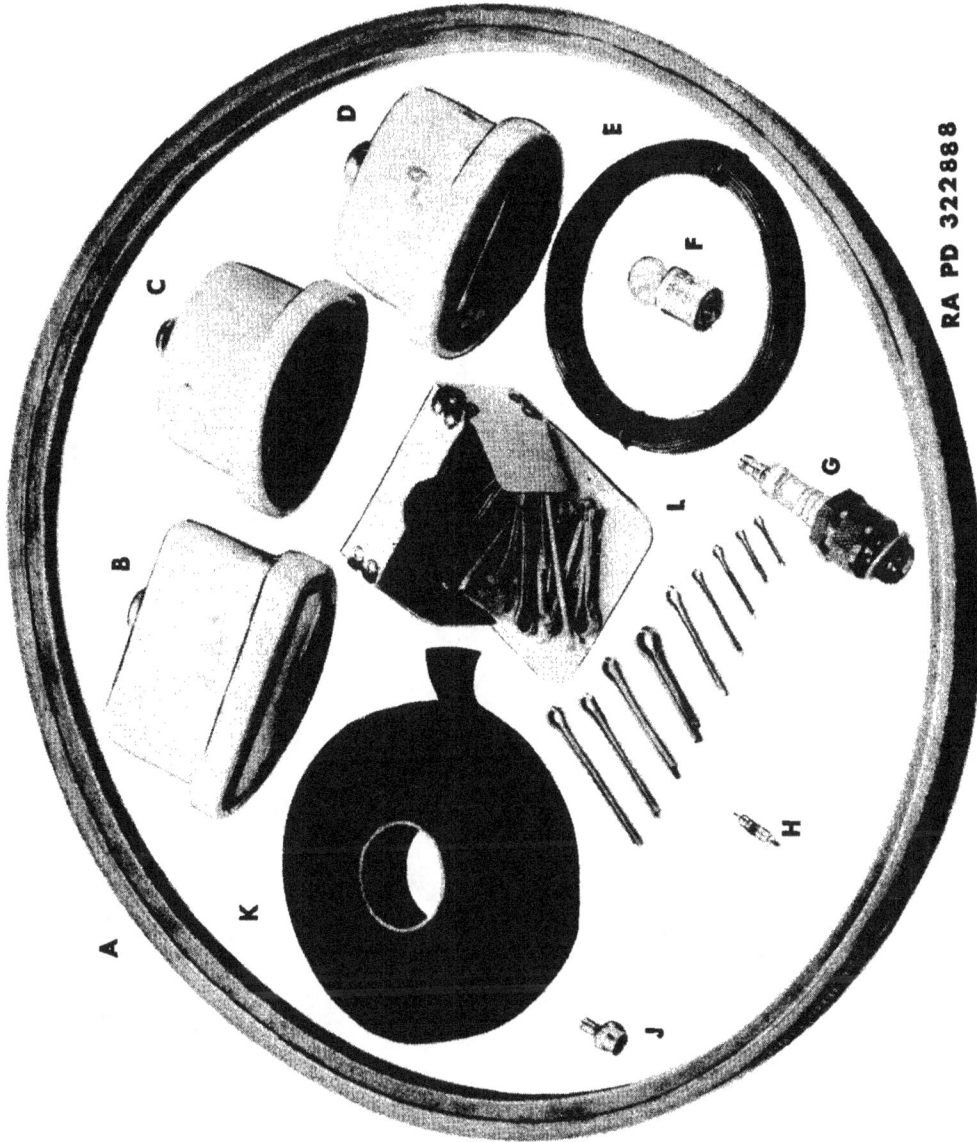

FIGURE 1—VEHICULAR SPARE PARTS

RA PD 322888

RA PD 322888

A	BELT	33-B-76
B	LAMP-UNIT	8-L-415
C	LAMP-UNIT	8-L-419
D	LAMP-UNIT	8-L-421
E	WIRE	22-W-650
F	LAMP	17-L-5215
G	PLUG	17-P-5365
H	CORE	8-C-6750
J	CAP	8-C-650
K	TAPE	17-T-805
L	PIN. COTTER, assorted	42-P-5347

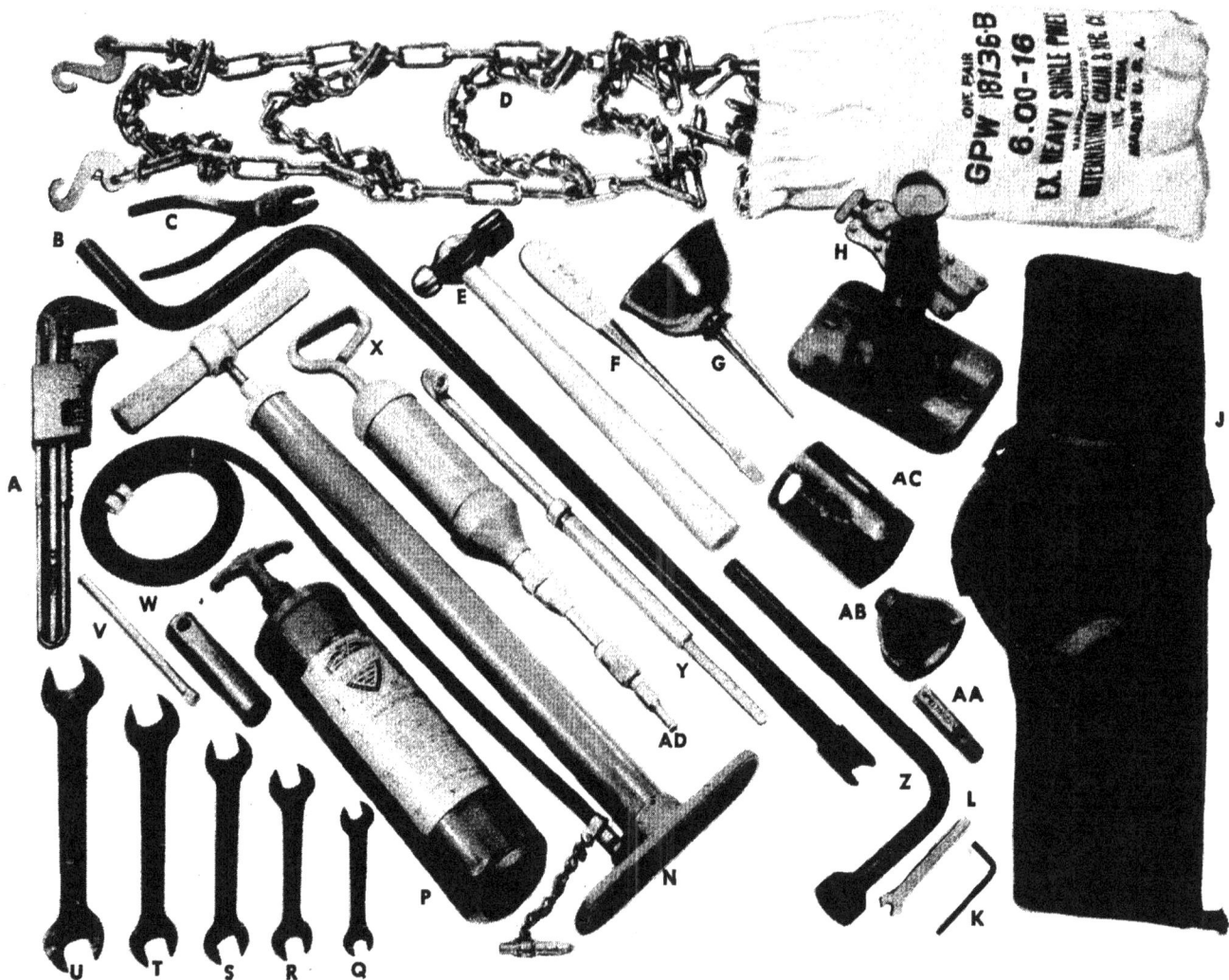

A	WRENCH	41-W-448	**K**	WRENCH	41-W-2459-500	**V**	HANDLE	FM-GPW-17011A
B	CRANK	FM-GPW-17036	**L**	WRENCH	FM-GPW-17030	**W**	WRENCH	41-W-3335-50
C	PLIERS	41-P-1650	**N**	PUMP	8-P-5000	**X**	GUN	41-G-1344
D	CHAINS	8-C-2538	**P**	EXTINGUISHER	58-E-202	**Y**	GAGE	8-G-615
E	HAMMER	41-H-523	**Q**	WRENCH	FM-GP-17043	**Z**	WRENCH	FM-GP-17035
F	SCREWDRIVER	41-S-1076	**R**	WRENCH	FM-GP-17044	**AA**	WRENCH	FM-GP-17062
G	OILER	13-O-1530	**S**	WRENCH	FM-GP-17045	**AB**	PULLER	41-P-2962-7
H	JACK	41-J-66	**T**	WRENCH	FM-GP-17046	**AC**	WRENCH	41-W-3825-200
J	BAG	41-B-15	**U**	WRENCH	FM-GP-17047	**AD**	ADAPTER	FM-GP-17126

RA PD 322889

FIGURE 2—VEHICULAR ACCESSORIES

RA PD 322889
SNL G-503

SECTION 4

ORGANIZATIONAL

SPARE PARTS AND

EQUIPMENT LIST

TABLE OF CONTENTS

Official Stockage Number	Ordnance Drawing Number	Mfr.'s Number Willys	Mfr.'s Number Ford	Item	Parts Allowances For: — Company			Parts Allowances For: — Regiment or Separate Battalion				
					Set No. 1 1-9 Vehicles	Set No. 2 10-25 Vehicles	Set No. 3 26 Up Vehicles	Set No. 4 1-9 Vehicles	Set No. 5 10-35 Vehicles	Set No. 6 36-75 Vehicles	Set No. 7 76-150 Vehicles	Set No. 8 151 Up Vehicles
Col. 1	Col. 2	Col. 3	Col. 4	Col. 5	Col. 6	Col. 7	Col. 8	Col. 9	Col. 10	Col. 11	Col. 12	Col. 13
				GROUP 01—ENGINE								
		WO-A-1236	FM-GPW-18662-A	ELEMENT, oil filter (PU-26637)	2	3	5	2	4	8	15	30
		WO-630299	FM-GPW-6649	GASKET, crankcase ventilator to valve spring cover	0	0	0	1	1	1	2	4
		WO-A-8558	FM-GPW-6051-B	GASKET, cylinder head	1	1	2	1	2	3	6	12
		WO-A-1235	FM-GPW-18688-A	GASKET, oil filter cover (PU-25802)	1	1	1	1	1	2	4	8
		WO-314338	FM-GPW-6734	GASKET, oil pan drain plug	1	2	3	1	3	5	10	20
		WO-A-1538	FM-GPW-18512	GASKET SET, oil pan (Consists of: 1 WO-630398 FM-GPW-6627 GASKET, oil float support; 1 WO-639980 FM-GPW-6710 GASKET, oil pan; 1 WO-314338 FM-GPW-6734 GASKET, oil pan drain plug)	1	1	1	1	1	2	4	8
		WO-A-7835	FM-GPW-18323	GASKET SET, manifold (Consists of: 1 WO-638640 FM-GPW-9448 GASKET, intake and exhaust manifold; 1 WO-634811 FM-GPW-9435 GASKET, intake to exhaust manifold; 1 WO-634814 FM-GPW-9450 GASKET, exhaust pipe flange)	0	0	0	1	1	1	2	4
		WO-51875	FM-GPW-6555	GASKET, valve cover bolt	0	1	1	1	1	2	3	6
		WO-630305	FM-GPW-6521	GASKET, valve spring cover	0	0	0	1	1	1	2	4
		WO-639650	FM-GPW-8255	GASKET, water outlet elbow	0	0	0	0	1	1	1	2
		WO-639979	FM-GPW-6727	PLUG, drain, oil pan (⅞ in.)	0	0	0	1	1	1	2	4
		WO-A-1197	FM-GPW-18667	TUBE, oil filter, inlet, assembly	1	1	1	1	1	2	4	8
		WO-A-1198	FM-GPW-18666	TUBE, oil filter, outlet, assembly	1	1	1	1	1	2	4	8
				GROUP 02—CLUTCH								
		WO-630593	FM-GPW-7523	SPRING, retracting, clutch and brake pedal	0	0	0	0	1	1	1	2

GROUP 03—FUEL SYSTEM

Qty 1	Qty 2	Qty 3	Qty 4	Qty 5	Qty 6	Qty 7	Qty 8	Description	Part No.	Ord. No.
2	1	1	1	0	0	0	0	CAP, fuel tank, w/safety CHAIN, assembly	FM-GPW-9030-B	WO-A-6333
2	1	1	1	0	0	0	0	CARBURETOR, assembly (CAR-539S)	FM-GPW-9510	WO-A-1223
4	2	2	1	1	0	0	0	CONNECTION, flexible (in fuel line) (FO-HA-8031)	FM-GPW-9288	WO-A-1325
2	1	1	1	0	0	0	0	GASKET, insulator, carburetor, and intake manifold DIFFUSER, assembly	FM-GPW-9445	WO-A-6357
6	3	2	1	1	1	1	0	GASKET, fuel pump bowl (AC-1523096)	FM-GPW-9364	WO-115656
4	2	2	1	1	0	0	0	GASKET, fuel pump to block (AC-838263)	FM-GPW-9417	WO-638737
8	4	2	1	1	1	1	1	GASKET SET, fuel strainer	FM-GPW-18337	WO-A-6883
								(Consists of:		
								1 WO-A-1257 FM-GPW-9184 GASKET, cover cap screw		
								1 WO-A-1259 FM-GPW-9160 GASKET, strainer bowl		
								1 WO-A-1260 FM-GPW-9186 GASKET, strainer unit)		
2	1	1	1	0	0	0	0	GAUGE fuel tank unit, assembly (AL-9979-A)	FM-GPW-9276	WO-A-1292
2	1	1	1	0	0	0	0	HOSE, flexible, air cleaner tube to cleaner	FM-GPW-9652	WO-A-1311
2	1	1	1	0	0	0	0	PUMP, fuel, assembly (AC-1538312)	FM-GPW-9350-B	WO-A-8323
2	1	1	1	0	0	0	0	SPRING, retracting, accelerator	FM-GPW-9799	WO-633011

GROUP 04—EXHAUST

Qty 1	Qty 2	Qty 3	Qty 4	Qty 5	Qty 6	Qty 7	Qty 8	Description	Part No.	Ord. No.
6	3	2	1	1	1	1	0	MUFFLER, assembly (rd. type)	FM-GPW-5230-A	WO-A-1146
8	4	2	1	1	1	1	1	PIPE, exhaust, assembly	FM-GPW-5246	WO-A-1296

GROUP 05—COOLING

Qty 1	Qty 2	Qty 3	Qty 4	Qty 5	Qty 6	Qty 7	Qty 8	Description	Part No.	Ord. No.
8	4	2	1	1	1	1	1	BELT, drive, fan and generator	FM-GPW-8620	WO-A-1495
2	1	1	1	0	0	0	0	CAP, radiator filler, assembly (AC-846709) (STN-6455-A)	FM-GPW-8100	WO-A-1215
16	8	4	2	1	3	2	2	CLAMP, hose, radiator, 1¹³⁄₁₆ in.	FM-60-8287	WO-52226
4	2	1	1	1	0	0	0	GASKET, radiator cap (AC-846732)	FM-GPW-8578	WO-A-1216
4	2	1	1	1	0	0	0	GASKET, water pump to cylinder block	FM-GPW-8543	WO-637053
20	10	5	3	3	3	2	1	HOSE, radiator, outlet	FM-GPW-8284	WO-A-592
2	1	1	1	0	0	0	0	PUMP, water, assembly	FM-GPW-8501	WO-639992
2	1	1	1	0	0	0	0	THERMOSTAT, assembly (HR-3108628)	FM-GPW-8575	WO-637646

GROUP 06—ELECTRICAL

Qty 1	Qty 2	Qty 3	Qty 4	Qty 5	Qty 6	Qty 7	Qty 8	Description	Part No.	Ord. No.
8	4	2	1	1	1	1	1	BATTERY, wet charged, 6 volt, 116 ampere assembly	FM-11AS-10655	WO-A-1238
2	1	1	1	0	0	0	0	CAP, distributor, assembly (AL-IG-1324)	FM-GPW-12106	WO-A-1655
2	1	1	1	0	0	0	0	COIL, ignition, w/BRACKET and WASHER, assembly (AL-IG-4070-L)	FM-GPW-12000-B	WO-A-7792
4	2	1	1	1	0	0	0	CONDENSER, distributor, assembly (AL-IGW-3139)	FM-GPW-12300	WO-A-1631
2	1	1	1	0	0	0	0	DISTRIBUTOR, assembly (AL-IGC-4705)	FM-GPW-12100	WO-A-1244
4	2	1	1	1	0	0	0	GASKET, spark plug (VC-2066C-C1)	FM-01A-12410	WO-637863
2	1	1	1	0	0	0	0	GENERATOR, assembly (AL-GEG-5002-D)	FM-GPW-10000A	WO-A-5992

ORGANIZATIONAL SPARE PARTS (Cont'd)

Official Stockage Number	Ordnance Drawing Number	Mfr's Number — Willys	Mfr's Number — Ford	Item	Parts Allowances For: Company — Set No. 1 — 1-9 Vehicles	Company — Set No. 2 — 10-25 Vehicles	Company — Set No. 3 — 26 Up Vehicles	Regiment or Separate Battalion — Set No. 4 — 1-9 Vehicles	Set No. 5 — 10-35 Vehicles	Set No. 6 — 36-75 Vehicles	Set No. 7 — 76-150 Vehicles	Set No. 8 — 151 Up Vehicles
Col. 1	Col. 2	Col. 3	Col. 4	Col. 5	Col. 6	Col. 7	Col. 8	Col. 9	Col. 10	Col. 11	Col. 12	Col. 13
				GROUP 06—ELECTRICAL (Cont'd)								
	C84908K	WO-A-6145	FM-GPW-13152	LAMP-UNIT, blackout driving, headlight sealed, 6-V8, assembly (CB-1267).	0	0	0	0	1	1	1	2
	C84934K (C91706C)	WO-A-1075	FM-GPW-13491-A	LAMP-UNIT, blackout tail, sealed, 6-8V., assembly (lower left and right) (CB-9225) (GL-9533078).	0	0	0	0	1	1	1	2
		WO-A-1033	FM-GPW-13007	LAMP-UNIT, head light, sealed, 6-V8, (Seelight unit) (CB-8494).	0	0	0	1	1	1	2	4
	C84908J	WO-A-1074	FM-GPW-13494	LAMP-UNIT, service tail and stop, sealed, 6-8V., 21-3 cp., assembly (upper, left) (CB-9218) (GL-5933104).	0	0	0	0	1	1	1	2
	C84934J	WO-A-1078	FM-GPW-13485-A	LAMP-UNIT, blackout stop, sealed, 6-8V., assembly (upper, right) (CB-9234) (GL-593121).	0	0	0	0	1	1	1	2
	DLAX1F	WO-51804	FM-B-13466	LAMP, elec., incand., min., 6-8 volt, sgle-tung-fil., No. 63, 3 cp. (blackout headlight) (MZ-63).	0	1	1	1	1	2	3	6
		WO-A-1245	FM-GPW-11001-A	MOTOR, cranking, assembly (AL-MZ-4113).	0	0	0	0	1	1	1	2
		WO-A-538	FM-GPW-12405	PLUG, spark, w/GASKET, assembly (CP-type-AN-7).	2	4	8	4	6	12	24	48
		WO-A-1687	FM-GPW-18354	CONTACT SET, contact, distributor (AL-IGP-3028FS).	1	1	2	1	2	3	6	12
		WO-A-1409	FM-GPW-10505	REGULATOR, generator, assembly.	0	0	0	0	1	1	1	2
		WO-A-1658	FM-GPW-12200	ROTOR, distributor (AL-IG-1657-R).	0	0	0	1	1	1	2	4
		WO-A-7225	FM-9N-11450-A	SWITCH, starting, assembly (AL-SW-4015).	0	0	0	1	1	1	2	4
		WO-638979	FM-GPW-13532	SWITCH, foot, headlight, assembly (CL-9634).	0	0	0	0	1	1	1	2
		WO-A-1271	FM-11A-13480	SWITCH, stop light, assembly.	0	0	0	0	1	1	1	2
				GROUP 11—REAR AXLE								
		WO-A-904	FM-GP-4032	GASKET, axle shaft (SP-17146).	1	1	1	1	1	2	4	8
				GROUP 12—BRAKE								
		WO-A-1484	FM-GPW-2061	CYLINDER, front wheel brake, assembly (LO-FD-7379).	1	1	1	1	1	2	4	8
		WO-A-556	FM-GP-2140	CYLINDER, master brake, assembly (LO-FE-1444).	0	0	0	0	1	1	1	2
		WO-A-6110	FM-GPW-2261	CYLINDER, rear wheel, ¾ in. diam., assembly (LO-FD-7568-A).	0	0	0	0	1	1	1	2
		WO-A-1460	FM-GPW-2079	HOSE, brake, front axle, assembly (LO-FC-8553).	0	0	0	1	1	1	2	4

Description	FM No.	WO No.	1	2	3	4	5	6	7	8
HOSE, brake, front, assembly (11 in.) (LO-FC-8502)	FM-GPW-2078	WO-A-1373	2	1	1	1	0	0	0	0
HOSE, brake, rear, assembly (15 in.) (LO-FC-5784)	FM-GP-2078	WO-637424	2	1	1	1	0	0	0	0
SHOE, brake, forward, w/LINING, assembly (BX-1141-S-1)	FM-GP-2018	WO-116549	2	1	1	1	0	0	0	0
SHOE, brake, reverse, w/LINING, assembly (BX-1141-S-2)	FM-GP-2019	WO-116550	2	1	1	1	0	0	0	0
SPRING, return, brake shoe (BX-41545)	FM-GP-2035	WO-637905	4	2	1	1	1	0	0	0
GROUP 13—WHEELS, HUBS & DRUM										
CONE and ROLLERS, tapered wheel bearing, bore, 1.625 x 1 1/16 (front and rear wheels, inner and outer) (TIM-18590) (BOW-BT-18590)	FM-GP-1201	WO-52942	2	1	1	1	0	0	0	0
CUP, tapered roller bearing, 2.875 x 1/2 (front and rear wheels, inner and outer) (TIM-18520) (BOW-BT-18520)	FM-GP-1202	WO-52943	2	1	1	1	0	0	0	0
OIL SEAL, wheel hub (SP-17004)	FM-GP-1177	WO-A-864	30	15	8	4	2	5	3	1
GROUP 16—SPRING & SHOCK ABSORBER										
ABSORBER, shock, front, assembly ....(Willys only)		WO-A-6902	2	1	1	1	0	0	0	0
SPRING, front, left, assembly	FM-GPW-5311	WO-A-612	2	1	1	1	0	0	0	0
SPRING, front, right, assembly	FM-GPW-5310	WO-A-613	2	1	1	1	0	0	0	0
GROUP 22—MISCELLANEOUS BODY										
ARM, windshield wiper, w/BLADE, assembly	FM-GP-17535	WO-A-2512	8	4	2	1	1	1	1	1
SHAFT, flexible, speedometer drive (53 5/32 in.) (AL-SPS-1105X)	FM-GPW-17262	WO-A-1344	4	2	1	1	1	0	0	0
GROUP 23—GENERAL USE & STANDARDIZED										
CAP SET, tire valve	FM-CPW-18322	WO-A-7681	8	4	2	1	1	1	1	1
CORE SET, tire valve	FM-GPW-18320	WO-A-7682	8	4	2	1	1	1	1	1
FITTING, grease, 90°		WO-638500	4	2	1	1	1	0	0	0
FITTING, grease, 45°		WO-638224	4	2	1	1	1	0	0	0
FITTING, grease	FM-350006-S8	WO-640038	4	2	1	1	1	0	0	0
FITTING, grease, straight, 1/8 in. (AD-1720)	FM-353027-AS7	WO-392909	6	3	2	1	1	1	1	0
FITTING, grease, straight, 1/4-28NF (AD-1641)	FM-353043-AS7	WO-638792	4	2	1	1	1	0	0	0

NOTE: Where the same part is common to two or more vehicles assigned to the same unit, the part should be issued on basis of the total number of vehicles to which the part applies.

ORGANIZATIONAL SPARE PARTS AND EQUIPMENT LIST

ORGANIZATIONAL ACCESSORIES

The issue of the following organizational tool kits and sets, the component, items of which are listed here-under in columnar form, are as follows:

TOOL KIT, MECHANICS: (Col. 6). 1 per Motor Mechanic (014) in Truck, ¼-Ton, 4 x 4, Command Reconnaissance , equipped unit.

TOOL-SET, COMPANY, (Col. 7). 1 per Truck, ¼-Ton, 4 x 4, Command Reconnaissance , company or troop equipped unit.

TOOL-SET, BATTALION CREW: (Col. 8). As authorized in applicable T/BA, T/A or T/O&E.

TOOL-SET, REGIMENTAL MAINTENANCE PLATOON: (Col. 9). As authorized in applicable T/BA, T/A, or T/O&E.

Figure Number	Official Stockage Number	Ordnance Drawing Number	Mfr.'s Number	Item	Tool Kit, Mechanics	Tool-Set, Company	Tool-Set, Battalion Crew	Tool Set, Regimental Maintenance Platoon	Note Symbol
Col. 1	Col. 2	Col. 3	Col. 4	Col. 5	Col. 6	Col. 7	Col. 8	Col. 9	Col. 10
	41-C-2554-400	B7076535	MAS-1148	COMPRESSOR, shock absorber rubber grommet				1	W
	41-W-3575		KM-J-4056	WRENCH, tappet, double end 11/32 and 17/32 in.				2	W

FIRE CONTROL EQUIPMENT

Issue Firing Tables pertinent to Gun authorized.

Figure Number	Official Stockage Number	Ordnance Drawing Number	Mfg.'s Number	Item	Quan. Per Vehicle	Note Symbol
Col. 1	Col. 2	Col. 3	Col. 4	Col. 5	Col. 6	Col. 7
				TABLE, firing, 0.30-A-4	Note 1	F-69
				TABLE, firing, 0.30-C-4	Note 1	F-69
				TABLE, firing, 0.30-J-1	Note 1	F-69
				TABLE, firing, 0.50-F-3 ....(Note 2).........	Note 1	F-69
				TABLE, firing, 0.50-H-1	Note 1	F-69

Note 1—One per gun. Additional issue as authorized in T/o&E.
Note 2—Under development.

APPLICABLE CLEANING AND PRESERVING MATERIALS

The following items extracted from SNL K-1 are requisitioned as required. See TM 9-850 for usage. (To be included when data become available).

ARTICLES FOR INSTRUCTIONAL PURPOSES; FOR:

The items listed hereunder are for training purposes only and will not be taken into the field. Upon permanent change of station involving movement into another Service Command or upon the departure for the theater of operations units will turn in all equipment held to the commanding officer of the station from which it departs. The receiving officer will make a report to the Commander of the Service Command without delay, showing number, type and conditions of items received. Issue only articles pertinent to Gun authorized.

GUN, MACHINE, CAL. .30, BROWNING, M2, M1919A4, FLEXIBLE

Stock No.	Dwg. No.	Description	Qty	Code
	B6252	CARTRIDGE, dummy, cal. .30, M1906 (corrugated)	25	T-1
		CHART, instruction, ammunition (chart No. 75) . . . (Note 1)	Note 2	A-6
		CHART, instruction, BMG, cal. .30	Note 3	A-6
		(Consisting of:		
		1 PLATE, No. 1-BMG, cal. .30, M1919A4 (ORD. 10987)		
		1 PLATE, No. 3-BMG, cal. .30, M1917 (ORD-10893)		
		1 PLATE, No. 5, BMG, cal. .30, M1917 (ORD-10895))		
		TRAINER, machine gun, cal. .22, M4	Note 4	A-48 W
		Accessories:		
A048-02-00010	B147548	ADAPTER, belt	500	A-48
A037-03-00005	D34338	BAG, metallic, belt link	1	A-19
	C64179	BRUSH, cleaning, cal. .22, M3	1	M-3 W

Note 1—Under development.

Note 2—One per organization; to schools as required.

Note 3—One per organization of fifty men or major fraction thereof equipped with machine guns; to schools as required.

Note 4—One per three vehicles equipped with machine gun.

Stock No.	Dwg. No.	Description	Qty	Code
A048-02-00020	D28201	CHEST, accessory and spare parts (STAX1AD modified)	1	M-8 W
	C64326	CHUTE, cartridge holder	1	A-19
A048-02-00030	A152916	CLIP, belt adapter	250	A-48
A048-02-00051	A147551	HOLDER, cartridge	500	A-48
	C3837	ROD, cleaning, cal. .22, M1	1	M-3
A048-02-00060	A152919	TOOL, cartridge ejecting	1	A-19

GUN, MACHINE, CAL. .50, BROWNING, M2, HB, FLEXIBLE

Stock No.	Dwg. No.	Description	Qty	Code
	B147421	ATTACHMENT, firing, blank ammunition	1	A-39
		CARTRIDGE, dummy, cal. .50, M2	25	T-1
	C56579	CHART, instruction, Browning Machine Gun, cal. .50, M2, Fixed and Flexible Set	Note 1	A-39
		(Consists of:		
		Plates, 5, 6, 7, 8, 15, 16, 17, 18 and 19)		

RIFLE, AUTOMATIC, CAL. .30, BROWNING, M1918A2

Stock No.	Dwg. No.	Description	Qty	Code
A004-02-00130	B6252	CARTRIDGE, dummy, cal. .30, M1906 (corrugated)	10	T-1
		CHART, instruction, Browning automatic rifle, cal. .30, M1918 and M1918A2	Note 2	A-4

AMMUNITION

Ammunition for use with these weapons is shown in SNL T-1.

Note 1—One set per organization equipped with BMG, cal. .50, M2; to schools as required.

Note 2—One per organization; to schools as required.

NOTES

Items listed in Section 4 of this SNL will be stored under SNL G-503 unless the note symbol indicates otherwise.

(a) Items indicated by (W) are non-expendable parts and the issue of such items must be in accordance with par. 3b (2), AR 35-6540.

(b) Items not marked by (W) are expendable in accordance with AR 35-6620.

W

SECTION 5

ORDNANCE
MAINTENANCE
UNIT
STOCKAGE
LIST

ORDNANCE MAINTENANCE UNIT STOCKAGE LIST

BASED ON 30 DAYS MAXIMUM ON HAND AND STOCKAGE

Figure Number	Official Stockage Number	Part Number		ITEM	Parts Allowances For:				
		Willys	Ford		1-9 Vehicles	10-35 Vehicles	36-75 Vehicles	76-150 Vehicles	151-300 Vehicles
Col. 1	Col. 2	Col. 3	Col. 4	Col. 5	Col. 6	Col. 7	Col. 8	Col. 9	Col. 10
				GROUP 01—ENGINE					
		WO-A-7233	FM-GPW-18330A	BEARING SET, Connecting Rod (Standard)	3	4	5	7	9
		WO-640070	FM-GPW-6065	BOLT, Engine Connecting Rod Bearing Cap	1	1	2	2	3
		WO-638635		BOLT, Cylinder Head	1	1	2	2	3
		WO-A-1198	FM-GPW-18666	CONNECTION, Flexible (For Oil Pressure Gauge Tube)	1	1	2	2	3
		WO-A-1236	FM-GPW-18662A	ELEMENT, Engine Oil Filter w/GASKET	14	18	24	32	43
		WO-A-5497	FM-GPW-6005	ENGINE, Assembly	0	1	1	1	2
		WO-A-1230	FM-GPW-18660A	FILTER, Oil, Engine, Assembly	0	1	1	1	2
		WO-A-8558	FM-GPW-6051B	GASKET, Engine, Cylinder Head	5	7	9	12	16
		WO-A-1235	FM-GPW-18688A	GASKET, Engine Oil Filter Cover	5	6	8	10	13
		WO-A-1233	FM-GPW-18675A	GASKET, Engine Oil Filter Cover Bolt	5	6	8	10	13
		WO-314338	FM-GPW-6734	GASKET, Engine Oil Pan Drain Plug	9	12	16	21	28
		WO-A-1538	FM-GPW-18512	GASKET SET, Oil Pan	4	5	6	8	11
		WO-630394	FM-GPW-6630	GASKET, Engine Oil Pump to Cylinder Block	1	1	1	1	2
		WO-51875	FM-GPW-6555	GASKET, Engine Valve Cover Screw	2	3	4	5	7
		WO-630305	FM-GPW-6521	GASKET, Engine Valve Spring Cover	2	3	4	5	7
		WO-630299	FM-GPW-6648	GASKET, Engine Crankcase Ventilator to Valve Spring Cover	2	2	3	4	5
		WO-A-1536	FM-GPW-18390	GASKET SET, Engine	2	2	3	4	5
		WO-A-7835	FM-GPW-18323	GASKET SET, Engine Manifold	2	2	3	4	5
		WO-A-6750	FM-GPW-18380	GASKET SET, Engine Oil Pump	2	2	3	4	5
		WO-A-1537	FM-GPW-18387	GASKET SET, Engine Valve Job	4	5	6	8	11
		WO-637045	FM-GPW-6511B	GUIDE, Engine Intake Valve Stem	1	2	2	2	3
		WO-375811	FM-GPW-6510B	GUIDE, Engine Exhaust Valve Stem	1	2	2	2	3
		WO-A-1534	FM-GPW-6050	HEAD, Engine Cylinder	0	1	1	1	2
		WO-A-5168	FM-GPW-6766B	INDICATOR and Breather CAP, Engine Oil Filler, Assembly	0	1	1	1	2
		WO-A-7498	FM-GPW-6038A	INSULATOR, Engine Support Front, Assembly	1	1	2	2	3
		WO-A-6156	FM-GPW-6040B	INSULATOR, Engine Support Rear	0	1	1	1	2
		WO-375994	FM-GPW-6546	LOCK, Engine Valve Spring Retainer, Lower	11	14	18	24	32
		WO-A-912	FM-GPW-9428	MANIFOLD, Exhaust, Assembly	0	1	1	1	2
		WO-637237	FM-GPW-6702	PACKING, Engine Crankshaft, Rear End	1	1	2	2	3
		WO-637098	FM-GPW-6700	PACKING, Engine Crankshaft, Front End	0	1	1	1	2
		WO-A-5105	FM-GPW-6770	PAN, Engine Oil, Assembly	0	1	1	1	2
		WO-639979	FM-GPW-6727	PLUG, Drain, Engine Oil Pan	1	2	2	3	4
		WO-637636	FM-GPW-6600	PUMP, Engine Oil, Assembly	0	1	1	1	2
		WO-637044	FM-GPW-6514	RETAINER, Engine Valve Spring, Lower	2	2	3	4	5
		WO-638636	FM-GPW-6513	SPRING, Engine Valve	3	4	5	7	9
		WO-349368	FM-GPW-6066	STUD, Engine Cylinder Head	1	1	2	2	3
		WO-A-1548	FM-GPW-6067	STUD, Engine Cylinder Head	0	1	1	1	2

GROUP 02—CLUTCH

WO No.	Ord. No.	Description					
WO-A-1197	FM-GPW-18667	TUBE, Engine Oil Filter Inlet, Assembly	12	9	7	5	4
WO-A-1198	FM-GPW-18666	TUBE, Engine Oil Filter, Outlet	12	9	7	5	4
WO-637183	FM-GPW-6505	VALVE, Exhaust, Engine	7	5	4	3	2
WO-637182	FM-GPW-6507	VALVE, Intake, Engine	3	2	2	1	1
WO-635529	FM-GPW-7580B	BEARING, Engine Clutch Release	3	2	2	1	1
WO-639578	FM-GPW-7600	BUSHING, Engine Clutch Pilot	2	1	1	1	0
WO-A-5102	FM-GPW-7530	CABLE, Engine Clutch Control Lever	2	1	1	1	0
WO-A-6751	FM-GPW-18358	FACING SET, Engine Clutch	3	2	2	1	1
WO-630112	FM-GPW-7515	FORK, Engine Clutch Release Shaft	2	1	1	1	0
WO-639654	FM-GPW-7561	CARRIER, Engine Clutch Release Bearing	2	1	1	1	0
WO-638992	FM-GPW-7563	PLATE, Pressure, Engine Clutch, Assembly	3	2	2	1	1
WO-A-7833	FM-GPW-18359B	KIT, Repair, Engine Clutch Cover	2	2	1	0	0
WO-636755	FM-GPW-7550	PLATE, Driven, Engine Clutch Assembly	3	2	1	1	1
WO-638152	FM-GPW-7566	PLATE, Pressure, Engine Clutch	2	1	1	1	0
WO-630593	FM-GPW-7523	SPRING, Retracting, Engine Clutch Pedal	3	2	2	2	1
WO-630117	FM-GPW-7562	SPRING, Engine Clutch Release Bearing Carrier	3	2	2	2	1

GROUP 03—FUEL SYSTEM

WO No.	Ord. No.	Description					
WO-A-6333	FM-GPW-9030	CAP, Fuel Tank, w/Safety CHAIN, Assembly	3	2	2	1	1
WO-A-1223	FM-GPW-9510	CARBURETOR, Assembly	2	1	1	1	0
WO-A-1325	FM-GPW-9288	CONNECTION, Flexible (in fuel line)	4	3	2	2	1
WO-A-5630	FM-GPW-9617	ELEMENT, Filter, Air Cleaner, Assembly	2	1	1	1	0
WO-A-6357	FM-GPW-9445	GASKET, Insulator, Carburetor and Intake Manifold DIFFUSER, Assembly	2	1	1	1	0
WO-638737	FM-GPW-9417	GASKET, Fuel Pump to Block	7	5	4	3	2
WO-115656	FM-GPW-9364	GASKET, Fuel Pump Bowl	8	6	5	4	3
WO-A-1293	FM-GPW-9276	GASKET, Fuel Tank Gauge	3	2	2	1	1
WO-A-1292	FM-GPW-9275	GAUGE, Fuel Tank Unit, Assembly	2	1	1	1	0
WO-A-6837	FM-GPW-18352	GASKET SET, Carburetor	5	4	3	2	2
WO-A-6883	FM-GPW-18337	GASKET SET, Fuel Strainer	13	10	8	6	5
WO-A-1311	FM-GPW-9652	HOSE, Flexible, Air Cleaner Tube to Carburetor	3	2	2	1	1
WO-A-6840	FM-GPW-18357B	KIT, Repair, Carburetor	4	3	2	2	1
WO-A-7834	FM-GPW-18373D	KIT, Repair, Fuel Pump	7	5	4	3	2
WO-A-7517	FM-GPW-9735B	KNOB, Throttle, Assembly	2	1	1	1	0
WO-A-6851	FM-GPW-9350	PEDAL, Accelerator, Assembly	2	1	1	1	0
WO-A-8323	FM-GPW-9799	PUMP, Fuel Assembly	2	3	1	1	0
WO-633011	FM-GPW-9140	SPRING, Retracting, Accelerator Pedal	4	3	2	2	1
WO-A-1261	FM-GPW-9289	STRAINER, Fuel	2	1	1	1	0
WO-A-1368		TUBE, Flexible Connection to Fuel Pump, Assembly	2	1	1	1	0

GROUP 04—EXHAUST

WO No.	Ord. No.	Description					
WO-634814	FM-GPW-9450	GASKET, Exhaust Pipe Flange	11	8	6	5	4
WO-A-658	FM-GPW-5283	INSULATOR, Muffler Support	2	1	1	1	0
WO-A-6118	FM-GPW-5230B	MUFFLER, Assembly (oval type)	11	8	6	5	4
WO-A-10198	FM-GPW-5246B	PIPE, Muffler, Assembly	11	8	6	5	4

Figure Number	Official Stockage Number	Part Number Willys	Part Number Ford	ITEM	1-9 Vehicles	10-35 Vehicles	36-75 Vehicles	76-150 Vehicles	151-300 Vehicles
Col. 1	Col. 2	Col. 3	Col. 4	Col. 5	Col. 6	Col. 7	Col. 8	Col. 9	Col. 10
				GROUP 05—COOLING					
		WO-A-1495	FM-GPW-8620A1	BELT, Drive, Fan and Generator	5	6	8	10	13
		WO-A-1215	FM-GPW-8100A	CAP, Radiator	1	1	2	2	3
		WO-52226	FM-60-8287	CLAMP, Hose, 1^{13}_{16} in. Dia.	7	9	12	16	21
		WO-A-1214	FM-GPW-8005	CORE, Radiator, w/TANK and SHROUD, Assembly	0	1	1	1	2
		WO-A-447	FM-GPW-8600	FAN, Assembly	0	1	1	1	2
		WO-A-1216	FM-K-7129-B	GASKET, Radiator Filler Cap	2	3	4	5	7
		WO-639650	FM-GPW-8255	GASKET, Engine Cylinder Head Water Outlet Elbow	1	1	2	2	3
		WO-637053	FM-GPW-8543	GASKET, Water Pump to Cylinder Block	2	2	3	4	5
		WO-A-592	FM-GPW-8284	HOSE, Radiator Water Outlet, Lower	9	12	16	21	28
		WO-A-6839	FM-GPW-18515B	KIT, Repair, Water Pump	2	2	3	4	5
		WO-636299	FM-GPW-8509A	PULLEY, Fan and Water Pump	0	1	1	1	2
		WO-639992	FM-GPW-8501	PUMP, Water, Assembly	0	1	1	1	2
		WO-637646	FM-GPW-8575	THERMOSTAT, Assembly	1	1	2	2	3
				GROUP 06—ELECTRICAL					
		WO-A-8186	FM-GPW-10850	AMMETER	0	1	1	1	2
		WO-A-1637	FM-GPW-10005	ARMATURE, Generator, Assembly	0	1	1	1	2
		WO-A-1568	FM-GPW-11005	ARMATURE, Cranking Motor, Assembly	0	1	1	1	2
		WO-109452	FM-GPW-11077	BAND, Head, Cranking Motor, Assembly	0	1	1	1	2
		WO-A-1767	FM-11AS-10658	BATTERY (Dry-Charged)	4	5	6	8	11
		WO-A-6299	FM-GPW-10094	BEARING, Ball, Generator (ND. 77503)	0	1	1	1	2
		WO-A-1651	FM-GPW-18274	BRUSH SET, Generator	1	1	1	2	3
		WO-A-1573	FM-GPW-11350	DRIVE, Bendix, Assembly	0	1	1	1	2
		WO-335912	FM-350343-S-16	BOLT, Battery Terminal Clamp w/NUT, Lead Plated, Assembly	2	3	4	5	7
		WO-109431	FM-GPW-11055	BRUSH, Insulated, Cranking Motor (AL MZ-12)	1	1	2	2	3
		WO-A-1452	FM-GPW-14300	CABLE, Battery to Starting Switch, Positive	0	1	1	1	2
		WO-A-1655	FM-GPW-12106	CAP, Distributor, Assembly	1	1	2	2	3
		WO-A-1733	FM-GPW-12250A	BREAKER, Circuit, 5 Amps	0	1	1	1	2
		WO-A-1734	FM-GPW-12250B	BREAKER, Circuit, 15 Amps	0	1	1	1	2
		WO-A-1349	FM-GPW-12250C	BREAKER, Circuit, 30 Amps	0	1	1	1	2
		WO-A-7792	FM-GPW-12000B	COIL, Ignition, w/BRACKET and WASHER, Assembly	0	1	1	1	2
		WO-A-1631	FM-GPW-12300	CONDENSER, Distributor, Assembly	2	2	3	4	5
		WO-635981	FM-GPW-14487B	CONNECTOR (2 Wire)	0	1	1	1	2
		WO-A-1244	FM-GPW-18354	CONTACT SET, Distributor	5	7	9	12	16
		WO-A-1687	FM-GPW-12100	DISTRIBUTOR, Assembly	0	1	1	1	2
		WO-A-11760	FM-GPW-18276	ELECTROLYTE, Battery, w/CONTAINER, 160 oz., Assembly	4	5	6	8	11
		WO-5980	FM-GPW-18960B	FILTER, and BRACKET, Radio Assembly	0	1	1	1	2
		WO-A-5337	FM-GPW-18935A	FILTER UNIT, Generator Ground	0	1	1	1	2
		WO-637863	FM-01A-12410	GASKET, Spark Plug	2	2	3	4	5

					Description	FM No.	WO No.
2	1	1	1	0	GAUGE, Fuel, Dash Unit	FM-GPW-9280	WO-A-8184
2	1	1	1	0	GAUGE, Oil Pressure, Dash Unit	FM-GPW-9273	WO-A-8190
2	1	1	1	0	GENERATOR, Assembly	FM-GPW-10000A	WO-A-5992
2	1	1	1	0	INDICATOR, Heat	FM-GPW-10883	WO-A-8188
2	1	1	1	0	KIT, Repair, Cranking Motor Field Coil	FM-GPW-18319	WO-A-7841
2	1	1	1	0	KIT, Repair, Distributor	FM-GPW-18343	WO-A-7843
2	1	1	1	0	KIT, Repair, Generator	FM-GPW-18363C	WO-A-9055
2	1	1	1	0	KIT, Repair, Generator Cutout Relay	FM-GPW-18298	WO-A-7794
2	1	1	1	0	KIT, Repair, Generator Current Regulator	FM-GPW-18299	WO-A-7805
2	1	1	1	0	KIT, Repair, Generator Voltage Regulator	FM-GPW-18311	WO-A-7810
3	2	2	1	1	KIT, Repair, Ignition Wiring	FM-GPW-18364B	WO-A-7844
7	5	4	3	2	LAMP, Electric 1 CP 6 Volt, Instrument Light	FM-48-15021	WO-52837
7	5	4	3	2	LAMP, Electric 3 CP 6 Volt, Blackout Marker Light	FM-B-13466	WO-51804
2	1	1	1	0	LIGHT, Head, L. H. Assembly	FM-GPW-13006	WO-A-1304
2	1	1	1	0	LIGHT, Head, R. H. Assembly	FM-GPW-13005	WO-A-1305
2	1	1	1	0	LAMP UNIT, Blackout Driving Headlight, Sealed, Assembly	FM-GPW-13152	WO-A-6145
3	2	2	2	1	LAMP UNIT, Blackout Tail, Sealed, Assembly	FM-GP-13491A	WO-A-1075
3	2	2	2	1	LAMP UNIT, Blackout Stop, Sealed, Assembly	FM-GP-13485A	WO-A-1078
5	4	3	3	2	LAMP UNIT, Head Light, Sealed, Assembly	FM-GPW-13007	WO-A-1033
3	2	2	2	1	LAMP UNIT, Service Tail and Stop, Sealed, Assembly	FM-GP-13494A	WO-A-1074
2	1	1	1	0	LIGHT, Driving, Blackout, Assembly	FM-GPW-13150	WO-A-6142
2	1	1	1	0	MOTOR, Cranking, Assembly	FM-GPW-11001	WO-A-1245
65	49	37	28	21	PLATE, Cranking Motor, Commutator End, Assembly	FM-GPW-11049	WO-A-1566
2	1	1	1	0	PLUG, Spark, w/Gasket Assembly	FM-GPW-12405	WO-A-538
2	1	1	1	0	PULLEY, Drive, Generator, w/SPACER, assembly	FM-GPW-10505	WO-A-1639
5	4	3	2	2	REGULATOR, Generator	FM-GPW-12200	WO-A-1409
2	1	1	1	0	ROTOR	FM-GPW-14301	WO-A-1658
5	4	3	2	2	STRAP, Battery Ground	FM-GPW-18812B	WO-635883
2	1	1	1	0	SUPPRESSOR, Spark Plug	FM-GPW-11649	WO-A-6320
2	1	1	1	0	SWITCH, Lighting, Assembly	FM-GTB-13739	WO-A-1332
4	3	2	2	1	SWITCH, Blackout Driving Light	FM-9N-11450A	WO-A-6149
3	2	2	2	1	SWITCH, Starting	FM-GPW-13532	WO-A-7225
3	2	2	1	1	SWITCH, Dimmer, Foot, Headlight, Assembly	FM-GPW-13532	WO-638979
2	1	1	1	0	SWITCH, Keyless, Ignition, Assembly	FM-GPW-3686B	WO-A-6811
2	1	1	1	0	SWITCH, Stop Light, Assembly	FM-11A-13480	WO-A-1271
2	1	1	1	0	TERMINAL, Battery Cable, "1/2"	FM-11A-14452	WO-371400
2	1	1	1	0	TERMINAL, Battery, Positive Cable Clamp	FM-19B-14451	WO-372668

GROUP 07—TRANSMISSION

					Description	FM No.	WO No.
4	3	2	2	1	GASKET, Transmission Case to Flywheel Housing	FM-GPW-7051B	WO-637495
5	4	3	2	2	GASKET SET, Transmission	FM-GPW-18356B	WO-A-7832
2	1	1	1	0	KIT, Repair, Speedometer Drive Gear	FM-GPW-18314	WO-A-7837
2	1	1	1	0	OIL, Seal Transmission Front Bearing Retainer	FM-GPW-7052	WO-640018
2	1	1	1	0	TRANSMISSION, Assembly	FM-GPW-7008	WO-A-7596

GROUP 08—TRANSFER CASE

					Description	FM No.	WO No.
2	1	1	1	0	CASE, Transfer Assembly	FM-GPW-7700	WO-A-1195
2	1	1	1	0	CONE and ROLLERS, Bearing, 1.3125 Diam. x .740 (Transfer Case and Output Clutch Shaft) (Bow-BT-14131) TIM-14131	FM-GP-7723	WO-51575

Figure Number	Official Stockage Number	Part Number		ITEM	Parts Allowances For:					
		Willys	Ford		1-9 Vehicles	10-35 Vehicles	36-75 Vehicles	76-150 Vehicles	151-300 Vehicles	
Col. 1	Col. 2	Col. 3	Col. 4	Col. 5	Col. 6	Col. 7	Col. 8	Col. 9	Col. 10	

GROUP 08—TRANSFER CASE—Cont'd

	WO-52883	FM-01Y-1202	CUP, Bearing, Transfer Case Output Shaft and Output Clutch Shaft (BOW-BT-14276) (TIM 14276)					
	WO-A-1106	FM-GP-7729	FLANGE AND SEALER, Transfer Case Front U Joint, Assembly	0	1	1	1	2
	WO-A-1509	FM-GPW-7707	GASKET, Transfer Case Cover Rear	0	1	1	1	2
	WO-A-7445	FM-GPW-18317	OIL SEAL SET, Transfer Case	1	1	2	2	3

GROUP 09—PROPELLER SHAFT, UNIVERSAL JOINTS

	WO-A-942	FM-GP-7077	CAP, Dust, Universal Joint Yoke Sleeve	0	1	1	1	2
	WO-A-1433	FM-GPW-18397	KIT, Repair, Universal Joint Journal	6	8	11	14	19
	WO-A-1326	FM-GPW-3365	SHAFT, Propeller, Front, Assembly	0	1	1	1	2
	WO-A-1327	FM-GPW-4602	SHAFT, Propeller, Rear, Assembly	0	1	1	1	2
	WO-A-490	FM-02Y-4529	U BOLT, Universal Joint	1	1	2	2	3
	WO-A-943	FM-GP-7097	WASHER, Universal Joint Yoke Sleeve	1	1	2	2	3
	WO-A-935	FM-GP-4841	YOKE, Universal Joint Sleeve	0	1	1	1	2

GROUP 10—FRONT AXLE

	WO-A-8249	FM-GPW-3131	BELL CRANK, Drag Link, Front	0	1	1	1	2
	WO-A-1712	FM-GPW-3113	ARM, Steering Knuckle, Upper L. H. Assembly	0	1	1	1	2
	WO-A-1710	FM-GPW-3112	ARM, Steering Knuckle, Upper R. H. Assembly	0	1	1	1	2
	WO-A-6029		AXLE, Front, Assembly	0	1	1	1	2
	WO-52940	FM-GP-3161	CONE and ROLLERS, Front Axle Spindle Bearing (Timken 11590)	0	1	1	1	2
	WO-52940	FM-GP-3161	CONE and ROLLERS, Front Axle King Pin Bearing Steering Pivot Arm	0	1	1	1	2
	WO-52941	FM-GP-3162	CUP, Front Axle Spindle Bearing (TIM-11320)	0	1	1	1	2
	WO-52941	FM-GP-3162	CUP, King Pin Bearing (Steering Pivot Arm)	0	1	1	1	2
	WO-A-782	FM-GPW-4035	GASKET, Front Axle Housing Cover	2	2	3	4	5
	WO-A-7830	FM-GPW-18365B	GASKET SET, Front Axle	1	1	2	2	3
	WO-A-820	FM-GP-1092	GASKET, Steering Knuckle Oil Seal, Assembly (Spicer 1704)	7	9	12	16	21
	WO-A-6816	FM-GPW-18384	KIT, Repair, Drive Gear and Pinion	0	1	1	1	2
	WO-A-6743	FM-GPW-18389	KIT, Repair, Front Axle Differential Gear	0	1	1	1	2
	WO-A-6882	FM-GPW-18338	SHIM SET, King Pin Bearing	0	1	1	1	2
	WO-A-6881	FM-GPW-18336	SHIM SET, Universal Joint Adjusting	0	1	1	1	2
	WO-A-812	FM-GP-3149A	KNUCKLE, Steering L. H.	0	1	1	1	2
	WO-A-811	FM-GP-3148A	KNUCKLE, Steering R. H.	0	1	1	1	2
	WO-A-A-819	FM-GP-3135	OIL SEAL, Felt (Half) (Spicer 17019)	8	11	14	18	25
	WO-A-818	FM-GP-3139	OIL SEAL, Felt Pressure Strip (Spicer 16983)	8	11	14	18	25
	WO-A-813	FM-GP-1088	OIL SEAL, Steering Knuckle	1	1	2	2	3
	WO-A-779	FM-GP-3034	OIL SEAL, Carrier, End	1	2	2	3	4
	WO-A-778	FM-GP-3031	RETAINER, Grease, Wheel, End	1	2	2	3	4

					Nomenclature	Ford Part No.	Willys Part No.
4	3	2	2	1	RETAINER, Pivot Oil Seal, Assembly	FM-GP-1089	WO-A-814
2	1	1	1	0	SHAFT, Axle, Left, w/Universal JOINT, Assembly	FM-GPW-3227A	WO-A-1728
2	1	1	1	0	SHAFT, Axle, Right, w/Universal JOINT, Assembly	FM-GPW-3206A	WO-A-809
2	1	1	1	0	SHAFT, Axle Front Left, Complete w/Universal JOINT, Assembly (Sp-17128-4X)		**WO-A-8836
2	1	1	1	0	SHAFT, Axle Front Right, Complete w/Universal JOINT, Assembly (Sp-17128-3X)		**WO-A-8637
3	2	2	1	1	SHAFT, Steering Bell Crank	FM-GPW-3165	WO-A-855
2	1	1	1	0	SPINDLE, Front Wheel Bearing, Assembly	FM-GP-3105	WO-A-851
4	3	2	2	1	STUD, Steering Arm Dowel	FM-GPW-3325	WO-A-5504
2	1	1	1	0	WASHER, Steering Arm, Lower	FM-GPW-3168	WO-A-859
2	1	1	1	0	WASHER, Steering Arm, Upper	FM-GPW-3169	WO-A-860
2	1	1	1	0	SPINDLE, Wheel Bearing, w/BUSHING, Assembly (Spicer 17202-X)	FM-GP-3105	WO-A-851

GROUP 11—REAR AXLE

					Nomenclature	Ford Part No.	Willys Part No.
2	1	1	1	0	DIFFERENTIAL, Assembly (Spicer 16968-X) Front and Rear		WO-A-788
5	4	3	2	2	GASKET, Axle Housing Cover	FM-GP-4035	WO-A-782
21	16	12	9	7	GASKET, Rear Axle Shaft	FM-GP-4032	WO-A-904
3	2	2	1	1	GASKET, Set, Rear Axle	FM-GPW-18366	WO-A-7831
2	1	1	1	0	KIT, Repair, Rear Axle Differential Gear	FM-GPW-18389	WO-A-6743
4	3	2	2	1	OIL SEAL, Axle Inboard (Spicer 17036)	FM-GP-3034	WO-A-779
3	2	2	1	1	RETAINER, Differential Pinion	FM-GP-4676	WO-639265
2	1	1	1	0	SHAFT, Rear Axle L. H.	FM-GPW-4235	WO-A-902
2	1	1	1	0	SHAFT, Rear Axle R. H.	FM-GPW-4234	WO-A-901
2	1	1	1	0	SHIM SET, Rear, Axle Drive Pinion	FM-GPW-18386	WO-A-6745
2	1	1	1	0	SHIM SET, Rear, Axle Differential Bearing	FM-GPW-18388	WO-A-6744

GROUP 12—BRAKES

					Nomenclature	Ford Part No.	Willys Part No.
2	1	1	1	0	BAND, Hand Brake Assembly	FM-GP-2648	WO-A-1009
2	1	1	1	0	BRAKE, Shoe and Lining Assembly Forward	FM-GP-2018	WO-116549
2	1	1	1	0	BRAKE, Shoe and Lining Assembly Rear	FM-GP-2019	WO-116550
2	1	1	1	0	CABLE, Hand Brake Ratchet, w/TUBE, Assembly	FM-GPW-2853	WO-A-1241
2	1	1	1	0	CYLINDER, Front Wheel Brake	FM-GPW-2061	WO-A-1484
2	1	1	1	0	CYLINDER, Master, Assembly	FM-GP-2140	WO-A-556
2	1	1	1	0	CYLINDER, Rear Brake Lo-FD-4665	FM-GP-2261	WO-637787
2	1	1	1	0	CYLINDER, Rear Wheel Brake Assembly ¾ in. Dia. (Lo-FD-7568) (After Serial No. 134356)	FM-GPW-2261	WO-A-6110
3	2	2	2	1	ECCENTRIC, Brake Shoe	FM-GP-2038	WO-A-754
12	9	7	5	4	GASKET, Fitting, Brake Hose (Lo-FC-5795)	FM-GP-2087	WO-637426
3	2	2	1	1	GASKET, Master Cylinder Cap (Lo-FC-6019)	FM-GP-2167	WO-637612
4	3	2	2	1	HOSE, Front Brake, 6 in. Long	FM-GPW-2079	WO-A-1460
3	2	2	2	1	HOSE, Front Brake, 11 in. Long	FM-GPW-2078	WO-A-1373
2	2	2	2	1	HOSE, Rear Brake, 15 in. Long	FM-GP-2078	WO-637424
3	2	2	2	1	LINING SET, Brake Shoe w/RIVETS	FM-GPW-18367	WO-116600
2	1	1	1	0	LINING SET, Hand Brake w/RIVETS	FM-GPW-18377	WO-A-6759
2	1	1	1	0	KIT, Repair, Master Cylinder	FM-GPW-18370B	WO-A-7838
3	2	2	1	1	KIT, Repair, Front Wheel Cylinder	FM-GPW-18371	WO-115962

**Willys only

237

ORDNANCE MAINTENANCE UNIT STOCKAGE LIST—Cont'd

Figure Number	Official Stockage Number	Part Number Willys	Part Number Ford	ITEM	1-9 Vehicles	10-35 Vehicles	36-75 Vehicles	76-150 Vehicles	151-300 Vehicles
Col. 1	Col. 2	Col. 3	Col. 4	Col. 5	Col. 6	Col. 7	Col. 8	Col. 9	Col. 10
				GROUP 12—BRAKES—Cont'd					
		WO-A-6133	FM-GPW-18368	KIT, Repair, Rear Wheel Cylinder	1	1	2	2	3
		WO-637899	FM-91A2027	PIN, Brake, Anchor	0	1	1	1	2
		WO-637540	FM-GP-2208	SCREW, Brake Wheel Cylinder Bleeder	1	1	2	2	3
		WO-630593	FM-GPW-7523	SPRING, Brake Pedal Return	1	1	2	2	3
		WO-637905	FM-GP-2035	SPRING, Return, Brake Shoe	2	2	3	4	5
		WO-A-5335	FM-GPW-2635	SPRING, Retracting, Hand Brake Lever	0	1	1	1	2
		WO-A-1017	FM-01T-2634	SPRING, Releasing Hand Brake	0	1	1	1	2
		WO-A-1488	FM-GPW-2298	TUBE, Brake, Assembly	0	1	1	1	2
				GROUP 13—WHEELS, HUBS AND DRUMS					
		WO-A-5470	FM-GPW-1029	BOLT, Wheel, Divided Rim	2	2	3	4	5
		WO-A-473	FM-GP-1108	BOLT, Wheel Hub, L. H.	1	1	2	2	3
		WO-A-474	FM-GP-1107	BOLT, Wheel Hub, R. H.	1	1	2	2	3
		WO-A-760	FM-GP-1110	BOLT, Wheel Hub to Flange, Front and Rear	2	2	3	4	5
		WO-52942	FM-GP-1201	CAP, Wheel Hub	0	1	1	1	2
				CONE and ROLLERS, Front and Rear Wheels Inner and Outer Bearing	1	1	2	2	3
		WO-52943	FM-GP-1202	CUP, Front and Rear Wheels Inner and Outer Bearing	1	1	2	2	3
		WO-A-1690	FM-GP-1103	HUB, Front and Rear, Left, w/DRUM, Assembly L. H.	0	1	1	1	2
		WO-A-1689	FM-GP-1102	HUB, Front and Rear, Right, w/DRUM, Assembly R. H.	0	1	1	1	2
		WO-A-866	FM-GP-4252	NUT, Front and Rear Axle Wheel Bearing	1	1	2	2	3
		WO-A-5471	FM-GPW-1030	NUT, Hex, S., $\frac{1}{2}$-20NF-2, Right Hand Thread (Wheel Hub Bolt)	1	2	2	3	4
		WO-A-475	FM-GP-1013	NUT, Wheel Bolt, L. H.	1	2	2	3	4
		WO-A-476	FM-GP-1012	NUT, Wheel Bolt, R. H.	1	2	2	3	4
		WO-A-864	FM-GP-1177	RETAINER, Grease, Wheel Hub, Assembly	15	20	26	32	43
		WO-A-865	FM-GP-1218	WASHER, Wheel Bearing	0	1	1	1	2
		WO-A-867	FM-GP-1124	WASHER, Lock, Wheel Bearing Nut	1	1	2	2	3
				GROUP 14—STEERING					
		WO-A-857	FM-GPW-3171	BEARING, Steering Bell Crank	0	1	1	1	2
		WO-A-847	FM-GP-3292	END, Steering Knuckle Tie Rod L. H., Assembly	0	1	1	1	2
		WO-A-838	FM-GP-3291	END, Steering Knuckle Tie Rod R. H., Assembly	0	1	1	1	2
		WO-A-6791	FM-GPW-18383	KIT, Repair, Steering Connecting Rod	0	1	1	1	2
		WO-A-8250	FM-GPW-3304	ROD, Connecting	0	1	1	1	2
		WO-A-1709	FM-GPW-3282	ROD, Tie Steering L. H., Assembly	0	1	1	1	2
		WO-A-1705	FM-GPW-3281	ROD, Tie Steering R. H., Assembly	0	1	1	1	2
		WO-A-861	FM-GPW-3170	SHIELD, Dust, Steering, Bell Crank Bearing (SP-17205)	0	1	1	1	2
		WO-A-1239	FM-GPW-3504	STEERING, Gear, w/WHEEL Assembly	0	1	1	1	2
		WO-A-6858	FM-GPW-3600A	WHEEL, Steering, Assembly	0	1	1	1	2

GROUP 16—SPRING

Part No.	Part No.	Description					
FM-GPW-18045	*	ABSORBER, Shock, Front, Assembly	3	2	2	1	1
FM-GPW-18080	*	ABSORBER, Shock, Rear, Assembly	3	2	2	1	1
	**WO-A-6902	ABSORBER, Shock, Front	2	1	1	1	0
	**WO-A-6903	ABSORBER, Shock, Rear	2	1	1	1	0
FM-GPW-5345A	WO-116609	BOLT, Front Spring, Center	16	12	9	7	5
FM-GPW-5345B	WO-116610	BOLT, Rear Spring, Center	7	5	4	3	2
FM-GPW-5468	WO-384228	BOLT, Pivot Spring Shackle	3	2	2	1	1
FM-GPW-5778	WO-A-513	BOLT U, Shackle, L. H.	2	1	1	1	0
FM-GPW-5779	WO-A-514	BOLT U, Shackle, R. H.	2	1	1	1	0
FM-GPW-18060	WO-637936	BUSHING, Shock Absorber Mounting Pin	5	4	3	2	2
FM-GPW-5463	WO-A-8255	BUSHING, Spring Shackle, L. H. Thread	3	2	2	1	1
FM-GPW-5464	WO-8256	BUSHING, Spring Shackle, R. H. Thread	5	4	3	2	2
FM-GPW-5724A	WO-116459	CLAMP, Rear Spring	4	3	2	2	1
FM-GPW-5482	WO-A-1252	RETAINER, Spring Shackle Grease Seal	3	3	2	1	1
FM-GPW-5481	WO-A-515	SEAL, Grease, Spring Shackle	7	5	4	3	2
FM-GPW-5310	WO-A-613	SPRING, Front Right	3	2	2	1	1
FM-GPW-5311	WO-A-612	SPRING, Left Front	3	2	2	1	1
FM-GPW-5560	WO-A-614	SPRING, Rear	2	1	1	1	0

GROUP 18—BODY AND HULL

Part No.	Part No.	Description					
FM-GP-17535	WO-A-2512	ARM, Windshield Wiper w/BLADE	11	8	6	5	4
FM-GP-1103027	WO-A-3197	CATCH, Hold Down w/SHIELD	2	1	1	1	0
FM-GP-1103100	WO-A-2478	GLASS, w/Shield, Assembly	3	2	2	1	1

GROUP 22—MISCELLANEOUS

Part No.	Part No.	Description					
FM-GPW-1509B	WO-A-7791	CASING, Pneumatic Tire 600 x 16—6 Ply	20	15	11	8	6
FM-GPW-17261	WO-A-1343	CASING, Speedometer Shaft	3	2	2	1	1
FM-21CS-17682-B	WO-A-2934	MIRROR, Rear View, Outside	2	1	1	1	0
FM-GPW-17262	WO-A-1344	SHAFT, Speedometer	5	4	3	2	2
FM-GPW-7255A	WO-A-8180	SPEEDOMETER, Assembly	2	1	1	1	0
FM-GPW-1655	WO-A-457	TUBE, Pneumatic Inner 600 x 16 (Heavy Duty)	20	15	11	8	6

GROUP 23—GENERAL USE STANDARDIZED PARTS

Part No.	Part No.	Description					
FM-GPW-18320	WO-A-7682	CORE, Kit Tire Valve	11	8	6	5	4
FM-353035-S7	WO-638224	FITTING, Grease 45°	4	3	2	2	1
FM-353023-S7	WO-638500	FITTING, Grease 90°	4	3	2	2	1
FM-350006-S8	WO-640038	FITTING, Grease (AD 1980)	4	3	2	2	1
FM-353027-AS7	WO-392909	FITTING, Grease Straight (AD 1720)	8	6	5	4	3
FM-353043-S7	WO-638792	FITTING, Grease Straight 1/4-28NF (AD 1641)	4	3	2	2	1

*Ford only.
**Willys only

SECTION 6

DEPOT STOCKAGE LIST

To determine the quantities of parts required by the various depots, the following procedures will be used:

1. FIELD DEPOT—The figures shown in the 12 Months Field Maintenance Column, No. 7, in Section 2, Maintenance Parts Procurement List, will be used as the basis for computing normal requirements. For example, 1 month of supply for 13 major items is the nearest whole equivalent of 1/12 x 13 items/100 items x the quantities listed in the 12 Months Field Maintenance Column.

2. BASE DEPOT—The figures shown in the Total First Year Procurement Column, No. 9, in Section 2, will be used as the basis for computing normal requirements. For example, 1 month of supply for 17 major items is the nearest whole equivalent of 1/12 x 17 items/100 items x the quantities listed in the Total First Year Procurement Column.

3. MOBILE DEPOT—The figures shown in the Ordnance Maintenance Unit Stockage List will be used as a basis for computing normal requirements. Using the proper column for the number of major items to be serviced and taking into consideration the time intervals being employed as the basis for computation, the quantities of parts required may be determined.

SECTION 7

GEOGRAPHICAL OR SEASONAL MAINTENANCE PARTS LISTS

Information to be included
when data becomes available

SECTION 8

INDEXES

FIGURE INDEX

Description	Gov't Group No.
BUSHING, reverse idler gear	0704C
BUSHING, rubber	2204
BUSHING, rubber (controls through dash)	0301B
BUSHING, shock absorber mounting pin, rubber	1603
BUSHING, shock absorber spring seat	1603
BUSHING, speedometer drive gear	0810
BUSHING, spring shackle, assembly	1602
BUSHING, spring shackle, assembly right thd	1602
BUSHING, spring shackle, inner	1602
BUSHING, spring shackle, outer	1602
BUSHING, starting motor	0602
BUSHING, steering column support	1405
BUSHING, torque reaction spring	1601C
BUSHING, wheel bearing spindle	1006
BUTTON, horn	0609B
CABLE, blackout driving lamp connector to switch	0606B
CABLE, blackout driving light, assembly	0607D
CABLE, blackout tail lamp to connector, left	0606B
CABLE, clutch control lever	0203
CABLE, coil primary	0604A
CABLE, coil, secondary	0604A
CABLE, engine stay, assembly	0110
CABLE, filter (reg. batt.)	0606B
CABLE, filter	0606B
CABLE, gasoline gauge to circuit breaker	0605B
CABLE, gasoline gauge	0605B
CABLE, ground, head light	0606B
CABLE, ground, trailer coupling socket to body	0606F
CABLE, ground, w/TERMINALS, assembly	2601
CABLE, hand brake ratchet, w/TUBE, assembly	1201
CABLE, horn assembly	0609B
CABLE, horn to junction block	0609B
CABLE, ignition switch to ammeter to blackout switch	0605B
CABLE, positive, battery to starting switch	0610B
CABLE, spark plug No. 1	0604A
CABLE, spark plug No. 2	0604A
CABLE, spark plug No. 3	0604A
CABLE, spark plug No. 4	0604A
CABLE, starter switch to starter motor	0610B
CABLE, starting switch to radio body, assembly	2601
CABLE, voltage regulator to generator	0606B
CABLE, voltage regulator to junction block	0606B
CAGE, universal joint	1007
CAM, anchor pin	1203B
CAM, hand brake	1201
CAMSHAFT	0106A
CAN, oil	2301
CAP, distributor, assembly	0603
CAP, dust universal joint yoke sleeve	0902
CAP, filler master cylinder	1205
CAP, gas tank well drain	0300
CAP, gasoline tank, w/safety CHAIN	0300
CAP, gasoline tank, w/safety CHAIN and BAR	0300
CAP, gearshift lever	0706A
CAP, hub	1302
CAP, insulator, spark plug	0604B
CAP, king pin bearing, lower	1006
CAP, output clutch, shaft bearing front	0801
CAP, output shaft bearing, w/BUSHING, rear	0801
CAP, radiator filler, assembly	0501
CARBURETOR, assembly	0301
CARRIER, clutch release bearing	0203
CASE, differential	1002
CASE, differential	1103
CASE, glove compartment lock, assembly	1809A

Description	Gov't Group No.
CASE, transfer	0801
CASE, transfer, assembly	0800
CASE, transmission, assembly	0701
CATCH, hold-down, windshield, assembly	1704
CATCH, hood, assembly	1701
CATCH, windshield clamp	1811
CHAIN, thumb screw	1811
CHAIN, thumb screw	2201A
CHAIN, tire, heavy duty, type "D"	2301B
CHAIN, tire, heavy duty, type "TS"	2301B
CHECK, intake ball, assembly	0301
CLAMP, axle brake tube	1101
CLAMP, axle brake tube	1209C
CLAMP, axle, front	2301A
CLAMP, axle rear, assembly	2301A
CLAMP, carburetor air horn, w/BOLT and NUT	0301C
CLAMP, exhaust pipe extension	0402
CLAMP, exhaust pipe to muffler	0402
CLAMP, front axle brake tube	1001
CLAMP, fuel pump valve	0302
CLAMP, gasoline tank hold-down, assembly	0300
CLAMP, hand brake cable tube	1201
CLAMP, hand brake cable tube (at transfer case bearing cap)	1201
CLAMP, hose	0505
CLAMP, oil filter, assembly	0107F
CLAMP, socket	1402
CLAMP, steering column, assembly	1405
CLAMP, steering column support	1405
CLAMP, support, muffler	0401
CLAMP, tail pipe	0402A
CLAMP, tube, assembly	0301
CLAMP, windshield, assembly	1811
CLEANER, air, assembly	0301C
CLIP	0606B
CLIP, Bond No. 4 to speedometer tube	0603
CLIP, Bond No. 4 to heat indicator cable	2603
CLIP, brake pipe	1209C
CLIP, breaker arm spring	0603
CLIP	2601
CLIP, 1 used cable to speedometer clip bolt	2601
CLIP	1201
CLIP, 1 used cable to speedometer clip bolt	2601
CLIP, front spring, left	1000
CLIP, front spring, left	1602
CLIP, front spring, right	1000
CLIP, front spring, right (was WO-A-1097)	1602
CLIP, lock, spring	1209C
CLIP, mounting	2204
CLIP, rear spring	1100
CLIP, retracting spring	0303
CLIP, retracting spring (on manifold)	0303
CLIP, spring leaf	1601
CLIP, spring leaf	1601A
CLIP, spring leaf (large for six leaves)	1601
CLIP, spring leaf, large	1601A
CLIP, support, hand brake cable	1201
CLIP, torque reaction	1601C
CLIP, tube	1209C
CLIP, tube (1 used tube to side rail enforcement)	1209C
CLIP	0304
CLIP	0107J
CLIP (tube to front engine plate, under generator brace bolt)	0304
COCK, drain, cylinder block	0101

Description	Gov't Group No.
SHAFT, axle, outer, w/universal JOINT, assembly	1007
SHAFT, axle, right, w/universal JOINT, assembly	1007
SHAFT, axle, right	1007
SHAFT, axle, right w/universal JOINT, assembly	1007
SHAFT, bell crank	1401
SHAFT, Bendix, drive w/PINION, assembly	0602
SHAFT, brake pedal, assembly	1204
SHAFT, cross, accelerator w/LEVER	0303
SHAFT, differential pinion	1003
SHAFT, drive, assembly	0603
SHAFT, drive, distributor, w/GOVERNOR, assembly	0603
SHAFT, heat control valve	0108C
SHAFT, intermediate	0804
SHAFT, main	0704
SHAFT, main, w/GEARS, assembly	0704
SHAFT, oil pump, assembly	0107
SHAFT, output, w/BUSHING, assembly	0803
SHAFT, output, w/NUT and WASHER, assembly	0803
SHAFT, output clutch	0803
SHAFT, output clutch w/NUT and WASHER, assembly	0803
SHAFT, pedal assembly	0204
SHAFT, pinion	1103
SHAFT, propeller, front, assembly	0901
SHAFT, propeller, rear, assembly	0901
SHAFT, propeller, rear, assembly	0902
SHAFT, rear axle, left	1102
SHAFT, rear axle, right	1102
SHAFT, reverse idler gear	0704C
SHAFT, right, axle, w/universal JOINT, assembly	1007
SHAFT, speedometer, assembly	2204
SHAFT, throttle, w/INNER. assembly	0301
SHEATH, axe	1802A
SHIELD, air cleaner, assembly	0301C
SHIELD, dust	0606F
SHIELD, dust	0803
SHIELD, dust, bell crank bearing dust	1401
SHIELD, dust cover	1401
SHIELD, dust, front universal joint end yoke	0902
SHIELD, dust universal joint end yoke	1003
SHIELD, dust, universal joint end yoke	1104
SHIELD. fan pulley	0503A
SHIELD, front seat to rear floor gas tank	0300
SHIELD, instrument, lamp assembly	0605C
SHIELD, master cylinder, assembly	1205
SHIM, adjusting (.003 thk.)	1103
SHIM, adjusting (.005 thk.)	1103
SHIM, adjusting (.010 thk.)	1103
SHIM, adjusting (.015 thk.)	1103
SHIM, adjusting (.030 thk.)	1103
SHIM, adjusting, bearing (small .003)	1104
SHIM, adjusting, bearing (large, .005)	1104
SHIM, adjusting, bearing (large, .003)	1104
SHIM, adjusting, bearing (large, .010)	1104
SHIM, adjusting, bearing (small, .010)	1104
SHIM, adjusting, bearing (small, .030)	1104
SHIM, adjusting, bearing (small, .005)	1104
SHIM, adjusting, differential bearing (.003 thk.)	1003
SHIM, adjusting, differential bearing (.005 thk.)	1003
SHIM, adjusting, differential bearing (.015 thk.)	1003
SHIM, adjusting, differential bearing (.010 thk.)	1003
SHIM, adjusting, differential bearing (.030 thk.)	1003
SHIM, adjusting, drive pinion, bearing (.003 thk.)	1003
SHIM, adjusting, drive pinion bearing (.005 thk.)	1003
SHIM, adjusting, drive pinion, bearing small (.010 thk.)	1003
SHIM, adjusting, drive pinion, bearing (.030 thk.)	1003

Description	Gov't Group No.
SHIM, adjusting, drive pinion, large (.003 thk.)	1003
SHIM, adjusting, drive pinion, large (.005 thk.)	1003
SHIM, adjusting, drive pinion, large (.010 thk.)	1003
SHIM, adjusting, king pin bearing (.003 thk.)	1006
SHIM, adjusting, king pin bearing (.005 thk.)	1006
SHIM, adjusting, king pin bearing (.010 thk.)	1006
SHIM, adjusting, king pin bearing (.030 thk.)	1006
SHIM, adjusting, universal joint (.010 thk.)	1007
SHIM, adjusting, universal joint (.030 thk.)	1007
SHIM, crankshaft (.002 thk.)	0102
SHIM, engine support insulator	0110
SHIM, mounting	1405
SHIM, oil relief spring	0107
SHIM, output shaft rear bearing, cap (.003 thk.)	0801
SHIM, output shaft rear bearing cap (.010 thk.)	0801
SHIM, output shaft rear bearing cap (.031 thk.)	0801
SHIM, radiator	0504
SHIM, radiator to support bracket	0501
SHIM, steering gear mounting	1403A
SHIM, upper cover, S. (.010 thk.)	1403A
SHIM, upper cover, S. (.003 thk.)	1403A
SHIM, upper cover, S. (.002 thk.)	1403A
SHIM SET, king pin bearing	1006
SHIM SET, universal joint adjusting	1007
SHIM SET, output shaft	0801
SHIM SET, rear axle drive pinion	1104
SHIM SET, axle differential bearing	1103
SHIM SET, steering gear housing upper cover	1403A
SHOE, brake, forward w/LINING, assembly	1202
SHOE, brake, reverse w/LINING, assembly	1202
SILL, cross, front floor pan, left, assembly	1802A
SILL, cross, front floor pan, rear, assembly	1802A
SILL, cross, front floor pan, rear, assembly	1802A
SILL, cross, front floor pan, right, assembly	1802A
SILL, cross, rear floor pan	1802A
SLEEVE, compression	0602
SLEEVE, second and direct speed clutch	0704B
SLEEVE, shock absorber, front, assembly	1603
SLEEVE, shock absorber, rear, assembly	1603
SLEEVE, speedometer driven gear	0810
SLINGER, drive pinion bearing	1104
SLINGER, oil, crankshaft	0102
SLINGER, oil, drive pinion bearing	1003
SNAP RING, axle shaft retainer	1007
SNAP RING, high and intermediate clutch hub	0704B
SNAP RING, main drive gear bearing	0703
SNAP RING, main shaft	0703
SNAP RING, main shaft	0704
SNAP RING, main shaft, pilot roller bearing	0703
SNAP RING, output clutch shaft bearing	0803
SNAP RING, output shaft	0802
SNAP RING, retaining, distributor cam	0603
SNAP RING, steering column	1403A
SNAP RING, trunnion bearing	0902
SNUBBER, transfer case support insulator	0809
SOCKET, coupling, electric trailer, assembly	0606F
SOCKET, coupling electric trailer, assembly (includes COVER, INTERNAL PARTS, ATTACHING PARTS)	0606F
SOCKET, curtain fastener	2201A
SOCKET, tie rod left, assembly	1402
SOCKET, tie rod right, assembly	1402
SOCKET, instrument lamp, w/CABLE, assembly	0605C
SPACER, cam	0603
SPACER, crankshaft sprocket	0106C

Description	Gov't Group No.
SPACER, countershaft bushing	0704C
SPACER, cross member intermediate front bracket	1500
SPACER, drive pinion bearing	1003
SPACER, drive pinion bearing	1104
SPACER, front bumper gusset	1500
SPACER, main shaft bearing	0704
SPACER, metering, shock absorber, front	1603
SPACER, shaft	0601
SPACER, shock absorber, metering, rear	1603
SPACER, steering gear to frame bolt	1500
SPACER, thrust, bearing	0602
SPACER, windshield adjusting arm	1811
SPARE PARTS SET	2309
SPEEDOMETER, luminous dial, assembly	2204
SPINDLE, wheel bearing, w/BUSHING, assembly	1006
SPRING, accelerator	0303
SPRING, accelerator cross shaft	0303
SPRING, accelerator treadle	0303
SPRING, anchor clip bolt	1201
SPRING, anti-rattle, distributor	0603
SPRING, ball seat	1401
SPRING, breaker arm	0603
SPRING, brush	0601
SPRING, brush	0602
SPRING, camshaft thrust plunger	0106
SPRING, choke pull-back	0301
SPRING, clutch control tube	0204A
SPRING, clutch release bearing carrier	0203
SPRING, connector link	0301
SPRING, connector rod	0301
SPRING, contact distributor	0603
SPRING, cut-out relay	0601B
SPRING, distributor cap	0603
SPRING, drive	0602
SPRING, friction, distributor shaft	0603
SPRING, front, L., assembly	1601
SPRING, front, R., assembly	1601
SPRING, front seat back, assembly	1804A
SPRING, front seat body side crash pad, left assembly	1804A
SPRING, front seat body side crash pad, right assembly	1804A
SPRING, front seat, cushion, assembly	1804A
SPRING, front wheel, cylinder cup	1207
SPRING, fuel pump diaphragm	0302
SPRING, fuel pump rocker arm	0302
SPRING, generator brace	0601A
SPRING, heat control valve	0108C
SPRING, horn button	0609B
SPRING, idle adjustment screw	0301
SPRING, intake needle	0301
SPRING, metering rod	0301
SPRING, needle, w/SEAT, assembly	0301
SPRING, oil filter cover bolt	0107E
SPRING, oil relief plunger	0107
SPRING, pedal retracting	0204
SPRING, pedal shank draft pad	0204
SPRING, pin	0301
SPRING, plunger	0301
SPRING, pressure	0202
SPRING, pressure plate return	0202
SPRING, pump	0301
SPRING, pump arm	0301
SPRING, ratchet tube	1201
SPRING, rear, assembly	1601A
SPRING, rear seat back assembly	1804A
SPRING, rear seat cushion assembly	1804A
SPRING, rear wheel, cylinder cup	1207

Description	Gov't Group No.
SPRING, releasing, hand brake	1201
SPRING, retaining, rear seat frame	1804
SPRING, retaining, water pump bearing	0503
SPRING, return, brake shoe	1203A
SPRING, return, master cylinder piston, w/RETAINER assembly	1205
SPRING, retracting, hand brake	1201
SPRING, shifter plate	0706C
SPRING, shift lever	0805B
SPRING, shift rail, poppet ball	0706B
SPRING, shift rod poppet	0805
SPRING, shock absorber, piston intake valve	1603
SPRING, shock absorber, rebound valve	1603
SPRING, steering column bearing	1403A
SPRING, strainer unit	0306B
SPRING, support, gearshift lever	0706A
SPRING, synchronizer	0704B
SPRING, tie rod socket stud	1402
SPRING, toggle	0301C
SPRING, torque reaction assembly	1601C
SPRING, valve	0105
SPRING, voltage regulator and current regulator coil	0601B
SPRING, water pump seal	0503
SPRING SET, governor weight	0603
SPROCKET, camshaft	0106C
SPROCKET, crankshaft	0102
STRAINER, body side panel to wheel house	1802A
STRAINER, gasoline assembly	0306B
STRAINER, pump check	0301
STRAINER, spare tire carrier, upper, assembly	1505
STRAINER, spare tire carrier, w/FILLER, lower, assembly	1505
STRAINER UNIT assembly	0306B
STRAP, axe assembly	2301A
STRAP, battery to front fender	0610A
STRAP, battery ground	0601B
STRAP, battery ground	0610B
STRAP, bond No. 1	2603
STRAP, bond No. 3	2603
STRAP, bond No. 4	2603
STRAP, bond No. 5	2603
STRAP, bond No. 6	2603
STRAP, bond No. 7	2603
STRAP, bond No. 8	2603
STRAP, bond No. 9	2603
STRAP, bond No. 10	2603
STRAP, bond No. 11	2603
STRAP, bond No. 12	2603
STRAP, bond No. 13	2603
STRAP, bond No. 15	2603
STRAP, bond No. 16	2603
STRAP, bond No. 21 and STRAP, bond No. 22	2603
STRAP, bond, radio box negative terminal to floor ground connection	2601
STRAP, hold-down, gasoline tank inner, assembly	0300
STRAP, hold-down, gasoline tank outer, assembly	0300
STRAP, hold-down, top, rear assembly	2201A
STRAP, lock, differential drive gear screw	1003
STRAP, lock, drive gear screw	1103
STRAP, muffler support	0401
STRAP, safety assembly	1804
STRAP, shovel front assembly	2301A
STRAP, shovel rear assembly	2301A
STRAP, spare gas can, w/BUCKLE, assembly	1817A
STRAP, spare gas can, w/TIP, assembly	1817A
STRAP, top lock, assembly	2201A

Vendor Number	Willys Part Number	Ford Part Number	Gov't Group No.	Vendor Number	Willys Part Number	Ford Part Number	Gov't Group No.
AETNA BALL BEARING MFG. CO. (AB)				AL-MU-28	WO-109433	FM-GPW-11103	0602
AB-A-935-1	WO-635529	FM-GPW-7580	0203	AL-IGH-28	WO-A-1671	FM-GPW-12133	0603
				AL-GEB-29	WO-A-1600	FM-GPW-10041	0601
A.C. SPARK PLUG DIV. (AC)				AL-M-29	WO-A-1556	FM-GPW-11120	0602
AC-119922-21	WO-A-1265	FM-GPW-9154	0306B	AL-MZ-30-A	WO-A-1557	FM-GPW-11089	0602
AC-127951	WO-A-1264	FM-GPW-9185	0306B	AL-IGP-30	WO-A-1587	FM-GPW-12169	0603
AC-132629	WO-51546	FM-26466-S7	0302	AL-GEW-31	WO-A-1638	FM-34709-S7	0601
AC-846709	WO-A-1215	FM-GPW-8100-A	0501	AL-GAU-31	WO-A-1644	FM-78-10212-A	0601
AC-846732	WO-A-1216	FM-GPW-8578	0501	AL-MU-31	WO-109436	FM-GPW-11107	0602
AC-850024	WO-A-1275	FM-GPW-9035-B	0300	AL-MAB-31	WO-A-1553	FM-GPW-11094	0602
AC-853558	WO-A-1259	FM-GPW-9160	0306B	AL-GBJ-32-A	WO-A-1626	FM-GPW-10118	0601
AC-853562	WO-A-1257	FM-GPW-9184	0306B	AL-GAA-32	WO-A-1591	FM-GPW-10202-A	0601
AC-853572	WO-A-1260	FM-GPW-9186	0306B	AL-MZ-32	WO-A-1558	FM-GPW-11090	0602
AC-854005	WO-113460	FM-GPW-9388	0302	AL-GBW-34	WO-A-1593	FM-GPW-10206-A	0601
AC-855064	WO-113440	FM-34803-S7	0302	AL-MU-37	WO-A-1555	FM-34706-S	0602
AC-855493	WO-113439	FM-31628-S7	0302	AL-MZ-38-E	WO-A-1559	FM-355385-S	0602
AC-855763	WO-113461	FM-GPW-9373	0302	AL-GBY-38-A	WO-A-1596	FM-355486-S7	0601
AC-1504117	WO-A-1263	FM-GPW-9162	0306B	AL-DA-39	WO-A-1622	FM-GPW-10124	0601
AC-1504118	WO-A-1262	FM-GPW-9182	0306B	AL-MU-39	WO-109437	FM-GPW-11095	0602
AC-1504212	WO-A-1258	FM-GPW-9149	0306B	AL-GEB-44	WO-A-1601	FM-GPW-10100	0601
AC-1521288	WO-A-1047	FM-GPW-9377	0302	AL-GAL-44	WO-A-1592	FM-OIA-10193	0601
AC-1521578	WO-A-1046	FM-GPW-9378	0302	AL-GBA-46	WO-A-1573	FM-GPW-11350	0602
AC-1521708	WO-115880	FM-INC-9381	0302	AL-MZ-51	WO-A-1569	FM-GPW-11053	0602
AC-1521880	WO-115870	FM-GPW-19469	0302	AL-MZ-52	WO-A-1584	FM-355164-S2	0602
AC-1521953	WO-115652	FM-GPW-9363	0302	AL-GCE-53	WO-A-1628	FM-GPW-10057	0601
AC-1521956	WO-115653	FM-11A-9361	0302	AL-GCE-54	WO-A-1629	FM-GPW-10105	0601
AC-1521960	WO-115641	FM-GPW-9399	0302	AL-MU-54	WO-109455	FM-GPW-11036-A	0602
AC-1521985	WO-115869	FM-GPW-9468	0302	AL-GEB-58	WO-A-1605	FM-GPW-10211-B	0601
AC-1522046	WO-115643	FM-GPW-9380	0302	AL-DA-60	WO-A-1623	FM-355260-S7	0601
AC-1523068	WO-116694	. . . .	0302	AL-GBW-66	WO-A-1594	FM-GPW-10106	0601
AC-1523084	WO-115650	FM-GPW-9354	0302	AL-SPK-66	WO-A-8242	. . . .	2204
AC-1523096	WO-115656	FM-GPW-9364	0302	AL-GBW-67	WO-A-1595	FM-GPW-10202-C	0601
AC-1523099	WO-115654	FM-GPW-9365	0302	AL-IGT-69	WO-A-1684	FM-GPW-12191	0603
AC-1523106	WO-115651	FM-11A-9352	0302	AL-8X-72-B	WO-A-8127	. . . .	2204
AC-1523231	WO-115657	FM-GPW-9387	0302	AL-MG-77-A	WO-A-1583	FM-GPW-11395	0602
AC-1537065	WO-A-1494	FM-GPW-9355	0302	AL-GT-78	WO-A-1646	FM-GPW-10212	0601
AC-1537812	WO-A-1045	FM-GPW-9386	0302	AL-MAB-88	WO-A-1586	FM-72798-S7-8	0602
AC-1538205	WO-116695	FM-GPW-9398	0302	AL-IG-90	WO-106740	FM-GPW-12193	0603
AC-1538312	WO-A-8323	. . . .	0302	AL-IG-94	WO-A-1652	FM-GPW-12006	0604A
AC-1542508	WO-A-1313	. . . .	0301C	AL-IGS-99	WO-A-1672	FM-GPW-12120	0603
AC-1542509	WO-A-1314	. . . .	0301C	AL-IGS-104	WO-A-1673	FM-GPW-12182	0603
AC-1542510	WO-A-1315	. . . .	0301C	AL-EB-108	WO-A-1574	FM-B-11379	0602
AC-1595235	WO-A-1255	FM-GPW-9155	0306B	AL-IGC-117	WO-A-1660	FM-GPW-12174	0603
AC-1595823	WO-A-1261	FM-GPW-9140	0306B	AL-GCE-125-A	WO-A-6301	FM-GPW-10139	0601
				AL-8X-140	WO-A-1610	FM-34051-S7-8	0601
ALEMITE DIV. (AD)				AL-X-156	WO-A-1640	FM-34019-S7	0601
AD-1636	WO-368224	FM-353035-S7	0805B	AL-8X-177	WO-A-1611	FM-355883-S	0601
AD-1641	WO-638792	FM-353043-A-S7	0204	AL-12X-193	WO-A-1612	FM-356263-S7	0601
AD-1980	WO-640038	FM-358006-S8	1204	AL-12X-194	WO-A-1613	FM-34801-S7	0601
				AL-X-195	WO-51532	FM-34802-S	0601
ELECTRIC AUTO-LITE (AL)				AL-X-196	WO-5168	FM-78-10212-A	0601
AL-GG-6	WO-106313	FM-GPW-10098	0601	AL-12X-196	WO-A-1614	FM-34803-S7	0601
AL-DH-7	WO-A-1624	FM-GPW-10116	0601	AL-IGB-199	WO-A-1659	FM-GPW-12195	0603
AL-MZ-12	WO-109431	FM-GPW-11055	0602	AL-X-199	WO-5045	FM-34805-S7-8	0601
AL-MU-14	WO-A-1554	FM-GPW-11102	0602	AL-IGB-202-S	WO-A-1677	FM-GPW-12084	0603
AL-MZ-16	WO-109442	FM-GPW-11061	0602	AL-12X-203	WO-A-1619	FM-34806-S2	0601
AL-MZ-19	WO-109445	FM-B-11059	0602	AL-XS-213	WO-A-1685	FM-72867-S	0603
AL-MN-21	WO-A-1609	FM-GPW-10218	0601	AL-TS-215	WO-A-5433	FM-11AS-10657	0610
AL-DA-22	WO-A-1621	FM-356208-S7	0601	AL-TSR-215	WO-A-1767	FM-11-AS-10658	0610
AL-DK-23	WO-A-1590	FM-GPW-10120	0601	AL-X-260	WO-A-1641	FM-74144-S	0601
AL-GCT-25	WO-A-1598	FM-GPW-10104	0601	AL-X-261	WO-5017	FM-74175-S	0602
AL-GCY-25	WO-A-1599	FM-GPW-10206-A	0601	AL-X-295	WO-51248	FM-GPW-10094	0601
AL-GBJ-25	WO-A-1625	FM-GPW-10119	0601	AL-8X-305	WO-A-1632	FM-26457-S7	0601
AL-GC-26	WO-A-1597	FM-GPW-10208-A	0601	AL-8X-311	WO-A-1647	FM-36800-S7-8	0601
AL-GEB-27	WO-A-1602	FM-GPW-10211-A	0601	AL-8X-323	WO-A-1633	FM-36787-S7	0603

Vendor Number	Willys Part Number	Ford Part Number	Gov't Group No.
ELECTRIC AUTO-LITE (AL) Cont'd			
AL-8X-349	WO-A-1615	FM-34703-S7-8	0601
AL-8X-361	WO-A-1616	FM-34705-S	0601
AL-X-404	WO-A-1642	FM-72034-S	0601
AL-SP-484-A	WO-A-1639	FM-GPW-10130	0601
AL-X-489	WO-302347	FM-B-10141	0601
AL-X-490	WO-107128	FM-B-10141	0603
AL-IG-514	WO-A-1656	FM-GPW-12011	0603
AL-IG-515	WO-A-1657	FM-GPW-12012	0603
AL-X-532	WO-A-1567	FM-GPW-11069	0602
AL-12X-544	WO-A-1635	FM-34803-S7	0601
AL-X-544	WO-A-1571	FM-34803-S7	0602
AL-IG-579-A	WO-A-1681	FM-GPW-12082	0603
AL-IG-676	WO-A-1663	FM-GPW-12217	0603
AL-IG-680	WO-A-1653	FM-GPW-12177	0603
AL-IG-694	WO-A-1682	FM-GPW-12144	0603
AL-X-714	WO-A-1588	FM-36954-S7-8	0602
AL-8X-715	WO-A-1650	FM-27161-S7-8	0601
AL-8X-794	WO-A-1589	FM-34141-S7-8	0601
AL-IG-816-C	WO-A-1654	FM-GPW-12267	0603
AL-X-847	WO-A-1603	....	0601
AL-8X-870	WO-A-1636	FM-31588-S7-8	0601
AL-8X-872	WO-A-1686	FM-31583-S	0603
AL-8X-884	WO-A-1668	FM-31027-S8-7	0603
AL-X-902	WO-A-1572	FM-31596-S	0602
AL-X-959	WO-A-1605	FM-GPW-10211-B	0601
AL-GEB-1005	WO-A-1604	FM-GPW-10175	0601
AL-GEB-1007-A	WO-A-1606	FM-GPW-10192	0601
AL-MZ-1007	WO-A-1560	FM-GPW-11083	0602
AL-GEB-1008-B	WO-A-1607	FM-GPW-10191	0601
AL-MZ-1008	WO-109428	FM-GPW-11084	0602
AL-MZ-1009	WO-109427	FM-GPW-11082	0602
AL-MZ-1010	WO-A-1563	FM-GPW-11085	0602
AL-GCE-1012	WO-A-1630	FM-GPW-10069	0601
AL-X-1012	WO-A-1669	FM-34801-S	0603
AL-12X-1014	WO-A-5288	FM-34806-S7	0601
AL-X-1014	WO-5051	FM-34806-S2	0602
AL-GCE-24	WO-A-1649	FM-GPW-10142	0601
AL-MZ-1024	WO-109452	FM-GPW-11077	0602
AL-MZ-1034	WO-109446	FM-GPW-11056	0602
AL-PS-1079-A	WO-A-1585	FM-GPW-11131	0602
AL-IGS-1080	WO-A-1647	FM-36800-S7-8	0601
AL-SPS-1100-R	WO-A-1343	FM-GPW-17261	2204
AL-SPS-1105-X	WO-A-1344	FM-GPW-17262	2204
AL-GCE-1125-A	WO-A-6300	FM-GPW-10138	0601
AL-IGC-1132-LB	WO-A-1661	FM-GPW-12176	0603
AL-IGS-1134-L	WO-A-1678	FM-GPW-12178	0603
AL-IGC-1148	WO-A-1664	FM-GPW-12010	0603
AL-IGC-1149	WO-A-1654	FM-GPW-12267	0603
AL-IG-1324	WO-A-1655	FM-GPW-12106	0603
AL-8X-1368	WO-A-6297	FM-37789-S7	0601
AL-5X-1376	WO-A-1565	FM-355944-S5	0602
AL-8X-1377	WO-A-1617	FM-350853-S7	0601
AL-8X-1420	WO-A-1618	FM-36009S7	0601
AL-X-1448	WO-A-1683	FM-GPW-12145	0603
AL-8X-1546	WO-A-1670	FM-31026-S7	0603
AL-X-1655	WO-A-6299	FM-B-10094	0601
AL-IG-1657	WO-A-1658	FM-GPW-12200	0604
AL-IG-1798-D	WO-A-1526	FM-GPW-12030	0604
AL-MZ-2012-S	WO-A-1552	FM-GPW-18535	0602
AL-GCG-2017-S	WO-A-1651	FM-GPW-18274	0601
AL-MAB-2040-A	WO-A-1582	FM-GPW-11394	0602
AL-MZ-2089	WO-A-1568	FM-GPW-11005	0602
AL-SPS-2099-M	WO-A-1267	FM-GPW-17260	2204
AL-GCE-2118-A	WO-A-6298	FM-GPW-10050	0601

Vendor Number	Willys Part Number	Ford Part Number	Gov't Group No.
AL-GCE-2118	WO-A-1627	FM-GPW-10050	0601
AL-GEG-2120-F	WO-A-1637	FM-GPW-10005	0601
AL-IGS-2134-L	WO-A-1675	FM-GPW-12175	0603
AL-IGS-2135	WO-A-1679	FM-GPW-12139	0603
AL-IGC-2148-C	WO-A-1662	FM-GPW-12151	0603
AL-MZ-2156	WO-A-1566	FM-GPW-11949	0601
AL-2229	WO-A-8129	....	0605
AL-2235	WO-6536	FM-34079-S7	0605
AL-IG-2456	WO-A-1676	FM-GPW-12188	0603
AL-SW-4015	WO-A-7225	FM-9N-11450-A	0605
AL-IGP-3028	WO-A-1570	FM-GPW-12162	0603
AL-IGP-3028-S	WO-A-1687	FM-GPW-18354	0603
AL-IGW-3139	WO-A-1631	FM-GPW-12300	0604
AL-SPK-4003	WO-A-8180	FM-GPW-17255-A	2200
AL-MZ-4113	WO-A-1245	FM-GPW-11001-A	0602
AL-VRY-4203-A	WO-A-1409	FM-GPW-10505	0601B
AL-EBA-4611	WO-A-1581	FM-GPW-11354	0602
AL-IGC-4705	WO-A-1244	FM-GPW-12100	0603
AL-GED-5002-D	WO-A-5992	FM-GPW-10000-A	0601
AL-E-7819-S	WO-A-1575	FM-B-11357-A	0602
AL-8170	WO-5848	FM-34079-S8	0605
AL-EB-8503	WO-A-1576	FM-B-11381	0601
AL-EB-8505	WO-A-1577	FM-GPW-11375	0602
AL-RB-8506	WO-A-1578	FM-GPW-11377	0602
AL-EB-8507	WO-A-1579	FM-GPW-11382	0602
AL-EB-8734	WO-A-1580	FM-B-11371	0602
AL-9288-A	WO-A-8130	....	0605
AL-9979-A	WO-A-1292	FM-GPW-9275	0300
AL-10310-A	WO-A-8190	FM-GPW-9273	0605
AL-10311-A	WO-A-8186	FM-GPW-10850	0605
AL-10312-A	WO-A-8188	FM-GPW-10883	0605
AL-10063	WO-A-8132	....	0605
AL-21608	WO-A-8131	....	0605
AL-IG-40700	WO-A-7792	FM-GPW-12000-B	0604
AINSWORTH MFG. CO. (AN)			
AN-S-384	WO-A-2483	....	1811
AN-X-598	WO-A-2482	FM-38192-S2	1811
AN-S-1044	WO-A-4260	....	1811
AN-X-1637	WO-A-2490	FM-356286-S	1811
AN-X-1700	WO-A-2486	FM-34805-S2	1811
AN-X-1747	WO-A-2239	FM-81W-811189	1811
AN-X-1750	WO-A-2480	FM-38259-S2	1811
AN-X-1751	WO-A-2481	FM-34176-S2	1811
AN-X-1755	WO-A-2487	FM-33249-S2	1811
ATWOOD VACUUM MFG. CO. (AVM)			
AVM-TP-28-7-1	WO-638992	FM-GPW-7563	0201B
AVM-TP-283	WO-638157	FM-GPW-7567	0202
AVM-TP-287	WO-638159	FM-GPW-7564	0202
AVM-TP-2811	WO-638151	FM-GPW-7580	0202
AVM-TP-2817	WO-638153	FM-GPW-7590	0202
AVM-TP-2818	WO-638154	FM-24325-S	0202
AVM-TP-2819	WO-638155	FM-33921-S7	0202
AVM-TP-2820	WO-638158	FM-GPW-7591	0202
AVM-TP-2827	WO-638305	FM-34745-S2	0202
AVM-TP-2831	WO-638993	FM-GPW-7572	0202
AVM-TP-2851	WO-638152	FM-GPW-7566	0202
BASSICK CO. (THE) (BA)			
BA-B-194	WO-A-3203	FM-GPW-1103028	1811
BA-SK-1670	WO-A-313	FM-GP-17037	2301

Vendor Number	Willys Part Number	Ford Part Number	Gov't Group No.	Vendor Number	Willys Part Number	Ford Part Number	Gov't Group No.
BORG AND BECK (B-B)				CAR-11B-125S	WO-116163	FM-GPW-9696	0301
B-B-2940	WO-371567	FM-GPW-7549	0201A	CAR-11B-127S	WO-116164	FM-GPW-9928	0301
B-B-4324	WO-636778	FM-GPW-7577	0201A	CAR-11B-129S	WO-116165	FM-GPW-9543	0301
B-B-11123	WO-636755	FM-GPW-7550	0201B	CAR-12-255	WO-116166	FM-GPW-9922	0301
				CAR-14-246S	WO-116545	FM-GPW-9546	0301
BOWER ROLLER BEARING CO. (BOW)				CAR-20-22	WO-116168	FM-GPW-9569	0301
BOW-BT-14131	WO-51575	FM-GP-7175	0803	CAR-20-26	WO-116169	FM-GPW-9608	0301
BOW-BT-14276	WO-52883	FM-O2Y-1202	0803	CAR-20-61	WO-116170	FM-GPW-9574	0301
BOW-BT-18250	WO-52942	FM-GP-1201	1301A	CAR-20-72	WO-116171	FM-GPW-9926	0301
BOW-BT-18590	WO-52943	FM-GP-1202	1301A	CAR-21-74S	WO-116172	FM-GPW-9550	0301
				CAR-24-23	WO-116173	FM-GPW-9558	0301
BUDD MFG. CO., EDW. G. (BUE)				CAR-25-93S	WO-116174	FM-GPW-9567	0301
BUE-5052751	WO-A-11319	FM-GPW-1153100	1820	CAR-30-20	WO-116175	FM-GPW-9575	0301
				CAR-30A-39	WO-116176	FM-GPW-9541	0301
BENDIX PRODUCTS CORP. (BX)				CAR-39-10	WO-116216	FM-GPW-9586	0301
BX-40-S-33	WO-5010	FM-34807-S2	1500	CAR-39-11	WO-116217	FM-GPW-9588	0301
BX-62-S-176	WO-51738	FM-20300-S7	1207	CAR-43-67	WO-116179	FM-GPW-9544	0301
BX-76-S-25	WO-52483	FM-34905-S2	1207	CAR-48-84	WO-116180	FM-GPW-9940	0301
BX-179-S-6	WO-374586	FM-351915-S	0201A	CAR-53A-168S	WO-116181	FM-GPW-9528	0301
BX-1067-S-3	WO-116551	FM-GP-2021	1202	CAR-53A-251S	WO-116537	FM-GPW-9529	0301
BX-1067-S-4	WO-116552	FM-GP-2022	1202	CAR-61-57	WO-116183	FM-GPW-9578	0301
BX-1141-S-1	WO-116549	FM-GP-2018	1202	CAR-61-119	WO-116184	FM-GPW-9624	0301
BX-1141-S-2	WO-116550	FM-GP-2019	1202	CAR-61-128	WO-116185	FM-GPW-9615	0301
BX-39953	WO-637899	FM-91A-2027	1203B	CAR-61-143	WO-116186	FM-GPW-9650	0301
BX-39956	WO-637901	FM-91A-2030	1203B	CAR-61-169	WO-116187	FM-GPW-9570	0301
BX-41545	WO-637905	FM-GP-2035	1203A	CAR-61-171	WO-116188	FM-GPW-9636	0301
BX-41665	WO-637923	FM-351466-S24	1203B	CAR-61-190	WO-116189	FM-GPW-9587	0301
BX-41708	WO-637924	FM-33846-S2	1203B	CAR-61-207	WO-116191	FM-GPW-9935	0301
BX-41876	WO-637900	FM-GP-2028	1203B	CAR-61-272	WO-116538	FM-GPW-9907	0301
BX-41887	WO-637787	FM-GP-2261	1207	CAR-62-108S	WO-116543	FM-GPW-6333-D	0301
BX-44517	WO-116600	FM-GPW-18367	1202	CAR-62-131S	WO-116548	. . . .	0301
BX-45322	WO-A-927	FM-GPW-2211	1200	CAR-62-134	WO-116587	FM-GPW-9595	0301
BX-45323	WO-A-924	FM-GP-7718-A	0804	CAR-62-135S	WO-116586	FM-GPW-9526	0301
BX-45771	WO-A-751	FM-GPW-3635	0609B	CAR-63-35	WO-116194	FM-GPW-9614	0301
BX-45908	WO-A-1484	FM-GPW-2061	1207	CAR-64-62S	WO-116195	FM-GPW-9631	0301
BX-46491	WO-A-6110	. . . .	1207	CAR-75-547	WO-116540	FM-GPW-9906	0301
BX-46752	WO-A-755	FM-33800-S7	1203A	CAR-86-10	WO-6330	FM-34802-S2	0301
BX-47047	WO-A-8894	FM-GPW-2011	1200	CAR-86-11	WO-5045	FM-34805-S2	0301
BX-47048	WO-A-8895	FM-GPW-2010	1200	CAR-86-15	WO-5010	FM-34807-S2	0301
BX-47050	WO-A-8896	FM-GPW-2211-B	1200	CAR-101-10	WO-116211	FM-31032-S7	0301
BX-47051	WO-A-8897	FM-GPW-2210-B	1200	CAR-101-28	WO-52290	FM-31588-S7	0301
BX-47054	WO-A-8898	FM-GP-2013	1203	CAR-101-82	WO-116213	FM-31061-S8	0301
				CAR-101-120	WO-116651	FM-GPW-9610	0301
COLUMBUS AUTO PARTS (CA)				CAR-101-122	WO-116215	FM-31662-S	0301
CA-SX-1109	WO-A-622	FM-GPW-3332-A	1401	CAR-101-142S	WO-116384	FM-355067-S7	0301
CA-SX-1110	WO-A-623	FM-GPW-3336	1401	CAR-101-150S	WO-116385	FM-355200-S7	0301
CA-S-5069	WO-630755	FM-GPW-3320	1401	CAR-105-11	WO-116218	. . . .	0301
CA-S-5070	WO-630756	FM-GPW-3323	1401	CAR-105-13	WO-116588	FM-355132-S	0301
CA-S-5143	WO-630757	FM-GPW-3328	1401	CAR-114-21S	WO-116197	FM-GPW-9583	0301
CA-S-5144	WO-630753	FM-O2W-3326	1401	CAR-115-59	WO-116198	FM-GPW-9531	0301
CA-S-5145	WO-630754	FM-GPW-3327	1401	CAR-117-58	WO-116199	FM-GPW-9527	0301
CA-D-8822	WO-A-8252	FM-GPW-3305-B	1401	CAR-117-106	WO-116542	FM-GPW-9598	0301
				CAR-120-151S	WO-116541	FM-GPW-9914	0301
CARTER CARBURETOR (CAR)				CAR-121-56	WO-116202	FM-GPW-9516	0301
CAR-1-407	WO-116584	FM-GPW-9518	0301	CAR-121-73	WO-116203	FM-GPW-9519	0301
CAR-2-89	WO-116154	FM-GPW-9585	0301	CAR-122-47S	WO-116204	FM-GPW-119594	0301
CAR-3-465-S	WO-116585	FM-GPW-9581	0301	CAR-122-64S	WO-116205	FM-GPW-9576	0301
CAR-3-4664	WO-A-5501	. . . .	0301	CAR-129-15	WO-116206	FM-GPW-9905	0301
CAR-6-312-S	WO-116544	FM-GPW-9520	0301	CAR-136-39	WO-116207	FM-34711-S	0301
CAR-7-116S	WO-116157	FM-GPW-9549	0301	CAR-146-95S	WO-116208	FM-GPW-9515	0301
CAR-11-180S	WO-116539	FM-GPW-9533	0301	CAR-150-97	WO-116209	FM-GPW-9930	0301
CAR-11B-35	WO-116159	FM-GPW-9522	0301	CAR-150-98	WO-116177	FM-GPW-9566	0301
CAR-11B-79	WO-116160	FM-GPW-9523	0301	CAR-150A-8	WO-116219	FM-355858-S	0301
CAR-11B-105	WO-116161	FM-GPW-9562	0301	CAR-150A-13	WO-50922	FM-33800-S2	0301
CAR-11B-108	WO-116162	FM-GPW-9579	0301	CAR-150A-19	WO-52615	FM-34051-S7	0301

Vendor Number	Willys Part Number	Ford Part Number	Gov't Group No.
CARTER CARBURETOR (CAR) Cont'd			
CAR-183-19	WO-116210	FM-GPW-9554	0301
CAR-595	WO-A-1223	FM-GPW-9510	0301
CHICAGO SCREW CO. (CCG)			
CCG-W-41-25	WO-637427	FM-78-2814A	1209C
CORCORAN BROWN (CB)			
CB-CB-300	WO-52221	FM-34803-S	0607
CB-CB-4057	WO-A-1032	FM-37206-S	0607
CB-CB-5099	WO-A-1037	FM-38095-S	0607
CB-CB-6012	WO-A-1361	FM-GPW-13022	0607
CB-CB-7783	WO-53071	FM-40631-S7	0607D
CB-CB-7784	WO-A-6313	FM-GPW-13180	0607D
CB-7861	WO-A-1072	FM-28378-S2	0607D
CB-CB-8010	WO-A-1031	FM-GPW-13022	0607
CB-CB-8494	WO-A-1033	FM-GPW-13007	0607A
CB-CB-8523	WO-A-1036	FM-GPW-13048	0607
CB-9204	WO-A-1070	FM-GP-13210-B	0607D
CB-CB-9212	WO-A-1073	FM-GPW-13408-B	0608
CB-CB-9218	WO-A-1074	FM-GPW-13494	0607A
CB-CB-9225	WO-A-1075	FM-GPW-13491-A	0607A
CB-CB-9231	WO-A-1076	FM-GPW-13448-B	0608
CB-CB-9232	WO-A-1079	FM-GPW-13349-A	0608
CB-CB-9233	WO-A-1077	FM-36931-S	0608
CB-CB-9234	WO-A-1078	FM-GPW-13485	0607A
CB-B-9276	WO-A-5806	. . . .	0607D
CB-CB-9281	WO-A-1071	FM-GP-13209	0607D
CB-CB-10436	WO-A-5586	FM-GPW-13012	0607
CB-CB-10461	WO-A-1439	FM-GPW-13217	0607D
CB-CB-10462	WO-A-1440	FM-GPW-13216-B	0607D
CB-CB-11193	WO-A-6144	FM-GPW-13162	0607D
CB-CB-11199	WO-A-6783	FM-GPW-13166	0607D
CB-CB-11200	WO-53070	. . . .	0607D
CB-CB-11261	WO-A-6145	FM-GPW-13153	0607A
CB-CB-11263	WO-A-6148	FM-131045-S2	0607D
CB-11495	WO-A-6142	FM-GPW-13150	0607D
CB-CB-11500	WO-A-6143	FM-GWP-13170	0607D
CB-CB-11503	WO-A-6146	F M-GPW-13175	0607D
CB-60142	WO-A-1064	FM-GPW-13405-B	0608
CB-60242	WO-A-1065	FM-GPW-13040-B	0608
CB-CB-215142	WO-A-1304	FM-GPW-13006	0607
CB-CB-215242	WO-A-1305	FM-GPW-13005	0607
CB-415342	WO-A-1346	. . . .	0606
CB-415442	WO-A-1347	. . . .	0606
CLUM MFG. CO. (CL)			
CL-9654	WO-638979	FM-GPW-13532	0606
COPELAND GIBSON PRODUCTS CORP. (CN)			
CN-310-0	WO-637237	FM-GPW-6702	0102A
CINCH MFG. CO. (CNC)			
CNC-0	WO-A-3113	FM-348148-S	2201A
CNC-12	WO-A-1312	FM-GPW-13802	0609
CNC-50	WO-A-3052	FM-95626-S	2201F
CNC-58	WO-A-1311	FM-GPW-9652	0301C
CNC-60	WO-A-3053	FM-95628-S	2201F
CNC-98	WO-A-3054	FM-95627-S	2201A
CNC-515	WO-A-3051	FM-95705-S	2201A
CNC-519	WO-A-4120	FM-352699-S15	1811
CHAMPION SPARK PLUG CO. (CP)			
CP-QM2	WO-A-538	FM-GPW-12405	0604B

Vendor Number	Willys Part Number	Ford Part Number	Gov't Group No.
DOUGLAS MFG. CO., H. A. (DM)			
DM-D-398	WO-A-6149	FM-GP-13739	0606
DM-5943	WO-A-1347	. . . .	0606
DM-5944	WO-A-1348	. . . .	0606
DM-5969	WO-A-1345	FM-34907-S7-8	0606
DM-5970	WO-A-1332	FM-GPW-11649	0606
DM-5995	WO-A-1333	FM-GPW-13740	0606
DM-6000	WO-A-1351	. . . .	0606
DM-5999	WO-A-1352	. . . .	0606
DM-29392	WO-A-1748	FM-GPW-13713	0605C
DM-5033	WO-52131	FM-26457-S7	0604C
DM-51104	WO-A-1350	FM-356229-S	0606
DM-52947	WO-52332	. . . .	0606
DM-53097-A	WO-A-1349	FM-GPW-12250-C	0606
DM-53170-A	WO-A-6813	. . . .	0604C
DM-53175	WO-A-1346	. . . .	0606
DM-53414	WO-A-1353	. . . .	0606
DEUTSCHMAN, TOBE, CORP. (DTC.)			
DTC-1107G	WO-A-1517	. . . .	2603
DTC-1126	WO-A-1287	FM-GPW-18936-A	2603
EATON MFG. CO. (EAT)			
EAT-LE-1212	WO-A-1126	FM-9N-8115	0501
ERIE RESISTOR CORP. (ER)			
ER-L7	WO-A-6320	. . . .	2603
FEDERAL BEARINGS CO. (FB)			
FB-1207MGF	WO-636885	FM-GPW-7025	0703

Ford Part Number	Willys Part Number	Gov't Group No
FORD MOTOR CO. (FM)		
FM-GP-1012	WO-A-476	1302A
FM-GP-1013	WO-A-475	1302A
FM-GP-1015	WO-A-465	1301
FM-GPW-1015	WO-A-5467	1301
FM-GPW-1016	WO-A-5468	1301
FM-GPW-1024	WO-A-5549	1301
FM-GPW-1025-C	WO-A-5488	1301
FM-GPW-1029	WO-A-5470	1301
FM-GPW-1030	WO-A-5471	1301
FM-GPW-1045	WO-A-5472	1301
FM-GP-1088	WO-A-813	1006
FM-GP-1089	WO-A-814	1006
FM-GP-1092	WO-A-820	1006
FM-GP-1102	WO-A-1690	1302
FM-GP-1103	WO-A-1689	1302
FM-GP-1107	WO-A-474	1302A
FM-GP-1108	WO-A-473	1302A
FM-GP-1110	WO-A-760	1007
FM-GP-1124	WO-A-867	1102
FM-GP-1139	WO-A-869	1302
FM-GP-1177	WO-A-864	1301B
FM-GP-1201	WO-52942	1301A
FM-O1Y-1202	WO-52883	0803
FM-GP-1202	WO-52943	1301A
FM-GP-1218	WO-A-865	1102
FM-GPW-1418	WO-A-11701	1505
FM-GPW-1433	WO-A-2359	1505
FM-GPW-2010	WO-A-8895	1200
FM-GPW-2011	WO-A-8894	1200
FM-GP-2013	WO-A-8898	1203
FM-GP-2018	WO-116549	1202
FM-GP-2019	WO-116550	1202

Ford Part Number	Willys Part Number	Gov't Group No.	Ford Part Number	Willys Part Number	Gov't Group No.
FORD MOTOR CO. (FM) Cont'd			FM-GPW-2263	WO-A-1501	1209C
FM-GP-2021	WO-116551	1202	FM-GPW-2264	WO-A-1377	1209C
FM-GP-2022	WO-116552	1202	FM-GPW-2265	WO-A-5224	1209C
FM-91A-2027	WO-637899	1203B	FM-GPW-2266	WO-A-1376	1209C
FM-GP-2028	WQ-637900	1203B	FM-GPW-2267	WO-A-5226	1209C
FM-91A-2030	WO-637901	1203B	FM-GPW-2268	WO-A-5225	1209C
FM-GP-2035	WO-637905	1203A	FM-GPW-2270	WO-A-1795	1201
FM-GP-2038	WO-A-754	1203A	FM-GPW-2270-B	WO-A-5393	1201
FM-GPW-2061	WO-A-1484	1207	FM-GPW-2272	WO-A-1735	1201
FM-GPW-2063	WO-A-1502	1207	FM-GPW-2274	WO-A-5227	1209C
FM-GPW-2073	WO-A-1283	1500	FM-GPW-2279	WO-638780	1201
FM-GP-2074	WO-637432	1209C	FM-GPW-2281	WO-A-5449	0606B
FM-GP-2076	WO-A-557	1205	FM-GPW-2298	WO-A-1488	1209C
FM-GP-2077	WO-637605	1205	FM-GPW-2452-B	WO-A-8253	1204
FM-GP-2078	WO-637424	1209B	FM-GPW-2454	WO-A-1359	1204
FM-GPW-2078	WO-A-1373	1209B	FM-GPW-2473	WO-A-495	0204
FM-GPW-2079	WO-A-1460	1209B	FM-GPW-2476-A	....	0204
FM-GPW-2082	WO-A-1487	1001	FM-GPW-2598	WO-A-1008	1201
FM-GPW-2084	WO-A-1486	1001	FM-GPW-2614	WO-A-1002	1201
FM-GP-2087	WO-637426	1209B	FM-O1T-2616	WO-A-1020	1201
FM-GPW-2096	WO-A-1457	1001	FM-GP-2620	WO-A-1014	1201
FM-GP-2133	WO-384710	1209C	FM-GPW-2630	WO-A-1005	1201
FM-GPW-2135	WO-A-6111	1207	FM-GPW-2632	WO-A-1003	1201
FM-GPW-2138	WO-A-1354	1205	FM-O1T-2634	WO-A-1017	1201
FM-GP-2140	WO-A-556	1205	FM-GPW-2635	WO-A-5335	1201
FM-GP-2143-A	WO-637599	1205	FM-O1T-2640	WO-A-1021	1201
FM-GP-2145	WO-637587	1205	FM-O1T-2642	WO-A-1016	1201
FM-91A-2151	WO-637606	1205	FM-GP-2648	WO-A-1009	1201
FM-91A-2152	WO-637604	1205	FM-GPW-2656	WO-A-1226	1201
FM-GP-2155	WO-637582	1205	FM-GPW-2659	WO-A-1228	1201
FM-GP-2160	WO-637583	1205	FM-GPW-2780	WO-A-1242	1201
FM-GP-2162	WO-637608	1205	FM-GPW-2782	WO-639244	0703
FM-GP-2167	WO-637612	1205	FM-O1T-2805	WO-A-1018	1201
FM-GP-2169	WO-637591	1205	FM-78-2814-A	WO-637427	1209C
FM-GP-2170	WO-637595	1205	FM-GPW-2848	WO-639010	1201
FM-GP-2173	WO-637590	1205	FM-GPW-2852	WO-A-2892	1201
FM-GP-2174	WO-637598	1205	FM-GPW-2853	WO-A-1241	1201
FM-GP-2175	WO-637584	1205	FM-GPW-3001	....	1000
FM-GP-2176	WO-637585	1205	FM-GP-3016-A	WO-A-1727	1007
FM-GP-2180	WO-637602	1205	FM-GP-3017-A	WO-A-1729	1007
FM-GP-2183	WO-637586	1205	FM-GP-3031-A	WO-A-778	1301B
FM-GP-2188	WO-637597	1205	FM-GP-3034	WO-A-779	1006
FM-GP-2192	WO-637789	1207	FM-GPW-3074	....	1001
FM-GP-2194	WO-637577	1207	FM-GP-3105	WO-A-851	1006
FM-GP-2196	WO-637541	1207	FM-GPW-3112	WO-A-1710	1006
FM-GPW-2196	WO-A-6113	1207	FM-GPW-3113	WO-A-1712	1006
FM-91A-2201	WO-637591	1205	FM-GP-3115	WO-A-824	1006
FM-WP-2201	WO-635544	1207	FM-GP-3117-A	WO-A-830	1006
FM-GPW-2201	WO-A-6116	1207	FM-GP-3117-B	WO-A-831	1006
FM-GP-2204	WO-637545	1207	FM-GP-3117-C	WO-A-832	1006
FM-GP-2205	WO-637580	1207	FM-GP-3117-D	WO-A-833	1006
FM-GPW-2206	WO-A-6117	1207	FM-GP-3122	WO-A-825	1006
FM-GP-2206-A	WO-637546	1207	FM-GPW-3131-B	WO-A-8249	1006
FM-GP-2208	WO-637540	1207	FM-GP-3135	WO-A-819	1006
FM-GPW-2210	WO-A-928	1200	FM-GP-3139	WO-A-818	1006
FM-GPW-2210-B	WO-A-8897	1200	FM-GP-3140	WO-A-828	1006
FM-GPW-2211	WO-A-927	1200	FM-GP-3148-A	WO-A-811	1006
FM-GPW-2211-B	WO-A-8896	1200	FM-GP-3149-A	WO-A-812	1006
FM-GPW-2223	....	1209C	FM-GP-3161	WO-52940	1006
FM-GPW-2244	WO-A-1378	1209C	FM-GP-3162	WO-52941	1006
FM-GPW-2250	WO-A-1515	1209C	FM-GPW-3165	WO-A-855	1401
FM-GP-2261	VENDOR-637787	1207	FM-GP-3166	WO-A-856	1401
FM-GPW-2261	WO-A-6110	1207	FM-GPW-3167	WO-A-858	1401
			FM-GPW-3168	WO-A-859	1401
			FM-GPW-3169	WO-A-860	1401

Ford Part Number	Willys Part Number	Gov't Group No.	Ford Part Number	Willys Part Number	Gov't Group No.
FORD MOTOR CO. (FM) Cont'd			FM-GPW-3631	WO-A-750	0609B
FM-GPW-3170	WO-A-861	1401	FM-GPW-3635-B	WO-A-751	0609B
FM-GPW-3171	WO-A-857	1006	FM-GPW-3646	WO-638885	0609B
FM-GP-3200-A	WO-A-1716	1007	FM-GPW-3652	WO-A-747	0609B
FM-GP-3204	WO-A-868	1007	FM-GPW-3655	WO-A-633	0609B
FM-GPW-3206-A1	WO-A-1715	1007	FM-GPW-3658	WO-A-2859	1405
FM-GPW-3207-A1	WO-A-1728	1007	FM-GPW-3682-A	WO-A-1277	1405
FM-GP-3208-A	WO-A-862	1007	FM-GPW-3685	WO-A-2518	0604C
FM-GP-3208-B	WO-A-863	1007	FM-GPW-3685-B	WO-A-6814	0604C
FM-GP-3215-A	WO-A-1719	1007	FM-GPW-3686-B	WO-A-6811	0604C
FM-GP-3216	WO-A-1726	1007	FM-GPW-4001	WO-A-5500	1100
FM-GP-3217	WO-A-1724	1007	FM-GPW-4004	WO-A-888	1101
FM-GP-3218	WO-A-1723	1007	FM-GP-4016	WO-A-781	1001
FM-GP-3219	WO-A-1722	1007	FM-GPW-4022	WO-A-870	1001
FM-GP-3221-A	WO-A-1720	1007	FM-GP-4032	WO-A-904	1102
FM-GPW-3279	WO-A-1708	1402	FM-GP-4035	WO-A-782	1001
FM-GPW-3280	WO-A-1704	1402	FM-GP-4206	WO-A-793	1103
FM-GPW-3281	....	1402	FM-GPW-4209	....	1103
FM-GPW-3282	WO-A-1709	1402	FM-GP-4211	WO-A-798	1103
FM-51-3287	WO-A-1706	1402	FM-GPW-4212	WO-A-5566	1103
FM-GP-3291	WO-A-838	1402	FM-GPW-4215	WO-A-796	1103
FM-GP-3292	WO-A-847	1402	FM-GP-4221	WO-52880	1003
FM-GPW-3304-B	WO-A-8250	1401	FM-GP-4222	WO-52881	1003
FM-GPW-3305-B	WO-A-8252	1401	FM-GP-4228	WO-A-795	1003
FM-GPW-3320	WO-630755	1401	FM-GP-4229-A	WO-A-784	1103
FM-GPW-3323	WO-630756	1401	FM-GP-4229-B	WO-A-785	1003
FM-GPW-3325	WO-A-5504	1006	FM-GP-4229-C	WO-A-786	1003
FM-GPW-3326	WO-630753	1401	FM-GP-4229-D	WO-A-787	1003
FM-GPW-3327	WO-630754	1401	FM-GP-4229-E	....	1103
FM-GPW-3328	WO-630757	1401	FM-GP-4230	WO-A-797	1003
FM-78-3332-A	....	1402	FM-GPW-4234	WO-A-901	1102
FM-GPW-3332-A	WO-A-622	1401	FM-GP-4235	WO-A-902	1102
FM-78-3336	WO-A-844	1402	FM-GPW-4236	WO-A-794	1103
FM-GPW-3336	WO-A-623	1401	FM-GP-4241	WO-636360	1103
FM-GPW-3365	WO-A-1326	0901	FM-GP-4252	WO-A-866	1301C
FM-GPW-3370	WO-A-1428	0901	FM-GPW-4259	WO-A-6439	1102
FM-GP-3374	WO-A-780	1007	FM-GP-4281	WO-A-792	1003
FM-GPW-3504	WO-A-1239	1403A	FM-O1Y-4529	WO-A-490	0902
FM-GPW-3506	WO-A-635	1405	FM-GPW-4602	WO-A-1327	0901
FM-GPW-3509	WO-A-1199	1403A	FM-GPW-4605	WO-A-1429	0902
FM-GPW-3511	WO-A-1276	1405	FM-86H-4616	WO-52877	1104
FM-GPW-3517	WO-639190	1403A	FM-GP-4619	WO-636566	1104
FM-GPW-3518	WO-639192	1403A	FM-86H-4621	WO-52876	1104
FM-GPW-3520	WO-639191	1403A	FM-GP-4628	WO-52879	1003
FM-GPW-3524	WO-A-742	1403A	FM-GP-4630	WO-52878	1104
FM-GPW-3548	WO-A-740	1403A	FM-GP-4659-A	WO-A-803	1003
FM-GPW-3552	WO-639102	1403A	FM-GP-4659-B	WO-A-804	1003
FM-GPW-3563	WO-638918	1405	FM-GP-4659-C	WO-A-805	1003
FM-GPW-3568	WO-A-1760	1403A	FM-GP-4659-D	WO-A-806	1003
FM-GPW-3571	WO-639104	1403A	FM-GP-4660-A	WO-A-800	1104
FM-GPW-3575	WO-A-745	1403A	FM-GP-4660-B	WO-A-801	1003
FM-GPW-3576	WO-639091	1403A	FM-GP-4660-C	WO-A-802	1003
FM-GPW-3577	WO-639118	1403A	FM-GP-4661	WO-636565	1104
FM-GPW-3581	WO-639119	1403A	FM-GP-4666	WO-636568	1104
FM-GPW-3583	WO-639116	1403A	FM-GP-4668	WO-A-799	1104
FM-GPW-3587	WO-639090	1403A	FM-GP-4676-A	WO-639265	1003
FM-GPW-3589	WO-639103	1403A	FM-GP-4841	WO-A-935	0902
FM-GPW-3590	WO-A-1116	1403A	FM-GP-4842	WO-A-1445	1104
FM-GPW-3591-A	WO-639095	1403A	FM-GP-4863	WO-A-1105	0803
FM-GPW-3593	WO-639108	1403A	FM-GP-4866	WO-A-950	0902
FM-GPW-3594	WO-639109	1403A	FM-GPW-5005	WO-A-1142	1500
FM-GPW-3595	WO-639110	1403A	FM-GPW-5019	....	1500
FM-GPW-3600-A3	WO-A-6858	1404	FM-GPW-5025	WO-A-5127	1500
FM-GPW-3626	WENO-638884	0609B	FM-GPW-5028	WO-A-1150	1500
FM-GPW-3627	WO-A-634	0609B	FM-GPW-5035	WO-A-547	1500

Ford Part Number	Willys Part Number	Gov't Group No.		Ford Part Number	Willys Part Number	Gov't Group No.
FORD MOTOR CO. (FM) Cont'd				FM-GPW-5566	WO-A-614-3	1601A
FM-GPW-5057	WO-A-1201	1500		FM-GPW-5567	WO-A-614-4	1601A
FM-GPW-5072	....	1500		FM-GPW-5568	WO-A-614-5	1601A
FM-GPW-5084	WO-A-6740	1500		FM-GPW-5569	WO-A-614-6	1601A
FM-GPW-5095	WO-A-1341	1500		FM-GPW-5570	WO-A-614-7	1601A
FM-GPW-5097	WO-A-534	1500		FM-GPW-5571	WO-A-614-8	1601A
FM-GPW-5106	WO-A-415	1500		FM-GPW-5572	WO-A-614-9	1601A
FM-GPW-5113	WO-A-185	1500		FM-GPW-5601	WO-A-6326	1601C
FM-GPW-5116	WO-A-1422	1500		FM-GPW-5602	WO-A-6068	1601C
FM-GPW-5125	WO-A-1151	1500		FM-GPW-5605	WO-A-6069	1601C
FM-GPW-5165	WO-A-1291	0610A		FM-GPW-5607	WO-A-6073	1601C
FM-GPW-5168	WO-A-1757	0610A		FM-GPW-5608	WO-A-6075	1601C
FM-GPW-5175	WO-A-1164	0610A		FM-GPW-5609	WO-A-6074	1601C
FM-GPW-5182	WO-A-593	1502		FM-GPW-5705-B	WO-A-575	1000
FM-GPW-5186	WO-A-6393	1502		FM-GPW-5724-A	WO-116459	1601A
FM-GPW-5196	WO-A-522	1502		FM-GPW-5724-B	WO-116589	1601A
FM-GPW-5230-A	WO-A-1146	0401		FM-GPW-5778-A	WO-A-513	1602
FM-GPW-5230-B	WO-A-6118	0401		FM-GPW-5779	WO-A-514	1602
FM-GPW-5246	WO-A-1296	0402		FM-GPW-5781	WO-359039	1601A
FM-GPW-5251	WO-A-1300	0402		FM-GPW-5783-A	WO-A-481	1601B
FM-GPW-5264-A	WO-A-655	0401		FM-GPW-6002	WO-A-5338	0100
FM-GPW-5264-B	WO-A-5753	0401		FM-GPW-6005	WO-A-5497	0100
FM-GPW-5270	WO-636004	0402		FM-GPW-6009	WO-A-5793	0101
FM-GPW-5274	WO-638058	0401		FM-GPW-6010	WO-A-1272	0101
FM-GPW-5283	WO-A-658	0401		FM-GPW-6016	WO-A-1190	0106
FM-GPW-5291-B	WO-A-1253	1500		FM-GPW-6020	WO-630359	0101A
FM-GPW-5298	WO-A-6119	0402A		FM-GPW-6031-A3	WO-A-1463	0101
FM-GPW-5310	WO-A-613	1601		FM-74-6038	WO-634758	0809
FM-GPW-5311	WO-A-612	1601		FM-GPW-6038-A	WO-A-7498	0110
FM-GPW-5313-A	WO-A-613-1	1601		FM-GPW-6040-B	WO-A-6156	0110
FM-GPW-5313-B	WO-A-612-1	1601		FM-GPW-6043	WO-A-146	0110
FM-GPW-5315-A	WO-A-613-2	1601		FM-GPW-6044	WO-A-5125	0110
FM-GPW-5315-B	WO-A-612-2	1601		FM-GPW-6050	WO-A-1534	0101A
FM-GPW-5316-A	WO-A-613-3	1601		FM-GPW-6051-B	WO-A-8558	0101A
FM-GPW-5316-B	WO-A-612-3	1601		FM-GPW-6065	WO-638635	0101A
FM-GPW-5317-A	WO-A-613-4	1601		FM-GPW-6066	WO-349368	0101A
FM-GPW-5317-B	WO-A-612-4	1601		FM-GPW-6067	WO-A-1548	0101
FM-GPW-5318-A	WO-A-613-5	1601		FM-GPW-6105-A	WO-637041	0103
FM-GPW-5318-B	WO-A-612-5	1601		FM-GPW-6105-C	WO-116019	0103
FM-GPW-5319-A	WO-A-613-6	1601		FM-GPW-6105-D	WO-116020	0103
FM-GPW-5319-B	WO-A-612-6	1601		FM-GPW-6135-A	WO-636961	0103B
FM-GPW-5320-A	WO-A-613-7	1601		FM-GPW-6149-A	WO-116110	0103A
FM-GPW-5320-B	WO-A-612-7	1601		FM-GPW-6149-C	WO-116112	0103A
FM-GPW-5321-A	WO-A-613-8	1601		FM-GPW-6149-D	WO-116113	0103A
FM-GPW-5321-B	WO-A-612-8	1601		FM-GPW-6149-E	WO-A-6794	0103A
FM-GPW-5330-A	WO-116458	1601		FM-GPW-6149-G	WO-A-6796	0103A
FM-GPW-5330-B	WO-116460	1601		FM-GPW-6149-H	WO-A-6797	0103A
FM-GPW-5337	WO-A-544	1500		FM-GPW-6150-A	WO-639864	0103A
FM-GPW-5341	WO-A-500	1500		FM-GPW-6150-C	WO-116502	0103A
FM-GPW-5345-A	WO-116609	1601		FM-GPW-6150-D	WO-116503	0103A
FM-GPW-5345-B	WO-116610	1601A		FM-GPW-6152-A	WO-116562	0103A
FM-GPW-5453	WO-A-574	1000		FM-GPW-6152-C	WO-116564	0103A
FM-GPW-5455	WO-A-6511	1000		FM-GPW-6152-D	WO-116565	0103A
FM-GPW-5456	WO-339372	1100		FM-GPW-6155-A	WO-637042	0103A
FM-GPW-5458	WO-A-568	1000		FM-GPW-6155-G	WO-116023	0103A
FM-GPW-5459	WO-A-572	1000		FM-GPW-6155-H	WO-116024	0103A
FM-GPW-5460	WO-A-571	1100		FM-GPW-6156-F	WO-116616	0103A
FM-GPW-5463	....	1602		FM-GPW-6156-H	WO-116116	0103A
FM-GPW-5464	....	1602		FM-GPW-6156-J	WO-116117	0103A
FM-GPW-5468	WO-384228	1602		FM-GPW-6159-A	WO-116566	0103A
FM-GPW-5481	WO-A-515	1602		FM-GPW-6159-C	WO-116568	0103A
FM-GPW-5482	WO-A-1252	1602		FM-GPW-6159-D	WO-116569	0103A
FM-GPW-5560	WO-A-614	1601A		FM-GPW-6200	WO-640071	0104
FM-GPW-5563	VENDOR-A-614-1	1601A		FM-GPW-6201	WO-640072	0104
FM-GPW-5565	WO-A-614-2	1601A		FM-GPW-6211-A	WO-639862	0104A

Ford Part Number	Willys Part Number	Gov't Group No.	Ford Part Number	Willys Part Number	Gov't Group No.
FORD MOTOR CO. (FM) Cont'd			FM-GPW-6617	WO-630397	0107A-1
FM-GPW-6211-B	WO-116534	0104A	FM-GPW-6619	WO-630392	0107
FM-GPW-6211-C	WO-116535	0104A	FM-GPW-6625	WO-375927	0107
FM-GPW-6243	WO-375907	0106	FM-GPW-6627	WO-630398	0107A-1
FM-GPW-6244	WO-375908	0106	FM-GPW-6628	WO-630389	0107
FM-GPW-6245	WO-375900	0106	FM-GPW-6630	WO-630394	0107
FM-GPW-6250	WO-637065	0106A	FM-GPW-6642	WO-634813	0107
FM-GPW-6256	WO-638458	0106C	FM-GPW-6644	WO-630390	0107
FM-GPW-6260	WO-638457	0106C	FM-GPW-6648	WO-630299	0105C
FM-GPW-6262-A	WO-639051	0106A	FM-GPW-6654	WO-356155	0107
FM-GPW-6269	WO-315932	0601C	FM-GPW-6659	WO-639870	0107
FM-GPW-6285	WO-630364	0106D	FM-GPW-6663	WO-630518	0107
FM-GPW-6286	WO-375917	0106D	FM-GPW-6664	WO-630387	0107
FM-GPW-6287	WO-375920	0102	FM-GPW-6673	WO-636600	0107
FM-GPW-6288	WO-630365	0106D	FM-GPW-6675	WO-A-7238	0107A
FM-GPW-6306	WO-638459	0102	FM-GPW-6684	WO-330964	0107
FM-GPW-6308	WO-634796	0102	FM-GPW-6700	WO-637098	0102A
FM-GPW-6310	WO-375877	0102	FM-GPW-6701	WO-637790	0102A
FM-GPW-6312	WO-638113	0102	FM-GPW-6702	WO-637237	0102A
FM-GPW-6319	WO-387633	0102	FM-GPW-6718	WO-A-6885	0107N
FM-GPW-6326	WO-630294	0102A	FM-GPW-6722	WO-A-6885	0107N
FM-GPW-6331-A	WO-638732	0102A	FM-GPW-6727	WO-639979	0107A
FM-GPW-6331-B	WO-639239	0102A	FM-GPW-6734	WO-314338	0107A
FM-GPW-6331-C	WO-116530	0102A	FM-GPW-6756	WO-A-6922	0107N
FM-GPW-6333-A	WO-637007	0102A	FM-GPW-6758	WO-A-1061	0107N
FM-GPW-6333-B	WO-637724	0102A	FM-GPW-6758-B	WO-A-6919	0107N
FM-GPW-6333-C	WO-116522	0102A	FM-GPW-6762	WO-630298	0107N
FM-GPW-6337-A	WO-638733	0102A	FM-GPW-6763-A	WO-639555	0107G
FM-GPW-6337-B	WO-639240	0102A	FM-GPW-6763-B	WO-A-5165	0107G
FM-GPW-6337-C	WO-116532	0102A	FM-GPW-6763-C	WO-A-6915	0107G
FM-GPW-6338-A	WO-637008	0102A	FM-GPW-6766-A	WO-639556	0107H
FM-GPW-6338-B	WO-637725	0102A	FM-GPW-6766-B	WO-A-5168	0107H
FM-GPW-6338-C	WO-116524	0102A	FM-GPW-6766-C	WO-A-6525	0107H
FM-GPW-6339-A	WO-638730	0102A	FM-GPW-6769	WO-A-6895	0107N
FM-GPW-6339-B	WO-639237	0102A	FM-GPW-6770	WO-A-5105	0107G
FM-GPW-6339-C	WO-116526	0102A	FM-GPW-6771	WO-A-6918	0301C
FM-GPW-6341-A	WO-638731	0102A	FM-GPW-6772	WO-53108	0301C
FM-GPW-6341-B	WO-639238	0102A	FM-GPW--6789	WO-A-7280	0107H
FM-GPW-6341-C	WO-116528	0102A	FM-GPW-7000	WO-A-1145	0700
FM-GPW-6342-A	WO-630727	0106C	FM-GPW-7005	WO-A-1148	0701
FM-GPW-6342-B	WO-630262	0102	FM-GPW-7008	WO-A-7596	0700
FM-GPW-6345	WO-381519	0102	FM-GPW-7015	WO-A-5553	0703
FM-GPW-6353	WO-334103	0102	FM-GPW-7017	WO-A-5554	0703
FM-GPW-6369	WO-635377	0102	FM-GPW-7023	WO-375217	0109C
FM-GPW-6384	WO-635394	0109B	FM-GPW-7025	WO-636885	0703
FM-GPW-6387	WO-632156	0109	FM-GPW-7050	WO-640017	0703
FM-GPW-6290	WO-116295	0109	FM-GPW-7051-B	WO-637495	0700
FM-GPW-6392	WO-A-439	0109C	FM-GPW-7052	WO-640018	0703
FM-GPW-6500-A	WO-637047	0105A	FM-GPW-7056	WO-637503	0701
FM-GPW-6500-B	WO-115948	0105A	FM-GPW-7059	WO-637835	0704B
FM-GPW-6510-B	WO-375811	0105	FM-GPW-7060	WO-A-6317	0704
FM-GPW-6511-B	WO-637045	0105	FM-GPW-7061	WO-A-519	0704
FM-GPW-6513	WO-638636	0105	FM-GPW-7062	WO-A-738	0704
FM-GPW-6514	WO-637044	0105	FM-GPW-7063	WO-639423	0703
FM-GPW-6520	WO-630303	0105C	FM-GPW-7064	WO-635844	0703
FM-GPW-6521	WO-630305	0105C	FM-GP-7065	WO-A-916	0704
FM-GPW-6546	WO-275994	0105	FM-B-7070	WO-635846	0703
FM-GPW-6549-B	WO-640020	0105C	FM-GP-7077	WO-A-942	0902
FM-GPW-6555	WO-51875	0105C	FM-O1T-7078-A	WO-A-941	0902
FM-GPW-6600	WO-637636	0107	FM-GPW-7080	WO-A-410	0704
FM-GPW-6604	WO-630384	0107	FM-O1Y-7083	WO-A-940	0902
FM-GPW-6608	WO-636599	0107	FM-O1Y-7096	WO-A-945	0902
FM-GPW-6610	WO-637425	0107	FM-GP-7097	WO-A-943	0902
FM-GPW-6614	VENDOR-343306	0107	FM-GPW-7100	WO-636879	0704B
FM-GPW-6615	WO-630396	0107A-1	FM-GPW-7102	WO-638798	0704B

Ford Part Number	Willys Part Number	Gov't Group No.		Ford Part Number	Willys Part Number	Gov't Group No.
FORD MOTOR CO. (FM) Cont'd				FM-GPW-7570	WO-638151	0202
FM-GPW-7104-A2	WO-640006	0704		FM-GPW-7572	WO-638993	0202
FM-GPW-7105	WO-A-6319	0704B		FM-GPW-7577	WO-636778	0201A
FM-GPW-7106	WO-637833	0704B		FM-74-7580-B	WO-635529	0203
FM-GPW-7107	WO-637834	0704B		FM-GPW-7590	WO-638153	0202
FM-GPW-7109	WO-637831	0704B		FM-GPW-7591	WO-638158	0202
FM-GPW-7111	WO-638948	0704C		FM-GPW-7600	WO-639578	0109
FM-GPW-7113	WO-A-739	0704B		FM-GPW-7700	WO-A-1195	0800
FM-GPW-7115	WO-A-880	0704C		FM-GPW-7705	WO-A-1503	0801
FM-GPW-7116	WO-637832	0704B		FM-GPW-7706	WO-A-1508	0801
FM-GPW-7119	WO-635812	0704C		FM-GPW-7707	WO-A-1509	0801
FM-GPW-7120	WO-639422	0703		FM-GP-7708	WO-A-953	0801
FM-GPW-7121	WO-A-878	0704C		FM-GP-7709	WO-A-954	0801
FM-GPW-7124	WO-A-6318	0704B		FM-GPW-7710	WO-A-1506	0805B
FM-GPW-7126	WO-A-879	0704C		FM-GP-7711	WO-A-960	0805A
FM-GPW-7129-A	WO-635811	0704C		FM-GP-7712	WO-A-959	0805A
FM-K-7129-B	WO-A-1216	0501		FM-GP-7718-A	WO-A-924	0804
FM-GPW-7140	WO-638952	0704C		FM-GP-7719	WO-A-1007	0803
FM-GPW-7141	WO-636882	0704B		FM-GP-7722	WO-A-1510	0802
FM-GPW-7143	WO-635804	0704C		FM-GP-7723	WO-51575	0803
FM-GPW-7155	WO-638949	0704C		FM-GP-7729	WO-A-1106	0803
FM-GPW-7206	WO-635836	0706C		FM-GP-7742	WO-A-999	0804
FM-GPW-7207	WO-635862	0706A		FM-GP-7743	WO-A-998	0804
FM-GPW-7208	WO-635839	0706C		FM-GP-7744	WO-A-1000	0804
FM-GPW-7210-A	WO-A-1380	0706A		FM-GP-7746	WO-A-1134	0803
FM-GPW-7211-B	WO-A-7260	0706C		FM-GP-7754	WO-A-934	0801
FM-GP-7213	WO-A-971	0805B		FM-GPW-7756	WO-A-1435	0801
FM-GP-7213	WO-A-971	0706A		FM-GPW-7761	WO-A-975	0803
FM-GPW-7214	WO-635859	0706A		FM-GP-7762	WO-A-992	0802
FM-GPW-7217	WO-A-1381	0706A		FM-GP-7763	WO-A-1764	0803
FM-BB-7220	WO-A-1379	0706A		FM-GP-7765	WO-A-988	0803
FM-GPW-7221	WO-A-1382	0706A		FM-GP-7766	WO-A-989	0803
FM-GPW-7223	WO-635861	0706A		FM-GP-7767	WO-A-1001	0804
FM-GPW-7227	WO-392328	0706A		FM-GPW-7768	WO-A-1507	0801
FM-GPW-7230	WO-636196	0706C		FM-GP-7770-A	WO-A-958	0803
FM-GPW-7231	WO-636197	0706C		FM-GP-7771	WO-A-990	0803
FM-GPW-7233	WO-A-1385	0706B		FM-GPW-7773	WO-A-957	0801
FM-GPW-7234	VENDOR-635837	0706A		FM-GPW-7774	WO-A-956	0801
FM-GPW-7240	WO-A-1156	0706B		FM-GPW-7776	WO-A-1111	0803
FM-GPW-7241	WO-A-1155	0706B		FM-GP-7777-A	WO-A-987	0803
FM-GPW-7245	WO-636200	0706B		FM-GPW-7781-A	WO-634759	0809
FM-GPW-7503	WO-A-1355	0204A		FM-GP-7782-A	WO-A-982	0801
FM-GPW-7507	WO-A-179	0204A		FM-GP-7782-B	WO-A-983	0801
FM-GPW-7508	WO-A-180	0204A		FM-GP-7782-C	WO-A-984	0801
FM-GPW-7512	WO-A-887	0204A		FM-GP-7783	WO-A-976	0803
FM-GPW-7514	WO-A-181	0204A		FM-GP-7784	WO-A-991	0802
FM-GPW-7515	WO-630112	0203		FM-GPW-7786	WO-A-1504	0805
FM-GPW-7516	WO-630068	0204A		FM-GP-7787	WO-A-962	0805
FM-GPW-7517	WO-A-177	0204A		FM-GP-7788	WO-A-966	0805
FM-GPW-7518	WO-630103	0109C		FM-GP-7789	WO-A-965	0805A
FM-GPW-7520	WO-A-405	0204		FM-GPW-7793	WO-A-1505	0805B
FM-GPW-7521	WO-A-499	0204A		FM-GP-7796	WO-A-972	0805B
FM-GPW-7525	WO-A-1360	0204		FM-GP-7798-A	WO-A-974	0805B
FM-GPW-7530	WO-A-5102	0203		FM-GP-7799	WO-A-970	0805B
FM-GPW-7532	WO-632177	0203		FM-GPW-8005	WO-A-1214	0501
FM-GPW-7539	WO-A-176	0204A		FM-GPW-8100-A	WO-A-1215	0501
FM-GPW-7545	WO-A-178	0204A		FM-GPW-8102	WO-A-3175	0501H
FM-GPW-7549	WO-371567	0201A		FM-GPW-8103	WO-A-3176	0501H
FM-GPW-7550	WO-636755	0201B		FM-9N-8115	WO-A-1126	0501
FM-GPW-7561	WO-639654	0203		FM-GPW-8125-A	WO-A-4413	0504
FM-GPW-7562	WO-630117	0203		FM-GPW-8133	WO-A-1217	0501
FM-GPW-7563	WO-638992	0201B		FM-GPW-8155	WO-A-3173	1803
FM-GPW-7565	WO-638159	0202		FM-GPW-8162	WO-A-2977	0501
FM-GPW-7566	VENDOR-638152	0202		FM-GPW-8166	WO-A-3574	1803
FM-GPW-7567	WO-638157	0202		FM-GPW-8222	WO-A-2979	1803

Ford Part Number	Willys Part Number	Gov't Group No.	Ford Part Number	Willys Part Number	Gov't Group No.
FORD MOTOR CO. (FM) Cont'd			FM-GPW-9273	WO-A-8190	0605
FM-GPW-8240	WO-A-1124	0503A	FM-GPW-9275	WO-A-1292	0300A
FM-GPW-8250	WO-A-1192	0101A	FM-GPW-9276	WO-A-1293	0300A
FM-GPW-8255	WO-639650	0101A	FM-GPW-9282	WO-A-1367	0304
FM-1GT-8260	WO-630512	0505	FM-GPW-9288	WO-A-1325	0304
FM-GPW-8264	WO-A-592	0505	FM-GPW-9289	WO-A-1368	0304
FM-GPW-8269	WO-636109	0505	FM-GPW-9295	WO-A-5450	0304
FM-GPW-8285	WO-A-6373	0505	FM-GPW-9315	WO-A-8131	0605
FM-60-8287	WO-52226	0505	FM-GPW-9316	WO-A-1450	0107J
FM-GPW-8290	WO-A-6374	0505	FM-GPW-9319-A	WO-662420	0605
FM-GPW-8307	WO-A-3615	2103	FM-GPW-9323	WO-A-1456	0107J
FM-GPW-8348	WO-A-3094	2103	FM-11A-9352	WO-115651	0302
FM-GPW-8349	WO-A-3095	2103	FM-GPW-9354	WO-115650	0302
FM-GPW-8501	WO-639992	0503	FM-GPW-9355	WO-A-1494	0302
FM-GPW-8505	WO-637052	0503	FM-11A-9361	WO-115653	0302
FM-GPW-8509-A	WO-636299	0503A	FM-GPW-9363	WO-115652	0302
FM-GPW-8512	WO-639993	0503	FM-GPW-9364	WO-115656	0302
FM-GPW-8524-A2	WO-640031	0503	FM-GPW-9365	WC-115654	0302
FM-GPW-8530	WO-636297	0503	FM-GPW-9369	WO-A-1369	0304
FM-GPW-8543	WO-637053	0503	FM-GPW-9373	WO-113461	0302
FM-GPW-8548-B	WO-640032	0503	FM-GPW-9377	WO-A-1047	0302
FM-GPW-8557-A	WO-640034	0503	FM-GPW-9378	WO-A-1046	0302
FM-GPW-8572-B	WO-640033	0503	FM-GPW-9380	WO-115643	0302
FM-GPW-8575	WO-637646	0502	FM-1NC-9381	WO-115880	0302
FM-GPW-8576	WO-636298	0503	FM-GPW-9386	WO-A-1045	0302
FM-GPW-8578	WO-639651	0502	FM-GPW-9387	WO-115657	0302
FM-GPW-8600	WO-A-447	0503A	FM-GPW-9388	WO-113460	0302
FM-GPW-8620	WO-A-1495	0503B	FM-GPW-9396	WO-116694	0302
FM-GPW-9001	WO-A-6897	0300	FM-GPW-9398	WO-113460	0302
FM-GPW-9002-A	WO-A-1221	0300	FM-GPW-9399	WO-115641	0302
FM-GPW-9002-B	WO-A-6618	0300	FM-GPW-9410-B	WO-A-1165	0108
FM-GPW-9019	WO-A-3195	0300	FM-GPW-9424-B	WO-A-1166	0108A
FM-GPW-9030-A	WO-A-1254	0300	FM-GPW-9428	WO-A-912	0108B
FM-GPW-9034-A	WO-A-6424	0300	FM-GPW-9435	WO-634811	0108
FM-GPW-9035-B	WO-A-1275	0300	FM-GPW-9443	WO-344732	0108
FM-GPW-9051	WO-A-2970	0300	FM-GPW-9448	WO-638640	0108
FM-GPW-9062	WO-A-2952	0300	FM-GPW-9450	WO-634814	0402
FM-GPW-9063	WO-A-2954	0300	FM-GPW-9456	WO-637206	0108C
FM-GPW-9065	WO-A-2953	0300	FM-GPW-9458	WO-637210	0108C
FM-GPW-9069	WO-A-1738	0300	FM-GPW-9460	WO-636439	0108C
FM-GPW-9071	WO-A-1741	0300	FM-GPW-9462	WO-636438	0108C
FM-GPW-9071	WO-A-3206	1803	FM-GPW-9463	WO-639743	0108C
FM-GPW-9074	WO-A-1476	0300	FM-GPW-9465	WO-637211	0108C
FM-GPW-9075	WO-A-1740	0300	FM-GPW-9467-A	WO-637208	0108C
FM-GPW-9078	WO-A-1480	0300	FM-GPW-9468	WO-115869	0302
FM-GPW-9079	WO-A-1739	0300	FM-GPW-9469	WO-115870	0302
FM-GPW-9082-A	WO-A-3207	1803	FM-GPW-9484	WO-637209	0108C
FM-GPW-9083-A	WO-A-3209	1803	FM-GPW-9510	WO-A-1223	0301
FM-GPW-9085-A	WO-A-3208	1803	FM-GPW-9515	WO-116208	0301
FM-GPW-9140-A	WO-A-1261	0306B	FM-GPW-9516	WO-116202	0301
FM-GPW-9149	WO-A-1258	0306B	FM-GPW-9518	WO-116584	0301
FM-GPW-9154	WO-A-1265	0306B	FM-GPW-9519	WO-116203	0301
FM-GPW-9155-A	WO-A-1255	0306B	FM-GPW-9520	WO-116544	0301
FM-GPW-9160	WO-A-1259	0306B	FM-GPW-9522	WO-116159	0301
FM-GPW-9162	WO-A-1263	0306B	FM-GPW-9523	WO-116160	0301
FM-GPW-9182	WO-A-1262	0306B	FM-GPW-9526	WQ-116586	0301
FM-GPW-9183	WO-A-1256	0306B	FM-GPW-9527	WO-116199	0301
FM-GPW-9184	WO-A-1257	0306B	FM-GPW-9528	WO-116181	0301
FM-GPW-9185	WO-A-1265	0306B	FM-GPW-9529	WO-116537	0301
FM-GPW-9186	WO-A-1260	0306B	FM-GPW-9531	WO-116198	0301
FM-GPW-9211	WO-A-1763	0300A	FM-GPW-9533	WO-116539	0301
FM-GPW-9237	WO-A-1366	0304	FM-GPW-9541	WO-116176	0301
FM-GPW-9267-A	WO-387249	0306B	FM-GPW-9543	WO-116165	0301
FM-GPW-9268	WO-384549	0306B	FM-GPW-9544	WO-116179	0301
FM-GPW-9273	WO-A-8184	0605	FM-GPW-9546	WO-116545	0301

Ford Part Number	Willys Part Number	Gov't Group No.
FORD MOTOR CO. (FM) Cont'd		
FM-GPW-9549	WO-116157	0301
FM-GPW-9550	WO-116172	0301
FM-GPW-9554	WO-116210	0301
FM-GPW-9558	WO-116173	0301
FM-GPW-9562	WO-116161	0301
FM-GPW-9566	WO-116177	0301
FM-GPW-9567	WO-116174	0301
FM-GPW-9569	WO-116168	0301
FM-GPW-9570	WO-116187	0301
FM-GPW-9574	WO-116170	0301
FM-GPW-9575	WO-116175	0301
FM-GPW-9576	WO-116205	0301
FM-GPW-9578	WO-116183	0301
FM-GPW-9579	WO-116162	0301
FM-GPW-9581	WO-116585	0301
FM-GPW-9583	WO-116197	0301
FM-GPW-9585	WO-116154	0301
FM-GPW-9586	WO-116216	0301
FM-GPW-9587	WO-116189	0301
FM-GPW-9588	WO-116217	0301
FM-GPW-9594	WO-116204	0301
FM-GPW-9595	WO-116587	0301
FM-GPW-9598	WO-116542	0301
FM-GPW-9599	WO-116178	0301
FM-GPW-9608	WO-116169	0301
FM-GPW-9609	WO-A-5629	0301C
FM-GPW-9610	WO-116651	0301
FM-GPW-9612	WO-A-7191	0301C
FM-GPW-9614	WO-116194	0301
FM-GPW-9615	WO-116185	0301
FM-GPW-9617	WO-A-5630	0301C
FM-GPW-9621	WO-A-5632	0301C
FM-GPW-9623	WO-A-5633	0301C
FM-GPW-9624	WO-116184	0301
FM-GPW-9628	WO-A-281	0301C
FM-GPW-9631	WO-116195	0301
FM-GPW-9632	WO-A-463	0301C
FM-GPW-9636	WO-116188	0301
FM-GPW-9637-A	WO-A-1290	0301C
FM-GPW-9637-B	WO-A-6911	0301C
FM-GPW-9647	WO-A-642	0301C
FM-GPW-9650	WO-116186	0301
FM-GPW-9652	WO-A-1311	0301C
FM-GPW-9653	WO-635097	0301C
FM-GPW-9656	WO-A-1278	0301C
FM-GPW-9657	WO-A-1279	0301C
FM-GPW-9658	WO-A-5631	0301C
FM-GPW-9686-A	WO-A-1451	0301C
FM-GPW-9696	WO-116163	0301
FM-GPW-9711	WO-650482	0303
FM-GPW-9716	WO-A-1225	0303
FM-GPW-9719	WO-A-6710	0303
FM-GPW-9726	WO-A-1910	0303
FM-GPW-9727	WO-A-1174	0303
FM-GPW-9728	WO-639607	0303
FM-GPW-9731-B	WO-651298	0303
FM-GPW-9732-A	WO-A-1084	0303
FM-GPW-9735-B	WO-A-6851	0303
FM-GPW-9737	WO-632174	0303
FM-GPW-9739	WO-A-1243	0303
FM-GPW-9742	WO-A-1175	0303
FM-GPW-9745	WO-639610	0303
FM-GPW-9751	WO-A-1173	0303
FM-GPW-9752	WO-633013	0303
FM-GPW-9775-B	WO-A-7517	0303
FM-GPW-9778	WO-A-7518	0303
FM-GPW-9795	WO-650483	0303
FM-GPW-9799	WO-633011	0303
FM-GPW-9905	WO-116206	0301
FM-GPW-9906	WO-116540	0301
FM-GPW-9907	WO-116538	0301
FM-GPW-9914	WO-116541	0301
FM-GPW-9922	WO-116166	0301
FM-GPW-9926	WO-116171	0301
FM-GPW-9928	WO-116164	0301
FM-GPW-9930	WO-116209	0301
FM-GPW-9935	WO-116191	0301
FM-GPW-9940	WO-116180	0301
FM-GPW-10000-A	WO-A-5992	0601
FM-GPW-10005	WO-A-1637	0601
FM-GPW-10041	WO-A-1600	0601
FM-GPW-10050	WO-A-6298	0601
FM-GPW-10057	WO-A-1628	0601
FM-GPW-10069	WO-A-1630	0601
FM-B-10094	WO-A-6299	0601
FM-GPW-10094	WO-51248	0601
FM-GPW-10098	WO-106313	0601
FM-GPW-10100	WO-A-1601	0601
FM-GPW-10104	WO-A-1598	0601
FM-GPW-10105	WO-A-1629	0601
FM-GPW-10116	WO-A-1624	0601
FM-GPW-10118	WO-A-1626	0601
FM-GPW-10119	WO-A-1625	0601
FM-GPW-10120	WO-A-1590	0601
FM-GPW-10124	WO-A-1622	0601
FM-GPW-10130	WO-A-1639	0601
FM-GPW-10134	WO-A-1638	0601
FM-GPW-10138	WO-A-6300	0601
FM-GPW-10139	WO-A-6301	0601
FM-B-10141	WO-107128	0603
FM-GPW-10142	WO-A-1649	0601
FM-GPW-10143	WO-A-1399	0601A
FM-GPW-10153-A	WO-A-1491	0601A
FM-GPW-10155	WO-A-1469	0601A
FM-GPW-10162	WO-A-1400	0601A
FM-GPW-10166	WO-A-1392	0601A
FM-GPW-10175	WO-A-1604	0601
FM-GPW-10176	WO-A-1468	0601A
FM-GPW-10177	WO-A-1470	0601A
FM-GPW-10178-A	WO-A-1395	0601A
FM-GPW-10191	WO-A-1607	0601
FM-GPW-10192	WO-A-1606	0601
FM-O1A-10193	WO-A-1592	0601
FM-GPW-10202-A	WO-A-1591	0601
FM-GPW-10202-C	WO-A-1594	0601
FM-GPW-10206-A	WO-A-1599	0601
FM-GPW-10208-A	WO-A-1597	0601
FM-GPW-10211-A	WO-A-1602	0601
FM-GPW-10211-B	WO-A-1605	0601
FM-78-10212-A	WO-A-1644	0601
FM-GPW-10212	WO-A-1646	0601
FM-GPW-10218	WO-A-1609	0601
FM-GPW-10505	WO-A-1409	0601B
FM-11AS-10655	WO-A-1238	0610
FM-11AS-10657	WO-A-5433	0610
FM-11AS-10658	WO-A-1767	0610
FM-GPW-10850	WO-A-8186	0605
FM-GPW-10883	WO-A-8188	0605
FM-GPW-11001-A	WO-A-1245	0602

Ford Part Number	Willys Part Number	Gov't Group No.		Ford Part Number	Willys Part Number	Gov't Group No.
FORD MOTOR CO. (FM) Cont'd				FM-GPW-12174	WO-A-1660	0603
FM-GPW-11005	WO-A-1568	0602		FM-GPW-12175	WO-A-1675	0603
FM-GPW-11036-B	WO-109455	0602		FM-GPW-12176	WO-A-1661	0603
FM-GPW-11049	WO-A-1566	0602		FM-GPW-12177	WO-A-1653	0603
FM-GPW-11053	WO-A-1569	0602		FM-GPW-12178	WO-A-1678	0603
FM-GPW-11055	WO-109431	0602		FM-GPW-12182	WO-A-1673	0603
FM-GPW-11056	WO-109446	0602		FM-GPW-12188	WO-A-1676	0603
FM-B-11059	WO-109445	0602		FM-GPW-12191	WO-A-1684	0603
FM-GPW-11061	WO-109442	0602		FM-GPW-12193	WO-106740	0603
FM-GPW-11069	WO-A-1567	0602		FM-GPW-12195	WO-A-1659	0603
FM-GPW-11077	WO-109452	0602		FM-GPW-12200	WO-A-1658	0603
FM-GPW-11082	WO-109427	0602		FM-GPW-12217	WO-A-1663	0603
FM-GPW-11083	WO-A-1560	0602		FM-GPW-12250-A	WO-A-1733	0606D
FM-GPW-11084	WO-109428	0602		FM-GPW-12250-B	WO-A-1734	0606D
FM-GPW-11085	WO-A-1563	0602		FM-GPW-12250-C	WO-A-1349	0606D
FM-GPW-11089	WO-A-1557	0602		FM-GPW-12267	WO-A-1654	0603
FM-GPW-11090	WO-A-1558	0602		FM-GPW-12283	WO-A-1416	0604A
FM-GPW-11094	WO-A-1553	0602		FM-GPW-12284	WO-A-1414	0604A
FM-GPW-11095	WO-109437	0602		FM-GPW-12286	WO-A-1418	0604A
FM-GPW-11102	WO-A-1554	0602		FM-GPW-12287	WO-A-1412	0604A
FM-GPW-11103	WO-109433	0602		FM-GPW-12298-B	WO-A-1420	0604A
FM-GPW-11107	WO-109436	0602		FM-GPW-12300	WO-A-1631	0603
FM-GPW-11120	WO-A-1556	0602		FM-GPW-12405	WO-A-538	0604B
FM-GPW-11130	WO-A-1582	0602		FM-01A-12410	WO-637863	0604B
FM-GPW-11131	WO-A-1585	0602		FM-11A-12425	WO-A-1096	0604B
FM-GPW-11134	WO-639734	0602		FM-GPW-13005-A	WO-A-1305	0607
FM-GPW-11135	WO-A-1583	0602		FM-GPW-13006-A	WO-A-1304	0607
FM-GPW-11140	WO-638646	0602		FM-GPW-13007	WO-A-1033	0607A
FM-GPW-11350	WO-A-1573	0602		FM-GPW-13012	WO-A-5586	0607
FM-GPW-11354	WO-A-1581	0602		FM-GPW-13015	WO-A-1031	0607
FM-B-11357-A	WO-A-1575	0602		FM-GPW-13020	WO-A-2873	0607
FM-B-11371	WO-A-1580	0602		FM-GPW-13022	WO-A-3161	0607
FM-GPW-11375	WO-A-1577	0602		FM-GPW-13032	WO-A-2878	0607
FM-GPW-11377	WO-A-1578	0602		FM-GPW-13043	WO-A-1036	0607
FM-B-11379	WO-A-1574	0602		FM-GPW-13070	WO-A-2871	0607
FM-B-11381	WO-A-1576	0602		FM-GPW-13071	WO-A-2870	0607
FM-GPW-11382	WO-A-1579	0602		FM-GPW-13075	WO-A-1363	0607
FM-9N-11450-A	WO-A-7225	0606		FM-GPW-13076	WO-A-1362	0607
FM-GPW-11474	WO-372438	0303A		FM-GPW-13150	WO-A-6142	0607D
FM-GPW-11514-A	WO-A-2992	1803		FM-GPW-13152	WO-A-6145	0607D
FM-GPW-11649	WO-A-1332	0606		FM-GPW-13162	WO-A-6144	0607D
FM-GPW-12000	WO-A-1424	0604		FM-GPW-13166	WO-A-6783	0607D
FM-GPW-12000-B	WO-A-7792	0604		FM-GPW-13174	WO-A-6147	0607D
FM-GPW-12006	WO-A-1652	0604A		FM-GPW-13175	WO-A-6146	0607D
FM-GPW-12010	WO-A-1664	0603		FM-GPW-13176	WO-A-4118	2103A
FM-GPW-12011	WO-A-1656	0603		FM-GPW-13180	WO-A-6313	0607D
FM-GPW-12012	WO-A-1657	0603		FM-GPW-13181	WO-A-6153	0606B
FM-GPW-12030	WO-A-1536	0604		FM-GPW-13200	WO-A-1437	0607D
FM-GPW-12064	WO-631105	0604		FM-GPW-13201	WO-A-1436	0607D
FM-GPW-12082	WO-A-1681	0603		FM-GPW-13209	WO-A-1071	0607D
FM-GPW-12083	WO-637615	0603		FM-GP-13210-B	WO-A-1070	0607D
FM-GPW-12084	WO-A-1677	0603		FM-GPW-13216-C	WO-A-1440	0607D
FM-GPW-12100	WO-A-1244	0603		FM-GPW-13217	WO-A-1439	0607D
FM-GPW-12106	WO-A-1655	0603		FM-GPW-13351	WO-A-2993	1802A
FM-9N-12113-A2	WO-327257	0604A		FM-GP-13404-B	WO-A-1065	0608
FM-GPW-12120	WO-A-1672	0603		FM-GP-13405-B	WO-A-1064	0608
FM-GPW-12133	WO-A-1671	0603		FM-GPW-13408-B	WO-A-1073	0608
FM-GPW-12139	WO-A-1679	0603		FM-GPW-13410	WO-A-719	0606B
FM-GPW-12144	WO-A-1682	0603		FM-GPW-13434-A	WO-345961	0606B
FM-GPW-12145	WO-A-1683	0603		FM-GPW-13434-A2	WO-345961	0606B
FM-GPW-12151	WO-A-1662	0603		FM-GPW-13437-A2	WO-662276	0606B
FM-GPW-12155	WO-A-1674	0603		FM-GP-13448-B	WO-A-1076	0608
FM-GPW-12162	WO-A-1570	0603		FM-GP-13449-A	WO-A-1079	0608
FM-GPW-12169	VENDOR-A-1587	0603		FM-B-13466	WO-51804	0607B
FM-GPW-12172	WO-A-1564	0603		FM-GP-13485-A	WO-A-1078	0607A

Ford Part Number	Willys Part Number	Gov't Group No.	Ford Part Number	Willys Part Number	Gov't Group No.
FORD MOTOR CO. (FM) Cont'd			FM-GPW-16082	WO-A-2842	1702
FM-GP-13491-A	WO-A-1075	0607A	FM-GPW-16083	WO-A-2841	1702
FM-GPW-13494-A	WO-A-1074	0607A	FM-GPW-16094	WO-A-2844	1701
FM-GPW-13532	WO-638979	0606	FM-GPW-16095	WO-A-2843	1701
FM-GPW-13549-A2	WO-A-2991	0606B	FM-GPW-16105	WO-A-2891	1701
FM-GPW-13704	WO-A-1334	0605C	FM-GPW-16128	WO-A-3096	1702
FM-GPW-13710	WO-A-1411	0605C	FM-GPW-16132	WO-A-2662	1701
FM-GPW-13713	WO-A-1748	0605C	FM-GPW-16133	WO-A-3159	1702
FM-GTB-13739	WO-A-6149	0606	FM-GPW-16159	WO-A-3158	1701
FM-GPW-13740	WO-A-1333	0606	FM-GPW-16610	WO-A-3225	1704
FM-GPW-13802	WO-A-1312	0609	FM-GPW-16684	WO-A-3059	1704
FM-GPW-13831	WO-A-1389	0609	FM-GPW-16802	WO-A-2188	1704
FM-GPW-13836	WO-A-302	0609B	FM-GTB-16847	WO-A-4680	1704
FM-GPW-14171	WO-A-752	0609B	FM-GTB-16848-A	WO-A-4683	1704
FM-GPW-14300	WO-A-1452	0610B	FM-9N-16868	WO-384569	0107J
FM-GPW-14301	WO-635883	0601B	FM-GPW-16892	WO-A-2896	1701
FM-GPW-14303	WO-A-1098	2603	FM-GPW-17003-B	WO-A-1162	2301
FM-GPW-14305	WO-A-5074	0606B	FM-GPW-17005	WO-A-372	2301
FM-GPW-14305-C	WO-A-7824	0606B	FM-GPW-17011-A	WO-306715	2301
FM-GPW-14320	WO-A-7645	2601	FM-GPW-17017-A	WO-637635	2301
FM-GPW-14321	WO-A-5083	0604A	FM-GP-17020	WO-A-375	2301
FM-09B-14362	WO-A-6356	0606F	FM-GP-17021	WO-A-377	2301
FM-GPW-14401-D	WO-A-5048	0606B	FM-GP-17023	WO-A-376	2301
FM-GPW-14406	WO-A-5070	0605B	FM-GP-17028	WO-A-574	2301
FM-GPW-14409	WO-A-5081	0609B	FM-GPW-17030	WO-A-5130	2301
FM-GPW-14416	WO-A-5080	0605B	FM-GP-17033	WO-A-692	2301
FM-GPW-14425	WO-A-1665	0606B	FM-GPW-17035	WO-A-348	2301
FM-GPW-14425-B	WO-A-7825	0606B	FM-GPW-17036	WO-A-289	2301
FM-GPW-14431	WO-A-1454	0610B	FM-GP-17037	WO-A-313	2301
FM-GPW-14432	WO-A-5981	0605B	FM-GP-17038	WO-A-379	2301
FM-GPW-14432-B	WO-A-7823	0606B	FM-GP-17042	WO-A-373	2301
FM-GPW-14436	WO-A-1731	0606B	FM-GP-17043	WO-A-596	2301
FM-GPW-14446	WO-A-5061	0606B	FM-GP-17044	WO-A-597	2301
FM-GPW-14448-C	WO-639599	0606E	FM-GP-17045	WO-A-598	2301
FM-GPW-14448-D	WO-A-1490	0606E	FM-GP-17046	WO-A-599	2301
FM-11A-14452	WO-371400	0610A	FM-GP-17047	WO-A-600	2301
FM-GPW-14456	WO-A-5079	0606B	FM-GPW-17052	WO-A-7511	2301
FM-GPW-14457	WO-A-5078	0606B	FM-GPW-17055	WO-A-4518	1804
FM-GPW-14458	WO-A-5072	0605B	FM-GPW-17062	WO-A-1100	2301
FM-GPW-14459	WO-A-5073	0606B	FM-GPW-17079	WO-A-4516	1804
FM-B-14463	WO-307556	0604A	FM-GPW-17080	WO-A-1240	2301
FM-91-BS-14463	WO-A-7466	2601	FM-GPW-17090	WO-A-1339	2301
FM-GPW-14464	WO-78318	0606B	FM-GPW-17091	WO-A-1492	2301
FM-GPW-14465	WO-A-5082	0606B	FM-GPW-17097	WO-A-693	2402
FM-B-14466	WO-314369	0604A	FM-GPW-17100	WO-A-616	2402
FM-GPW-14480-B	WO-A-8113	2601	FM-GP-17125	WO-A-213	2301
FM-GPW-14481-B	WO-A-8116	2601	FM-GPW-17126-B	WO-A-11765	2301
FM-GPW-14487-A	WO-635985	0606B	FM-GPW-17255-A	WO-A-8180	2204
FM-GPW-14487-B	WO-635981	0606B	FM-GPW-17260	WO-A-1267	2204
FM-GPW-14513	WO-A-7640	2601	FM-GPW-17261	WO-A-1343	2204
FM-GPW-14521	WO-A-7180	2601	FM-GPW-17262	WO-A-1344	2204
FM-GPW-14531	WO-A-7636	2601	FM-GPW-17271	WO-A-1512	0810
FM-GPW-14532	WO-A-7182	2601	FM-GP-17277	WO-A-985	0810
FM-GPW-14534	WO-A-7638	2601	FM-GP-17285	WO-A-1511	0810
FM-GPW-14535	WO-A-7637	2601	FM-GP-17333	WO-636396	0810
FM-GPW-14536	WO-A-7181	2601	FM-GPW-17500	WO-A-11433	1811D
FM-GPW-14537	WO-A-7947	2601	FM-GP-17535	WO-A-2512	1811D
FM-GPW-14561	WO-A-1694	2603	FM-GP-17541	WO-A-2513	1811D
FM-GPW-14561	WO-A-5598	0606B	FM-21CS-17682-B	WO-A-2934	2202
FM-GPW-14566	WO-78932	0606B	FM-GPW-17750	WO-A-1117	2101
FM-GPW-14585	WO-A-1289	0304	FM-GPW-17751	WO-A-1147	2101
FM-GPW-14589	WO-A-1289	0107J	FM-GPW-17752	WO-A-1130	1500
FM-48-15021	WO-52837	0607B	FM-GPW-17753	WO-A-1129	1500
FM-GPW-16005	VENDOR-A-2942	1701	FM-GPW-17754	WO-A-1128	1500
FM-GPW-16006	WO-52031	0300	FM-GPW-17755	WO-A-1127	1500

Ford Part Number	Willys Part Number	Gov't Group No.
FORD MOTOR CO. (FM) Cont'd		
FM-GPW-17759	WO-A-1120	1500
FM-GPW-17775	WO-A-1157	2101
FM-GPW-1151266	WO-A-2897	2201A
FM-GPW-1151270	WO-A-2901	2201A
FM-GPW-1151272	WO-A-2898	2201A
FM-GPW-1151289	WO-A-2501	1811
FM-GPW-1151297	WO-A-3189	1811
FM-GPW-1152700	WO-A-3216	2201
FM-GPW-1152720	WO-A-3110	2201A
FM-GPW-1152730	WO-A-3109	2201A
FM-GPW-1152950	WO-A-3141	2201A
FM-GPW-1153030	WO-A-2754	2201A
FM-GPW-1153142	WO-A-2900	1804
FM-GPW-1154810-A	WO-A-2168	1802A
FM-GPW-146211	WO-A-1693	2603
FM-GPW-18040	....	1500
FM-GPW-18045	....	1603
FM-GPW-18060	WO-637936	1603
FM-GPW-18075	....	1500
FM-GPW-18080	....	1500
FM-GPW-18136-B	WO-A-7687	2301B
FM-11YS-18142-B	WO-A-6019	0606F
FM-11YS-18143-B	WO-A-6589	0606F
FM-11YS-18144-B	WO-A-6588	0606F
FM-11YS-18149-B	WO-A-6587	0606F
FM-11YS-18151-B	WO-A-6586	0606F
FM-GPW-18205-B	WO-A-5621	0301C
FM-GPW-18274	WO-A-1650	0601
FM-GPW-18288	WO-A-7568	0102
FM-GPW-18289	WO-A-7503	0109
FM-GPW-18308	WO-A-8279	2302
FM-GPW-18314	WO-A-7837	0810
FM-GPW-18317	WO-A-7445	0800
FM-GPW-18323	WO-A-7835	0108
FM-GPW-18325	WO-A-6855	2301
FM-GPW-18330-A	WO-A-7233	0104A
FM-GPW-18330-B	WO-A-7234	0104A
FM-GPW-18330-C	WO-7235	0104A
FM-GPW-18336	WO-A-6881	1007
FM-GPW-18337	WO-6883	0306B
FM-GPW-18338	WO-A-6882	1006
FM-GPW-18342	WO-A-7840	0601
FM-GPW-18343	WO-A-7843	0603
FM-GPW-18347	WO-A-6798	0102A
FM-GPW-18348	WO-A-6746	0102A
FM-GPW-18349	WO-A-6747	0102A
FM-GPW-18352	WO-A-6837	0301
FM-GPW-18353	WO-A-7680	2309
FM-GPW-18354	WO-A-1687	0603
FM-GPW-18355-B	WO-A-7443	0800
FM-GPW-18356-B	WO-A-7832	0700
FM-GPW-18357-B	WO-A-6840	0301
FM-GPW-18358	WO-A-6751	0201A
FM-GPW-18359-B	WO-A-7833	0202
FM-GPW-18360	WO-A-6753	0801
FM-GPW-18361-B	WO-A-7845	0606B
FM-GPW-18363-B	WO-A-7844	0604A
FM-GPW-18363-C	WO-A-9055	0601
FM-GPW-18365-B	WO-A-7830	1001
FM-GPW-18366-B	WO-A-7831	1100
FM-GPW-18367	WO-116600	1202
FM-GPW-18368	VENO-A-6133	1207
FM-GPW-18370-B	WO-A-7838	1205

Ford Part Number	Willys Part Number	Gov't Group No.
FM-GPW-18371	WO-115962	1207
FM-GPW-18372	WO-115963	1207
FM-GPW-18374	WO-A-6760	1403A
FM-GPW-18376-B	WO-A-7836	0602
FM-GPW-18377	WO-A-6759	1201
FM-GPW-18359	WO-A-6749	0107
FM-GPW-18380	WO-A-6750	0107
FM-GPW-18382	WO-A-6742	0609B
FM-GPW-18383	WO-A-6791	1401
FM-GPW-18384	WO-A-6816	1003
FM-GPW-18386	WO-A-6745	1104
FM-GPW-18387	WO-A-1537	0101
FM-GPW-18388	WO-A-6744	1103
FM-GPW-18389	WO-A-6743	1103
FM-GPW-18390	WO-A-1536	0101
FM-21C-18397-B	WO-A-1433	0902
FM-GPW-18512	WO-A-1538	0107A
FM-GPW-18515-B	WO-A-6839	0503
FM-GPW-18535	WO-A-1552	0602
FM-GPW-18660	WO-A-1230	0107E
FM-GPW-18662-A	WO-A-1236	0107E
FM-GPW-18663	WO-A-1247	0107F
FM-GPW-18664	WO-A-1251	0107F
FM-GPW-18666	WO-A-1198	0107E
FM-GPW-18667	WO-A-1197	0107E
FM-GPW-18675-A	WO-A-1233	0107E
FM-9N-18679	WO-387891	0107J
FM-GPW-18685	WO-A-1234	0107E
FM-9N-18686	WO-384569	0107J
FM-GPW-18687	WO-A-1231	0117E
FM-GPW-18688-A	WO-A-1235	0107E
FM-GPW-18691-A	WO-A-1232	0107E
FM-GPW-18812-B	WO-A-6320	2603
FM-GPW-18840	WO-A-7826	2601
FM-GPW-18841	WO-A-7600	2601
FM-GPW-18846	WO-A-5041	0606B
FM-GPW-18849	WO-A-5035	2603
FM-GPW-18850	WO-A-5040	2603
FM-GPW-18852	WO-A-1699	2603
FM-GPW-18853	WO-A-5039	2603
FM-GPW-18854	WO-A-5027	2603
FM-GPW-18857	WO-A-5037	2603
FM-GPW-18858	WO-A-5033	2603
FM-GAA-18861-A	WO-A-8114	2601
FM-GPW-18871	WO-A-5030	2603
FM-GPW-18872	WO-A-5031	2603
FM-GPW-18873	WO-A-5034	2603
FM-GPW-18874	WO-A-5036	2603
FM-GPW-18876	WO-A-5038	2601
FM-GPW-18876-B	WO-A-11766	2601
FM-GPW-18877-B	....	2601
FM-GPW-18935-A	WO-A-5337	2603
FM-GPW-18936-A	WO-A-8883	2603
FM-GPW-18937-A	WO-A-8884	2603
FM-GPW-18938-C	WO-A-7848	2603
FM-GPW-18960-B	WO-A-5980	2603
FM-GPW-1100000	WO-A-3565	1800
FM-GPW-1100001	WO-A-3563	1800
FM-GPW-1101610	WO-A-3005	1809
FM-GPW-1101621-A	WO-A-1330	2203
FM-GPW-1101627-A	WO-A-1331	2203
FM-GPW-1101670	WO-A-3132	1804
FM-GPW-1101698-A	WO-A-2931	1803
FM-GPW-1101735-A	WO-A-3784	0805B
FM-GPW-1102015	WO-A-2747	1802A

Ford Part Number	Willys Part Number	Gov't Group No.	Ford Part Number	Willys Part Number	Gov't Group No.
FORD MOTOR CO. (FM) Cont'd			FM-GPW-1160314	WO-A-2515	1804
FM-GPW-1102038	WO-A-3007	1802A	FM-GPW-1150326	WO-A-2925	1804
FM-GPW-1102039	WO-A-3008	1802A	FM-GPW-1160327-B	WO-A-2886	1804
FM-GPW-1102130	WO-A-2190	1802A	FM-GPW-1160780	WO-A-2788	1804A
FM-GPW-1102330	WO-A-3943	1803	FM-GPW-1161300	WO-A-2783	1804
FM-GPW-1102396	WO-A-2138	1802A	FM-GPW-1161302	WO-A-2830	1804
FM-GPW-1102510	WO-A-2816	1802	FM-GPW-1161326	WO-A-2782	1804
FM-GPW-1102511	WO-A-2815	1802	FM-GPW-1162180	WO-A-2911	1804
FM-GPW-1103010	WO-A-3210	1811	FM-GPW-1162181	WO-A-2910	1804
FM-GP-1103014	WO-A-2226	1811A	FM-GPW-1162400	WO-A-3029	1804
FM-GP-1103020	WO-A-2255	1811B	FM-GPW-1162402	WO-A-2787	1804
FM-GPW-1103027	WO-A-3197	1811	FM-GPW-1162410	WO-A-2945	1804
FM-GPW-1103028	WO-A-3203	1811	FM-GPW-1162414	WO-A-2833	1804
FM-GPW-1103030	WO-A-2246	1811B	FM-GPW-1162552	WO-A-2453	1804
FM-GPW-1103050	WO-A-2798	1811	FM-GPW-1162606	WO-A-3144	1802A
FM-GPW-1103080-A	WO-A-2476	1811B	FM-GPW-1162900-B	WO-A-2968	1802A
FM-GP-1103100	WO-A-2478	1811A	FM-GPW-1163846-B	WO-A-3983	1804A
FM-GP-1103106	WO-A-2479	1811B	FM-GPW-1166400	WO-A-3115	1804A
FM-GP-1103110	WO-A-2250	1811B	FM-GPW-1166401	WO-A-3114	1804A
FM-GPW-1103162	WO-A-2213	1811	FM-GPW-1166800-B	WO-A-3982	1804A
FM-GPW-1103214	WO-A-3211	2201	FM-G8T-8103074-B	WO-A-2239	1811
FM-GPW-1103268	WO-A-3204	1811B	FM-99W-8103311	WO-A-2483	1811
FM-GP-1103302	WO-A-2235	1811	FM-B-45482-C	WO-A-2238	1811
FM-GPW-1103304	WO-A-2232	1811	FM-GPW-1120040	WO-A-2999	2201F
FM-GP-1103310	WO-A-4300	1811	FM-GPW-1120041	WO-A-2998	2201F
FM-GP-1103334	WO-A-2234	1811	FM-GPW-1127846	WO-A-2757	1802A
FM-GP-1103350	WO-A-4303	1811	FM-GPW-1127847	WO-A-2756	1802A
FM-GPW-1103446-A	WO-A-2939	1802A	FM-GPW-1127848	WO-A-2772	1802A
FM-GP-1103482-A	WO-A-2227	1811	FM-GPW-1127849	WO-A-2771	1802A
FM-GPW-1103488	WO-A-2791	1811	FM-GPW-1127850	WO-A-2770	1802A
FM-GPW-1104320	WO-A-3578	1809	FM-GPW-1127851	WO-A-2769	1802A
FM-GPW-1104364	WO-A-3155	1809	FM-GPW-1128237	WO-A-3137	2301A
FM-GPW-1106024-B	WO-A-3434	1809A	FM-GPW-1128242	WO-A-2940	1802A
FM-GPW-1106050	WO-A-3436	1809A	FM-GPW-1128244	WO-A-2986	1804A
FM-GPW-1106064	WO-A-3818	1809A	FM-GPW-1128247	WO-A-3135	2301A
FM-GPW-1106083	WO-A-3532	1809A	FM-GPW-1128254	WO-A-2984	2301A
FM-GPW-1106084-A	WO-A-3531	1809A	FM-GPW-1128256	WO-A-2995	2301A
FM-GPW-1106084-B	WO-A-3823	1809A	FM-GPW-1128258	WO-A-3082	2301A
FM-GPW-1106085	WO-A-3536	1809A	FM-GPW-1128267	WO-A-3139	2301A
FM-GPW-1106087-A	WO-A-3652	1809A	FM-GPW-1128286	WO-A-2325	1802A
FM-GPW-1110610	WO-A-2838	1505	FM-GPW-1129068	WO-A-2745	2201A
FM-GPW-1110625	WO-A-2935	1802A	FM-GPW-1129069	WO-A-2744	2201A
FM-GPW-1110750	WO-A-2948	1802A	FM-GPW-1129200	WO-A-2946	2301
FM-GPW-1110895	WO-A-2950	1802A	FM-GPW-1129670	WO-A-2390	1800A
FM-GPW-1111137-A	WO-A-2932	1803	FM-GPW-1129672	WO-A-2389	1800A
FM-GPW-1111140	WO-A-2768	1802A	FM-GPW-1131414	WO-A-2883	1804
FM-GPW-1111155	WO-A-3222	1802A	FM-GPW-1140326	WO-A-4607	1802A
FM-GPW-1111218	WO-A-2773	1802A	FM-GPW-1140327-B	WO-A-4606	1802A
FM-GPW-1111238	WO-A-2879	1802A	FM-GPW-1140330	WO-A-4123	1817A
FM-GPW-1111286-A	WO-A-3783	1803	FM-GPW-1140334	WO-A-4127	1817A
FM-GPW-1111286-A3	WO-A-3783	1803	FM-GPW-1140344	WO-A-4128	1817A
FM-GPW-1111299-A	WO-A-3116	1803	FM-GPW-1140447	WO-A-2470	1802A
FM-GPW-1111322	WO-A-3055	0300	FM-GPW-1140449	WO-A-2853	1802A
FM-GPW-1111323	WO-A-3056	1802A	FM-GPW-1140474	WO-A-2930	0300
FM-GPW-1111324	WO-A-3497	0300	FM-GPW-1143501	WO-A-2895	1817
FM-GPW-1111330	WO-A-2760	1803	FM-GPW-1144598	WO-A-2820	1505
FM-GPW-1111331	WO-A-2759	1803	FM-GPW-1144600	WO-A-2823	1505
FM-GPW-1111520	WO-A-3782	1803	FM-GPW-1144620	WO-A-3120	1802A
FM-GPW-1112110	WO-A-2982	1803	FM-GPW-1146100	WO-A-3227	1817
FM-GPW-1112115	WO-A-2990	1803	FM-GPW-1146132	WO-A-3226	1817
FM-GPW-1112116	WO-A-2994	1803	FM-GPW-1146152	WO-A-3198	1817
FM-GPW-1112117	WO-A-2918	1803	FM-20027-S2	WO-52132	
FM-GPW-1112119	WO-A-2919	1803	FM-20046-S	WO-51523	
FM-GPW-1112158	WO-A-2917	1803	FM-20046-S2	WO-51523	
FM-GPW-1160026	WO-A-3108	1804A	FM-20066-S2	WO-6428	

Ford Part Number	Willys Part Number	Gov't Group No.		Ford Part Number	Willys Part Number	Gov't Group No.
FORD MOTOR CO. (FM) Cont'd				FM-24427-S7	WO-6609	
FM-20300-S7	WO-51738			FM-24428-S	WO-51406	
FM-20308-S2	WO-51514			FM-24429-S7	WO-50878	
FM-20308-S2	WO-51763			FM-24430-S	WO-6184	
FM-20308-S2	WO-52170			FM-24449-S2	WO-52600	
FM-20308-S7-8	WO-51514			FM-24454-S2	WO-A-2485	
FM-20309-S7	WO-51732			FM-24489-S	WO-52379	
FM-20310-S2	WO-51391			FM-24505-S7	WO-50992	
FM-20311-S7	WO-A-1648			FM-24555-S	WO-51371	
FM-20324-S	WO-51514			FM-24622-S2	WO-1725	
FM-20324-S2	WO-51514			FM-24650-S2	WO-52987	
FM-20324-S2	WO-52168			FM-24916-S2	WO-A-1707	
FM-20325-S7	WO-5920			FM-26147-S7	WO-52036	
FM-20326-S	WO-51485			FM-26147-S7	WO-51040	
FM-20326-S2	WO-51485			FM-26457-S7	WO-1632	
FM-20326-S7	WO-51485			FM-26457-S7	WO-51040	
FM-20328-S2	WO-52189			FM-26457-S7	WO-52131	
FM-20344-S2	WO-52167			FM-26466-S7	WO-51546	
FM-20344-S2	WO-52168			FM-26480-S2	WO-6287	
FM-20346-S2	WO-51523			FM-26483-S	WO-51492	
FM-20347-S2	WO-51396			FM-26483-S2	WO-51492	
FM-20348-S2	WO-53031			FM-26483-S7	WO-51492	
FM-20349-S2	WO-50163			FM-26496-S2	WO-52131	
FM-20364-S7	. . . .			FM-26498-S2	WO-53267	
FM-20366-S	WO-6428			FM-27063-S7	WO-1620	
FM-20366-S	WO-639689			FM-27068-S	WO-52236	
FM-20366-S2	WO-6468			FM-27145-S2	WO-6383	
FM-20367-S7	WO-50929			FM-27161-S7	WO-1650	
FM-20384-S2	WO-6188			FM-27698-S	WO-52994	
FM-20386-S7	WO-639120			FM-28378-S2	WO-A-1072	
FM-20387-S2	WO-5934			FM-31026-S7	WO-A-1670	
FM-20388-S2	WO-51405			FM-31027-S8	WO-A-1668	
FM-20388-S7	WO-51405			FM-31032-S7	WO-116211	
FM-20389-S	. . . .			FM-31037-S7	WO-A-5197	
FM-20406-S7	WO-6660			FM-31045-S2	WO-A-6148	
FM-20411-S4	WO-6923			FM-31061-S8	WO-116213	
FM-20486-S2	. . . .			FM-31079-S	WO-51819	
FM-20514-S2	WO-51545			FM-31583-S7	WO-A-1686	
FM-21492-S2	WO-51308			FM-31588-S	WO-A-1636	
FM-23017-S16	WO-323397			FM-31588-S7	WO-A-1636	
FM-23393-S2	WO-52983			FM-31596-S	WO-A-1572	
FM-23395-S	WO-52945			FM-31628-S7	WO-113439	
FM-23498-S2	WO-52372			FM-31662-S	WO-116215	
FM-24308-S	WO-51763			FM-31866-S	WO-5580	
FM-24308-S2	WO-52170			FM-32475-S2	WO-53103	
FM-24325-S	WO-638154			FM-32852-S	WO-52993	
FM-24327-S2	WO-A-872			FM-32866-S2	WO-52142	
FM-24327-S2	WO-52132			FM-32866-S2	WO-52142	
FM-24327-S7	WO-52132			FM-32866-S7	WO-52889	
FM-24328-S2	WO-52189			FM-32924-S2	WO-52809	
FM-24346-S2	WO-A-2943			FM-32924-S7-8	WO-52809	
FM-24347-S2	WO-51396			FM-32989-S2	WO-52990	
FM-24348-S	WO-6412			FM-33184-S2	WO-A-2489	
FM-24349-S2	WO-50163			FM-33249-S2	WO-A-2487	
FM-24366-S2	. . . .			FM-33784-S2	WO-A-491	
FM-24367-S2	WO-50929			FM-33786-S2	WO-636575	
FM-24368-S	WO-6299			FM-33795-S	WO-5790	
FM-24369-S	WO-50151			FM-33795-S2	WO-5790 / WO-52217	
FM-24386-S2	WO-51486					
FM-24389-S2	WO-5919			FM-33795-S7	WO-5790 / WO-52847	
FM-24407-S	WO-5922					
FM-24408-S	WO-52911			FM-33796-S2	WO-5914	
FM-24409-S7	WO-6606			FM-33797-S	WO-6167	
FM-24426-S	WO-6157			FM-33797-S2	WO-6167	
FM-24426-S2	WO-6157			FM-33797-S2	WO-52350	

Ford Part Number	Willys Part Number	Gov't Group No.		Ford Part Number	Willys Part Number	Gov't Group No
FORD MOTOR CO. (FM) Cont'd				FM-34707-S2	WO-5455	
FM-33798-S	WO-5910			FM-34707-S2	WO-52819	
FM-33798-S2	WO-5910			FM-34707-S7	WO-A-1680	
FM-33798-S7	WO-5910			FM-34707-S7	WO-50921	
FM-33798-S7	WO-111062			FM-34708-S	WO-A-1547	
FM-33799-S	WO-5544			FM-34711-S	WO-116207	
FM-33799-S2	WO-5544			FM-34745-S	WO-52702	
FM-33799-S2	WO-6169			FM-34745-S2	WO-52702	
FM-33799-S2	WO-53033			FM-34745-S2	WO-638305	
FM-33799-S7	WO-5544			FM-34746-S2	WO-A-2490	
FM-33799-S7	WO-53033			FM-34746-S2	WO-52274	
FM-33800-S	WO-5901			FM-34746-S7	WO-A-1533	
FM-33800-S2	WO-5901			FM-34747-S	WO-5455	
FM-33800-S2	WO-50922			FM-34747-S	WO-52101	
FM-33800-S7	WO-A-755			FM-34747-S2	WO-A-1680	
FM-33801-S2	WO-5336			FM-34747-S2	WO-52101	
FM-33802-S	WO-5939			FM-34747-S7	WO-A-1680	
FM-33802-S2	WO-5939			FM-34753-S	WO-A-8087	
FM-33832-S2	WO-52804			FM-34801-S7	WO-A-1613	
FM-33845-S2	WO-6163			FM-34801-S7	WO-A-1669	
FM-33846-S	WO-5916			FM-34801-S7	WO-52614	
FM-33846-S2	WO-5916			FM-34802-S	WO-51532	
FM-33846-S2	WO-637924			FM-34802-S	WO-52961	
FM-33880-S	WO-5901			FM-34802-S2	WO-6330	
FM-33881-S2	WO-638539			FM-34802-S2	WO-52652	
FM-33896-S2	WO-A-2466			FM-34802-S2	WO-52705	
FM-33896-S2	WO-A-300329			FM-34803-S	WO-5168	
FM-33909-S7	WO-52954			FM-34803-S2	WO-52221	
FM-33911-S2	WO-A-873			FM-34803-S2	WO-52960	
FM-33921-S7	WO-638155			FM-34803-S7	WO-A-1571	
FM-33925-S7	WO-5901			FM-34803-S7	WO-5186	
FM-33927-S7	WO-52925			FM-34803-S7	WO-52221	
FM-34032-S7	WO-A-1640			FM-34803-S7	WO-52960	
FM-34033-S18	WO-6436			FM-34803-S7	WO-113440	
FM-34051-S7	WO-A-1610			FM-34804-S	WO-52164	
FM-34051-S7	WO-52615			FM-34804-S2	WO-51969	
FM-34052-S2	WO-6214			FM-34805-S	WO-52706	
FM-34052-S2	WO-52182			FM-34805-S2	WO-A-2486	
FM-34054-S	WO-6989			FM-34805-S2	WO-5045	
FM-34054-S2	WO-A-2586			FM-34805-S2	WO-52031	
FM-34054-S2	WO-6989			FM-34805-S2	WO-52706	
FM-34056-S16	WO-A-9094			FM-34805-S2	WO-53058	
FM-34079-S7	WO-6536			FM-34805-S7	WO-5045	
FM-34079-S7	WO-53061			FM-34805-S7	WO-52706	
FM-34079-S8	WO-5848			FM-34805-S8	WO-52707	
FM-34083-S2	WO-636575			FM-34806-S2	WO-A-1619	
FM-34084-S	WO-A-1546			FM-34806-S2	WO-51833	
FM-34129-S2	WO-52165			FM-34806-S2	WO-51840	
FM-34130-S2	WO-A-3550			FM-34806-S2	WO-52045	
FM-34130-S2	WO-6167			FM-34806-S2	WO-53047	
FM-34141-S2	WO-6273			FM-34806-S7	WO-A-5288	
FM-34141-S7	WO-1589			FM-34806-S7	WO-51833	
FM-34176-S2	WO-2481			FM-34806-S7-8	WO-51840	
FM-34701-S7	WO-A-1667			FM-34807-S	WO-5010	
FM-34702-S2	WO-A-1666			FM-34807-S2	WO-5010	
FM-34703-S7	WO-A-1615			FM-34807-S2	WO-51304	
FM-34703-S7	WO-A-1702			FM-30807-S2	WO-52046	
FM-34705-S	WO-52702			FM-34807-S7	WO-5010	
FM-34705-S2	WO-5121			FM-34807-S7	WO-52046	
FM-34706-S2	WO-A-1555			FM-34808-S	WO-5059	
FM-34706-S2	WO-5121			FM-34808-S	WO-52349	
FM-34706-S2	WO-5437			FM-34808-S2	WO-5039	
FM-34706-S2	WO-52274			FM-34809-S	WO-5009	
FM-34706-S2	VENDOR-52768			FM-34809-S	WO-5009	
FM-34706-S7	WO-5437					

Ford Part Number	Willys Part Number	Gov't Group No.	Ford Part Number	Willys Part Number	Gov't Group No.
FORD MOTOR CO. (FM) Cont'd			FM-72071-S	WO-5397	
FM-34809-S2	WO-5009		FM-72798-S7-8	WO-A-1586	
FM-34811-S2	WO-5038		FM-72809-S	WO-5134	
FM-34824-S2	WO-2587		FM-72867-S7	WO-A-1685	
FM-34836-S7	WO-5051		FM-73880-S	WO-339043	
FM-34846-S2	WO-52045		FM-73889-S7	WO-A-1006	
FM-34848-S2	WO-5938		FM-73904-S7	WO-311003	
FM-34886-S	WO-52968		FM-73928-S7	WO-A-1004	
FM-34902-S	WO-52754		FM-74019-S	WO-650484	
FM-34903-S2	WO-53194		FM-74113-S	WO-51921	
FM-34905-S	WO-352760		FM-74121-S	WO-51091	
FM-34905-S2	WO-52483		FM-74127-S	WO-15460	
FM-34906-S	WO-52428		FM-74144-S	WO-A-1641	
FM-34907-S2	WO-52332		FM-74175-S7	WO-5017	
FM-34909-S2	WO-52330		FM-74178-S	WO-5036	
FM-34921-S	WO-52874		FM-74182-S	WO-50917	
FM-34922-S	WO-636528		FM-88022-S	WO-384958	
FM-34941-S	WO-52510		FM-88032-S	WO-332515	
FM-34941-S2	WO-52510		FM-88042-S	WO-300143	
FM-36002-S2	WO-51541		FM-88057-S7	WO-632159	
FM-36009-S	WO-A-1618		FM-88082-S	WO-349712	
FM-36009-S7	WO-A-1618		FM-88141-S	WO-375981	
FM-36046-S2	WO-52963		FM-88350-S	WO-337304	
FM-36787-S7	WO-A-1633		FM-92047-S	WO-51904	
FM-36800-S	WO-A-1647		FM-95626-S	WO-A-3052	
FM-36845-S2	WO-6988		FM-95627-S	WO-A-3054	
FM-36868-S2	WO-51539		FM-95628-S	WO-A-3053	
FM-36931-S2	WO-A-1077		FM-95633-S	WO-A-1312	
FM-36954-S7	WO-A-1588		FM-95705-S	WO-A-3051	
FM-37206-S	WO-A-1032		FM-348148-S	WO-A-3113	
FM-36364-S7	WO-A-1572		FM-350343-S16	WO-A-9073	
FM-37621-S7	WO-53020		FM-350744-S2	WO-51612	
FM-37789-S7	WO-A-6297		FM-350796-S2	WO-A-2375	
FM-38095-S	WO-A-1037		FM-350850-SA2	WO-A-4679	
FM-38192-S2	WO-2482		FM-350853-S7	WO-A-1617	
FM-38259-S2	WO-A-2480		FM-350976-S2	WO-A-2375	
FM-39114-S2	....		FM-351015-S7	WO-A-2263	
FM-40631-S7	WO-53071		FM-351023-S	WO-53287	
FM-60093-S	....		FM-351025-S7	WO-A-1550	
FM-60332-S2	WO-5247		FM-351025-S8	WO-638539	
FM-60371-S	WO-52207		FM-351027-S	WO-53289	
FM-60275-S	WO-51561		FM-351193-S	WO-53057	
FM-60416-S	WO-5267		FM-351274-S7	WO-53024	
FM-60446-S	WO-52832		FM-351303-S2	WO-A-2584	
FM-60466-S	WO-52832		FM-351355-S7	WO-5051	
FM-60470-S	WO-5216		FM-351359-S7	WO-4116	
FM-60472-S	WO-50769		FM-351359-S7	WO-52274	
FM-62216-S	WO-635860		FM-351370-S2	WO-2902	
FM-62218-S	WO-51977		FM-351370-S2	WO-71633	
FM-62627-S3	WO-51182		FM-351838-S2	WO-5455	
FM-63848-S	WO-53243		FM-351466-S24	637923	
FM-64647-S	WO-A-1015		FM-351915-S	WO-374586	
FM-72003-S	WO-5067		FM-351926-S	WO-374681	
FM-72004-S	WO-5354		FM-352101-S	....	
FM-72016-S	WO-5020		FM-352103-S	....	
FM-72016-S	WO-52967		FM-352104-S	....	
FM-72017-S	WO-5020		FM-352126-S	WO-5215	
FM-72025-S	WO-5152		FM-352127-S	....	
FM-72034-S	WO-A-1642		FM-352692-S15	WO-A-4120	
FM-72035-S	WO-5021		FM-353023-S7	WO-638500	
FM-72037-S	WO-52946		FM-353043-A-S7	WO-638792	
FM-72043-S	WO-5802		FM-353051-S	WO-636538	
FM-72053-S	WO-5108		FM-353052-S	WO-52525	
FM-72062-S	WO-52527		FM-353053-S	WO-A-1104	
FM-72063-S	WO-52944		FM-353055-S	WO-5138	

Ford Part Number	Willys Part Number	Gov't Group No.	Ford Part Number	Willys Part Number	Gov't Group No.
FORD MOTOR CO. (FM) Cont'd			FM-355752-S	WO-A-1227	
FM-353055-S7	WO-A-5120		FM-355835-S	WO-A-1701	
FM-353055-S7	WO-5138		FM-355836-S7	WO-5352	
FM-353064-S	WO-5140		FM-355836-S7	WO-6352	
FM-353075-S	WO-5599		FM-355836-S7	WO-52651	
FM-353081-S	WO-635838		FM-355858-S	WO-116219	
FM-353104-S	WO-3111		FM-355883-S	WO-A-1611	
FM-353202-S	WO-662010		FM-355900-S2	WO-654934	
FM-355035-S7	WO-638224		FM-355909-S	WO-A-2933	
FM-355067-S7	WO-116384		FM-355943-S7	WO-52845	
FM-355095-S7	....		FM-355944-S5	WO-A-1565	
FM-355130-S7	WO-A-1700		FM-355965-S2	WO-A-2607	
FM-355130-S7	WO-5113		FM-355982-S2	WO-53285	
FM-355131-S	WO-5066		FM-356016-S7	WO-52893	
FM-355132-S	WO-5272		FM-356021-S	WO-636962	
FM-355132-S	WO-116588		FM-356028-S	WO-52825	
FM-355151-S	WO-5064		FM-356077-S8	WO-639115	
FM-355158-S2	WO-5182		FM-356123-S7	WO-A-3537	
FM-355160-S2	WO-5272		FM-356124-S	WO-A-876	
FM-355162-S2	WO-6290		FM-356125-S	WO-A-980	
FM-355162-S2	WO-51537		FM-356126-S	WO-636569	
FM-355163-S2	....		FM-356134-S18	WO-A-520	
FM-355164-S	WO-A-1584		FM-356200-S7	WO-53026	
FM-355165-S2	....		FM-356201-S	WO-51662	
FM-355200-S7	WO-116385		FM-356205-S	WO-A-1135	
FM-355253-S	WO-115905		FM-356208-S7	WO-A-1621	
FM-355260-S7	WO-A-1623		FM-356229-S	WO-A-1350	
FM-355262-S	WO-380197		FM-356235-S8	WO-8129	
FM-355319-S	WO-A-4590		FM-356263-S7	WO-A-1612	
FM-355351-S	....		FM-356264-S7	WO-A-1616	
FM-355351-S7	....		FM-356299-S	WO-A-1089	
FM-355352-S7	WO-A-1019		FM-356303-S	WO-639121	
FM-355378-S	WO-A-973		FM-356305-S	WO-52768	
FM-355398-S2	WO-51798		FM-356309-S7	WO-53025	
FM-355403-S2	....		FM-356361-S	WO-A-4260	
FM-355426-S	WO-639107		FM-356371-S	WO-A-1401	
FM-355433-S2	WO-51612		FM-356373-S2	WO-2220	
FM-355442-S2	WO-51858		FM-356376-S	WO-A-1125	
FM-355444-S	WO-52836		FM-356394-S2	WO-52835	
FM-355449-S2	WO-A-2214		FM-356436-S	WO-A-1396	
FM-355451-S	WO-632158		FM-356439-S	WO-638381	
FM-355452-S	WO-639052		FM-356504-S	WO-A-1028	
FM-355455-S	WO-A-1397		FM-356504-S	WO-636570	
FM-355456-S	WO-51823		FM-356519-S	WO-A-1410	
FM-355470-S2	WO-A-2473		FM-356522-S2	WO-A-2639	
FM-355476-S	WO-630129		FM-356524-S	WO-634762	
FM-355485-S7	WO-A-1559		FM-356561-S	WO-A-498	
FM-355486-S7	WO-A-1596		FM-357008-S	....	
FM-355496-S	WO-633949		FM-357016-S2	....	
FM-355497-S	WO-632157		FM-357074-S	....	
FM-355498-S7	WO-50163		FM-357076-S	....	
FM-355499-S	WO-634850		FM-357100-S	WO-52832	
FM-355511-S	WO-A-871		FM-357155-S	WO-53056	
FM-355528-S	WO-A-821		FM-357202-S	WO-636571	
FM-355533-S	WO-A-1137		FM-357417-S	WO-A-524	
FM-355550-S	WO-A-963		FM-357418-S	WO-635840	
FM-355551-S	WO-A-997		FM-357419-S2	WO-2875	
FM-355552-S	WO-A-877		FM-357420-S	WO-337112	
FM-355554-S	WO-A-1136		FM-357553-S18	WO-392468	
FM-355569-S2	WO-A-2983		FM-357574-S	WO-28023	
FM-355578-S	WO-A-903		FM-357699-S2	WO-A-834	
FM-355597-S	WO-630101		FM-357703-S	WO-A-1714	
FM-355698-S	WO-A-967		FM-358006-S7	WO-640038	
FM-355699-S2	WO-636527		FM-358006-S8	WO-640038	
FM-355741-S	WO-A-147		FM-358019-S	WO-A-5120	

FORD MOTOR CO. (FM) Cont'd

Ford Part Number	Willys Part Number	Gov't Group No.
FM-358040-S	WO-A-1237	
FM-358048-S	WO-636577	
FM-358063-S	WO-376373	
FM-358064-S	WO-5085	
FM-358074-S	WO-A-1721	

Vendor Number	Willys Part Number	Ford Part Number	Gov't Group No.
FLEX-O-TUBE CO. (FO)			
FO-HA-8031	WO-A-1325	FM-GPW-9288	0304
FYR-FIGHTER CO. (FYR)			
FYR D-10-A	WO-A-616	FM-GPW-17100	2402
GENERAL MOTORS CORP. (GM)			
GM-100014	WO-5934	FM-2046-S2	0401
GM-100025	WO-50163	FM-20349-S2	0106D
GM-100026	WO-5919	FM-24389-S2	0106D
GM-100027	WO-50878	FM-24449-S7	0402
GM-100044	WO-51371	FM-24555-S	1403A
GM-100121	WO-51523	FM-20346-S2	1802A
GM-100768	WO-6383	FM-27145-S2	0301C
GM-102634	WO-6167	FM-33797-S2	0601A
GM-103024	WO-5914	FM-33796-S2	0301C
GM-103026	WO-5901	FM-33800-S2	0601A
GM-106262	WO-5437	FM-34706-S2	0601B
GM-106279	WO-52132	FM-20027-S2	0301C
GM-106281	WO-5922	FM-24407	0402
GM-106331	WO-52911	FM-24408-S	0101A
GM-106333	WO-6486	FM-24534-S	0402
GM-107322	WO-52909	....	0110
GM-107381	WO-52825	FM-356028-S	0104
GM-110347	WO-638792	FM-353043-S7	0902
GM-112657	WO-53050	FM-20387-S7	2103A
GM-113844	WO-51612	FM-355433-S2	1403A
GM-114493	WO-52945	FM-23395-S	1500
GM-115093	WO-52046	FM-34807-S2	1817A
GM-115729	WO-50802	FM-33798-S2	2103A
GM-118613	WO-52217	FM-33799-S2	1804
GM-118772	WO-53048	FM-24347-S7	2103A
GM-119034	WO-52983	FM-23393-S2	0402
GM-122122	WO-53031	FM-20348-S2	1817A
GM-123228	WO-53033	FM-33799-S2	1817A
GM-123499	WO-53131	FM-20407-S7	2601
GM-127753	WO-52839	....	1209C
GM-128854	WO-6352	FM-355836-S8	0303
GM-136837	WO-53303	....	2601
GM-138485	WO-52725	FM-34706-S7	2601
GM-138489	WO-53135	....	1802A
GM-174916	WO-53024	FM-351274-57	0301C
GM-178378	WO-53023	....	2601
GM-178532	WO-53025	FM-356309-S7	2601
GM-178551	WO-53036	....	2601
GM-263549	WO-A-302	FM-GPW-13836	0609B
OAKES PRODUCTS DIV. (HH)			
HH-613300	WO-A-5621	FM-GPW-18205-A	0301C
HH-613306	WO-A-5631	FM-GPW-9658	0301C
HH-613313	WO-A-5632	FM-GPW-9621	0301C
HH-613314	WO-A-5633	FM-GPW-9623	0301C
HH-613380	WO-A-7191	FM-GPW-9612	0301C
HH-613387	WO-A-5630	FM-GPW-9617	0301C
HH-613455	WO-A-5629	FM-GPW-9609	0301C

Vendor Number	Willys Part Number	Ford Part Number	Gov't Group No.
HOLLAND HITCH CO. (HLH)			
HLH-T-60-B	WO-A-593	FM-GPW-5182	1502
HOOVER BALL AND BEARING CO. (HO)			
HO-88541	WO-636297	FM-GPW-8530	0503
HARRIS PRODUCTS CO. (HP)			
HP-770	WO-637936	FM-GPW-18060	1603
HARRISON RADIATOR CORP. (HR)			
HR-3108628	WO-637646	FM-GPW-8575	0502
HYATT BEARINGS DIV. (HY)			
HY-94322	WO-A-924	FM-GP-7718-A	0804
KELSEY HAYES WHEEL CO. (KHW)			
KHW-24382	WO-A-348	FM-GPW-17035	2301
KHW-24562	WO-A-465	FM-GPW-1015	1301
KHW-24566	WO-A-472	....	1302
KHW-24568	WO-A-473	FM-GP-1108	1302A
KHW-24575	WO-A-475	FM-GP-1013	1302A
KHW-24576	WO-A-474	FM-GP-1107	1302A
KHW-25646	WO-A-1690	FM-GP-1102	1302
KHW-25647	WO-A-1689	FM-GP-1103	1302
KHW-25649	WO-A-1691	....	1302
KHW-25692	WO-A-5467	FM-GPW-1015	1301
KHW-25693	WO-A-5468	FM-GPW-1016	1301
KHW-25695	WO-A-5470	FM-GPW-1029	1301
KHW-25696	WO-A-5472	FM-GPW-1045	1301
KHW-25779	WO-A-5471	FM-GPW-1030	1301
KHW-25917	WO-A-5539	FM-GPW-1024	1301
KHW-25930	WO-A-5488	FM-GPW-1025-C	1301
KING SEELEY CORP. (KS)			
KS-513	WO-53194	FM-34903-S2	0606F
KS-40333	WO-A-8242	....	2204
KS-40350	WO-A-8125	....	2204
KS-40355	WO-A-8180	FM-GPW-17255-A	2204
LINK-BELT CO. (LK)			
LK-S-35116-1	WO-634796	FM-GPW-6308	0102
LK-S-35117-1	WO-638458	FM-GPW-6256	0106C
LK-S-40936	WO-638457	FM-GPW-6260	0106C
WAGNER ELECTRIC CO. (LO)			
LO-21-11	WO-A-8279	....	2302
LO-S-FC-602	WO-637604	FM-91A-2152	1205
LO-S-FC-603	WO-637606	FM-90A-2151	1205
LO-FE-1444	WO-A-556	FM-GP-2140	1205
LO-S-FC-1499	WO-637579	FM-91A-2201	1207
LO-S-FD-2108-E	WO-637590	FM-GP-2173	1205
LO-S-FD-2109-B	WO-637595	FM-GP-2170	1205
LO-S-FC-2917	WO-637584	FM-GP-2175	1205
LO-S-FC-2918-A	WO-637585	FM-GP-2176	1205
LO-S-FC-2919-E	WO-637586	FM-GP-2183	1205
LO-S-FC-2926	WO-637597	FM-GP-2188	1205
LO-S-FC-2927	WO-637598	FM-GP-2174	1205
LO-S-FC-3023	WO-638544	....	1207
LO-FC-3052	WO-637427	FM-78-2814-A	1209C
LO-FC-4158	WO-A-6116	FM-GPW-2201	1207
LO-S-FD-4564	WO-637582	FM-GP-2155	1205
LO-FD-4664	WO-637789	FM-GP-2192	1207
LO-FD-4665	WO-637787	FM-GP-2261	1207
LO-FD-5381	WO-115962	FM-GPW-18371	1207
LO-S-FC-5727-A	WO-A-557	FM-GP-2076	1205

Vendor Number	Willys Part Number	Ford Part Number	Gov't Group No.
WAGNER ELECTRIC CO. (LO) Cont'd			
LO-FC-5778	WO-637432	FM-GP-2074	1209C
LO-FC-5784	WO-637424	FM-GP-2078	1209B
LO-S-FC-5992	WO-637580	FM-GP-2205	1207
LO-S-FC-5993	WO-637540	FM-GP-2208	1207
LO-FC-5994	WO-637546	FM-GP-2206	1207
LO-FD-5997	WO-637577	FM-GP-2194	1207
LO-FD-5998	WO-637541	FM-GP-2196	1207
LO-FC-6003	WO-637545	FM-GP-2204	1207
LO-S-FC-6007	WO-637591	FM-GP-2169	1205
LO-S-FC-6009	WO-637587	FM-GPW-2145	1205
LO-S-FC-6010	WO-637583	FM-GP-2160	1205
LO-S-FC-6011	WO-637602	FM-GP-2180	1205
LO-S-FC-6014	WO-637599	FM-GP-2143-A	1205
LO-S-FC-6018-E	WO-637608	FM-GP-2162	1205
LO-S-FC-6019	WO-637612	FM-GP-2167	1205
LO-S-FD-7379	WO-A-1484	GM-GPW-2061	1207
LO-FD-7568-A	WO-A-6110	....	1207
LO-FC-8502	WO-A-1373	FM-GPW-2078	1209B
LO-FD-8547	WO-A-1502	FM-GPW-2063	1502
LO-FC-8553	WO-A-1460	FM-GPW-2079	1209B
LO-FC-8772	WO-A-6111	FM-GP-2135	1207
LO-FC-8779	WO-A-6117	FM-GPW-2206	1207
LO-FC-8782	WO-A-6111	FM-GP-2135	1207
MONROE AUTO EQUIPMENT (MAE)			
MAE-T-317	WO-A-7778	....	2301
MAE-T-347	WO-A-7779	....	2301
MAE-8301	WO-116624	....	1603
MAE-10639-B	WO-637803	....	1603
MAE-10640-B	WO-637804	....	1603
MAE-10855	WO-116625	....	1603
MAE-10856-8	WO-116626	....	1603
MAE-10863-1	WO-116627	....	1603
MAE-10863:2	WO-116628	....	1603
MAE-10875	WO-637810	....	1603
MAE-10906-A	WO-116629	....	1603
MAE-10966	WO-116630	....	1603
MAE-11465	WO-A-6902	....	1603
MAE-11466	WO-A-6903	....	1603
MAE-12448	WO-116631	....	1603
MAE-12449	WO-116632	....	1603
MAE-12463	WO-116633	....	1603
MAE-12464	WO-116634	....	1603
MAE-12468	WO-116635	....	1603
MAE-12469	WO-116636	....	1603
MAE-12507	WO-116637	....	1603
MAE-12620	WO-116638	....	1603
MAE-12621	WO-116639	....	1603
MAE-12626	WO-116640	....	1603
MAE-12627	WO-116641	....	1603
MAE-12628	WO-116642	....	1603
MAE-12629	WO-116643	....	1603
MAE-12631	WO-116644	....	1603
MAE-108496	WO-638343	....	1603
McCORD RADIATOR & MFG. CO. (MDR)			
MDR-AM-504K	WO-637863	FM-O1A-12410	0604B
MALLORY & CO., P. R. (MLL)			
MLL-205244	WO-A-7600	FM-GPW-18841	2601
MARLIN ROCKWELL CORP. (MRC)			
MRC-206	WO-A-1007	FM-GP-7777-A	0803
MRC-307	WO-A-916	FM-GP-7065	0704
MRC-DZ-13567	WO-636297	FM-GPW-8530	0503

Vendor Number	Willys Part Number	Ford Part Number	Gov't Group No.
MIDLAND STEEL PRODUCTS CO. (MSP)			
MSP-F-2005	WO-A-668	....	1500
MSP-F-2009	WO-A-549	....	1500
MAZDA (GENERAL ELECTRIC) (MZ)			
MZ-51	WO-52837	FM-48-15021	0607E
MZ-1245	WO-51804	FM-B-13466	0607E
NEW DEPARTURE MFG. CO. (ND)			
ND-1203	WO-51248	FM-B-13466	0607E
ND-3206	WO-A-1007	FM-GPW-7777-A	0803
ND-7607	WO-A-916	FM-GP-7065	0704
ND-77503	WO-A-6299	FM-B-10094	0601
ND-885141	WO-636297	FM-GPW-8530	0503
NEW PROCESS GEAR, INC. (NP)			
NP-33873	WO-A-6382	....	1007
NP-38487	WO-A-6361	....	1007
NP-38491	WO-A-6383	....	1007
NP-38492	WO-A-6384	....	1007
NP-38493	WO-A-6362	....	1007
PUROLATOR PRODUCTS (PU)			
PU-25755	WO-A-1232	FM-GPW-18691-A	0107J
PU-25756	WO-A-1233	FM-GPW-18675-A	0107J
PU-25757	WO-A-1234	FM-GPW-18685-A	0107J
PU-25791	WO-A-1231	FM-GPW-18687-A	0107J
PU-25795	WO-A-1237	FM-358040-S	0107J
PU-25802	WO-A-1235	FM-GPW-18688-A	0107J
PU-26637	WO-A-1236	FM-GPW-18866	0107J
PU-27078	WO-A-1230	FM-GPW-18660-A	0107J
PU-27081	WO-A-1251	FM-GPW-18664-A	0107J
RICH MFG. CORP. (RMC)			
RMC-A-365	WO-637183	FM-GPW-6505	0105
RMC-A-366	WO-637182	FM-GPW-6507	0105
ROSS GEAR & TOOL CO. (RG)			
RG-7698-4-5/8	WO-A-745	FM-GPW-3575	1403
RG-8287-32	WO-A-752	FM-GPW-14171	0609
RG-T-13086	WO-A-1239	FM-GPW-3504	1403
RG-021116	WO-639118	FM-GPW-3577	1403
RG-025060	WO-639115	FM-356077-S8	1403
RG-026083	WO-A-633	FM-GPW-3655	0609
RG-028093	WO-639992	FM-GPW-8501	0503
RG-029021	WO-639121	FM-356303-S	1403
RG-029046	WO-638885	FM-GPW-3646	0609
RG-029049	WO-A-750	FM-GPW-3631	0609
RG-032075	WO-639095	FM-GPW-3591-A	1403
RG-032087	WO-A-302	FM-GPW-13836	0609
RG-033046	WO-639108	FM-GPW-3593	1403
RG-033047	WO-639109	FM-GPW-3594	1403
RG-033048	WO-639110	FM-GPW-3595	1403
RG-051035	WO-A-751	FM-GPW-3635	0609
RG-063011	WO-639090	FM-GPW-3587	1403
RG-063012	WO-639091	FM-GPW-3576	1403
RG-063991	WO-A-747	FM-GPW-3652	0609
RG-063996	WO-639190	FM-GPW-3517	1403
RG-T-126000	WO-A-1760	FM-GPW-3568	1403
RG-T-129001	WO-639119	FM-GPW-3581	1403
RG-400013	WO-639104	FM-GPW-3571	1403
RG-400025	WO-639102	FM-GPW-3552	1403
RG-401090	WO-639191	FM-GPW-3520	1403
RG-401100	WO-639103	FM-GPW-3589	1403
RG-401107	WO-638884	FM-GPW-3626	0609
RG-450054	WO-A-634	FM-GPW-3627	0609

Vendor Number	Willys Part Number	Ford Part Number	Gov't Group No.
ROSS GEAR & TOOL CO. (RG) Cont'd			
RG-502282	WO-A-635	FM-GPW-3506	1405
RG-5025401	WO-639116	FM-GPW-3583	1403A
RG-503284	WO-A-740	FM-GPW-3548	1403A
RG-503308	WO-A-742	FM-GPW-3524	1403A
STANT MFG. CO. (SA)			
SA-6455-A	WO-A-6424	FM-GPW-9034-A	0300
SA-6935-A	WO-A-1215	FM-GPW-8100-A	0501
SKF INDUSTRIES, INC. (SKF)			
SKF-6206	WO-A-1007	FM-GP-7777-A	0803
SKF-6307-Z	WO-A-916	FM-GP-7065	0704
SPICER MFG. CORP. (SP)			
SP-LW-1	WO-A-964	. . . .	0805
SP-42-G	WO-5354	FM-72004-S	0204A
SP-S-58	WO-A-784	FM-GP-4229-A	1103
SP-S-59	WO-A-785	FM-GP-4229-B	1003
SP-S-74	WO-A-786	FM-GP-4229-C	1003
SP-S-75	WO-A-787	FM-GP-4229-D	1003
SP-S-112	WO-A-800	FM-GP-4660-A	1003
SP-S-113	WO-A-801	FM-GP-4660-B	1003
SP-S-114	WO-A-802	FM-GP-4660-C	1003
SP-122-D	WO-A-1136	FM-355554-S	0801
SP-124-J	WO-A-980	FM-356125-S	0803
SP-145-W	WO-A-1028	FM-356504-S	0803
SP-153-SP	WO-A-1006	FM-73889-S7-8	1201
SP-S-171	WO-636565	FM-GP-4661	1104
SP-232-SP	WO-A-1004	FM-73928-S7-8	1201
SP-238-R	WO-A-1015	FM-64647-S	1201
SP-363-SP	WO-A-972	FM-GP-7796	0805B
SP-373-J	WO-A-1018 •	FM-O1T-2805	1201
SP-433-D	WP-A-973	FM-355378-S	0805F
SP-477-W	WO-A-1135	FM-356205-S	0801
SP-529-W	WO-52510	FM-34941-S	0607D
SP-634-D	WO-A-1137	FM-355533-S	0801
SP-690-W	WO-A-1000	FM-GP-7744	0804
SP-710-W	WO-A-990	FM-GP-7771	0803
SP-745-D	WO-A-963	FM-355550-S	0805A
SP-S-780	WO-636577	FM-358048-S	1101
SP-854-D	WP-A-967	FM-355698-S	0805
SP-886-D	WO-A-1019	FM-355352-S7	1201
SP-887-D	WO-A-1020	FM-O1T-2616	1201
SP-S-953	WO-A-780	FM-GP-3374	1007
SP-S-957	WO-A-825	FM-GP-3122	1006
SP-S-962	WO-A-1714	FM-357703-S	1006
SP-S-1056	WO-636570	FM-356504-S	1301C
SP-S-1106	WO-A-876	FM-356124-S	1401
SP-S-1107	WO-10558	FM-351059-S7	1402
SP-S-1135	WO-636569	FM-356126-S	1301C
SP-IS-1310	WO-A-877	FM-355552-S	1203
SP-K2-62-210-212-1718	WO-A-1429	FM-GPW-4605	0902
SP-K2-62-210-212-1907	WO-A-1428	FM-GPW-3370	0901
SP-5815-X	WO-A-1008	FM-GPW-2598	1201
SP-5900-X	WO-A-1009	FM-GPW-2648	1201
SP-5951-X	WO-A-934	FM-GP-7754	0801
SP-5962-X	WO-A-1111	FM-GP-7776	0803
SP-8996-SF	WP-A-1326	FM-GPW-3365	0901
SP-8997-S7	WO-A-1327	FM-GPW-4602	0901
SP-11586	WO-636528	FM-34922-S	1001
SP-13439	WO-A-1706	FM-51-3287	1402
SP-13449	WO-636360	FM-GP-4241	1003
SP-13575	WO-636566	FM-GP-4619	1003
SP-14223	WO-639265	FM-GP-4676A	1104
SP-15099	WO-636568	FM-GP-4666	1104
SP-15367	WO-A-799	FM-GP-4668	1104
SP-15926	WO-A-796	FM-GPW-4215	1103
SP-16067	WO-A-844	FM-78-3336	1402
SP-16075	WO-A-798	FM-GP-4211	1003
SP-16383	WO-A-793	FM-GP-4206	1002
SP-16385	WO-A-794	FM-GPW-4236	1003
SP-16409	WO-A-782	FM-GP-4035	1001
SP-16412-X	WO-A-789	FM-GPW-4209	1003
SP-16866	WO-A-792	FM-GP-4281	1103
SP-16968-X	WO-A-788	. . . .	1103
SP-16976	WO-A-781	FM-GP-4016	1101
SP-16979	WO-A-870	FM-GP-4022	1001
SP-16983	WO-A-818	FM-GP-3139	1006
SP-16991	WO-A-824	FM-GP-3115	1006
SP-17004	WO-A-864	FM-GP-1177	1301B
SP-17015	WO-A-864	FM-GP-1218	1301C
SP-17016	WO-A-866	FM-GP-4252	1007
SP-17017	WO-A-867	FM-GP-1124	1102
SP-17018	WO-A-778	FM-GP-3031-A1	1301B
SP-17019	WO-A-819	FM-GP-3135	1006
SP-17036	WO-A-779	FM-GP-3034	1006
SP-17041	WO-A-820	FM-GP-1092	1006
SP-17046-X	WO-A-838	FM-GP-3289	1402
SP-17047-X	WO-A-847	FM-GP-3290	1402
SP-17048-X	WO-A-828	FM-GP-3140	1006
SP-17071	WO-A-869	FM-GP-1139	1302
SP-17121-X	WO-A-1716	FM-GP-3200-A1	1007
SP-17133-X	WO-A-814	FM-GP-1089	1006
SP-17135-X	WO-A-813	FM-GPW-1088	1006
SP-17146	WO-A-904	FM-GP-4032	1102
SP-17153	WO-A-868	FM-GPW-3204	1007
SP-17202-X	WO-A-851	FM-GP-3105	1006
SP-17205	WO-A-861	FM-GPW-3170	1401
SP-17210	WO-A-855	FM-GPW-3165	1401
SP-17211	WO-A-856	FM-GPW-3166	1401
SP-17213	WO-A-858	FM-GPW-3167	1401
SP-17214	WO-A-859	FM-GPW-3168	1401
SP-17215	WO-A-860	FM-GPW-3169	1401
SP-17216	WO-A-1724	FM-GP-3217	1007
SP-17217	WO-A-1725	FM-24622-S	1007
SP-17218	WO-A-1726	FM-GP-3216	1007
SP-17221	WO-A-811	FM-GP-3148-A2	1006
SP-17222	WO-A-812	FM-GP-3149-A2	1006
SP-17224	WO-A-1723	FM-GP-3218	1007
SP-17226-X	WO-A-888	FM-GPW-4004	1101
SP-17230	WO-A-1719	FM-GP-3215-B	1007
SP-17231	WO-A-1720	FM-GP-3221-A	1007
SP-17232	WO-A-1721	FM-358074-S	1007
SP-17233	WO-A-1722	FM-GP-3219	1007
SP-17301-X	WO-A-1710	FM-GPW-3112	1006
SP-17302-X	WO-A-1712	FM-GPW-3113	1006
SP-17307	WO-A-8249	FM-GPW-3131-13	1006
SP-17308-X	WO-A-1704	FM-GPW-3280	1402
SP-17309-X	WO-A-1708	FM-GPW-3279	1402
SP-17310-X	WO-A-1703	FM-GPW-3074	1001
SP-17347	WO-A-6305	. . . .	1402
SP-17356-X	WO-A-6361	. . . .	1007
SP-17359	WO-A-6382	. . . .	1007
SP-17361	WO-A-6362	. . . .	1007
SP-17377	WO-A-6439	FM-GPW-4259	1102
SP-SKA-34172	WO-A-1765	FM-GPW-1101670	1809
SP-10-B-18	WO-A-1016	FM-O1T-2642	1201
SP-12-B-5	WO-A-1017	FM-O1T-2634	1201
SP-14-B-6	WO-A-1021	FM-O1T-2640	1201

Vendor Number	Willys Part Number	Ford Part Number	Gov't Group No.
SPICER MFG. CORP. (SP) Cont'd			
SP-20-B-62	WO-A-1002	FM-GP-2614	1201
SP-3-O-74	WO-A-1005	FM-GPW-2630	1201
SP-3-Q-76	WO-311003	FM-73904-S	1201
SP-4-B-26	WO-A-1016	FM-GP-2620	1201
SP-5-B-15	WO-A-1003	FM-GPW-2632	1201
SP-5-B-16	WO-A-1226	FM-GPW-2656	1201
SP-K-5-21-X	WO-A-1433	FM-21C-18397-B	0902
SP-39-Q-20	WO-A-971	FM-GP-7213	0706A
SP-97-Q-13	WO-A-970	FM-GP-7799	0805B
SP-475-2	WO-A-985	FM-GP-17277	0810
SP-S-638	WO-A-803	FM-GPW-4659-A	1104
SP-S-638-1	WO-A-804	FM-GP-4659-B	1104
SP-S-638-2	WO-A-805	FM-GP-4659-C	1104
SP-S-638-3	WO-A-806	FM-GP-4659-D	1104
SP-2058-1	WO-A-1212	. . . .	1000
SP-2058-2	WO-A-1387	. . . .	1000
SP-16322-2	WO-A-797	FM-GP-4230	1003
SP-16323-2	WO-A-795	FM-GP-4228	1103
SP-16992-1	WO-S-830	FM-GP-3117-A	1006
SP-16992-2	WO-A-831	FM-GP-3117-B	1006
SP-16992-3	WO-A-832	FM-GP-3117-C	1006
SP-16992-4	WO-A-833	FM-GP-3117-D	1006
SP-17120-3-X	WO-A-1715	FM-GPW-3206-A1	1007
SP-17120-4-X	WP-A-1728	FM-GPW-3207-A1	1007
SP-17123-3	WO-A-1727	FM-GPW-3016-A	1007
SP-17122-4	WO-A-1729	FM-GPW-3017-A	1007
SP-17128-2-X	WO-A-810	. . . .	1007
SP-17128-3-X	WO-A-809	. . . .	1007
SP-17144-3	WO-A-901	FM-GPW-4234	1102
SP-17144-4	WO-A-902	FM-GPW-4235	1102
SP-17155-1	WO-A-862	FM-GPW-3208-A	1007
SP-17155-2	WO-A-863	FM-GPW-3208-B	1007
SP-17295-1	WO-A-1705	FM-GPW-3281	1402
SP-17295-2	WO-A-1709	FM-GPW-3282	1402
SP-17360-1	WO-A-6384	. . . .	1007
SP-17360-2	WO-A-6383	FM-27145-S2	1007
SP-K2-1-28	WO-A-1105	FM-GP-4863	0803
SP-K2-4-88-X	WO-A-1106	FM-GP-7729	0803
SP-K2-7-29	WO-A-945	FM-O1Y-7096	0902
SP-K2-14-69	WO-A-942	FM-GP-7077	0902
SP-K2-15-63	WO-A-943	FM-GP-7097	0902
SP-K2-76-17	WO-A-940	FM-O1Y-7083	0902
SP-K2-94-29-V	WO-A-490	FM-O1Y-4529	0902
SP-K2-3-198-X	WO-A-935	FM-GP-4841	0902
SP-K2-4-108-X	WO-A-1445	FM-GP-4842	1104
SP-K3-86-89	WO-A-941	FM-O1T-7078-A	0902
SP-5-74-11	WO-A-491	FM-33784-S2	0902
SP-5-75-19	WO-51833	FM-34806-S7	0902
SP-5-73-310	WO-A-872	FM-24327-S2	1006
SP-6-74-11	WO-636575	FM-33786-S2	1402
SP-6-73-1117	WO-A-821	FM-355526-S	1006
SP-6-73-124	WO-A-760	FM-GP-1110	1007
SP-6-73-218	WO-A-997	FM-355551-S	1201
SP-6-73-414	WO-A-871	FM-355511-S	1003
SP-6-73-513	WO-A-903	FM-355578-S	1203
SP-6-74-101	WO-A-873	FM-33911-S	1006
SP-7-72-39	WO-636571	FM-357202-S	1003
SP-13-463-1	WO-A-974	. . . .	0805B
SP-22-381-19	WO-A-976	FM-GP-7783	0803
SP-31-246-2	WO-A-1001	FM-GP-7767	0804
SP-50-39-2	WO-5140	FM-353064-S	0701
SP-50-39-4	WO-636538	FM-353051-S	1001
SP-50-80-1	WO-5599	FM-353075-S	0804
SP-62-463-2	WO-A-958	. . . .	0803

Vendor Number	Willys Part Number	Ford Part Number	Gov't Group N..
SP-18-5-1	WO-A-999	FM-GP-7742	0804
SP-18-8-1	WO-A-989	FM-GP-7766	0803
SP-18-8-2	WO-A-988	FM-GP-7765	0803
SP-18-8-7	WO-A-1510	FM-GP-7722	0802
SP-18-12-1	WO-A-965	FM-GP-7789	0805A
SP-18-15-9	WO-A-1503	FM-GPW-7705	0801
SP-18-16-3	WO-A-953	FM-GP-7708	0801
SP-18-19-8	WO-A-956	FM-GPW-7774	0801
SP-18-19-18	WO-A-1507	FM-GPW-7769	0801
SP-18-24-1	WO-A-987	FM-GP-7777-A2	0803
SP-18-66-1	WO-A-959	FM-GP-7712	0805A
SP-18-66-7	WO-A-960	FM-GPW-7711	0805A
SP-18-67-2	WO-A-962	FM-GP-7787	0805
SP-18-67-5	WO-A-1504	FM-GPW-7786	0805
SP-18-155-2	WO-A-954	FM-GP-7709	0801
SP-18-187-1	WO-A-998	FM-GP-7743	0804
SP-18-223-2	WO-A-957	FM-GPW-9773	0801
SP-18-223-3	WO-A-1134	FM-GP-7746	0803
SP-18-267-4	WO-A-1508	FM-GPW-7706	0801
SP-18-324-2	WO-A-1509	FM-GPW-7707	0801
SP-18-352-5	WO-A-1435	FM-GPW-7756	0801
SP-18-362-3	WO-A-975	FM-GP-7761	0803
SP-18-363-4-566-X	WO-A-1764	FM-GP-7763	0803
SP-18-381-2	WO-A-991	FM-GP-7784	0802
SP-18-452-2	WO-A-1511	FM-GP-17285	0810
SP-18-453-3	WO-A-1512	FM-GPW-17271	0810
SP-18-454-4	WO-636396	FM-GP-17333	0810
SP-18-466-1	WO-A-992	FM-GP-7762	0802
SP-25-Q-1055	WO-A-1505	FM-GPW-7793	0805F
SP-25-Q-1056	WO-A-1506	FM-GPW-7710	0805F
SPARKS WITHINGTON CO. (SPW)			
SPW-B-9427	WO-A-1312	FM-GPW-13082	0609
SCHREADER'S SON, A. (SV)			
SV-7188-BT	WO-A-6855	FM-GPW-6722	0107I
STEWART-WARNER CORP. (SW)			
SW-5585	WO-A-213	FM-GP-17125	2301
SCHWARZE ELECTRIC CO. (SZE)			
SZE-61400	WO-A-1312	FM-GPW-13082	0609
TIMKEN ROLLER BEARING CO. (TM)			
TM-02820	WO-52879	FM-GP-4628	1104
TM-02872	WO-52878	FM-GP-4630	1104
TM-11520	WO-52941	FM-GP-3162	1006
TM-11590	WO-52940	FM-GP-3161	1006
TM-14131	WO-51575	FM-O1Y-1202	0803
TM-14276	WO-52883	FM-O1Y-1202	0803
TM-18250	WO-52942	FM-GP-1201	1301A
TM-18590	WO-52943	FM-GP-1202	1301A
TM-24721	WO-52881	FM-GP-4222	1003
TM-24780	WO-52880	FM-GP-4221	1003
TM-31530	WO-52877	FM-86H-4616	1003
TM-31593	WO-52876	FM-86H-4621	1003
TORRINGTON CO. (TR)			
TR-B-1210	WO-A-857	FM-GPW-3171	1006
TRICO PRODUCTS CO. (TRI)			
TRI-80540	WO-A-11432	FM-GPW-17534	1811I

Vendor Number	Willys Part Number	Ford Part Number	Gov't Group No.
AMERICAN CHAIN & CABLE (TRU)			
TRU-SA-2844-1	WO-630068	FM-GPW-7516	0204A
TAYLOR SALES ENGINEERING CO. (TSE)			
TSE-215-B	WO-630396	FM-GPW-6615	0107A
TOMPSON PRODUCTS (TP)			
TP-6MN27	WO-A-1707	FM-24916-S2	1402
TP-14DS20	WO-A-6305	. . . .	1402
TP-14DM-43	WO-A-844	FM-78-3336	1402
TP-14SV79-A-7	WO-A-838	FM-GP-3291	1402
TP-148V90-A-7	WO-A-847	FM-GP-3292	1402
TP-16MN8	WO-A-1706	FM-51-3287	1402
USL BATTERY CORP. (USL)			
USL-461389	WO-635097	FM-GPW-9653	0301C
VICTOR MFG. & GASKET CO. (VG)			
VG-2066-C-C1	WO-637863	FM-O1A-12410	0604B
WILLARD STORAGE BATTERY CO. (WB)			
WB-SW-2-119	WO-A-1238	FM-11AS-10655	0610
WB-SR-2-119	WO-A-1767	FM-11AS-10658	0610
WARNER ELECTRIC BRAKE (WEB)			
WEB-3529	WO-A-6586	FM-11YS-18151B	0606F
WEB-3604	WO-A-6019	FM-11YS-18142-B	0606F
WEB-11935-B	WO-A-6587	. . . .	0606F
WEB-20098	WO-A-6589	FM-11YS-18193-B	0606F
WEB-20099	WO-A-6588	FM-11YS-18198-B	0606F
WEB-110242	WO-A-8088	. . . .	0606F
WEB-110477	WO-53061	. . . .	0606F
WARNER GEAR (WG)			
WG-X-802	WO-639689	FM-20366-S	0706A
WG-X-2136	WO-635838	FM-353081-S	0706B
WG-X-2428-A	WO-A-1019	FM-355352-S7-8	1201
WG-X-3204-ML	WO-636885	FM-GPW-7025	0703
WG-4418-M	WO-636200	FM-GPW-7245	0706B
WG-4496-K	WO-A-1379	FM-BB-7220	0706A
WG-4498-K	WO-392328	FM-GPW-7227	0706A
WG-4682-K	WO-637831	FM-GPW-7109	0704B
WG-4686	WO-637835	FM-GPW-7059	0704B
WG-B-7070	WO-635846	FM-B-7070	0703
WG-C-8-2-½	WO-A-1381	FM-GPW-7217	0706A
WG-T-9-50P	WO-A-520	FM-356134-S18	0802
WG-ASI-T-84-J	WO-A-1145	FM-GPW-7000	0700
WG-AT-84-J-1A	WO-A-1148	FM-GPW-7005	0701
WG-2AT-84-H-2	WO-A-6317	FM-GPW-7060	0704
WG-T-84-H-2	WO-A-519	FM-GPW-7061	0704
WG-T-84-J-2½	WO-A-6319	FM-GPW-7105	0704B
WG-AT-84-F-2½A	WO-A-6318	FM-GPW-7124	0704B
WG-AC-84-J-2A	WO-A-1380	FM-GPW-7210-A	0706A
WG-T-84-C-3	WO-638948	FM-GPW-7111	0704C
WG-T-84-J-6	WO-640017	FM-GPW-7050	0703
WG-T-84-J-8	WO-A-739	FM-GPW-7113	0704B
WG-AT-84-F-10A	WO-636882	FM-GPW-7141	0704B
WG-T-84-F-11-A	WO-638798	FM-GPW-7102	0704B
WG-T-84-F-12-A	WO-636879	FM-GPW-7100	0704B
WG-C-84-B-12-A	WP-A-1382	FM-GPW-7221	0706A
WG-T-84-F-13	WO-637832	FM-GPW-7116	0704B
WG-T-84-F-14	WO-637834	FM-GPW-7107	0704B
WG-T-84-F-15	WO-637833	FM-GPW-7106	0704B
WG-AT-84-J-16-A	WO-A-5553	FM-GPW-7015	0703
WG-T-84-J-16-A	WO-A-5554	FM-GPW-7017	0703

Vendor Number	Willys Part Number	Ford Part Number	Gov't Group No.
WG-T-84-17	WO-635844	FM-GPW-7064	0703
WG-T-84-C-19	WO-640006	FM-GPW-7104-A2	0704
WG-T-84-J-20-A	WO-A-1155	FM-GPW-7241	0706B
WG-T-84-J-21	WO-A-1156	FM-GPW-7240	0706B
WG-T-84-22	WO-635836	FM-GPW-7206	0706C
WG-T-84-C-23-A	WO-636196	FM-GPW-7230	0706C
WG-T-84-C-24-A	WO-636197	FM-GPW-7231	0706C
WG-T-84-G-25	WO-639423	FM-GPW-7063	0703
WG-AT-84-J-25	WO-A-7260	FM-GPW-7211-B	0706C
WG-T-84-G-26	WO-639422	FM-GPW-7120	0703
WG-T-84-J-28	WO-A-738	FM-GPW-7062	0704
WG-T-84-J-28	WO-A-880	FM-GPW-7115	0704C
WG-T-84-J-29	WO-A-879	FM-GPW-7126	0704C
WG-T-84-29	WO-635811	FM-GPW-7129A	0704C
WG-T-84-B-30-A	WO-635812	FM-GPW-7119	0704C
WG-T-84-30	WO-635840	FM-357418-S	0706C
WG-T-84-31	WO-635839	FM-GPW-7208	0706C
WG-T-84-J-31	WO-A-524	FM-357417-S	0704C
WG-T-84-B-32	WO-635859	FM-GPW-7214	0706A
WG-T-84-G-35	WO-638952	FM-GPW-7140	0704C
WG-T-84-42	WO-635837	FM-GPW-7234	0706A
WG-T-84-C-48	WO-638949	FM-GPW-7155	0704C
WG-T-84-J-50½A	WO-A-1410	FM-356519-S	0802
WG-T-84-J-54	WO-640018	FM-GPW-7052	0703
WG-T-84-85A	WO-635804	FM-GPW-7143	0704C
WG-T-84-J-86A	WO-A-1385	FM-GPW-7233	0706B
WG-T-84-J-100	WO-A-1492	FM-GPW-17091	2301
WG-T-84-115	WO-635861	FM-GPW-7223	0706A
WG-T-84-H-137	WO-A-410	FM-GPW-7080	0704
WG-T-84-J-167	WO-A-878	FM-GPW-7121	0704C
WEATHERHEAD CO. (WH)			
WH-145-A	WO-A-1126	FM-9N-8115	0501
WILCOX RICH (WIL)			
WIL-ST-1265	WO-637047	FM-GPW-6500-A	0105A

	Willys Part Number	Ford Part Number	Gov't Group No
WILLYS-OVERLAND MOTORS, INC. (WO)			
	WO-A-146	FM-GPW-6043	0110
	WO-A-147	FM-355741-S	0809
	WO-A-168	FM-34747-S7	0610B
	WO-A-176	FM-GPW-7539	0204A
	WO-A-177	FM-GPW-7517	0204A
	WO-A-178	FM-GPW-7545	0204A
	WO-A-179	FM-GPW-7507	0204A
	WO-A-180	FM-GPW-7508	0204A
	WO-A-181	FM-GPW-7514	0204A
	WO-A-183	FM-GPW-2462	1205
	WO-A-185	FM-GPW-5113	1500
	WO-A-213	FM-GP-17125	2301
	WO-A-227	FM-356525	1603
	WO-A-281	FM-GPW-9628	0301C
	WO-A-302	FM-GPW-13836	0609B
	WO-A-313	FM-GP-17037	2301
	WO-A-348	FM-GPW-17035	2301
	WO-A-371	. . . .	2301
	WO-A-372	FM-GPW-17005	2301
	WO-A-373	FM-GP-17042	2301
	WO-A-374	FM-GP-17028	2301
	WO-A-375	FM-GP-17020	2301
	WO-A-376	FM-GP-17023	2301
	WO-A-377	FM-GP-17021	2301

Willys Part Number	Ford Part Number	Gov't Group No.	Willys Part Number	Ford Part Number	Gov't Group No.
WILLYS-OVERLAND MOTORS, INC. (WO) Cont'd			WO-A-612-4	FM-GPW-5317-B	1601
WO-A-378	. . . .	2301	WO-A-612-5	FM-GPW-5318-B	1601
WO-A-379	FM-GP-17038	2301	WO-A-612-6	FM-GPW-5319-B	1601
WO-A-405	FM-GPW-7250	0204	WO-A-612-7	FM-GPW-5320-B	1601
WO-A-410	FM-GPW-7080	0704	WO-A-612-8	FM-GPW-5321-B	1601
WO-A-415	FM-GPW-5106	1500	WO-A-613	FM-GPW-5310	1601
WO-A-416	. . . .	1500	WO-A-613-1	FM-GPW-5313-A	1601
WO-A-417	. . . .	1500	WO-A-613-2	FM-GPW-5315-A	1601
WO-A-418	. . . .	1500	WO-A-613-3	FM-GPW-5316-A	1601
WO-A-419	. . . .	1500	WO-A-613-4	FM-GPW-5317-A	1601
WO-A-420	. . . .	1500	WO-A-613-5	FM-GPW-5318-A	1601
WO-A-421	. . . .	1500	WO-A-613-6	FM-GPW-5319-A	1601
WO-A-439	FM-GPW-6392	1091C	WO-A-613-7	FM-GPW-5320-A	1601
WO-A-445	. . . .	1100	WO-A-613-8	FM-GPW-5321-A	1601
WO-A-447	FM-GPW-8600	0503A	WO-A-614	FM-GPW-5560	1601A
WO-A-463	FM-GPW-9632	0301C	WO-A-614-1	FM-GPW-5563	1601A
WO-A-465	FM-GPW-1015	1301	WO-A-614-2	FM-GPW-5565	1601A
WO-A-472	. . . .	1302	WO-A-614-3	FM-GPW-5566	1601A
WO-A-473	FM-GP-1108	1302A	WO-A-614-4	FM-GPW-5567	1601A
WO-A-474	FM-GP-1107	1302A	WO-A-614-5	FM-GPW-5568	1601A
WO-A-475	FM-GP-1013	1302A	WO-A-614-6	FM-GPW-5569	1601A
WO-A-476	FM-GP-1012	1505	WO-A-614-7	FM-GPW-5570	1601A
WO-A-481	FM-GPW-5783-A	1601B	WO-A-614-8	FM-GPW-5571	1601A
WO-A-484	. . . .	1500	WO-A-614-9	FM-GPW-5572	1601A
WO-A-485	. . . .	1500	WO-A-616	FM-GPW-17100	2402
WO-A-490	FM-O1Y-4529	0902	WO-A-622	FM-GPW-3332-A	1401
WO-A-491	FM-33784-S2	0902	WO-A-623	FM-GPW-3336	1401
WO-A-493	. . . .	1500	WO-A-630	. . . .	1209C
WO-A-495	FM-GPW-2473	1204	WO-A-631	. . . .	1209C
WO-A-498	FM-356561-S	0204	WO-A-633	FM-GPW-3655	0609B
WO-A-499	FM-GPW-7521	0204A	WO-A-634	FM-GPW-3627	0609B
WO-A-500	FM-GPW-5341	1500	WO-A-635	FM-GPW-3506	1405
WO-A-508	. . . .	1500	WO-A-637	FM-GPW-5077	1500
WO-A-510	. . . .	1500	WO-A-642	FM-GPW-9647	0301C
WO-A-513	FM-GPW-5778	1602	WO-A-647	FM-GPW-5118	1205
WO-A-514	FM-GPW-5779	1602	WO-A-655	FM-GPW-5264-A	0401
WO-A-515	FM-GPW-5481	1602	WO-A-657	. . . .	0401
WO-A-519	FM-GPW-7061	0704	WO-A-658	FM-GPW-5283	0401
WO-A-520	FM-356134-S18	0802	WO-A-668	. . . .	1500
WO-A-524	FM-357417-S	0704C	WO-A-683	. . . .	1209C
WO-A-534	FM-GPW-5097	1500	WO-A-692	FM-GP-17033	2301
WO-A-538	FM-GPW-12405	0604B	WO-A-693	FM-GPW-17097	2402
WO-A-544	FM-GPW-5337	1500	WO-A-719	FM-GPW-13410	0606B
WO-A-547	FM-GPW-5035	1500	WO-A-738	FM-GPW-7062	0704
WO-A-548	. . . .	1500	WO-A-739	FM-GPW-7113	0704B
WO-A-549	. . . .	1500	WO-A-740	FM-GPW-3548	1403A
WO-A-552	. . . .	1502	WO-A-742	FM-GPW-3524	1403A
WO-A-556	FM-GP-2140	1205	WO-A-745	FM-GPW-3575	1403A
WO-A-557	FM-GP-2076	1205	WO-A-747	FM-GPW-3652	0609B
WO-A-568	FM-GPW-5458	1000	WO-A-750	FM-GPW-3631	0609B
WO-A-571	FM-GPW-5460	1100	WO-A-751	FM-GPW-3635	0609B
WO-A-572	FM-GPW-5459	1000	WO-A-752	FM-GPW-14171	0609B
WO-A-574	FM-GPW-5453	1602	WO-A-754	FM-GP-2038	1203A
WO-A-575	FM-GPW-5705-B	1602	WO-A-755	FM-33800-S7	1203A
WO-A-592	FM-GPW-8284	0505	WO-A-760	FM-GP-1110	1007
WO-A-593	FM-GPW-5182	1502	WO-A-778	FM-GP-3031-A	1301B
WO-A-596	FM-GP-17043	2301	WO-A-779	FM-GP-3034	1006
WO-A-597	FM-GP-17044	2301	WO-A-780	FM-GP-3374	1007
WO-A-598	FM-GP-17045	2301	WO-A-781	FM-GP-4015	1001
WO-A-599	FM-GP-17046	2301	WO-A-782	FM-GF-4035	1101
WO-A-600	FM-GP-17047	2301	WO-A-784	FM-GP-4229-A	1003
WO-A-612	FM-GPW-5311	1601	WO-A-785	FM-GP-4229-B	1003
WO-A-612-1	FM-GPW-5313-B	1601	WO-A-786	FM-GP-4229-C	1003
WO-A-612-2	FM-GPW-5315-B	1601	Cont'A-787	FM-GP-4229-D	1003
WO-A-612-3	FM-GPW-5316-B	1601	WO-A-788	. . . .	1002

Willys Part Number	Ford Part Number	Gov't Group No.	Willys Part Number	Ford Part Number	Gov't Group No.
WILLYS-OVERLAND MOTORS, INC. (WO) Cont'd			WO-A-901	FM-GPW-4234	1102
WO-A-789	FM-GP-4209	1103	WO-A-902	FM-GPW-4235	1102
WO-A-792	FM-GP-4281	1003	WO-A-903	FM-355578-S	1203
WO-A-793	FM-GP-4206	1103	WO-A-904	FM-GP-4032	1102
WO-A-794	FM-GPW-4236	1003	WO-A-912	FM-GPW-9428	0108B
WO-A-795	FM-GP-4228	1003	WO-A-916	FM-GP-7065	0704
WO-A-796	FM-GPW-4215	1003	WO-A-924	FM-GP-7718-A	0804
WO-A-797	FM-GP-4230	1103	WO-A-927	FM-GPW-2211	1200
WO-A-798	FM-GP-4211	1003	WO-A-928	FM-GPW-2210	1200
WO-A-799	FM-GP-4668	1104	WO-A-934	FM-GP-7754	0801
WO-A-800	FM-GP-4660-A	1104	WO-A-940	FM-O1Y-7083	0902
WO-A-801	FM-GP-4660-B	1003	WO-A-941	FM-O1T-7078-A	0902
WO-A-802	FM-GP-4660-C	1003	WO-A-942	FM-GP-7077	0902
WO-A-803	FM-GP-4659-A	1003	WO-A-943	FM-GP-7097	0902
WO-A-804	FM-GP-4659-B	1003	WO-A-945	FM-OIY-7096	0902
WO-A-805	FM-GP-4659-C	1003	WO-A-953	FM-GPW-7708	0801
WO-A-806	FM-GP-4659-D	1003	WO-A-954	FM-GP-7709	0801
WO-A-809	. . . .	1007	WO-A-956	FM-GPW-7774	0801
WO-A-810	. . . .	1007	WO-A-957	FM-GPW-7773	0801
WO-A-811	FM-GP-3148-A	1006	WO-A-958	. . . .	0803
WO-A-812	FM-GP-3149-A	1006	WO-A-959	FM-GP-7712	0805A
WO-A-813	FM-GP-1088	1006	WO-A-960	FM-GP-7711	0805A
WO-A-814	FM-GP-1089	1006	WO-A-962	FM-GP-7787	0805
WO-A-818	FM-GP-3139	1006	WO-A-963	FM-355550-S	0805A
WO-A-819	FM-GP-3135	1006	WP-A-964	. . . .	0805
WO-A-820	FM-GP-1092	1006	WO-A-965	FM-GP-7789	0805A
WO-A-821	FM-355526-S	1006	WO-A-966	FM-GP-7788	0805
WO-A-824	FM-GP-3115	1006	WO-A-967	FM-355698-S	0805
WO-A-825	FM-GP-3112	1006	WO-A-970	FM-GP-7799	0805B
WO-A-828	FM-GP-3140	1006	WO-A-971	FM-GPW-7213	0706A
WO-A-830	FM-GP-3117-A	1006	WO-A-972	FM-GP-7796	0805B
WO-A-831	FM-GP-3117-B	1006	WO-A-973	FM-355378-S	0805B
WO-A-832	FM-GP-3117-C	1006	WO-A-974	. . . .	0805B
WO-A-833	FM-GP-3117-D	1006	WO-A-975	FM-GP-7761	0803
WO-A-838	FM-GP-3291	1402	WO-A-976	FM-GP-7783	0803
WO-A-844	FM-78-3336	1402	WO-A-980	FM-356125-S	0803
WO-A-847	FM-GP-3292	1402	WO-A-982	FM-GP-7782-A	0801
WO-A-851	FM-GP-3105	1006	WO-A-983	FM-GP-7782-B	0801
WO-A-853	. . . .	1006	WO-A-984	FM-GP-7782-C	0801
WO-A-855	FM-GPW-3165	1401	WO-A-985	FM-GP-17277	0810
WO-A-856	FM-GP-3166	1401	WO-A-987	FM-GP-7777-A	0803
WO-A-857	FM-GPW-3171	1401	WO-A-988	FM-GP-7765	0803
WO-A-858	FM-GPW-3167	1401	WO-A-989	FM-GP-7766	0803
WO-A-859	FM-GPW-3168	1401	WO-A-990	FM-GP-7771	0803
WO-A-860	FM-GPW-3169	1401	WO-A-991	FM-GP-7784	0802
WO-A-861	FM-GPW-3170	1401	WO-A-992	FM-GP-7762	0802
WO-A-862	FM-GP-3208-A	1007	WO-A-997	FM-355551-S	1201
WO-A-863	FM-GP-3208-B	1007	WO-A-998	FM-GP-7743	0804
WO-A-864	FM-GP-1177	1007	WO-A-999	FM-GP-7742	0804
WO-A-865	FM-GP-1218	1102	WO-A-1000	FM-GP-7744	0804
WO-A-866	FM-GP-4252	1007	WO-A-1001	FM-GP-7767	0804
WO-A-867	FM-GP-1124	1102	WO-A-1002	FM-GP-2614	1201
WO-A-868	FM-GP-3204	1007	WO-A-1003	FM-GPW-2632	1201
WO-A-869	FM-GP-1139	1302	WO-A-1004	FM-73928-S7	1201
WO-A-870	FM-GP-4022	1001	WO-A-1005	FM-GPW-2630	1201
WO-A-871	FM-355511-S	1103	WO-A-1006	FM-73889-S7	1201
WO-A-872	FM-24327-S2	1006	WO-A-1007	GP-7777-A	0803
WO-A-873	FM-33911-S2	1006	WO-A-1008	FM-GPW-2598	1201
WO-A-876	FM-33911-S2	1401	WO-A-1009	FM-GP-2648	1201
WO-A-877	FM-355552-S	1203	WO-A-1014	FM-GP-2620	1201
WO-A-878	FM-GPW-7121	0704C	WO-A-1015	FM-64647-S	1201
WO-A-879	FM-GPW-7126	0704C	WO-A-1016	FM-O1T-2642	1201
WO-A-880	FM-GPW-7115	0704C	WO-A-1017	FM-O1T-2634	1201
WO-A-887	FM-GPW-7512	0204A	WO-A-1018	FM-O1T-2805	1201
WO-A-888	FM-GPW-4004	1101	WO-A-1019	FM-355352-S7	1201

Willys Part Number	Ford Part Number	Gov't Group No.		Willys Part Number	Ford Part Number	Gov't Group No.
WILLYS-OVERLAND MOTORS, INC. (WO) Cont'd				WO-A-1159	. . . .	1500
WO-A-1020	FM-GP-2616	1201		WO-A-1160	. . . .	1500
WO-A-1021	FM-O1T-2640	1201		WO-A-1162	FM-GPW-17003	2301
WO-A-1028	FM-356504-S	0803		WO-A-1164	FM-GPW-5175	0610
WO-A-1031	FM-GPW-13022	0607		WO-A-1165	FM-GPW-9410-B	0108
WO-A-1032	FM-37206-S	0607		WO-A-1166	FM-GPW-9424-B	0108
WO-A-1033	FM-GPW-13007	0607A		WO-A-1173	FM-GPW-9751	0303
WO-A-1036	FM-GPW-13043	0607		WO-A-1174	FM-GPW-9727	0303
WO-A-1037	FM-38095-S	0607		WO-A-1175	FM-GPW-9742	0303
WO-A-1045	FM-GPW-9386	0302		WO-A-1190	FM-GPW-6016	0106
WO-A-1046	FM-GPW-9378	0302		WO-A-1192	FM-GPW-8250	0101
WO-A-1047	FM-GPW-9377	0302		WO-A-1195	FM-GPW-7700	0800
WO-A-1051	FM-GPW-6098	0110		WO-A-1197	FM-GPW-18667	0107
WO-A-1061	FM-GPW-6758	0107N		WO-A-1198	FM-GPW-18666	0107
WO-A-1064	FM-GPW-13405-B	0608		WO-A-1199	FM-GPW-3509	1403
WO-A-1065	FM-GPW-13404-B	0608		WO-A-1201	FM-GPW-5057	1500
WO-A-1070	FM-GP-13210-B	0607D		WO-A-1202	. . . .	1500
WO-A-1071	FM-GP-13209-B2	0607D		WO-A-1203	FM-GPW-5019	1500
WO-A-1072	FM-28378-S2	0607D		WO-A-1204	. . . .	1500
WO-A-1073	FM-GPW-13408-B	0608		WO-A-1205	. . . .	1500
WO-A-1074	FM-GPW-13494	0607A		WO-A-1207	. . . .	1500
WO-A-1075	FM-GPW-13491-A	0607A		WO-A-1210	. . . .	1500
WO-A-1076	FM-GPW-13448-B	0608		WO-A-1212	. . . .	1000
WO-A-1077	FM-36931-S	0608		WO-A-1214	FM-GPW-8005	0501
WO-A-1078	FM-GPW-13485	0607A		WO-A-1215	FM-GPW-8100-A	0501
WO-A-1079	FM-GPW-13449-A	0608		WO-A-1216	FM-GPW-8578	0501
WO-A-1082	. . . .	0607D		WO-A-1217	FM-GPW-8133	0501
WO-A-1084	FM-GPW-9732-A	0303		WO-A-1221	FM-GPW-9002-A	0300
WO-A-1089	FM-356299	0606E		WO-A-1223	FM-GPW-9510	0301
WO-A-1096	FM-11A-12425	0604B		WO-A-1224	. . . .	0301
WO-A-1097	. . . .	1602		WO-A-1225	FM-GPW-9716	0303
WO-A-1098	FM-GPW-14303	2603		WO-A-1226	FM-GPW-2656	1201
WO-A-1100	FM-GPW-17062	2301		WO-A-1227	FM-355752-S	1201
WO-A-1104	FM-353053	0801		WO-A-1228	FM-GPW-2659	1201
WO-A-1105	FM-GP-4863	0803		WO-A-1230	FM-GPW-18660-A	0107
WO-A-1111	FM-GP-7776	0803		WO-A-1231	FM-GPW-18687-A	0107
WO-A-1116	FM-GPW-3590	1403A		WO-A-1232	FM-GPW-18691-A	0107
WO-A-1117	FM-GPW-17750	2101		WO-A-1233	FM-GPW-18675-A	0107
WO-A-1120	FM-GPW-17759	1500		WO-A-1234	FM-GPW-18685-A	0107
WO-A-1122	. . . .	1500		WO-A-1235	FM-GPW-18688-A	0107
WO-A-1124	FM-GPW-8240	0503A		WO-A-1236	FM-GPW-18662-A	0107
WO-A-1125	FM-356376-S	0503A		WO-A-1237	FM-358040-S	0107
WO-A-1126	FM-9N-8115	0501		WO-A-1238	FM-11AS-10655	0610
WO-A-1127	FM-GPW-17755	1500		WO-A-1239	FM-GPW-3504	1403
WO-A-1128	FM-GPW-17754	1500		WO-A-1240	FM-GPW-17080	2301
WO-A-1129	FM-GPW-17753	1500		WO-A-1241	FM-GPW-2853	1201
WO-A-1130	FM-GPW-17752	1500		WO-A-1242	FM-GPW-2780	1201
WO-A-1134	FM-GP-7746	0803		WO-A-1243	FM-GPW-9738	0303
WO-A-1135	FM-356205-S	0801		WO-A-1244	FM-GPW-12100	0603
WO-A-1136	FM-355554-S	0801		WO-A-1245	FM-GPW-11001-A	0602
WO-A-1137	FM-355533	0801		WO-A-1247	FM-GPW-18663	0107
WO-A-1138	. . . .	1500		WO-A-1251	FM-GPW-18664-A	0107
WO-A-1142	. . . .	1500		WO-A-1252	FM-GPW-5482	1602
WO-A-1145	FM-GPW-7000	0700		WO-A-1253	FM-GPW-5291-B	1500
WO-A-1146	FM-GPW-5230-A	0401		WO-A-1254	FM-GPW-9030-A	0300
WO-A-1147	FM-GPW-17751	2101		WO-A-1255	FM-GPW-9155	0306
WO-A-1148	FM-GPW-7005	0701		WO-A-1256	FM-GPW-9183	0306
WO-A-1150	FM-GPW-5028	1500		WO-A-1257	FM-GPW-9184	0306
WO-A-1151	FM-GPW-5125	1500		WO-A-1258	FM-GPW-9149	0306
WO-A-1152	. . . .	1500		WO-A-1259	FM-GPW-9160	0306
WO-A-1153	. . . .	1500		WO-A-1260	FM-GPW-9186	0306
WO-A-1154	. . . .	1500		WO-A-1261	FM-GPW-9140	0306
WO-A-1155	FM-GPW-7241	0706B		WO-A-1262	FM-GPW-9182	0306
WO-A-1156	FM-GPW-7240	0706B		WO-A-1263	FM-GPW-9162	0306
WO-A-1157	FM-GPW-17775	2101		WO-A-1264	FM-GPW-9185	0306

Willys Part Number	Ford Part Number	Gov't Group No.	Willys Part Number	Ford Part Number	Gov't Group No.
WILLYS-OVERLAND MOTORS, INC. (WO) Cont'd			WO-A-1377	FM-GPW-2264	1209C
WO-A-1265	FM-GPW-9154	0306B	WO-A-1378	FM-GPW-2244	1209C
WO-A-1267	FM-GPW-17260	2204	WO-A-1379	FM-BB-7220	0706A
WO-A-1272	FM-GPW-6010	0101	WO-A-1380	FM-GPW-7210-A	0706A
WO-A-1275	FM-GPW-9035-B	0300	WO-A-1381	FM-GPW-7217-A	0706A
WO-A-1276	FM-GPW-3511	1405	WO-A-1382	FM-GPW-7221	0706A
WO-A-1277	FM-GPW-3682-A	1405	WO-A-1385	FM-GPW-7233	0706B
WO-A-1278	FM-GPW-9656	0301C	WO-A-1386	FM-GPW-2452-A	1204
WO-A-1279	FM-GPW-9657	0301C	WO-A-1387	. . . .	1000
WO-A-1283	FM-GPW-2073	1500	WO-A-1389	FM-GPW-13831	0609
WO-A-1287	FM-GPW-18936-A	2603	WO-A-1392	FM-GPW-10166	0601A
WO-A-1289	FM-GPW-14589	0606B	WO-A-1395	FM-GPW-10178-A	0601A
WO-A-1290	FM-GPW-9637-A	0301C	WO-A-1396	FM-356436-S	0601A
WO-A-1291	FM-GPW-5165	0610A	WO-A-1397	FM-355455-S	0601A
WO-A-1292	FM-GPW-9275	0300A	WO-A-1399	FM-GPW-10143	0601A
WO-A-1293	FM-GPW-9276	0300A	WO-A-1400	FM-GPW-10162	0601A
WO-A-1296	FM-GPW-5246	0402	WO-A-1401	FM-356371-S7	0601A
WO-A-1300	FM-GPW-5251	0402	WO-A-1409	FM-GPW-10505	0601B
WO-A-1302	FM-GPW-9775	0303	WO-A-1410	FM-356519-S	0802
WO-A-1304	FM-GPW-13006	0607	WO-A-1411	FM-GPW-13710	0605C
WO-A-1305	FM-GPW-13005	0607	WO-A-1412	FM-GPW-12287	0604A
WO-A-1306	FM-GPW-13380-A	2203A	WO-A-1414	FM-GPW-12284	0604A
WO-A-1307	FM-GPW-97303	0301B	WO-A-1416	FM-GPW-12283	0604A
WO-A-1311	FM-GPW-9652	0301C	WO-A-1418	FM-GPW-12286	0604A
WO-A-1312	FM-GPW-13802	0609	WO-A-1420	FM-GPW-12298-B	0604A
WO-A-1313	. . . .	0301C	WO-A-1422	FM-GPW-5116	1500
WO-A-1314	. . . .	0301C	WO-A-1423	. . . .	1500
WO-A-1315	. . . .	0301C	WO-A-1424	FM-GPW-12030	0604
WO-A-1320	FM-GPW-14301	0610B	WO-A-1428	FM-GPW-3370	0901
WO-A-1325	FM-GPW-9288	0304	WO-A-1429	FM-GPW-4605	0902
WO-A-1326	FM-GPW-3335	0901	WO-A-1431	. . . .	1500
WO-A-1327	FM-GPW-4602	0901	WO-A-1433	FM-21C-18397-B	0902
WO-A-1330	FM-GPW-1101621-A	2203	WO-A-1434	FM-GPW-7074	0902
WO-A-1331	FM-GPW-1101627-A	2203	WO-A-1435	FM-GPW-7756	0801
WO-A-1332	FM-GPW-11649	0606	WO-A-1436	FM-GPW-13021	0607D
WO-A-1333	FM-GPW-13740	0606	WO-A-1437	FM-GPW-13200	0607D
WO-A-1334	FM-GPW-13704	0605C	WO-A-1439	FM-GPW-13217	0607D
WO-A-1339	FM-GPW-17090	2301	WO-A-1440	FM-GPW-13216-B	0607D
WO-A-1341	FM-GPW-5095	1500	WO-A-1443	FM-GPW-6375	0109
WO-A-1343	FM-GPW-17261	2204	WO-A-1445	FM-GP-4842	1003
WO-A-1344	FM-GPW-17262	2204	WO-A-1450	FM-GPW-9316	0107J
WO-A-1345	FM-34907-S7-8	0606	WO-A-1451	FM-GPW-9686-A	0301C
WO-A-1346	. . . .	0606	WO-A-1452	FM-GPW-14300	0610B
WO-A-1347	. . . .	0606	WO-A-1454	FM-GPW-14431	0610B
WO-A-1348	. . . .	0606	WO-A-1456	FM-GPW-9323	0107J
WO-A-1349	FM-GPW-12250-C	0606D	WO-A-1457	FM-GPW-2096	1001
WO-A-1350	FM-356229-S	0604C	WO-A-1460	FM-GPW-2079	1209B
WO-A-1351	. . . .	0606	WO-A-1463	FM-GPW-6031	0110
WO-A-1352	. . . .	0606	WO-A-1468	FM-GPW-10176	0601A
WO-A-1353	. . . .	0606	WO-A-1469	FM-GPW-10155	0601A
WO-A-1354	FM-GPW-2138	1205	WO-A-1470	FM-GPW-10177	0601A
WO-A-1355	FM-GPW-7503	0204A	WO-A-1472	FM-GPW-9095	0300
WO-A-1359	FM-GPW-2454	1204	WO-A-1476	FM-GPW-9074	0300
WO-A-1360	FM-GPW-7525	0204	WO-A-1477	FM-GPW-9057	0300
WO-A-1361	FM-GPW-13022	0607	WO-A-1480	FM-GPW-9078	0300
WO-A-1362	FM-GPW-13076	0607	WO-A-1481	FM-GPW-9066	0300
WO-A-1363	FM-GPW-13075	0607	WO-A-1483	FM-24505-S7	0300
WO-A-1366	FM-GPW-9237	0304	WO-A-1484	FM-GPW-2061	1207
WO-A-1367	FM-GPW-9282	0304	WO-A-1486	FM-GPW-2084	1001
WO-A-1368	FM-GPW-9289	0304	WO-A-1487	FM-GPW-2082	1001
WO-A-1369	FM-GPW-9369	0304	WO-A-1488	FM-GPW-2298	1209C
WO-A-1373	FM-GPW-2078	1209B	WO-A-1490	FM-GPW-14448-D	0606E
WO-A-1376	FM-GPW-2266	1209C	WO-A-1491	FM-GPW-10153-A	0601A
			WO-A-1492	FM-GPW-17091	2301
			WO-A-1493	. . . .	0100

Willys Part Number	Ford Part Number	Gov't Group No.	Willys Part Number	Ford Part Number	Gov't Group No.
WILLYS-OVERLAND MOTORS, INC. (WO) Cont'd			WO-A-1586	FM-72798-S7-8	0602
WO-A-1494	FM-GPW-9355	0302	WO-A-1587	FM-GPW-12169	0603
WO-A-1495	FM-GPW-8260	0503B	WO-A-1588	FM-36954-S7	0602
WO-A-1501	FM-GPW-2263	1209C	WO-A-1589	FM-34141-S2	0602
WO-A-1502	FM-GPW-2063	1207	WO-A-1590	FM-GPW-10120	0601
WO-A-1503	FM-GPW-7705	0801	WO-A-1591	FM-GPW-10202-A	0601
WO-A-1504	FM-GPW-7786	0805	WO-A-1592	FM-01A-10193	0601
WO-A-1505	FM-GPW-7793	0805B	WO-A-1593	FM-GPW-10202-A	0601
WO-A-1506	FM-GPW-7710	0805B	WO-A-1594	FM-GPW-10202-C	0601
WO-A-1507	FM-GPW-7768	0801	WO-A-1595	FM-GPW-10208-A	0601
WO-A-1508	FM-GPW-7706	0801	WO-A-1596	FM-355486-S7	0601
WO-A-1509	FM-GPW-7707	0801	WO-A-1597	FM-GPW-10208-A	0601
WO-A-1510	FM-GP-7722	0802	WO-A-1598	FM-GPW-10104	0601
WO-A-1511	FM-GP-17285	0810	WO-A-1599	FM-GPW-10206-A	0601
WO-A-1512	FM-GPW-17271	0810	WO-A-1600	FM-GPW-10041	0601
WO-A-1514	. . . .	1500	WO-A-1601	FM-GPW-10100	0601
WO-A-1515	FM-GPW-2250	1101	WO-A-1602	FM-GPW-10211-A	0601
WO-A-1517	. . . .	2603	WO-A-1603	. . . .	0601
WO-A-1526	FM-GPW-12030	0604	WO-A-1604	FM-GPW-10175	0601
WO-A-1532	FM-33798-S7	2601	WO-A-1605	FM-GPW-10211-B	0601
WO-A-1533	FM-34746-S7	1201	WO-A-1606	FM-GPW-10192	0601
WO-A-1534	FM-GPW-6050	0101A	WO-A-1607	FM-GPW-10191	0601
WO-A-1536	FM-GPW-18390	0101	WO-A-1608	. . . .	0601
WO-A-1537	FM-GPW-18387	0101	WO-A-1609	FM-GPW-10218	0601
WO-A-1538	FM-GPW-18512	0107A	WO-A-1610	FM-34051-S7	0601
WO-A-1542	FM-GPW-18356	0700	WO-A-1611	FM-35583-S	0601
WO-A-1543	FM-GPW-18355	0800	WO-A-1612	FM-356263-S7	0601
WO-A-1545	FM-GPW-18366	1100	WO-A-1613	FM-34801-S7	0601
WO-A-1546	FM-34084-S	0501	WO-A-1614	FM-34803-S7	0601
WO-A-1547	FM-34708-S	0501	WO-A-1615	FM-34703-S7	0601
WO-A-1548	FM-GPW-6067	0101	WO-A-1616	FM-34705-S2	0601
WO-A-1549	. . . .	0101	WO-A-1617	FM-350853-S7	0601
WO-A-1550	FM-356025-S	0101A	WO-A-1618	FM-36009-S	0601
WO-A-1552	FM-GPW-18535	0602	WO-A-1619	FM-34806-S2	0601
WO-A-1553	FM-GPW-11094	0602	WO-A-1620	FM-27063-S7	0601
WO-A-1554	FM-GPW-11102	0602	WO-A-1621	FM-356208-S7	0601
WO-A-1555	FM-34706-S2	0602	WO-A-1622	FM-GPW-10124	0601
WO-A-1556	FM-GPW-11120	0602	WO-A-1623	FM-355267-S7	0601
WO-A-1557	FM-GPW-11089	0602	WO-A-1624	FM-GPW-10116	0601
WO-A-1558	FM-GPW-11090	0602	WO-A-1625	FM-GPW-10119	0601
WO-A-1559	FM-355485-S7	0602	WO-A-1626	FM-GPW-10118	0601
WO-A-1560	FM-GPW-11083	0602	WO-A-1628	FM-GPW-10057	0601
WO-A-1563	FM-GPW-11085	0602	WO-A-1629	FM-GPW-10105	0601
WO-A-1564	FM-GPW-12172	0603	WO-A-1630	FM-GPW-10069	0601
WO-A-1565	FM-355944-S5	0602	WO-A-1631	FM-GPW-12300	0603
WO-A-1566	FM-GPW-11049	0602	WO-A-1632	FM-26457-S7	0601
WO-A-1567	FM-63543-S	0602	WO-A-1633	FM-36787-S7	0603
WO-A-1568	FM-GPW-11005	0602	WO-A-1635	FM-34803-S7	0601
WO-A-1569	FM-GPW-11053	0602	WO-A-1636	FM-31588-S	0601
WO-A-1570	FM-GPW-12164	0603	WO-A-1637	FM-GPW-10005	0601
WO-A-1571	. . . .	0602	WO-A-1638	FM-GPW-10134	0601
WO-A-1572	FM-31596-S	0602	WO-A-1639	FM-GPW-10130	0601
WO-A-1573	FM-GPW-11350	0602	WO-A-1640	FM-34032-S7	0601
WO-A-1574	FM-B-11379	0602	WO-A-1642	FM-72034-S	0601
WO-A-1575	FM-B-11357-A	0602	WO-A-1644	FM-78-10212-A	0601
WO-A-1576	FM-B-11381	0602	WO-A-1645	FM-GPW-10138	0601
WO-A-1577	FM-GPW-11375	0602	WO-A-1646	FM-GPW-10212	0601
WO-A-1578	FM-GPW-11377	0602	WO-A-1647	FM-36800-S	0601
WO-A-1579	FM-GPW-11382	0602	WO-A-1648	FM-20311-S7	2603
WO-A-1580	FM-B-11371	0602	WO-A-1649	FM-GPW-10142	0601
WO-A-1581	FM-GPW-11354	0602	WO-A-1650	FM-27161-S7	0601
WO-A-1582	FM-GPW-11394	0602	WO-A-1651	FM-GPW-18274	0601
WO-A-1583	FM-GPW-11395	0602	WO-A-1652	FM-GPW-12006	0604
WO-A-1584	FM-355164-S	0602	WO-A-1653	FM-GPW-12177	0603
WO-A-1585	FM-GPW-11131	0602	WO-A-1654	FM-GPW-12267	0603

Willys Part Number	Ford Part Number	Gov't Group No.	Willys Part Number	Ford Part Number	Gov't Group No.
WILLYS-OVERLAND MOTORS, INC. (WO) Cont'd			WO-A-1729	FM-GP-3017-A	1007
WO-A-1655	FM-GPW-12106	0603	WO-A-1731	FM-GPW-14436	0606B
WO-A-1656	FM-GPW-12011	0603	WO-A-1733	FM-GPW-12250-A	0606D
WO-A-1657	FM-GPW-12012	0603	WO-A-1734	FM-GPW-12250-B	0606D
WO-A-1658	FM-GPW-12200	0603	WO-A-1735	FM-GPW-2272	1201
WO-A-1659	FM-GPW-12195	0603	WO-A-1738	FM-GPW-9069	0300
WO-A-1660	FM-GPW-12174	0603	WO-A-1739	FM-GPW-9079	0300
WO-A-1661	FM-GPW-12176	0603	WO-A-1740	FM-GPW-9075	0300
WO-A-1662	FM-GPW-12151	0603	WO-A-1741	FM-GPW-9071	0300
WO-A-1663	FM-GPW-12217	0603	WO-A-1746	. . . .	0602
WO-A-1664	FM-GPW-12010	0603	WO-A-1748	FM-GPW-13713	0605C
WO-A-1665	FM-GPW-14425	0606B	WO-A-1755	. . . .	1500
WO-A-1666	FM-34702-S2	0603	WO-A-1756	. . . .	1500
WO-A-1667	FM-34701-S7	0603	WO-A-1757	FM-GPW-5168	0610A
WO-A-1668	FM-31027-S8	0603	WO-A-1760	FM-GPW-3568	1403A
WO-A-1669	FM-34801-S7	0603	WO-A-1763	FM-GPW-9211	0300A
WO-A-1670	FM-31026-S7	0603	WO-A-1764	FM-GP-7763	0803
WO-A-1671	FM-GPW-12133	0603	WO-A-1765	FM-GPW-1101670	1809
WO-A-1672	FM-GPW-12120	0603	WO-A-1767	FM-11AS-10658	0610
WO-A-1673	FM-GPW-12182	0603	WO-A-1795	. . . .	1201
WO-A-1674	FM-GPW-12155	0603	WO-A-1798	. . . .	2301B
WO-A-1675	FM-GPW-12175	0603	WO-A-1799	. . . .	1301
WO-A-1676	FM-GPW-12188	0603	WO-A-1910	. . . .	0303
WO-A-1677	FM-GPW-42084	0603	WO-A-2138	FM-GPW-1102396	1802A
WO-A-1678	FM-GPW-12178	0603	WO-A-2168	FM-GPW-1154810-A	1800A
WO-A-1679	FM-GPW-12139	0603	WO-A-2188	FM-GPW-16802	1704
WO-A-1680	FM-34707-S7	2603	WO-A-2190	FM-GPW-1102130	1802A
WO-A-1681	FM-GPW-12082	0603	WO-A-2190	FM-GPW-1102130	1802A
WO-A-1682	FM-GPW-12144	0603	WO-A-2213	FM-GPW-1103162	1811
WO-A-1683	FM-GPW-12145	0603	WO-A-2214	FM-GPW-1150498	1811
WO-A-1684	FM-GPW-12191	0603	WO-A-2220	. . . .	1811
WO-A-1685	FM-72867-S7	0603	WO-A-2226	FM-GP-1103014	1811A
WO-A-1686	FM-31583-S7	0603	WO-A-2227	FM-GP-1103482-A	1811
WO-A-1687	FM-GPW-18354	0603	WO-A-2232	FM-GPW-1103304	1811
WO-A-1689	FM-GP-1103	1302	WO-A-2234	FM-GP-1103334	1811
WO-A-1690	FM-GP-1102	1302	WO-A-2235	FM-GP-1103302	1811
WO-A-1691	. . . .	1302	WO-A-2238	FM-B-45482-C	1811
WO-A-1693	FM-GPW-14621	2603	WO-A-2239	FM-GBT-8103074-B	1811
WO-A-1694	FM-GPW-14561	2603	WO-A-2246	FM-GPW-1103030	1811B
WO-A-1699	FM-GPW-18852	2603	WO-A-2250	FM-GP-1103110	1811B
WO-A-1700	FM-355130-S7	2603	WO-A-2263	FM-351015-S7	1802A
WO-A-1701	FM-355835-S	2603	WO-A-2278	. . . .	1802A
WO-A-1702	FM-34703-S7	2603	WO-A-2301	. . . .	1811
WO-A-1703	. . . .	1001	WO-A-2311	. . . .	1802A
WO-A-1704	FM-GPW-3280	1402	WO-A-2312	. . . .	1802A
WO-A-1705	FM-GPW-3281	1402	WO-A-2325	FM-GPW-1128286	1802A
WO-A-1706	FM-51-3287	1402	WO-A-2336	FM-GPW-1144606	1802A
WO-A-1707	FM-24916-S2	1402	WO-A-2359	FM-GPW-1433	1505
WO-A-1708	FM-GPW-3279	1402	WO-A-2375	FM-350976-S2	1802A
WO-A-1709	FM-GPW-3282	1402	WO-A-2386	FM-GPW-16892	1811
WO-A-1710	FM-GPW-3112	1006	WO-A-2389	FM-GPW-1129672	1800A
WO-A-1712	FM-GPW-3113	1006	WO-A-2390	FM-GPW-1129670	1800A
WO-A-1714	FM-357703-S	1006	WO-A-2453	FM-GPW-1162552	1804
WO-A-1715	FM-GPW-3206	1007	WO-A-2466	FM-33896	1802A
WO-A-1716	FM-GP-3200-A	1007	WO-A-2470	FM-GPW-1140447	1802A
WO-A-1719	FM-3215-A	1007	WO-A-2476	FM-GPW-1103080-A	1811B
WO-A-1720	FM-GP-3221-A	1007	WO-A-2478	FM-GPW-1103100	1811A
WO-A-1721	FM-358074-S	1007	WO-A-2479	FM-GP-1103106	1811B
WO-A-1722	FM-GP-3219	1007	WO-A-2480	FM-38259-S2	1811
WO-A-1723	FM-GP-3218	1007	CO'-A-2481	FM-34176-S2	1811
WO-A-1724	FM-GP-3217	1007	WO-A-2482	FM-38192-S2	1811
WO-A-1725	FM-24622-S2	1007			
WO-A-1726	FM-GP-3216	1007			
WO-A-1727	FM-GP-3016-A	1007			
WO-A-1728	FM-GPW-3207-A1	1007			

Willys Part Number	Ford Part Number	Gov't Group No.		Willys Part Number	Ford Part Number	Gov't Group
WILLYS-OVERLAND MOTORS, INC. (WO) Cont'd				WO-A-2871	FM-GPW-16005	0607
WO-A-2483	FM-99N-8103311	1811		WO-A-2873	FM-GPW-13020	0607
WO-A-2485	FM-24454-S2	1811		WO-A-2875	FM-357419-S2	0607
WO-A-2486	FM-34805-S2	1811		WO-A-2878	FM-GPW-13032	0607
WO-A-2487	FM-33249-S2	1811		WO-A-2879	FM-GPW-1111238	1802
WO-A-2489	FM-33184-S2	1811B		WO-A-2883	FM-GPW-1131414	1804
WO-A-2490	FM-356286-S	1811		WO-A-2886	FM-GPW-1160327-B	1804
WO-A-2501	FM-GPW-1151289	1811		WO-A-2891	FM-GPW-16105	1701
WO-A-2512	FM-GP-17535	1811D		WO-A-2892	FM-GPW-2852	1201
WO-A-2513	FM-GP-17531	1811D		WO-A-2895	FM-GPW-1143501	1817
WO-A-2515	FM-GPW-1160314	1804		WO-A-2896	FM-GPW-16892	1701
WO-A-2518	FM-GPW-3685	0604C		WO-A-2897	FM-GPW-1151266	2201
WO-A-2584	FM-351303-S2	1811D		WO-A-2898	FM-GPW-1151272	2201
WO-A-2587	. . . .	1811D		WO-A-2900	FM-GPW-1153142	2201
WO-A-2588	FM-34054-S2	1811D		WO-A-2901	FM-GPW-1151270	2201
WO-A-2601	. . . .	2301A		WO-A-2902	FM-351370-S2	2201
WO-A-2607	FM-355965-S2	1804		WO-A-2909	. . . .	2201
WO-A-2639	FM-356522-S2	2201A		WO-A-2910	FM-GPW-1162181	1804
WO-A-2662	FM-GPW-16132	1701		WO-A-2911	FM-GPW-1162180	1804
WO-A-2744	FM-GPW-1129069	2201A		WO-A-2917	FM-GPW-1112158	1803
WO-A-2745	FM-GPW-1129068	2201A		WO-A-2918	FM-GPW-1112117	1803
WO-A-2747	FM-GPW-1102015	1802A		WO-A-2919	FM-GPW-1112119	1803
WO-A-2754	FM-GPW-1153030	2201A		WO-A-2925	FM-GPW-1160326	1804
WO-A-2756	FM-GPW-1127847	1802A		WO-A-2930	FM-GPW-1140474	0300
WO-A-2757	FM-GPW-1127846	1802A		WO-A-2931	FM-GPW-1101698-A	1803
WO-A-2758	FM-GPW-1140324	1802A		WO-A-3932	FM-GPW-1111137-A	1803
WO-A-2759	FM-GPW-1111331	1803		WO-A-2933	FM-355909-S2	1802.
WO-A-2760	FM-GPW-1111330	1803		WO-A-2934	FM-21CS-17682-B	2202
WO-A-2768	FM-GPW-1111140	1802A		WO-A-2935	FM-GPW-1110625	1802.
WO-A-2769	FM-GPW-1127851	1802A		WO-A-2939	FM-GPW-1103446-A	1802.
WO-A-2769	FM-GPW-1127851	1802A		WO-A-2940	FM-GPW-1128242	1802.
WO-A-2770	FM-GPW-1127850	1802A		WO-A-2942	FM-GPW-16006	1701
WO-A-2771	FM-GPW-1127849	1802A		WO-A-2943	FM-GPW-16005	1701
WO-A-2772	FM-GPW-1127848	1802A		WO-A-2945	FM-GPW-1162410	1804
WO-A-2773	FM-GPW-1111218	1802A		WO-A-2948	FM-GPW-1110750	1802.
WO-A-2774	. . . .	1817		WO-A-2950	FM-GPW-1110895	1802.
WO-A-2775	. . . .	1817		WO-A-2952	FM-GPW-9062	0300
WO-A-2776	. . . .	1811B		WO-A-2953	FM-GPW-9065	0300
WO-A-2782	FM-GPW-1161326	1804		WO-A-2954	FM-GPW-9063	0300
WO-A-2783	FM-GPW-1161300	1804		WO-A-2968	FM-GPW-1128244	1802.
WO-A-2787	FM-GPW-1162402	1804		WO-A-2970	FM-GPW-9051	0300
WO-A-2788	FM-GPW-1160780	1804A		WO-A-2977	FM-GPW-8162	0501
WO-A-2791	FM-GPW-1103488	1811		WO-A-2979	FM-GPW-8222	1803
WO-A-2796	. . . .	1811		WO-A-2982	FM-GPW-1112110	1803
WO-A-2798	FM-GPW-1103050	1811		WO-A-2983	FM-355569-S2	1804
WO-A-2810	. . . .	1817		WO-A-2984	FM-GPW-1128254	2301.
WO-A-2811	. . . .	1817		WO-A-2986	FM-GPW-1162900-B	1804.
WO-A-2815	FM-GPW-1102511	1802		WO-A-2989	. . . .	1704
WO-A-2816	FM-GPW-1102510	1802		WO-A-2990	FM-GPW-1112115	1803
WO-A-2820	FM-GPW-1144598	1505		WO-A-2992	FM-GPW-11514-A	1803
WO-A-2823	FM-GPW-1144600	1505		WO-A-2993	FM-GPW-13351	1802.
WO-A-2830	FM-GPW-1161302	1804		WO-A-2994	FM-GPW-1112116-A	1803
WO-A-2832	FM-GPW-1146126	1817		WO-A-2995	FM-GPW-1128256	2301.
WO-A-2833	FM-GPW-1162414	1804		WO-A-2998	FM-GPW-1120041	2201
WO-A-2836	. . . .	1704		WO-A-2999	FM-GPW-1120040	2201
WO-A-2837	. . . .	1802A		WO-A-3005	FM-GPW-1101610	1809
WO-A-2838	FM-GPW-1110610	1802A		WO-A-3007	FM-GPW-1102038	1802.
WO-A-2841	FM-GPW-16083	1702		WO-A-3008	FM-GPW-1102039	1802.
WO-A-2842	FM-GPW-16082	1702		WO-A-3029	FM-GPW-1162400	1804
WO-A-2843	FM-GPW-16095	1701				
WO-A-2844	FM-GPW-16094	1701				
WO-A-2853	FM-GPW-1140449	1802A				
WO-A-2858	. . . .	2103				
WO-A-2859	FM-GPW-3658	1405				
WO-A-2870	FM-GPW-16006	0607				

Willys Part Number	Ford Part Number	Gov't Group No.	Willys Part Number	Ford Part Number	Gov't Group No.
			WO-A-3563	FM-GPW-1100001	1800
WILLYS-OVERLAND MOTORS, INC. (WO) Cont'd			WO-A-3565	FM-GPW-1000000	1800
WO-A-3051	FM-95705-S	2201A	WO-A-3574	FM-GPW-8166	1803
WO-A-3052	FM-95626-S	2201F	WO-A-3575	. . . .	1803
WO-A-3053	FM-95628-S	2201F	WO-A-3578	FM-GPW-1104320	1809
WO-A-3054	FM-95627-S	2201A	WO-A-3615	FM-GPW-8307	2103
WO-A-3055	FM-GPW-1111322	0300	WO-A-3652	FM-GPW-1106087-A	1809A
WO-A-3056	FM-GPW-1111323	1802A	WO-A-3728	. . . .	0610A
WO-A-3059	FM-GPW-16684	1704	WO-A-3782	FM-GPW-1111520-A	1803
WO-A-3070	FM-GPW-1102980	2201	WO-A-3783	FM-GPW-1111286-A	1803
WO-A-3073	. . . .	2201	WO-A-3784	FM-GPW-1101735-A	0805B
WO-A-3082	FM-GPW-1128258	2301A	WO-A-3818	FM-GPW-1106064	1809A
WO-A-3094	FM-GPW-8348	2103	WO-A-3823	FM-GPW-1106084-A3	1809A
WO-A-3095	FM-GPW-8349	2103	WO-A-3835	. . . .	1809A
WO-A-3096	FM-GPW-16128	1702	WO-A-3933	. . . .	1817
WO-A-3108	FM-GPW-1160026	1804A	WO-A-3934	. . . .	1817
WO-A-3109	FM-GPW-1152730	2201A	WO-A-3940	. . . .	1811D
WO-A-3110	FM-GPW-1152720	2201A	WO-A-3943	FM-GPW-1102330-A	1809A
WO-A-3111	FM-353104-S	2201A	WO-A-3980	. . . .	1804A
WO-A-3112	FM-95633-S	2201A	WO-A-3981	FM-GPW-1163206	1804A
WO-A-3113	FM-348148-S	2201A	WO-A-3982	FM-GPW-1166800-B	1804A
WO-A-3114	FM-GPW-1166401	1804A	WO-A-3983	FM-GPW-1163846-B	1804A
WO-A-3115	FM-GPW-1166400	1804A	WO-A-3984	. . . .	1804A
WO-A-3116	FM-GPW-1111299-A	1803	WO-A-3985	. . . .	1804A
WO-A-3120	FM-GPW-1144620	1802A	WO-A-4116	. . . .	2103A
WO-A-3135	FM-GPW-1128247	2301A	WO-A-4118	FM-GPW-13176	2103A
WO-A-3137	FM-GPW-1128237	2301A	WO-A-4120	FM-352699-S15	1811
WO-A-3139	FM-GPW-1128267	2301A	WO-A-4123	FM-GPW-1140330	1817A
WO-A-3141	FM-GPW-1152950	2201A	WO-A-4127	FM-GPW-1140334	1817A
WO-A-3144	FM-GPW-1162606	1802A	WO-A-4128	FM-GPW-1140344	1817A
WO-A-3155	FM-GPW-1104364	1809	WO-A-4260	. . . .	1811
WO-A-3158	FM-GPW-16159	1701	WO-A-4300	FM-GP-1103310	1811
WO-A-3159	FM-GPW-16133	1702	WO-A-4303	FM-GP-1103350	1811
WO-A-3173	FM-GPW-8155	1803	WO-A-4413	FM-GPW-8125-A	0501
WO-A-3175	FM-GPW-8102	0501H	WO-A-4414	. . . .	1801
WO-A-3176	FM-GPW-8103	0501H	WO-A-4415	. . . .	1801
WO-A-3182	. . . .	1803	WO-A-4416	FM-GPW-17079	1804
WO-A-3189	FM-GPW-1151297	1811	WO-A-4518	FM-GPW-17055	1804
WO-A-3190	. . . .	1811	WO-A-4592	FM-GPW-1111162	1803
WO-A-3197	FM-GPW-1103027	1811	WO-A-4606	FM-GPW-1140327-B	1802A
WO-A-3198	FM-GPW-1146152	1817	WO-A-4607	FM-GPW-1140326-B	1802A
WO-A-3203	FM-GPW-1103028	1811	WO-A-4679	. . . .	1704
WO-A-3204	FM-GPW-1103268	1811B	WO-A-4680	FM-GTB-16847	1704
WO-A-3206	FM-GPW-9071	1803	WO-A-4683	FM-GTB-16848-A	1704
WO-A-3207	FM-GPW-9082-A	1803	WO-A-4687	. . . .	1811D
WO-A-3208	FM-GPW-9085-A	1803	WO-5009	FM-34809-S	. . . .
WO-A-3209	FM-GPW-9083-A	1803	WO-5010	FM-34807-S2	. . . .
WO-A-3210	FM-GPW-1103010	1811	WO-5020	FM-72017-S	. . . .
WO-A-3211	FM-GPW-1103214	2201	WO-5021	FM-72035-S	. . . .
WO-A-3216	FM-GPW-1152700	2201	WO-5027	FM-GPW-18874	2603
WO-A-3222	FM-GPW-1111155	1802A	WO-A-5030	FM-GPW-13871	2603
WO-A-3225	FM-GPW-16610	1704	WO-A-5031	FM-GPW-18872	2603
WO-A-3226	FM-GPW-1146132	1817	WO-A-5032	FM-GPW-18859	2603
WO-A-3227	FM-GPW-1146100	1817	WO-A-5033	FM-GPW-18858	2603
WO-A-3434	FM-GPW-1106024	1809A	WO-A-5034	FM-GPW-18873	2603
WO-A-3436	FM-GPW-1106050	1809A			
WO-A-3497	FM-GPW-1111324	0300			
WO-A-3531	FM-GPW-1106084-A	1809A			
WO-A-3532	FM-GPW-1106068	1809A			
WO-A-3536	FM-GPW-1106085	1809A			
WO-A-3537	FM-356123-S7	1809A			
WO-A-3538	. . . .	1809A			
WO-A-3549	. . . .	2103			
WO-A-3550	FM-34130-S2	2103			

WILLYS-OVERLAND MOTORS, INC. (WO) Cont'd

Willys Part Number	Ford Part Number	Gov't Group No.	Willys Part Number	Ford Part Number	Gov't Group No.
WO-A-5035	FM-GPW-18849	2603	WO-A-5449	FM-GPW-14561	2204
WO-A-5036	FM-GPW-18874	2603	WO-A-5450	. . . .	2601
WO-5036	FM-74178-S	. . . .	WO-5455	FM-34747-S	. . . .
WO-A-5037	FM-GPW-18857	2603	WO-A-5467	FM-GPW-1015	1301
WO-A-5038	FM-GPW-18876	2601	WO-A-5468	FM-GPW-1016	1301
WO-A-5039	FM-GPW-18853	2603	WO-A-5470	FM-GPW-1029	1301
WO-A-5040	FM-GPW-18850	2603	WO-A-5471	FM-GPW-1030	1301
WO-A-5041	FM-GPW-18846	2603	WO-A-5472	FM-GPW-1045	1301
WO-5045	FM-34805-S2	. . . .	WO-A-5488	FM-GPW-1025-C	1301
WO-A-5048	FM-GPW-14401-C	0606B	WO-A-5497	FM-GPW-6005	0100
WO-5051	FM-34836-S-7	. . . .	WO-A-5498	. . . .	1000
WO-A-5061	FM-GPW-14446	0606B	WO-A-5499	. . . .	1000
WO-5064	FM-355130-S7	. . . .	WO-A-5500	FM-GPW-4001	1100
WO-5067	FM-72003-S	. . . .	WO-A-5501	. . . .	0301
WO-A-5070	FM-GPW-14406	0605B	WO-A-5504	FM-GPW-3325	1006
WO-A-5072	FM-GPW-14458	0605B	WO-5544	FM-33799-S	. . . .
WO-A-5073	FM-GPW-14459	0606B	WO-A-5549	FM-GPW-1024	1301
WO-A-5074	FM-GPW-14305	0606B	WO-A-5553	FM-GPW-7015	0703
WO-A-5078	FM-GPW-14457	0606B	WO-A-5554	FM-GPW-7017	0703
WO-A-5079	FM-GPW-14456	0606B	WO-A-5586	FM-GPW-13012	0607
WO-A-5080	FM-GPW-14416	0605B	WO-A-5598	FM-GPW-14561	0606B
WO-A-5081	FM-GPW-14409	0609B	WO-5599	FM-353075-S	. . . .
WO-A-5082	FM-GPW-14465	0606B	WO-A-5621	FM-GPW-18205-A	0301C
WO-A-5083	FM-GPW-14321	0604A	WO-A-5629	FM-GPW-9609	0301C
WO-5085	FM-358064-S	. . . .	WO-A-5630	FM-GPW-9617	0301C
WO-A-5102	FM-GPW-7530	0203	WO-A-5631	FM-GPW-9658	0301C
WO-A-5105	FM-GPW-6770	0107G	WO-A-5632	FM-GPW-9621	0301C
WO-5108	FM-72053	. . . .	WO-A-5633	FM-GPW-9623	0301C
WO-5113	FM-355130-S7	. . . .	WO-A-5753	FM-GPW-5264-B	0401
WO-A-5120	FM-358019-S	1802A	WO-5790	FM-33795-S	. . . .
WO-5121	FM-34705-S2	. . . .	WO-A-5806	. . . .	0607D
WO-A-5125	FM-GPW-6044	0110	WO-5901	FM-33800-S	. . . .
WO-A-5127	FM-GPW-5025	1500	WO-5910	FM-33798-S2	. . . .
WO-A-5130	FM-GPW-17030	2301	WO-5914	FM-33796-S	. . . .
WO-5138	FM-353055	. . . .	WO-5916	FM-33846-S	. . . .
WO-5140	FM-353064-S	0801	WO-5919	. . . .	. . . .
WO-A-5165	FM-GPW-6763-B	0107G	WO-5920	FM-20325-S7	. . . .
WO-A-5168	FM-GPW-6766-B	0107H	WO-5922	FM-2407-S	. . . .
WO-5168	FM-34703-S	. . . .	WO-5934	. . . .	. . . .
WO-A-5181	. . . .	. . . .	WO-5938	FM-34838-S	. . . .
WO-5182	FM-355158-S2	. . . .	WO-5939	FM-33802-S2	. . . .
WO-A-5197	FM-31037-S7	0606	WO-A-5980	FM-GPW-18960-B	2603
WO-5215	. . . .	. . . .	WO-A-5981	FM-GPW-14432	0605B
WO-52221	FM-GPW-34803-S7	. . . .	WO-A-5992	FM-GPW-10000-A	0601
WO-A-5224	FM-GPW-2265	1209C	WO-A-6019	FM-11YS-18142-B	0606F
WO-A-5225	FM-GPW-2268	1209C	WO-A-6029	. . . .	1000
WO-A-5226	FM-GPW-2267	1209C	WO-A-6030	. . . .	1007
WO-A-5227	FM-GPW-2274	1209C	WO-A-6031	. . . .	1007
WO-5247	. . . .	1802A	WO-A-6066	FM-GPW-5588-A	1601C
WO-A-5256	. . . .	0601B	WO-A-6067	FM-GPW-5610	1601C
WO-A-5260	. . . .	0601B	WO-A-6068	FM-GPW-5602	1601C
WO-A-5262	. . . .	0601B	WO-A-6069	FM-GPW-5605	1601C
WO-5267	. . . .	. . . .	WO-A-6072	FM-HPW-5604	1601C
WO-5272	FM-355160-S2	. . . .	WO-A-6073	FM-GPW-5607	1601C
WO-A-5288	FM-34806-S7	0601	WO-A-6074	FM-GPW-5609	1601C
WO-A-5335	FM-GPW-2635	1201	WO-A-6075	FM-GPW-5608	1601C
WO-A-5337	FM-GPW-18935-A	2603	WO-A-6110	. . . .	1207
WO-A-5338	FM-GPW-6002	0100	WO-A-6111	FM-GP-2135	1207
WO-5354	FM-72004-S	. . . .	WO-A-6113	FM-GPW-2196	1207
WO-A-5393	FM-GPW-2270-B	1201	WO-A-6116	FM-GPW-2201	1207
WO-5397	FM-72071	. . . .	WO-A-6117	FM-GPW-2206	1207
WO-A-5415	. . . .	1500	WO-A-6118	FM-GPW-5230-B	0401
WO-A-5433	FM-11AS-10657	0610	WO-A-6119	FM-GPW-5298	0402A
WO-5437	FM-34706-S2	. . . .	WO-A-6133	FM-GPW-18368	1207
			WO-A-6142	FM-GPW-13150	0607D

Willys Part Number	Ford Part Number	Gov't Group No.	Willys Part Number	Ford Part Number	Gov't Group No.
WILLYS-OVERLAND MOTORS, INC. (WO) Cont'd			WO-6606	FM-24409-S7	....
WO-A-6143	FM-GPW-13170	0607D	WO-6609	FM-24427-S7	....
WO-A-6144	FM-GPW-13162	0607D	WO-A-6618	FM-GPW-9002-B	0300
WO-A-6145	FM-GPW-13153	0607A	WO-A-6701	....	0503A
WO-A-6146	FM-GPW-13175	0607D	WO-A-6710	....	0303
WO-A-6147	FM-GPW-13174	0607D	WO-A-6740	FM-GPW-5084	1500
WO-A-6148	FM-131045-S2	0607D	WO-A-6742	FM-GPW-18382	0609B
WO-A-6149	FM-GT-13739	0606	WO-A-6743	FM-GPW-18389	1003
WO-A-6152	....	0606	WO-A-6744	FM-GPW-18388	1103
WO-A-6153	FM-GPW-13181	0606B	WO-A-6745	FM-GPW-18386	1104
WO-A-6154	FM-GPW-14402	0606B	WO-A-6746	FM-GPW-18348	0102A
WO-A-6156	FM-GPW-6040-B	0110	WO-A-6747	FM-GPW-18349	0102A
WO-6157	FM-24426-S2	....	WO-A-6749	FM-GPW-18379	0107
WO-6167	FM-33797-S2	....	WO-A-6750	FM-GPW-18380	0107
WO-A-6168	FM-GPW-5590-B	1601C	WO-A-6751	FM-GPW-18358	0201A
WO-A-6169	FM-GPW-5611	1601C	WO-A-6752	....	0202
WO-6184	FM-24430-S	....	WO-A-6753	FM-GPW-18360	0801
WO-6188	FM-20384-S2	....	WO-A-6756	FM-GPW-18376	0602
WO-6273	FM-34141-S2	....	WO-A-6759	FM-GPW-18377	1201
WO-6290	FM-GPW-355162-S2	....	WO-A-6760	FM-GPW-18374	1403A
WO-A-6297	FM-37789-S7	0601	WO-A-6783	FM-GPW-13166	0607D
WO-A-6298	FM-GPW-10050	0601	WO-A-6791	FM-GPW-18383	1401
WO-6299		....	WO-A-6793	FM-GPW-6009	0101
WO-A-6299	FM-B-10094	0601	WO-A-6794	FM-GPW-6149-E	0103A
WO-A-6300	....	0601	WO-A-6796	FM-GPW-6149-G	0103A
WO-A-6301	FM-GPW-10139	0601	WO-A-6797	FM-GPW-6149-H	0103A
WO-A-6305	....	1402	WO-A-6798	FM-GPW-18347	0102A
WO-A-6313	FM-GPW-13180	0607D	WO-A-6809	....	2601
WO-A-6317	FM-GPW-7060	0704	WO-A-6811	FM-GPW-3686-B	0604C
WO-A-6318	FM-GPW-7124	0704B	WO-A-6813	....	0604C
WO-A-6319	FM-GPW-7105	0704B	WO-A-6814	FM-GPW-3685-B	0604C
WO-A-6320	FM-GPW-18812	2603	WO-A-6816	FM-GPW-18384	1003
WO-A-6321	....	0604A	WO-A-6816	FM-GPW-18384	1103
WO-A-6326	FM-GPW-5601	1602	WO-A-6837	FM-GPW-18352	0301
WO-A-6333	FM-GPW-9030-B	0300	WO-A-6839	FM-GPW-18515	0503
WO-6352	FM-355836-S7	....	WO-A-6840	FM-GPW-18357-B	0301
WO-A-6356	FM-O9B-14362	0606F	WO-A-6851	....	0303
WO-A-6357	FM-GPW-9445	0301A	WO-A-6855	FM-GPW-18325	2301
WO-A-6359	....	0204	WO-A-6858	FM-GPW-3600-A3	1404
WO-A-6360	....	0204	WO-A-6861	FM-GPW-5587	1601C
WO-A-6361	....	1007	WO-A-6881	FM-GPW-18336	1007
WO-A-6362	....	1007	WP-A-6882	FM-GPW-18388	1006
WO-A-6373	FM-GPW-8285	0505	WO-A-6883	FM-GPW-18337	0306B
WO-A-6374	FM-GPW-8290	0505	WO-A-6895	FM-GPW-6769	0107N
WO-A-6382	....	1007	WO-A-6897	FM-GPW-9001	0300
WO-A-6383	....	1007	WO-A-6902	....	1603
WO-6383	FM-27145-S2	....	WO-A-6903	....	1603
WO-A-6384	....	1007	WO-A-6911	FM-GPW-9637-B	0301C
WO-A-6393	FM-GPW-5186	1502	WO-A-6915	FM-GPW-6763-C	0107G
WO-6412	FM-24348-S	....	WO-A-6918	FM-GPW-6771	0301C
WO-A-6424	FM-GPW-9034-A	0300	WO-A-6919	FM-GPW-6758-B	0107N
WO-6428	FM-20366-S	....	WO-A-6922	FM-GPW-6756	0107N
WO-6436	FM-34033-S-18	....	WO-6923	FM-24411-S	....
WO-A-6439	FM-GPW-4259	1102	WO-6989	FM-34054-S	2402
WO-A-6442	....	1000	WO-A-7180	FM-GPW-14521	2601
WO-6470	....	....	WO-A-7181	FM-GPW-14536	2601
WO-A-6472	....	1007	WO-A-7182	FM-GPW-14532	2601
WO-6486	FM-GP-106333	....	WO-A-7191	FM-GPW-9612	0301C
WO-A-6511	FM-GPW-5455	1602	WO-A-7225	....	0606
WO-A-6525	....	0107H	WO-A-7233	FM-GPW-18330-A	0104A
WO-A-6586	FM-11YS-18151-B	0606F	WO-A-7234	FM-GPW-18330-B	0104A
WO-A-6587	....	0606F	WO-A-7235	FM-GPW-18330-C	0104A
WO-A-6588	FM-11YS-18198-B	0606F	WO-A-7238	....	0107A
WO-A-6589	FM-AAYS-18193-B	0606F	Cont-A-7260	FM-GPW-7211-B	0706C
			WO-A-7280	FM-GPW-6789	0107H

VENDOR'S NUMERICAL INDEX (Cont'd)

Willys Part Number	Ford Part Number	Gov't Group No.		Willys Part Number	Ford Part Number	Gov't Group
WILLYS-OVERLAND MOTORS, INC. (WO) Cont'd				WO-A-8116	FM-GPW-14481-B	2601
WO-A-7443	FM-GPW-18355-B	0800		WO-A-8124	FM-GPW-9273	0605
WO-A-7445	FM-GPW-18317-B	0800		WO-A-8125	....	2204
WO-A-7466	FM-91BS-14463	2601		WO-A-8126	....	2204
WO-A-7498	FM-GPW-6038-A	0110		WO-A-8127	....	2204
WO-A-7503	FM-GPW-18289	0109		WO-A-8129	....	2204
WO-A-7511	FM-GPW-17052	2301		WO-A-8130	....	0605
WO-A-7517	FM-GPW-9775-B	0301B		WO-A-8132	....	0605
WO-A-7518	FM-GPW-9778	0301B		WO-A-8180	FM-GPW-17255-A	2204
WO-A-7568	FM-GPW-18288	0102		WO-A-8186	FM-GPW-10850	0605
WO-A-7596	....	0700		WO-A-8188	FM-GPW-10883	0605
WO-A-7600	FM-GPW-18841	2601		WO-A-8190	FM-GPW-9273	0604
WO-A-7636	FM-GPW-14531	2601		WO-A-8242	....	220
WO-A-7637	FM-GPW-14535	2601		WO-A-8249	....	1000
WO-A-7638	FM-GPW-14534	2601		WO-A-8250	FM-GPW-3304-B	1401
WO-A-7640	FM-GPW-14513	2601		WO-A-8252	FM-GPW-3305-B	1401
WO-A-7645	FM-GPW-14320	2601		WO-A-8253	FM-GPW-2452-B	1204
WO-A-7680	FM-GPW-18353	2309		WO-A-8255	....	160
WO-A-7687	FM-GPW-18136-B	2301B		WO-A-8256	....	160
WO-A-7715	....	2601		WO-A-8279	....	230
WO-A-7718	....	2601		WO-A-8322	....	060
WO-A-7778	....	2301		WO-A-8323	....	030
WO-A-7779	....	2301		WO-A-8498	....	060
WO-A-7792	FM-GPW-12000-B	0604		WO-A-8558	FM-GPW-6051-B	010
WO-A-7794	....	0601B		WO-A-8809	....	160
WO-A-7795	....	0601B		WO-A-8810	....	160
WO-A-7796	....	0601B		WO-A-8834	....	030
WO-A-7797	....	0601B		WO-A-8835	....	080
WO-A-7798	....	0601B		WO-A-8841	FM-GPW-7736	080
WO-A-7799	....	0601B		WO-A-8842	....	060
WO-A-7800	....	0601B		WO-A-8883	FM-GPW-18958-C	260
WO-A-7801	....	0601B		WO-A-8884	FM-GPW-18937	260
WO-A-7802	....	0601B		WO-A-8894	FM-GPW-2011	120
WO-A-7803	....	0601B		WO-A-8895	FM-GPW-2010	120
WO-A-7805	....	0601B		WO-A-8896	FM-GPW-2211-B	120
WO-A-7806	....	0601B		WO-A-8897	FM-GPW-2210-B	120
WO-A-7807	....	0601B		WO-A-8898	FM-GPW-2013	120
WO-A-7808	....	0601B		WO-A-8914	....	060
WO-A-7809	....	0601B		WO-A-8993	....	060
WO-A-7810	....	0601B		WO-A-8997	....	060
WO-A-7823	....	0606B		WO-A-9040	....	060
WO-A-7824	....	0606B		WO-A-9041	....	060
WO-A-7830	FM-GPW-18365-B	1001		WO-A-9042	....	060
WO-A-7831	FM-GPW-18366-B	1100		WO-A-9043	....	060
WO-A-7832	FM-GPW-18356-B	0700		WO-A-9044	....	060
WO-A-7833	FM-GPW-18359-B	0202		WO-A-9046	....	060
WO-A-7834	FM-GPW-18373-C	0302		WO-A-9047	....	060
WO-A-7835	FM-GPW-18323	0108		WO-A-9048	....	060
WO-A-7836	FM-GPW-18376-B	0602		WO-A-9049	....	060
WO-A-7837	FM-GPW-18314	0810		WO-A-9050	....	060
WO-A-7838	FM-GPW-18370-B	1205		WO-A-9051	....	060
WO-A-7840	FM-GPW-18342	0601		WO-A-9052	....	060
WO-A-7841	FM-GPW-18319	0602		WO-A-9053	....	060
WO-A-7842	FM-GPW-18329	0602		WO-A-9054	....	060
WO-A-7843	FM-GPW-18343	0603		WO-A-9055	....	060
WO-A-7844	FM-GPW-18363-B	0604A		WO-A-9220	....	060
WO-A-7845	....	0606B		WO-A-9490	FM-GPW-8620-A2	050
WO-A-7848	FM-GPW-18938-C	2603		WO-A-9492	FM-GPW-10130	060
WO-A-7895	FM-GPW-18363-B	0601		WO-A-11432	FM-GPW-17534	181
WO-A-7947	FM-GPW-14537	2601		WO-A-11433	FM-GPW-17500	181
WO-A-7956	....	0302		WO-A-11519	....	181
WO-A-8087	FM-34753-S	0606F		WO-A-11701	....	150
WO-A-8088	....	0606F		WO-A-11729	....	180
WO-A-8113	FM-GPW-14480-B	2601		WO-A-11730	....	180
WO-A-8114	FM GPW-18861-B	2601		WO-A-11731	....	180

Willys Part Number	Ford Part Number	Gov't Group No.		Willys Part Number	Ford Part Number	Gov't Group No.
WILLYS-OVERLAND MOTORS, INC. (WO) Cont'd				WO-52168	FM-20324-S2	
WO-A-11732	....	1804A		WO-52170	FM-24308-S2	
WO-A-11757	FM-GPW-1101629-A	2203		WO-52189	FM-24328-S2	
WO-A-11765	FM-GPW-17126	2301		WO-52217	FM-33799-S2	
WO-A-11768	....	0607D		WO-52221	FM-34803-S2	
WO-A-11770	....	0607D		WO-52226	FM-60-8287	
WO-A-11850	FM-34804-S2	2202		WO-52236	FM-27068-S	
WO-A-11861	....	2202		WO-52274	FM-34746-S2	
WO-A-11862	....	2202		WO-52332	....	
WO-A-12025	....	1802A		WO-52350	FM-33797-S2	
WO-A-12054	....	1803		WO-52379	FM-24489-S	
WO-A-12055	....	1803		WO-52424	FM-34806-S7	
WO-A-12056	....	0606		WO-52510	FM-34941-S2	
WO-A-2940	FM-GPW-1128242	1802A		WO-52600	FM-24449-S2	
WO-A-28023	FM-357574-S			WO-52615	FM-34051-S7	
WO-50151	....			WO-52702	FM-34745-S	
WO-50163	FM-355498-S7			WO-52705	....	
WO-50769	....			WO-52706	FM-34805-S7	
WO-50878	FM-24449-S7			WO-52700	FM-34805-S2	
WO-50921	....			WO-52706	FM-34805-S2	
WO-50922	FM-33800-S2			WO-52768	FM-34706-S2	
WO-50929	FM-20367-S7			WO-52781	....	
WO-50992	FM-24505-S7			WO-52809	FM-32924-S2	
WO-51040	FM-26457-S			WO-52832	....	
WO-51091	FM-74121-S			WO-52836	FM-355444-S	
WO-51248	FM-GPW-10094			WO-52837	FM-48-15021	
WO-51304	FM-34807-S2			WO-52839	....	
WO-51308	FM-21492-S2			WO-52857	....	
WO-51371	FM-24555-S			WO-52863	....	
WO-51391	FM-24310-S			WO-52876	FM-86H-4621	
WO-51396	FM-24347-S2			WO-52877	FM-86H-4616	
WO-51405	....			WO-52877	FM-86H-4616	
WO-51406	FM-24428-S			WO-52878	FM-GP-4630	
WO-51485	FM-20326-S2			WO-52879	FM-GP-4628	
WO-51486	FM-24386-S2			WO-52880	FM-GP-4221	
WO-51492	FM-26483-S2			WO-52881	FM-GP-4222	
WO-51514	FM-20324			WO-52883	FM-O1Y-1202	
WO-51523	FM-20346-S2			WO-52893	FM-356016-S7	
WO-51532	....			WO-52909	....	
WO-51545	....			WO-52911	FM-24408-S	
WO-51546	FM-26466-S7			WO-52921	....	
WO-51575	FM-O1Y-1202			WO-52925	FM-33927-S7	
WO-51612	FM-350744-S2			WO-52940	FM-GP-3161	
WO-51662	FM-356201-S			WO-52941	FM-GP-3162	
WO-51732	FM-20309-S7			WO-52942	FM-GP-1201	
WO-51738	FM-20300-S7			WO-52943	FM-GP-1202	
WO-51763	FM-20308-S2			WO-52944	FM-72063-S	
WO-51798	FM-355398-S2			WO-52945	....	
WO-51804	FM-B-13466			WO-52954	FM-33909-S	
WO-51823	FM-355456-S			WO-53023	FM-351274-S7	
WO-51833	FM-34806-S2			WO-53024	....	
WO-51840	FM-34806-S2			WO-53025	FM-356309-S7	
WO-51858	FM-355442-S			WO-53026	....	
WO-51875	FM-GPW-6555			WO-53031	FM-20348-S2	
WO-51921	FM-74113-S			WO-53048	FM-24347-S7	
WO-51954	....			WO-53069	....	
WO-51969	FM-34804-S2			WO-53070	....	
WO-52031	FM-34805-S2			WO-53071	FM-40631-S7	
WO-52045	FM-34846-S2			WO-53135	....	
WO-52046	FM-34807-S2			WO-53194	FM-34903-S2	
WO-52101	....			WO-53024	....	
WO-52131	FM-26457-S7			WO-53036	....	
WO-52132	FM-24327-S7			WO-53135	....	
WO-52142	FM-32866-S2			WO-53285	FM-33798-S7	
WO-52167	FM-20344-S2			WO-53303	....	

Ford Part Number	Willys Part Number	Gov't Group No.	Willys Part Number	Ford Part Number	Gov't Group No.
WILLYS-OVERLAND MOTORS, INC. (WO) Cont'd			WO-116175	FM-GPW-9575	0301
WO-71633	FM-351370-S2	1804	WO-116176	FM-GPW-9541	0301
WO-78932	FM-GPW-14566	0606B	WO-116177	FM-GPW-9566	0301
WO-106313	FM-GPW-10098	0601	WO-116178	FM-GPW-9599	0301
WO-106740	FM-GPW-12193	0603	WO-116179	FM-GPW-9544	0301
WO-107128	FM-B-10141	0603	WO-116180	FM-GPW-9940	0301
WO-109427	. . . .	0602	WO-116181	FM-GPW-9528	0301
WO-109428	. . . .	0602	WO-116183	FM-GPW-9578	0301
WO-109431	FM-GPW-11055	0602	WO-116184	FM-GPW-9624	0301
WO-109433	FM-GPW-11103	0602	WO-116185	FM-GPW-9615	0301
WO-109436	FM-GPW-11107	0602	WO-116186	FM-GPW-9650	0301
WO-109437	FM-GPW-11071	0602	WO-116187	FM-GPW-9570	0301
WO-109442	FM-GPW-11061	0602	WO-116188	FM-GPW-9636	0301
WO-109445	FM-GPW-11036-B	0602	WO-116189	FM-GPW-9587	0301
WO-109446	FM-GPW-11056	0602	WO-116191	FM-GPW-9935	0301
WO-109452	FM-GPW-11077	0602	WO-116194	FM-GPW-9614	0301
WO-109453	. . . .	0603	WO-116195	FM-GPW-9631	0301
WO-111063	FM-33798-S7	0606	WO-116197	FM-GPW-9583	0301
WO-113440	FM-34803-S7	0302	WO-116198	FM-GPW-9531	0301
WO-113460	FM-GPW-9388	0302	WO-116199	FM-GPW-9527	0301
WO-113461	FM-GPW-9373	0302	WO-116202	FM-GPW-9516	0301
WO-115641	FM-GPW-9399	0302	WO-116203	FM-GPW-9519	0301
WO-115643	FM-GPW-9380	0302	WO-116204	FM-GPW-119594	0301
WO-115650	FM-GPW-9354	0302	WO-116205	FM-GPW-9576	0301
WO-115651	FM-11A-9352	0302	WO-116206	FM-GPW-9905	0301
WO-115652	FM-GPW-9363	0302	WO-116207	FM-34711-S	0301
WO-115653	FM-11A-9361	0302	WO-116208	FM-GPW-9515	0301
WO-115654	FM-GPW-9365	0302	WO-116209	FM-GPW-9930	0301
WO-115656	FM-GPW-9364	0302	WO-116210	FM-GPW-9554	0301
WO-115657	FM-GPW-9387	0302	WO-116211	FM-31032-S7	0301
WO-115869	FM-GPW-9468	0302	WO-116213	FM-31061-S8	0301
WO-115870	FM-GPW-19469	0302	WO-116215	FM-31662-S	0301
WO-115880	FM-INC-9381	0302	WO-116216	FM-GPW-9586	0301
WO-115905	FM-355253-S	0301C	WO-116217	FM-GPW-9588	0301
WO-115948	FM-GPW-6552-C	0105A	WO-116218	. . . .	0301
WO-115962	FM-GPW-18371	1207	WO-116219	FM-355858-S	0301
WO-115963	FM-GPW-18372	1207	WO-116295	FM-GPW-6390	0109
WO-116017	. . . .	0103	WO-116384	FM-355067-S7	0301
WO-116018	. . . .	0103	WO-116385	FM-355200-S7	0301
WO-116019	FM-GPW-6105-C	0103	WO-116458	FM-GPW-5330-A	1601
WO-116020	FM-GPW-6105-D	0103	WO-116459	FM-GPW-5724-A	1601
WO-116023	FM-GPW-6155-G	0130A	WO-116460	FM-GPW-5330-B	1601
WO-116024	FM-GPW-6155-H	0103A	WO-116502	FM-GPW-6150-C	0103
WO-116110	FM-GPW-6149-A	0103A	WO-116503	FM-GPW-6150-D	0103
WO-116112	FM-GPW-6149-C	0103A	WO-116522	FM-GPW-6333-C	0102
WO-116113	FM-GPW-6149-D	0103A	WO-116524	FM-GPW-6338-C	0102
WO-116116	FM-GPW-6156-H	0103A	WO-116526	FM-GPW-6339-C	0102
WO-116117	FM-GPW-6156-J	0103A	WO-116528	FM-GPW-6341-C	0102
WO-116154	FM-GPW-9585	0301	WO-116530	FM-GPW-6331-C	0102
WO-116157	FM-GPW-9549	0301	WO-116532	FM-GPW-6337-C	0102
WO-116159	FM-GPW-9522	0301	WO-116534	FM-GPW-6211-B	0104
WO-116160	FM-GPW-9523	0301	WO-116535	FM-GPW-6211-C	0104
WO-116161	FM-GPW-9562	0301	WO-116537	FM-GPW-9529	0301
WO-116162	FM-GPW-9579	0301	WO-116538	FM-GPW-9907	0301
WO-116163	FM-GPW-9696	0301	WO-116539	FM-GPW-9553	0301
WO-116164	FM-GPW-9928	0301	WO-116540	FM-GPW-9906	0301
WO-116165	FM-GPW-9543	0301	WO-116541	FM-GPW-9914	0301
WO-116166	FM-GPW-9922	0301	WO-116542	FM-GPW-9598	0301
WO-116168	FM-GPW-9569	0301	WO-116543	FM-GPW-6333-D	0301
WO-116169	FM-GPW-9608	0301	WO-116544	FM-GPW-9520	0301
WO-116170	FM-GPW-9574	0301	WO-116545	FM-GPW-9546	0301
WO-116171	FM-GPW-9926	0301	WO-116548	. . . .	0301
WO-116172	FM-GPW-9550	0301	WO-116549	FM-GP-2018	1202
WO-116173	FM-GPW-9558	0301	WO-116550	FM-GP-2019	1202
WO-116174	FM-GPW-9567	0301	WO-116551	FM-GP-2021	1202

Willys Part Number	Ford Part Number	Gov't Group No		Willys Part Number	Ford Part Number	Gov't Group No.
WILLYS-OVERLAND MOTORS, INC. (WO) Cont'd				WO-337112	FM-357420-S	0102
WO-116552	FM-GP-2022	1202		WO-337304	FM-88350-S	0303
WO-116558	....	0103		WO-339043	FM-73880-S	0203
WO-116560	....	0103		WO-339372	FM-GPW-5456	1602
WO-116561	....	0103		WO-343306	FM-GPW-6614	0107
WO-116562	FM-GPW-6152-A	0103A		WO-344732	FM-GPW-9443	0108
WO-116564	FM-GPW-6152-C	0103A		WO-345961	FM-GPW-13434-A	0606B
WO-116565	FM-GPW-6152-D	0103A		WO-349368	FM-GPW-6066	0101A
WO-116566	FM-GPW-6159-A	0103A		WO-349712	FM-88082-S	0108
WO-115667	FM-GPW-6159-B	0103A		WO-352760	FM-34905-S	0608
WO-116568	FM-GPW-6159-C	0103A		WO-356155	FM-GPW-6654	0107
WO-116569	FM-GPW-6159-D	0103A		WO-359039	FM-GPW-5781	1601A
WO-116584	FM-GPW-9518	0301		WO-371400	FM-11A-14452	2601
WO-116585	FM-GPW-9581	0301		WO-371567	FM-GPW-7549	0201A
WO-116586	FM-GPW-9526	0301		WO-372438	FM-GPW-11474	0303A
WO-116587	FM-GPW-9595	0301		WO-374586	FM-361915-S	0201A
WO-116588	FM-355132-S	0301		WO-375217	FM-GPW-7023	0109C
WO-116589	FM-GPW-5724-B	1601A		WO-375811	FM-GPW-6510-B	0105
WO-116600	FM-GPW-18367	1202		WO-375877	FM-GPW-6310	0102
WO-116609	FM-GPW-5345-A	1601		WO-375900	FM-GPW-6245	0106
WO-116610	FM-GPW-5345-B	1601		WO-375907	FM-GPW-6243	0106
WO-116616	FM-GPW-6156-F	0103A		WO-375908	FM-GPW-6244	0106
WO-116624	....	1603		WO-375917	FM-GPW-6286	0106D
WO-116625	....	1603		WO-375920	FM-GPW-6287	0102
WO-116626	....	1603		WO-375927	FM-GPW-6625	0107
WO-116627	....	1603		WO-375981	FM-88141-S	0107
WO-116628	....	1603		WO-375994	FM-GPW-6546	0105
WO-116629	....	1603		WO-376373	FM-358063-S	0101
WO-116630	....	1603		WO-380197	FM-355262	0107
WO-116631	....	1603		WO-381519	FM-GPW-6345	0102
WO-116632	....	1603		WO-384228	FM-GPW-5468	1602
WO-116633	....	1603		WO-384549	FM-GPW-9268	0107N
WO-116634	....	1603		WO-384710	FM-GPW-2133	1209C
WO-116635	....	1603		WO-384958	FM-88022-S	0106D
WO-116636	....	1603		WO-387249	FM-GPW-9267-A	0306B
WO-116637		1603		WO-387633	FM-GPW-6319	0102
WO-116638	....	1603		WO-387891	FM-9N-18679	0107J
WO-116639	....	1603		WO-390510	....	1403A
WO-116640	....	1603		WO-392328	FM-GPW-7227	0706A
WO-116641	....	1603		WO-392468	FM-357553-S-18	1201
WO-116642	....	1603		WO-393594	....	0603
WO-116643	....	1603		WO-630068	FM-GPW-7516	0203
WO-116644	....	1603		WO-630101	FM-355597	0109C
WO-116651	FM-GPW-9610	0301		WO-630103	FM-GPW-7518	0109C
WO-116694	FM-GPW-9396	0302		WO-630112	FM-GPW-7515	0203
WO-116695	FM-GPW-9398	0302		WO-630117	FM-GPW-7562	0203
WO-52217	FM-33795-S2	0501		WO-630129	FM-355476-S	0201B
WO-300143	FM-88042	0108		WO-630262	FM-GPW-6342-B	0102
WO-300329	FM-33896-S2	0610A		WO-630294	FM-GPW-6326	0102A
WO-301232	FM-31037-S7	0301		WO-630298	FM-GPW-6762	0107N
WO-302347	FM-GPW-10141	0601		WO-630299	FM-GPW-6648	0105C
WO-303922	....	1500		WO-630303	FM-GPW-6519	0105C
WO-306715	FM-GPW-17011-A	2301		WO-630305	FM-GPW-6521	0105C
WO-307556	FM-B-14463	0604A		WO-630359	FM-GPW-6020	0101A
WO-311003	FM-73904-S7	1201		WO-630364	FM-GPW-6285	0106D
WO-314338	FM-GPW-6734	0107A		WO-630365	FM-GPW-6288	0106D
WO-314369	FM-B-14466	0604A		WO-630384	FM-GPW-6604	0107
WO-315932	FM-GPW-6269	0106C		WO-630387	FM-GPW-6664	0107
WO-323397	FM-23017-S16	0610B		WO-630389	FM-GPW-6628	0107
WO-323457	FM-34056-S16	0610B		WO-630390	FM-GPW-6644	0107
WO-327257	FM-9N-12113-A	0604A		WO-630392	FM-GPW-6619	0107
WO-330964	FM-GPW-6684	0107		WO-630394	FM-GPW-6630	0107
WO-332515	FM-88032-S	0108B		WO-630396	FM-GPW-6615	0107A-1
WO-334103	FM-GPW-6353	0102		WO-630397	FM-GPW-6617	0107A-1
WO-335912	FM-350343-S-16	0610B		WO-630398	FM-GPW-6627	0107A-1

WILLYS-OVERLAND MOTORS, INC. (WO) Cont'd

Willys Part Number	Ford Part Number	Gov't Group No.	Willys Part Number	Ford Part Number	Gov't Group No.
WO-630512	FM-IGT-8260	0505	WO-636538	FM-353051-S	1001
WO-630518	FM-GPW-6663	0107	WO-636565	FM-GP-4661	1003
WO-630526	FM-GPW-5269	0402	WO-636566	FM-GP-4619	1003
WO-630593	FM-GPW-7523	0204	WO-636568	FM-GP-4666	0902
WO-630727	FM-GPW-6342-A	0106C	WO-636569	FM-356126-S	1104
WO-630753	FM-GPW-3326	1401	WO-636570	FM-356504-S	1003
WO-630754	FM-GPW-3327	1401	WO-636571	FM-357202-S	1104
WO-630755	FM-GPW-3320	1401	WO-636575	FM-33786-S2	1201
WO-630756	FM-GPW-3323	1401	WO-636577	FM-358048-S	1101
WO-630757	FM-GPW-3328	1401	WO-636599	FM-GPW-6608	0107
WO-631105	FM-GPW-12064	0604	WO-636600	FM-GPW-6673	0107
WO-632156	FM-GPW-6387	0109	WO-636755	FM-GPW-7550	0201
WO-632157	FM-355497-S	0109	WO-636778	FM-GPW-7577	0201
WO-632158	FM-355451-S	0105C	WO-636796	FM-355396-S	0107A
WO-632159	FM-88057-S7	0301A	WO-636879	FM-GPW-7100	0704
WO-632174	FM-GPW-9737	0303	WO-636882	FM-GPW-7141	0704
WO-632177	FM-GPW-7532	0203	WO-636885	FM-GPW-7025	0703
WO-633011	FM-GPW-9799	0303	WO-636961	FM-GPW-6135-A	0103
WO-633013	FM-GPW-9752	0303	WO-636962	FM-356021-S	0104
WO-633949	FM-355496-S	0601A	WO-637007	FM-GPW-6333-A	0102
WO-634758	FM-74-6038	0809	WO-637008	FM-GPW-6338-A	0102
WO-634759	FM-GPW-7781-A	0809	WO-637037	. . . .	0103
WO-634762	FM-356524-S	0809	WO-637041	FM-GPW-6105-A	0103
WO-634796	FM-GPW-6308	0102	WO-637042	FM-GPW-6155-A	0103
WO-634811	FM-GPW-9435	0108	WO-637044	FM-GPW-6514	0105
WO-634813	FM-GPW-6642	0107	WO-637045	FM-GPW-6511-B	0105
WO-634814	FM-GPW-9450	0402	WO-637047	FM-GPW-6500-A	0105
WO-634850	FM-355499-S	0106C	WO-637052	FM-GPW-8505	0503
WO-635377	FM-GPW-6369	0102	WO-637053	FM-GPW-8543	0503
WO-635394	FM-GPW-6384	0109B	WO-637065	FM-GPW-6250	0106
WO-635529	FM-GPW-7580	0203	WO-637098	FM-GPW-6700	0102
WO-635681	FM-GPW-7291	1201	WO-637107	FM-355426-S	1403
WO-635804	FM-GPW-7143	0704C	WO-637182	FM-GPW-6507	0105
WO-635811	FM-GPW-7129	0704C	WO-637183	FM-GPW-6505	0105
WO-635812	FM-GPW-7119	0704C	WO-637206	FM-GPW-9456	0108
WO-635836	FM-GPW-7206	0706C	WO-637208	FM-GPW-9467-A	0108
WO-635837	FM-GPW-7234	0706B	WO-637209	FM-GPW-9484	0108
WO-635839	FM-GPW-7208	0706C	WO-637210	FM-GPW-9458	0108
WO-635840	FM-357418-S	0706C	WO-637211	FM-GPW-9465	0108
WO-635844	FM-GPW-7064	0703	WO-637237	FM-GPW-6702	0102
WO-635846	. . . .	0703	WO-637424	FM-GP-2078	1209
WO-635859	FM-GPW-7214	0706A	WO-637425	FM-GPW-6610	0107
WO-635861	FM-GPW-7223	0706A	WO-637426	FM-GP-2087	1209
WO-635862	FM-GPW-7267	0706A	WO-637427	FM-78-2814-A	1209
WO-635863	FM-GPW-7227	0706A	WO-637432	FM-GP-2074	1209
WO-635868	. . . .	0706A	WO-637439	. . . .	1209
WO-635883	. . . .	0601B	WO-637495	FM-GPW-7051-B	0700
WO-635886	FM-357689-S	0101	WO-637503	FM-GPW-7056	0701
WO-635981	FM-GPW-11487-B	0606B	WO-637540	FM-GP-2208	1207
WO-635985	FM-GPW-11487-A	0606B	WO-637541	FM-GP-2196	1207
WO-636004	FM-GPW-5270	0402	WO-637544	. . . .	1207
WO-636109	FM-GPW-8269	0505	WO-637545	FM-GP-2204	1207
WO-636196	FM-GPW-7230	0706C	WO-637546	FM-GP-2206-A	1207
WO-636197	FM-GPW-7231	0706C	WO-637577	FM-GP-2194	1207
WO-636200	FM-GPW-7245	0706B	WO-637579	FM-91A-2201	1207
WO-636297	FM-GPW-8530	0503	WO-637580	FM-GP-2205	1207
WO-636298	FM-GPW-8576	0503	WO-637582	FM-GP-2155	1205
WO-636299	FM-GPW-8509-A	0503A	WO-637583	FM-GP-2160	1205
WO-636360	FM-GP-4241	1003	WO-637584	FM-GP-2175	1205
WO-636396	FM-GP-17333	0810	WO-637585	FM-GP-2176	1205
WO-636438	FM-GPW-9462	0108C	WO-637586	FM-GP-2183	1205
WO-636439	FM-GPW-9460	0108C	WO-637587	FM-GP-2145	1205
WO-636527	FM-355699-S2	1001	WO-637590	FM-GP-2173	1205
WO-636528	FM-34922-S	1001	WO-637591	FM-GP-2169	1205
			WO-637595	FM-GP-2170	1205

Willys Part Number	Ford Part Number	Gov't Group No-	Willys Part Number	Ford Part Number	Gov't Group No-
WILLYS-OVERLAND MOTORS, INC. (WO) Cont'd			WO-638884	FM-GPW-3626	0609B
WO-637597	FM-GP-2188	1205	WO-638885	FM-GPW-3646	0609B
WO-637599	FM-GP-2143-A	1205	WO-638918	FM-GPW-3563	1403A
WO-637602	FM-GP-2180	1205	WO-638918	FM-GPW-3563	1405
WO-637604	FM-91A-2152	1205	WO-638948	FM-GPW-7111	0704C
WO-637605	FM-GP-2077	1205	WO-638949	FM-GPW-7155	0704C
WO-637606	FM-91A-2151	1205	WO-638952	FM-GPW-7140	0704C
WO-637608	FM-GP-2162	1205	WO-638979	FM-GPW-13532	0606
WO-637612	FM-GP-2167	1205	WO-638992	FM-GPW-7563	1201
WO-637615	FM-GPW-12083	0603	WO-638993	FM-GPW-7572	0202
WO-637635	FM-GPW-17017-A	2301	WO-639010	FM-GPW-28482	1201
WO-637636	FM-GPW-6600	0107	WO-639051	FM-GPW-6252-A	0106A
WO-637646	FM-GPW-8575	0502	WO-639052	FM-355452-S	0105C
WO-637724	FM-GPW-6333-B	0102A	WO-639090	FM-GPW-3587	1403A
WO-637725	FM-GPW-6338-B	0102A	WO-639091	FM-GPW-3576	1403A
WO-637787	FM-GP-2261	1207	WO-639095	FM-GPW-3591-A	1403A
WO-637789	FM-GP-2192	1207	WO-639102	FM-GPW-3552	1403A
WO-637790	FM-GPW-6701	0102A	WO-639103	FM-GPW-3589	1403A
WO-637803	. . . .	1603	WO-639104	FM-GPW-3571	1403A
WO-637804	. . . .	1603	WO-639108	FM-GPW-3593	1403A
WO-637810	. . . .	1603	WO-639109	FM-GPW-3594	1403A
WO-637831	FM-GPW-7109	0704B	WO-639110	FM-GPW-3595	1403A
WO-637832	FM-GPW-7116	0704B	WO-639116	FM-GPW-3583	1403A
WO-637833	FM-GPW-7106	0704B	WO-639118	FM-GPW-3577	1403A
WO-637834	FM-GPW-7107	0704B	WO-639119	FM-GPW-3581	1403A
WO-637835	FM-GPW-7059	0704B	WO-639120	FM-20386-S7	1403A
WO-637863	FM-O1A-12410	0604B	WO-639121	FM-356303-S	1403A
WO-637899	FM-91A-2027	1203B	WO-639190	FM-GPW-3517	1403A
WO-637900	FM-GP-2028	1203B	WO-639191	FM-GPW-3520	**1403A**
WO-637901	FM-91A-2030	1203B	WO-639192	FM-GPW-3518	**1403A**
WO-637905	FM-GP-2035	1203A	WO-639237	FM-GPW-6339-B	0102A
WO-637923	FM-351466-S-24	1203B	WO-639238	FM-GPW-6341-B	0102A
WO-637924	FM-33846-S2	1203B	WO-639239	FM-GPW-6331-B	0102A
WO-637936	FM-GPW-18060	1603	WO-639240	FM-GPW-6337-B	0102A
WO-638058	FM-GPW-5274	0401	WO-639244	FM-GPW-2782	1201
WO-638113	FM-GPW-6312	0102	WO-639265	FM-GP-4676-A	1003
WO-638121	FM-GPW-6303-A	0102	WO-639422	FM-GPW-7120	0703
WO-638151	FM-GPW-7580	0202	WO-639423	FM-GPW-7063	0703
WO-638152	FM-GPW-7566	0202	WO-639555	FM-GPW-6763-A	0107G
WO-638153	FM-GPW-7590	0202	WO-639556	FM-GPW-6766-A	0107H
WO-638154	FM-24325-S	0202	WO-639578	FM-GPW-7600	0109
WO-638155	FM-33921-S7	0202	WO-639599	FM-GPW-14448-C	0606E
WO-638157	FM-GPW-7567	0202	WO-639607	FM-GPW-9728	0303
WO-638158	FM-GPW-7591	0202	WO-639610	FM-GPW-9745	0303
WO-638159	FM-GPW-7564	0202	WO-639650	FM-GPW-8255	0101A
WO-638305	FM-34745-S2	0202	WO-639651	FM-GPW-8578	0502
WO-638343	. . . .	1603	WO-639654	FM-GPW-7561	0203
WO-638381	FM-356439-S	1403A	WO-639689	FM-20366-S	0706A
WO-638458	FM-GPW-6256	0106A	WO-639734	FM-GPW-11134	0601B
WO-638459	FM-GPW-6306	0102	WO-639743	FM-GPW-9463	0108C
WO-638500	FM-353023-S7	1602	WO-639862	FM-GPW-6211-A	0104A
WO-638539	FM-33881-S2	1602	WO-639864	FM-GPW-6150-A	0103A
WO-638635	FM-GPW-6065	0101A	WO-639870	FM-GPW-6659	0107
WO-638636	FM-GPW-6513	0105	WO-639979	FM-GPW-6727	0107A
WO-638640	FM-GPW-9448	0108	WO-639980	FM-GPW-6710 .	0107A
WO-638646	FM-GPW-11140	0602	WO-639992	FM-GPW-8501	0503
WO-638730	FM-GPW-6339-A	0102A	WO-639993	FM-GPW-8512	0503
WO-638731	FM-GPW-6341-A	0102A	WO-640006	FM-GPW-7104-A2	0704
WO-638732	FM-GPW-6331-A	0102A	WO-640017	FM-GPW-7050	0703
WO-638733	FM-GPW-6337-A	0102A	WO-640018	FM-GPW-7052	0703
WO-638737	FM-GPW-9417	0302A	WO-640020	FM-GPW-6549-B	0105A
WO-638780	FM-GPW-2279	1201	WO-640031	FM-GAA-8524-A2	0503
WO-638792	FM-353043-A-57	0204	WO-640032	FM-GPW-8549	0503
WO-638798	FM-VGPW-7102	0704B	WO-640033	FM-GPW-8572-B	0503
WO-638809	. . . .	1500	WO-640034	FM-GPW-8557-A	0503

VENDOR'S NUMERICAL INDEX (Cont'd)

Willys Part Number	Ford Part Number	Gov't Group No.		Willys Part Number	Ford Part Number	Gov't Group No.
WILLYS-OVERLAND MOTORS, INC. (WO) Cont'd				WO-650484	FM-74019-S	0303
WO-640038	FM-358006-S8	1602		WO-650684	. . . .	0204
WO-640066	. . . .	0104		WO-651298	FM-GPW-9731-B	0303
WO-640067	. . . .	0104		WO-662010	FM-353202-S	1802A
WO-640070	. . . .	0104		WO-662276	FM-GPW-13437-A	0606B
WO-640071	FM-GPW-6200	0104		WO-662420	FM-GPW-9319-A	0605
WO-640072	FM-GPW-6201	0104				
WO-635838	FM-353081-S	0706B		**YALE & TOWNE MFG. CO. (YA)**		
WO-650482	FM-GPW-9711	0303				
WO-650483	FM-GPW-9795	0303		YA-OP-528	WO-A-2895	FM-GPW-1143501	1817

www.ingramcontent.com/pod-product-compliance
Lightning Source LLC
Chambersburg PA
CBHW080416030426
42335CB00020B/2474